Experimental Thermodynamics Volume X
Non-equilibrium Thermodynamics with Applications

Experimental Thermodynamics Series

Titles in this Series:

1: **Calorimetry of Non-Reacting Systems**
Edited by J. P. McCullough and D. W. Scott
Butterworths, London, 1968.

2: **Experimental Thermodynamics of Non-Reacting Fluids**
Edited by B. Le Neindre and B. Vodar
Butterworths, London, 1975

3: **Measurement of the Transport Properties of Fluids**
Edited by W. A. Wakeham, A. Nagashima, and J. V. Sengers
Blackwell Scientific Publications, Oxford, 1991

4: **Solution Calorimetry**
Edited by K. N. Marsh and P. A. G. O'Hare
Blackwell Scientific Publications, Oxford, 1994

5: **Equations of State for Fluids and Fluid Mixtures**
Edited by J. V. Sengers, R. F. Kayser, C. J. Peters, and H. J. White, Jr.
Elsevier, Amsterdam, 2000

6: **Measurement of the Thermodynamic Properties of Single Phases**
Edited by A. R. H. Goodwin, K. N. Marsh, W. A. Wakeham
Elsevier, Amsterdam, 2003

7: **Measurement of the Thermodynamic Properties of Multiple Phases**
Edited by R. D. Weir and T. W. de Loos
Elsevier, Amsterdam, 2005

8: **Applied Thermodynamics of Fluids**
Edited by A. R. H. Goodwin, J. V. Sengers, C. J. Peters
Royal Society of Chemistry, Cambridge, 2010

9: **Experimental Thermodynamics Volume IX: Advances in Transport Properties of Fluids**
Edited by M. J. Assael, A. R. H. Goodwin, V. Vesovic and W. A. Wakeham
Royal Society of Chemistry, Cambridge, 2014

10: **Experimental Thermodynamics Volume X: Non-equilibrium Thermodynamics with Applications**
Edited by Dick Bedeaux, Signe Kjelstrup and Jan V. Sengers
Royal Society of Chemistry, Cambridge, 2016

Experimental Thermodynamics Volume X

Non-equilibrium Thermodynamics with Applications

Edited by

Dick Bedeaux
Norwegian University of Science and Technology, Trondheim, Norway
Email: dick.bedeaux@chem.ntnu.no

Signe Kjelstrup
Norwegian University of Science and Technology, Trondheim, Norway
Email: signe.kjelstrup@ntnu.no

Jan V. Sengers
University of Maryland, College Park, MD, USA
Email: sengers@umd.edu

ROYAL SOCIETY OF CHEMISTRY

IUPAC

ISBN: 978-1-78262-024-2
PDF eISBN: 978-1-78262-254-3

A catalogue record for this book is available from the British Library

Published by The Royal Society of Chemistry,
Thomas Graham House, Science Park, Milton Road,
Cambridge CB4 0WF, UK

Registered Charity Number 207890

Visit our website at www.rsc.org/books

Printed in the United Kingdom by CPI Group (UK) Ltd, Croydon,
CR0 4YY, UK

List of Contributors

J. Armstrong
Department of Chemistry, Imperial College, Room 160, Exhibition Road, London SW7 2AZ, UK
j.armstrong@imperial.ac.uk

A. Bardow
Lehrstuhl für Technische Thermodynamik, RWTH Aachen University, Schinkelstr. 8, 52062 Aachen, Germany
andre.bardow@ltt.rwth-aachen.de

V. M. Barragán
Department of Applied Physics, Universidad Complutense, 28040 Madrid, Spain
vmabarra@ucm.es

D. Bedeaux
Department of Chemistry, Norwegian University of Science and Technology, 7491 Trondheim, Norway
dick.bedeaux@chem.ntnu.no

F. Bresme
Department of Chemistry, Imperial College, Room 160, Exhibition Road, London SW7 2AZ, UK
f.bresme@imperial.ac.uk

F. E. Genceli Güner
İstanbul Teknik Üniversitesi Kimya Mühendisliği Bölümü, 34469, İstanbul, Turkey
egenceli@hotmail.com

W. Hendriksen
Delft University of Technology, Department of Chemical Technology, Julianalaan 136, 2628 BL Delft, The Netherlands
W.E.J.Hendriksen@tudelft.nl

E. Johannessen
Department of Chemistry, Norwegian University of Science and Technology, 7491 Trondheim, Norway
eivjohannessen@gmail.com

T. R. Kirkpatrick
Institute for Physical Science and Technology, University of Maryland, College Park, Maryland, MD 20742-2431, USA
tedkirkp@umd.edu

S. Kjelstrup
Department of Chemistry, Norwegian University of Science and Technology, 7491 Trondheim, Norway
signe.kjelstrup@ntnu.no

G. J. M. Koper
Delft University of Technology, Department of Chemical Technology, Julianalaan 136, 2628 BL Delft, The Netherlands
G.J.M.Koper@tudelft.nl

A. Lervik
Department of Chemistry, Norwegian University of Science and Technology, 7491 Trondheim, Norway
anders.lervik@gmail.com

J. M. Ortiz de Zárate
Department of Applied Physics, Universidad Complutense, 28040 Madrid, Spain
jmortizz@fis.ucm.es

I. Pagonabarraga
Department of Fundamental Physics, Universitat de Barcelona, Diagonal 647, 08028 Barcelona, Spain
ipagonabarraga@ub.edu

C. Peters
Lehrstuhl für Experimentalphysik III, RWTH Aachen University, Schinkelstr. 8, 52062 Aachen, Germany
christine.peters@ltt.rwth-aachen.de

D. Reguera
Department of Physics, Universitat de Barcelona, Diagonal 647, 08028 Barcelona, Spain
dreguera@ub.edu

J. M. Rubí
Department of Physics, Universitat de Barcelona, Diagonal 647, 08028 Barcelona, Spain
mglrb2@gmail.com

L. Sagis
Physics and Physical Chemistry of Foods, Agrotechnology and Food Sciences Group, Wageningen University, Bomenweg 2, 6703 HD Wageningen, The Netherlands
Leonard.Sagis@wur.nl

J. V. Sengers
Institute for Physical Science and Technology, University of Maryland, College Park, Maryland, MD 20742-2431, USA
sengers@umd.edu

J.-M. Simon
Laboratoire Interdisciplinaire Carnot de Bourgogne, Université de Bourgogne, BP 47870, 21078 Dijon, France
jmsimon@u-bourgogne.fr

J. P. G. Villaluenga
Department of Applied Physics, Universidad Complutense, 28040 Madrid, Spain
juanpgv@fis.ucm.es

T. J. H. Vlugt
Delft University of Technology, Process & Energy Laboratory, Leeghwaterstraat 39, 2628 CA Delft, The Netherlands
T.J.H.Vlugt@tudelft.nl

Ø. Wilhelmsen
Department of Chemistry, Norwegian University of Science and Technology, 7491 Trondheim, Norway
oivind.wilhelmsen@ntnu.no

L. Wolff
Lehrstuhl für Technische Thermodynamik, RWTH Aachen University, Schinkelstr. 8, 52062 Aachen, Germany
ludger.wolff@ltt.rwth-aachen.de

Dedication

Anthony Robert Holmes Goodwin (1961–2014)

Our friend and colleague Anthony (Tony) Goodwin completed the Foreword to this volume in November 2014 but passed away suddenly in December 2014 before he could see its completion.

Tony had been the inspiration for the latest volumes on the Transport Properties of Fluids as well as for several earlier volumes in the long-running IUPAC Series on Experimental Thermodynamics. In that series he had been variously, author, editor and driving-force. His activities within science in the broad field of Thermodynamics are widely recognised and his role for the International Union of Pure and Applied Chemistry within the Physical Chemistry Division and beyond were much appreciated. His work on this particular volume illustrated vividly his commitment to the completion of a task undertaken in a timely manner and I know that he would have been pleased by the adherence of the current editors to his timescale.

Tony Goodwin will be sadly missed but we hope that this volume will represent one of the reminders of the quality and breadth of his scientific contribution.

W. A. Wakeham

Experimental Thermodynamics Volume X: Non-equilibrium Thermodynamics with Applications
Edited by Dick Bedeaux, Signe Kjelstrup and Jan V. Sengers

Published by the Royal Society of Chemistry, www.rsc.org

Foreword

Monographs concerned with the general field of thermophysics were initiated by Commission 1.2 of the International Union of Pure and Applied Chemistry (IUPAC) in 1956. Its former Subcommittee on Transport Properties continued and extended the coverage from 1991. The intention of the books was that they should summarize the state of knowledge with regard to experimental and theoretical methods in thermodynamics, thermochemistry and transport properties. The texts have appeared in two series, the first[1,2] reporting methods in thermochemistry. The present volume is the tenth in the second series.[3–11] Many of the texts in the second series have been concerned with topics in equilibrium thermodynamics; for example, the first volume was concerned with the experimental calorimetry of non-reacting systems,[3] the fourth monograph was concerned with the calorimetry of reacting fluids[6] and also provided updates to the first series.[1,2] In a complementary fashion the fifth volume presented the theoretical basis for equations of state of both fluids and fluid mixtures.[7] The sixth and seventh volumes[8,9] were concerned with the measurement of the thermodynamic properties of single and multiple phases, respectively, and were updates to the second volume[4] reporting measurements of a broader class of thermodynamic properties including techniques with industrial applications for chemically non-reacting systems. The eighth volume[10] was an update of ref. 7 and included subject matter of importance to the practitioner including equations of state for chemically reacting and non-equilibrium fluids, and others, which have undergone significant developments. It is noteworthy that this eighth volume was the first published in conjunction with the International Association of Chemical Thermodynamics (IACT)[†],

---

[†]http://www.iactweb.org/

Experimental Thermodynamics Volume X: Non-equilibrium Thermodynamics with Applications
Edited by Dick Bedeaux, Signe Kjelstrup and Jan V. Sengers

Published by the Royal Society of Chemistry, www.rsc.org

which is the successor body to the former Subcommittee on Thermodynamic Tables of Commission 1.2.

The former Subcommittee on Transport Properties of Commission 1.2, now known as the International Association for Transport Properties (IATP),[‡] made its first contribution to the series with the third volume.[5] It described measurements of the transport properties of fluids characterizing the relaxation of a fluid from a non-equilibrium state, while another volume,[12] also produced under the auspices of IUPAC, discussed the correlation, prediction and estimation of transport properties. Since the publication of ref. 5, there have been significant developments in the measurement and theory of non-equilibrium states of matter. First, several new measurement techniques have been developed and other, older techniques, have received a new lease of life because of technological developments. Secondly, the advent of high-performance computing machinery has enabled theoretical calculations that were not possible hitherto. These considerations led to the decision by IATP to initiate an update of ref. 5 and ref. 12. However, it was recognized by IATP that the importance of the non-equilibrium state and its areas of application have increased during the last two decades so that a single update would not encompass all that is new and useful in the field of non-equilibrium thermodynamics. This realization led to the decision to produce three separate volumes in the Experimental Thermodynamics series entitled *Advances in Transport Properties* Vol. IX; *Non-Equilibrium Thermodynamics with Applications* Vol. X; and another one entitled *Applied Transport Properties* Vol. XI. Volume IX, already published,[11] deals with the experimental and theoretical developments on transport properties. The current Vol. X deals with conceptual developments in non-equilibrium thermodynamics and applications. Together with Vol. IX,[11] it will provide background information for the forthcoming Vol. XI.

This volume compliments other recent publications associated with IUPAC that have covered a range of diverse issues reporting applications of solubility data,[13] to the topical issue of alternate sources of energy,[14] heat capacities of liquids and vapours[15] and the application of chemical thermodynamics to other matters of current industrial and scientific research including separation technology, biology, medicine and petroleum in one[16] of eleven monographs of an IUPAC series entitled *Chemistry for the 21st Century*.[17]

Anthony R. H. Goodwin
Titular Member of Physical and Bio-Physical Division (I), Interdivisional Committee on Terminology, Nomenclature and Symbols (ICTNS) and Commission I.1 on Physicochemical Symbols, Terminology, and Units of the International Union of Pure and Applied Chemistry

‡http://transp.cheng.auth.gr/index.php/iatp/terms

Schlumberger Technology Corporation,
150 Gillingham Lane,
Sugar Land, TX 77478, USA

Prof. Sir William A. Wakeham
Chair International Association of Transport Properties,
Royal Academy of Engineering, London, UK

References

1. *Experimental Thermochemistry*, ed. F. D. Rossini, for IUPAC, Interscience, New York, 1956.
2. *Experimental Thermochemistry*, ed. H. A. Skinner, for IUPAC, Interscience, New York, 1962, vol. II.
3. *Experimental Thermodynamics, Calorimetry of Non-Reacting Systems*, ed. J. P. McCullough and D. W. Scott, for IUPAC, Butterworths, London, 1968, vol. I.
4. *Experimental Thermodynamics, Experimental Thermodynamics of Non-Reacting Fluids*, ed. B. Le Neindre and B. Vodar, for IUPAC, Butterworths, London, 1975, vol. II.
5. *Experimental Thermodynamics, Measurement of the Transport Properties of Fluids*, ed. W. A. Wakeham, A. Nagashima and J. V. Sengers, for IUPAC, Blackwell Scientific Publications, Oxford, 1991, vol. III.
6. *Experimental Thermodynamics, Solution Calorimetry*, ed. K. N. Marsh and P. A. G. O'Hare, for IUPAC, Blackwell Scientific Publications, Oxford, 1994, vol. IV.
7. *Experimental Thermodynamics, Equations of Sate for Fluids and Fluid Mixtures, Parts I and II*, ed. J. V. Sengers, R. F. Kayser, C. J. Peters and H. J. White, Jr., for IUPAC, Elsevier, Amsterdam, 2000, vol. V.
8. *Experimental Thermodynamics, Measurement of the Thermodynamic Properties of Single Phases*, ed. A. R. H. Goodwin, K. N. Marsh and W. A. Wakeham, for IUPAC, Elsevier, Amsterdam, 2003, vol. VI.
9. *Experimental Thermodynamics, Measurement of the Thermodynamic Properties of Multiple Phases*, ed. R. D. Weir and T. W. de Loos, for IUPAC, Elsevier, Amsterdam, 2003, vol. VII.
10. *Experimental Thermodynamics, Applied Thermodynamics of Fluids*, ed. A. R. H. Goodwin, J. V. Sengers and C. J. Peters, for IUPAC, Royal Society of Chemistry, Cambridge, UK, 2010, vol. VIII.
11. *Experimental Thermodynamics, Advances in Transport Properties of Fluids*, ed. M. J. Assael, A. R. H. Goodwin, V. Vesovic and W. A. Wakeham, Royal Society of Chemistry, Cambridge, UK, 2014, Vol. IX.
12. *Transport Properties of Fluids: Their Correlation, Prediction and Estimation*, ed. J. Millat, J. Dymond and C. A. Nieto de Castro, for IUPAC, Cambridge University Press, Cambridge, UK, 1996.
13. *Developments and Applications of Solubility*, ed. T. M. Letcher, for IUPAC, Royal Society of Chemistry, Cambridge, UK, 2007.

14. *Future Energy: Improved, Sustainable and Clean Options for our Planet*, ed. T. M. Letcher, for IUPAC, Elsevier, Amsterdam, 2008.
15. *Heat Capacities of Liquids and Vapours*, ed. E. Wilhelm and T. M. Letcher, for IUPAC, Royal Society of Chemistry, Cambridge, UK, 2009.
16. *Chemical Thermodynamics*, ed. T. M. Letcher, for IUPAC, Blackwell Scientific Publications, Oxford, UK, 2000.
17. C. L. Watkins, *J. Chem. Educ.*, 2000, 77, 973.

Preface

The field of non-equilibrium thermodynamics originates in the work of Lars Onsager who gave a more specific formulation of the second law of thermodynamics in 1931, useful beyond the inequality given in equilibrium thermodynamics. His formulation, in terms of conjugate thermodynamic fluxes and forces, has in the last decade been explored on mesoscopic time- and

L. Onsager

J. Meixner

I. Prigogine

S.R. de Groot

P. Mazur

Experimental Thermodynamics Volume X: Non-equilibrium Thermodynamics with Applications
Edited by Dick Bedeaux, Signe Kjelstrup and Jan V. Sengers

Published by the Royal Society of Chemistry, www.rsc.org

length-scales, when surfaces and heterogeneous systems are involved, and when processes are activated. The theory has furthermore been extended to describe long-range thermal fluctuations in non-equilibrium systems. This has considerably extended what has been called the classical line of development, starting with the work of Onsager, Meixner, Prigogine, de Groot and Mazur. The aim of the present book is to show how the important new developments, can be beneficial in practice. Several chapters have been collected as an inspiration for further applications in new contexts and new fields.

Non-equilibrium thermodynamics is as versatile as thermodynamics. Areas of applications can be found everywhere. The theory can be simplified to more well-known versions when the set of fluxes remain uncoupled. It becomes indispensable, when the fluxes are coupled, most importantly to ensure that the symmetry of the matrix of transport coefficients, proven by Onsager, is obeyed. In order for experimentalists or modellers to make sure that possible approximations are done consistently, the complete set of force-flux relations must be known. It is then important to have the overarching complete theory, the framework where approximations can be introduced.

Practical problems often pose fruitful questions or challenges to the theory. It is a hope that experimental verification of theoretical assumptions and predictions, as well as the development of the theory on the basis of experiments, can continue to guide us in the work to specify the second law even further. The regime of small systems away from equilibrium, for instance, is as of yet practically unexplored.

Dick Bedeaux and Signe Kjelstrup
Norwegian University of Science and Technology

Jan V. Sengers
University of Maryland

Contents

Experimental Thermodynamics Volume X: Non-equilibrium Thermodynamics with Applications
Edited by Dick Bedeaux, Signe Kjelstrup and Jan V. Sengers

Published by the Royal Society of Chemistry, www.rsc.org

CHAPTER 1

Basis and Scope

DICK BEDEAUX,*[a] SIGNE KJELSTRUP[a] AND JAN V. SENGERS[b]

[a] Department of Chemistry, Norwegian University of Science and Technology, 7491 Trondheim, Norway; [b] Institute for Physical Science and Technology, University of Maryland, College Park MD 20742, USA
*Email: dick.bedeaux@chem.ntnu.no

1.1 Short Historic Overview

Non-equilibrium thermodynamics describes transport processes in systems that are not in global equilibrium. The now classical field, which we briefly review here, resulted from efforts of many scientists to find a more explicit formulation of the second law of thermodynamics. This had started already in 1856 with Thomson's studies of thermoelectricity.[1] Onsager is, however, counted as the founder of the field with his papers published in 1931,[2–4] because these put earlier research by Thomson, Boltzmann, Nernst, Duhem, Jauman and Einstein into a systematic framework. Onsager was given the Nobel prize in chemistry in 1968 for this work.

The second law is reformulated in terms of the entropy production, σ. In Onsager's formulation, the entropy production is given by the product sum of so-called conjugate fluxes,[1,†] J_i, and forces, X_i, in the system. The second law then becomes

$$\sigma = \sum_i J_i X_i \geq 0. \tag{1.1}$$

†In non-equilibrium thermodynamics it is standard to use the name flux rather than flux density, as not all of these variables are vectors.

Experimental Thermodynamics Volume X: Non-equilibrium Thermodynamics with Applications
Edited by Dick Bedeaux, Signe Kjelstrup and Jan V. Sengers

Published by the Royal Society of Chemistry, www.rsc.org

The entropy production is per unit of volume. Each flux is taken to be a linear combination of all forces,

$$J_i = \sum_j L_{ij} X_j. \tag{1.2}$$

Onsager showed that the reciprocal relations

$$L_{ji} = L_{ij}, \tag{1.3}$$

apply. They now bear his name. In order to use the theory, one first has to identify a complete set of extensive *independent* variables, α_i, like, for instance, the internal energy and the mass densities per unit of volume. The resulting conjugate fluxes and forces are $J_i = \mathrm{d}\alpha_i/\mathrm{d}t$ and $X_i = \partial S/\partial\alpha_i$ respectively. Here t is the time and S is the entropy of the system. The three equations above contain then all information on the non-equilibrium behaviour of the system. For cases where $\mathrm{d}\alpha_i/\mathrm{d}t$ is equal to minus the divergence of a flux density, this flux density replaces J_i and the gradient of $\partial S/\partial\alpha_i$ replaces $\partial S/\partial\alpha_i$. For surfaces and contact lines the densities are per unit of surface area or length, respectively.

Following Onsager, a consistent theory of non-equilibrium processes in continuous systems was set up in the forties by Meixner[5–8] and Prigogine.[9] They calculated the entropy production for a number of physical problems. Prigogine received the Nobel prize for his work on dissipative structures in systems that are out of equilibrium in 1977, and Mitchell the year after for his application of the (driving) force concept to transport processes in biology.[10]

The most general description of classical non-equilibrium thermodynamics is still the 1962 monograph of de Groot and Mazur[11] reprinted in 1985.[12] Haase's book,[13] also reprinted,[14] contains many results for electrochemical systems and systems with temperature gradients. Katchalsky and Curran developed the theory for biophysical systems.[15] Their analysis was carried further by Caplan and Essig.[16] Førland and co-workers gave various applications in electrochemistry and biology, and they treated frost heave.[17,18] Their book presented the theory in a way suitable for chemists. Newer books on equilibrium thermodynamics or statistical thermodynamics often include chapters on non-equilibrium thermodynamics, see, *e.g.*, Carey.[19] Kondepudi and Prigogine[20] presented a textbook which integrated texts on basic equilibrium and non-equilibrium thermodynamics. Jou *et al.*[21] published a book on extended non-equilibrium thermodynamics. Öttinger gave a non-equilibrium description which also extends to the nonlinear regime.[22]

Non-equilibrium thermodynamics is constantly being applied in new contexts. Fitts gave an early presentation of viscous phenomena.[23] In 1994 Kuiken[24] wrote the most general treatment of multicomponent diffusion and rheology of colloidal systems. Rubí and co-workers[25–27] used internal (molecular) degrees of freedom to explore the development towards equilibrium within a system. This allows us to deal with chemical reactions within the framework of non-equilibrium thermodynamics.[12,26] Bedeaux and Mazur[28] extended the theory to quantum mechanical systems. Kjelstrup and Bedeaux[29] wrote a book dealing with transport into and across surfaces, presenting non-equilibrium thermodynamics for heterogeneous systems. Doi[30] used the variational principle of Onsager[2,3] at constant temperature to derive equations of motion for colloids.

A fundamental assumption in non-equilibrium thermodynamics is that of local equilibrium. In recent years the nature and conditions of local equilibrium have been investigated with computer simulations. Kjelstrup *et al.*[32] pointed out that there can be local equilibrium in volume elements exposed to large fields, with as few as 10 particles. We have reasons to expect that the same holds true in surfaces and in systems at the mesoscale (see Chapter 4). Non-equilibrium thermodynamics has also been extended to include thermal non-equilibrium fluctuations as reviewed by Ortiz de Zárate and Sengers.[31] Local equilibrium is no longer valid for hydrodynamic fluctuations, however. Correlations of the densities and temperature are much larger and longer-ranged in a system exposed, for instance, to temperature gradients than at equilibrium.[31]

The linear relations in eqn (1.2) apply to a volume element, which according to the assumption of local equilibrium, can be very small. Mesoscopic non-equilibrium thermodynamics has been developed to deal with shorter space and time scales (*cf.*, Chapters 14 to 16). Integration from the local to the macroscopic level leads to highly non-linear flux–force relations. To refer to non-equilibrium thermodynamics as a linear theory is misleading. Quoting the preface of the Dover edition of de Groot and Mazur,[12] the theory is non-linear for a variety of reasons, because it includes (i) the presence of convection terms, (ii) quadratic source terms in the energy equation, (iii) the non-linear character of the equation of state, and (iv) the dependence of the Onsager coefficients on the state variables, and so on.

Chemical and mechanical engineering needs theories of transport for systems with gradients in velocity, pressure, concentrations, and temperature, see Denbigh[33] and Bird *et al.*[34] In isotropic systems there is no coupling between tensorial (viscous), vectorial (diffusional) and scalar (reactions) phenomena, so they can usually be dealt with separately.[12] Simple vectorial transport laws have long worked well in engineering, but there is now an increased effort to be more precise. The need for more accurate flux equations in the modelling of non-equilibrium processes[35] increases the need for non-equilibrium thermodynamics. The books by Taylor and Krishna,[36] Cussler,[37] Demirel,[38] and Kjelstrup *et al.*,[29,35] which present Maxwell–Stefan's formulation of the flux equations, are important books in this context. Krishna and Wesselingh[39] and Kuiken,[24] analysing an impressive amount of experimental data, have shown that the diffusion coefficients in the Maxwell–Stefan equations depend less on the concentrations.

Non-equilibrium thermodynamics is necessary for a precise description of all systems that have transport of heat, mass, charge and momentum.[29] There is also a need in mechanical and chemical engineering to design systems that waste less work.[35,40–44] Fossil energy sources, as long as they last, lead to global warming. Better and more efficient use of energy resources is therefore central. It is, then, not enough to only optimize the first-law efficiency. One should minimize the entropy production, which defines the second-law efficiency. The total entropy production should be used as an appropriate measure of sustainability. Through non-equilibrium thermodynamics, one can obtain control[44] of the local entropy production, develop more precise descriptions of the transport processes, and improve the second-law efficiency.[35]

We believe that we are now in a situation where potential new users are looking to the field and asking themselves how non-equilibrium thermodynamics possibly can add to the understanding, the description, or the experimental design of their transport or energy conversion problems. The present collection of chapters is meant to help such readers, by giving them inspiration for additional applications, similar to the ones presented here. We expect that the reader is familiar with the basic elements that are used. For a pedagogical presentation of non-equilibrium thermodynamics theory we refer to textbooks, *e.g.* ref. 29 and 35. For the basis of hydrodynamic fluctuations, we refer to Ortiz de Zárate and Sengers.[31] The seventeen chapters of this book, will update published works, and give state-of-the art mini-reviews of the field of non-equilibrium thermodynamics. We have chosen topics where the theory can add to the present description and open new lines of research and application.

In the remainder of Chapter 1 we provide the essence of the foundations for eqn (1.1) to (1.3) and give advice on how to derive the equations and use them in particular situations.

1.2 The Entropy Production in a Homogeneous Phase

In order to illustrate the points raised in the end of the previous section, we repeat how the entropy production in a homogeneous system can be derived for transport of heat, mass, charge, momentum and chemical reactions,[12] aiming to make eqn (1.1) as concrete as possible. A homogeneous system can be described by the same variable set over its entire extension. The examples covered by this book extend from coupled transport of heat and mass, coupled transport of heat, mass and charge and chemical reactions, to all phenomena present in hydrodynamic flow. We will therefore give the most general expressions for hydrodynamic flow, before we introduce simplifying conditions that apply for the other cases.

The start in any case is to write the entropy balance. In the presence of flow, the best frame of reference is the barycentric (center-of-mass) frame of reference. The change in the entropy density in a volume element by net flow of entropy into the volume element and entropy production inside it is then given by:

$$\frac{\partial\rho(\boldsymbol{r},t)s(\boldsymbol{r},t)}{\partial t} = -\boldsymbol{\nabla}\cdot[\rho(\boldsymbol{r},t)s(\boldsymbol{r},t)\boldsymbol{v}(\boldsymbol{r},t)+\boldsymbol{J}_{\mathrm{s}}(\boldsymbol{r},t)]+\sigma(\boldsymbol{r},t). \tag{1.4}$$

Here $\rho(\boldsymbol{r},t)$ is the mass density, $s(\boldsymbol{r},t)$ is the specific entropy, $\boldsymbol{v}(\boldsymbol{r},t)$ is the barycentric velocity, $\boldsymbol{J}_s(\boldsymbol{r},t)$ is the entropy flux and $\sigma(\boldsymbol{r},t)$ is the entropy production. The position vector is denoted $\boldsymbol{r}=(x,y,z)$ and t is the time. We use italic bold symbols for vectors. The entropy density and flux depend on position as well as time, so we use partial derivatives. The volume element has a sufficient number of particles to give a statistical basis for thermodynamic calculations, *i.e.*, local equilibrium. The state is given by the temperature $T(\boldsymbol{r},t)$, the specific chemical potential, $\mu_j(\boldsymbol{r},t)$ for n neutral components and the pressure $p(\boldsymbol{r},t)$. We suppress the explicit dependence on $(\boldsymbol{r},t)$ from now on. It is straightforward to use standard IUPAC symbols and units for homogeneous systems, see however the special situation for heterogeneous systems (*cf.*, Chapters 4, 8 and 12).

The explicit expression for σ is always found by introducing relevant balance equations into the Gibbs equation and comparing the resulting expression with eqn (1.4). This will identify the entropy flux as well as the entropy production, provided that the Gibbs equation holds, or in other words, provided that the system is in *local equilibrium*, see Chapter 4 for a discussion. The Gibbs equation for a co-moving volume element is

$$\frac{\mathrm{d}u}{\mathrm{d}t} = T\frac{\mathrm{d}s}{\mathrm{d}t} - p\frac{\mathrm{d}(1/\rho)}{\mathrm{d}t} + \sum_j \mu_j \frac{\mathrm{d}w_j}{\mathrm{d}t}, \tag{1.5}$$

where u is the specific internal-energy and $w_j \equiv \rho_j/\rho$ are the mass fractions. The total differentials are time derivatives in the barycentric frame of reference, which will be defined below.

The balance equation for the mass density of species j is

$$\frac{\partial\rho_j}{\partial t} = -\boldsymbol{\nabla}\cdot(\rho_j\boldsymbol{v}+\boldsymbol{J}_j)+M_j\nu_j r \quad \text{for } j=1,\ldots,n, \tag{1.6}$$

where $\boldsymbol{J}_j=\rho_j(\boldsymbol{v}_j-\boldsymbol{v})$ are the diffusion fluxes in the barycentric frame of reference, and M_j and ν_j are the molar mass and the stoichiometric coefficient of species j in the chemical reaction, respectively. Furthermore, $r(\boldsymbol{r},t)$ is the rate of the reaction. The coefficient ν_j is positive if j is a product and negative if j is a reactant. For simplicity, we consider only one reaction.[12] The reaction Gibbs energy is $\Delta_{\mathrm{r}}G(\mathbf{r},t)=\sum M_j\nu_j\mu_j$. Independent components are, according to the phase rule, the number of species minus the number of restrictions between them. A chemical reaction at equilibrium may pose such a restriction.

We sum eqn (1.6) over all components and use conservation of mass in the chemical reaction. This gives the continuity equation

$$\frac{\partial \rho}{\partial t} = -\nabla \cdot \rho \boldsymbol{v}, \tag{1.7}$$

where $\rho = \sum \rho_j$ and $\rho \boldsymbol{v} = \sum \rho_j \boldsymbol{v}_j$. It follows from the definitions that the sum of all diffusive fluxes $\boldsymbol{J}_j = \rho_j(\boldsymbol{v}_j - \boldsymbol{v})$ is zero. The continuity equation for an arbitrary density, a, is

$$\frac{\partial \rho a}{\partial t} + \nabla \cdot \rho a \boldsymbol{v} = \rho \left(\frac{\partial a}{\partial t} + \boldsymbol{v} \cdot \nabla a \right) \equiv \rho \frac{\mathrm{d}a}{\mathrm{d}t}. \tag{1.8}$$

The last identity defines the time derivative of a quantity that moves with the flow, a co-moving quantity. This defines then the derivatives in Gibbs eqn (1.5).

With the help of eqn (1.8) we can write terms in eqn (1.4) and (1.6) alternatively as

$$\rho \frac{\mathrm{d}s}{\mathrm{d}t} = -\nabla \cdot \boldsymbol{J}_{\mathrm{s}} + \sigma, \tag{1.9}$$

$$\rho \frac{\mathrm{d}w_j}{\mathrm{d}t} = -\nabla \cdot \boldsymbol{J}_j + M_j \nu_j r \quad \text{for } j = 1, \ldots, n. \tag{1.10}$$

The momentum balance (the equation of motion) in the absence of external forces is given by

$$\rho \frac{\mathrm{d}\boldsymbol{v}}{\mathrm{d}t} = -\nabla p - \nabla \cdot \boldsymbol{\Pi}, \tag{1.11}$$

where $\boldsymbol{\Pi}$ is the viscous pressure tensor. The momentum balance enters the Gibbs equation *via* the change in internal-energy density. According to the first law of thermodynamics,[12] the change in internal-energy density per unit of time is equal to:

$$\rho \frac{\mathrm{d}u}{\mathrm{d}t} = -\nabla \cdot \boldsymbol{J}_q - p \nabla \cdot \boldsymbol{v} - \boldsymbol{\Pi} : \nabla \boldsymbol{v} + \boldsymbol{E} \cdot \boldsymbol{j}, \tag{1.12}$$

where $\boldsymbol{J}_q$ in the absence of external forces is the total heat (energy) flux in the barycentric frame of reference. The total heat flux across the volume element is the sum of the measurable heat flux, $\boldsymbol{J}'_q$, and a latent heat flux (the specific partial enthalpies, h_j, carried by the diffusive fluxes, $\boldsymbol{J}_j$ of the neutral components)

$$\boldsymbol{J}_q = \boldsymbol{J}'_q + \sum_{j=1}^{n} h_j \boldsymbol{J}_j. \tag{1.13}$$

The electric field $\boldsymbol{E} = -\nabla \phi$ is minus the derivative of ϕ, the electric potential. The product $\boldsymbol{E} \cdot \boldsymbol{j}$ is the electrical work per unit of volume. The systems we consider are electro-neutral. It follows that the electric current density, $\boldsymbol{j}$, is independent of the position. The electric current can be externally controlled, and is a measure of the relative motion of the charge carriers in the system.

By introducing all balance equations and definitions in the Gibbs eqn (1.5), we can identify the terms in eqn (1.4), giving the entropy flux

$$\boldsymbol{J}_s = \frac{1}{T}\left(\boldsymbol{J}_q - \sum_{j=1}^{n} \mu_j \boldsymbol{J}_j\right) = \frac{1}{T}\boldsymbol{J}'_q - \sum_{j=1}^{n} s_j \boldsymbol{J}_j, \tag{1.14}$$

and the entropy production

$$\sigma = \boldsymbol{J}_q \cdot \boldsymbol{\nabla}\frac{1}{T} + \sum_{j=1}^{n} \boldsymbol{J}_j \cdot \left(-\boldsymbol{\nabla}\frac{\mu_j}{T}\right) + \boldsymbol{j}\cdot\frac{\boldsymbol{E}}{T} + r\left(-\frac{\Delta_\mathrm{r}G}{T}\right) + \boldsymbol{\Pi} : \left(-\frac{\boldsymbol{\nabla v}}{T}\right). \tag{1.15}$$

In eqn (1.14) s_j is the specific partial entropy of component j. By replacing the total heat flux, $\boldsymbol{J}_q$, by the measurable heat flux, $\boldsymbol{J}'_q$, using eqn (1.13), we obtain

$$\sigma = \boldsymbol{J}'_q \cdot \boldsymbol{\nabla}\frac{1}{T} + \sum_{j=1}^{n} \boldsymbol{J}_j \cdot \left(-\frac{1}{T}\boldsymbol{\nabla}\mu_{j,T}\right) + \boldsymbol{j}\cdot\frac{\boldsymbol{E}}{T} + r\left(-\frac{\Delta_\mathrm{r}G}{T}\right) + \boldsymbol{\Pi} : \left(-\frac{\boldsymbol{\nabla v}}{T}\right). \tag{1.16}$$

Here $\boldsymbol{\nabla}\mu_{j,T} \equiv \boldsymbol{\nabla}\mu_j + s_j\boldsymbol{\nabla}T$ is the gradient of the chemical potential keeping the temperature constant. These gradients do not depend on a choice of standard state. The expressions apply for fluxes which are measured in the barycentric frame of reference. We also emphasize that:

(1) The results apply to systems in *local equilibrium*. This need not mean *local chemical equilibrium*, which is a special case of local equilibrium.[12,45] The important condition is that Gibbs eqn (1.5) applies, see Chapter 4 for further comments.

(2) The entropy production contains pairs of fluxes and forces. Each pair has a so-called *conjugate flux and force*. We see that the conjugate flux–force pairs in eqn (1.15) and (1.16) differ. A change in one pair, will lead to a change in another pair, as is seen by comparing the two first terms on the right-hand side of eqn (1.15) and (1.16). It is hard to know the conjugate flux–force pairs without deriving the entropy production from actual balance equations.

(3) The diffusion fluxes depend on the *frame of reference*. The total heat flux depends also on such a reference. The measurable heat flux is, however, independent of the frame of reference. The measurable heat flux differs from the reduced heat flux introduced by de Groot and Mazur by a contribution proportional to the electric current density, see ref. 43 for a discussion. The electric current density does also not depend on the frame of reference. De Groot and Mazur[12] used charged and uncharged components; appropriate for systems that are not electro-neutral. For electro-neutral systems, the number of uncharged components is smaller and therefore more convenient, *cf.*, Chapter 12.

(4) Chemical potentials and enthalpies are not absolute quantities; they depend on the choice of a standard state. This can be the state of pure

components at 0.1 MPa at the temperature in question. The standard state value of a chemical potential depends on the temperature. The total heat flux in eqn (1.15) depends on the standard state, while all fluxes and driving forces in eqn (1.16) are absolute and can be measured.

(5) The different sets of conjugate flux–force pairs are equivalent and describe the same physical situation. The entropy production is invariant for transformations and has an absolute value. This property can be used if a change in variables is needed, *cf.*, Chapter 12.

(6) It is always possible to find driving forces and conjugate fluxes which are absolute. In eqn (1.16) all forces and fluxes are absolute. The sum of the diffusive fluxes is zero. This can be used to reduce the number of conjugate pairs by one, replacing μ_j by $\mu_j - \mu_n$ everywhere. A procedure that applies in mechanical equilibrium is discussed in the next section.

(7) The problem of interest should guide the choice of flux–force pairs to use. In descriptions of measurements, the measurable heat flux is a useful variable. In descriptions of molecular simulations, the total heat flux (which often is the energy flux) is a good variable. In the first case, we can find the Onsager coefficients from measurements, see the next section. In the second case, we can only calculate them, as they depend on the choice of a standard state.

(8) Ross and Mazur[46] showed that the contribution to σ from the chemical reaction is also equal to the product of r and the driving force, $\Delta_r G/T$, when the reaction rate is a non-linear function of the driving force, provided that the ensemble of particles is nearly Maxwellian. Prigogine showed that the Gibbs relation was valid for such conditions, see ref. 46.

The separate products in the *entropy production do not individually give pure dissipation* of energy. It is the sum that does so. For instance, the electric power per unit of volume does not necessarily give only an Ohmic contribution to the entropy production; there may also be electric work terms included in the product. Each product normally contains work terms as well as energy storage terms. It is *their combination* which gives the entropy production rate, and the work that is lost per unit of time, see Chapter 13 for further comments. The entropy production can be used to find the second-law efficiency of a process.[35]

De Groot and Mazur[11,12] used the affinity, A, of the chemical reaction, rather than the reaction Gibbs energy. According to Kondepudi and Prigogine[20] (page 111), the reaction Gibbs energy is primarily used in connection with equilibrium states and reversible processes, while De Donder's affinity concept is more general. We dispute that there is a principle difference between the two concepts. *The affinity is simply equal to minus the reaction Gibbs energy.* We use the reaction Gibbs energy, because chemists are more familiar with this concept.

Haase[12,13] defined the dissipation function in his monograph, and the dissipation function is still used in many books.[30,31,38] For homogeneous systems, Haase defined $\psi = T\mathrm{d}S_{\mathrm{irr}}/\mathrm{d}t$, where $\mathrm{d}S_{\mathrm{irr}}/\mathrm{d}t$ is the rate of increase of the entropy due to processes which occur inside a system. For a continuous system, in which the temperature can vary from point to point, he used the definition $\psi = T\sigma$,[13] see ref. 13, page 83. This last definition is analogous to Rayleigh's dissipation function for hydrodynamic flow.[30] For non-isothermal conditions, neither of the above definitions are correct, see ref. 29 page 56. *Correct linear force–flux relations can only be obtained from the entropy production* and not from the dissipation function.

1.3 Linear Flux–Force Relations

As mentioned in Section 1.1, the linear relation between the fluxes and forces, refers to a volume element, and not to the overall behaviour of the system. An expression for the entropy production in the presence of flow through the element was given in Section 1.2. The barycentric frame of reference was used, motivated by our wish to include applications where hydrodynamic flow takes place. As soon as we no longer are interested in flow, other frames of reference become relevant. We explain here why this is so (Prigogine's theorem), and give one other option for a frame of reference (the laboratory frame of reference), as well as a formulation that is independent on the frame (the Maxwell–Stefan formulation). We intend to illustrate how the theory can be adopted to particular needs. The phenomenological coefficients may depend on the frame of reference for the fluxes. They may depend on a standard state as well.

We start with the flux–force relations that derive from eqn (1.16). The measurable heat flux is independent of the frame of reference and is therefore convenient from an experimental point of view. All forces and fluxes in eqn (1.16), except the last two pairs, have a direction and are thus vectors. The chemical reaction has a scalar flux and force. The viscous contribution has a tensorial and a scalar force and flux. For isotropic systems, the (viscous) pressure tensor is symmetric. According to the Curie's principle, tensors of different order do not couple. We can thus discuss the vectorial contributions, and subsequently the reaction rate and the viscous pressure tensor. We further show how a substitution of flux–force relations into the balance equations yields differential equations for the variables.

1.3.1 Vectorial Contributions

The diffusive fluxes are given by $\boldsymbol{J}_j = \rho_j(\boldsymbol{v}_j - \boldsymbol{v})$. We introduce these into eqn (1.16) and obtain:

$$\sigma = \boldsymbol{J}'_q \cdot \boldsymbol{\nabla}\frac{1}{T} + \sum_{j=1}^{n} (\boldsymbol{v}_j - \boldsymbol{v}) \cdot \left(-\frac{\rho_j}{T}\boldsymbol{\nabla}\mu_{j,T}\right) + \boldsymbol{j} \cdot \frac{\boldsymbol{E}}{T} + r\left(-\frac{\Delta_\mathrm{r} G}{T}\right) + \boldsymbol{\Pi} : \left(-\frac{\boldsymbol{\nabla}\boldsymbol{v}}{T}\right). \tag{1.17}$$

For a one-component system, the second term in eqn (1.17) is zero. At least two components are needed to observe diffusion. The first three terms in eqn (1.17) contain vectors. In an isotropic system, the linear relations for the vectorial forces are:

$$\begin{aligned}
\nabla\frac{1}{T} &= r_{qq}\boldsymbol{J}'_q + \sum_{j=1}^{n} r_{qj}(\boldsymbol{v}_j - \boldsymbol{v}) + r_{q\phi}\boldsymbol{j}, \\
-\frac{\rho_k}{T}\nabla\mu_{k,T} &= r_{kq}\boldsymbol{J}'_q + \sum_{j=1}^{n} r_{kj}(\boldsymbol{v}_j - \boldsymbol{v}) + r_{k\phi}\boldsymbol{j}, \\
\frac{1}{T}\boldsymbol{E} &= r_{\phi q}\boldsymbol{J}'_q + \sum_{j=1}^{n} r_{\phi j}(\boldsymbol{v}_j - \boldsymbol{v}) + r_{\phi\phi}\boldsymbol{j}.
\end{aligned} \tag{1.18}$$

The Onsager resistivities r_{ij} in eqn (1.18) form a symmetric matrix, which is the inverse of the conductivity matrix L_{ij} in eqn (1.2).

In many cases without flow, the pressure equilibrates fast and we achieve mechanical equilibrium, $\nabla p = 0$. In that case, the Gibbs–Duhem equation gives

$$\sum_{k=1}^{n} \frac{\rho_k}{T}\nabla\mu_{k,T} = 0. \tag{1.19}$$

It follows that

$$\sum_{k=1}^{n} r_{kq} = \sum_{k=1}^{n} r_{kj} = \sum_{k=1}^{n} r_{k\phi} = 0. \tag{1.20}$$

Using the Onsager symmetry relations it follows that

$$\sum_{k=1}^{n} r_{qk} = \sum_{k=1}^{n} r_{jk} = \sum_{k=1}^{n} r_{\phi k} = 0. \tag{1.21}$$

These identities imply that the barycentric velocity in eqn (1.18) may be replaced by an arbitrary reference velocity. This is Prigogine's theorem.[12] We chose the laboratory frame of reference so that $\boldsymbol{v} = 0$.

For our further analysis we write eqn (1.18) in the form

$$\begin{aligned}
\nabla T &= -\frac{1}{\lambda}\left[\boldsymbol{J}'_q - \sum_{j=1}^{n} c_j q_j^* \boldsymbol{v}_j - \frac{\Pi}{F}\boldsymbol{j}\right], \\
\frac{\rho_k}{T}\nabla\mu_{k,T} &= -\frac{c_k q_k^*}{T^2}\nabla T - \sum_{j=1}^{n} R_{kj}\boldsymbol{v}_j - R_{k\phi}\boldsymbol{j}, \\
\frac{1}{T}\boldsymbol{E} &= \frac{\Pi}{FT^2}\nabla T + \sum_{j=1}^{n} R_{\phi j}\boldsymbol{v}_j + R_{\phi\phi}\boldsymbol{j}.
\end{aligned} \tag{1.22}$$

where the thermal conductivity λ, the measurable heats of transfer q_k^*, the Peltier coefficient Π and the resistivities for transport at constant temperature R_{jk} are defined by

$$\lambda \equiv \frac{1}{T^2 r_{qq}}, \quad c_j q_j^* \equiv -\frac{r_{qj}}{r_{qq}}, \quad \Pi \equiv -F\frac{r_{q\phi}}{r_{\phi\phi}}$$
$$R_{jk} \equiv r_{jk} - \frac{r_{jq} r_{qk}}{r_{qq}} \quad \text{for } j, k = 1, \ldots, n, \phi. \tag{1.23}$$

F is Faraday's constant. The density of component k is given by $c_k \equiv \rho_k / M_k$. Superscript * is used for the heat of transfer to emphasize that it is a transport property. It is related to the thermal diffusion coefficient *via* a symmetry relation. It follows from these definitions and eqn (1.20) and (1.21) that the R_{jk} matrix is symmetric and that

$$\sum_{j=1}^{n} c_j q_j^* = \sum_{j=1}^{n} R_{jk} = \sum_{j=1}^{n} R_{kj} = \sum_{j=1}^{n} R_{j\phi} = \sum_{j=1}^{n} R_{\phi j} = 0. \tag{1.24}$$

The Maxwell–Stefan diffusion coefficients are defined by

$$R_{jk} \equiv -cR\frac{x_j x_k}{\mathrm{D}_{jk}}, \tag{1.25}$$

where the mole fractions are defined by $x_j = c_j / c$ and c is the total molar density, and R is the universal gas constant. The Maxwell–Stefan diffusion coefficients, D_{jk}, are also symmetric. Eqn (1.24) gives the diagonal coefficients in terms of the cross coefficients. By using this with eqn (1.22), we obtain

$$\nabla T = -\frac{1}{\lambda}\left[\boldsymbol{J}'_q - \sum_{j=1}^{n-1} c_j q_j^* (\boldsymbol{v}_j - \boldsymbol{v}_n) - \frac{\Pi}{F}\boldsymbol{j}\right],$$
$$\frac{\rho_k}{T}\nabla \mu_{k,T} = -\frac{c_k q_k^*}{T^2}\nabla T + cR\sum_{j \neq k}^{n} \frac{x_k x_j}{\mathrm{D}_{kj}}(\boldsymbol{v}_j - \boldsymbol{v}_k) - R_{k\phi}\boldsymbol{j}, \tag{1.26}$$
$$\frac{1}{T}\boldsymbol{E} = \frac{\Pi}{FT^2}\nabla T + \sum_{j=1}^{n-1} R_{\phi j}(\boldsymbol{v}_j - \boldsymbol{v}_n) + R_{\phi\phi}\boldsymbol{j}.$$

Due to the identities in eqn (1.24), all fluxes are now independent of the frame of reference. The gradients of the chemical potentials do not depend on the choice of the standard state. The standard chemical potential depends only on the temperature. All the constitutive coefficients in eqn (1.26) are therefore independent of both a choice of standard state and of a frame of reference for the fluxes. The Maxwell–Stefan diffusion coefficients are found to be less dependent on concentrations than Fick's diffusion coefficients,[24] see also ref. 29, 35 and 36. We can now emphasize that:

1. The Onsager coefficients (resistivity) in eqn (1.22) depend on the frame of reference for the fluxes. Choosing eqn (1.26) as the starting point, they neither depend on a standard state for the driving forces, as all driving forces in eqn (1.26) are absolute, nor on the frame of reference chosen for the fluxes.
2. A description exists that makes all variables including the Onsager coefficients free of choice for frame of reference or standard states. The example above is such a case. Transformations to other sets of fluxes and forces, and also to sets that depend on a standard state, can be done using the invariance of the entropy production.
3. The barycentric frame of reference is convenient for flow problems. At mechanical equilibrium we are free to choose any other convenient frame of reference. In the present example, we first chose the laboratory frame of reference. We next used the Gibbs–Duhem equation to obtain a description independent of the choice of the frame of reference (the Maxwell–Stefan description).
4. The linear force–flux relations obey Onsager symmetry and can be interpreted for any choice of frame of reference and/or standard state. Any such dependence should be stated along with their values, as the interpretation will rest on the choice taken.
5. The number of independent Maxwell–Stefan diffusion coefficients is equal to the number of independent Fick diffusion coefficients. The relation between the two sets contains the so-called thermodynamic factors (the derivatives of the chemical potentials with respect to the densities) and is discussed in Chapter 5.

The general form of the force–flux relations, eqn (1.18), contains the barycentric reference velocity as well as the component velocities. A tempting interpretation of eqn (1.18) is to explain the part connected to $\boldsymbol{v}$, the co-moving part, as a reversible contribution to the equation, while the remaining part is a dissipative part. This interpretation is not correct. The remaining diffusive parts are not purely dissipative. This was explained below item (8) in the previous section. Another way to understand this point is to diagonalise the Onsager resistivity matrix. With m force–flux pairs, there are m positive eigenvalues and eigenvectors. These are the m independent dissipative processes in the system. The resistivity matrix has, because of Onsager symmetry, $m(m+1)/2$ independent coefficients. The $m(m-1)/2$ remaining coefficients are crucial for the conversion of one kind of energy into another kind of energy. Above, we took the example of conversion between electric and chemical energy. Another example is the Soret effect, in which a temperature gradient causes separation of components (work). The concentration gradients represent work stored. The coupling coefficient responsible for this can be connected to a reversible process. In the important effort to reduce the entropy production in the process industry such insights may enable wanted reductions.

1.3.2 Scalar Contributions

There are two scalar contributions in eqn (1.17) for the entropy production. The first one is due to the reaction and the second one is due to the trace of the viscous pressure tensor

$$\sigma_{\mathrm{s}} = r\left(-\frac{\Delta_{\mathrm{r}}G}{T}\right) + \mathrm{tr}\boldsymbol{\Pi}\left(-\frac{\boldsymbol{\nabla}\cdot\boldsymbol{v}}{T}\right). \tag{1.27}$$

The viscous pressure tensor can be written in terms of its trace, which is scalar, and a symmetric traceless part, which is truly tensorial. The scalar trace couples in principle to the scalar reaction rate. The coupling coefficient is negligible, however.[24] The scalar flux–force pairs therefore give as flux–force relations,

$$r = L_{rr}\left(-\frac{\Delta_{\mathrm{r}}G}{T}\right), \tag{1.28}$$

$$\mathrm{tr}\,\boldsymbol{\Pi} = -\,l_{vv}\frac{\boldsymbol{\nabla}\cdot\boldsymbol{v}}{T} \equiv -\eta_{\mathrm{v}}\boldsymbol{\nabla}\cdot\boldsymbol{v}, \tag{1.29}$$

where $\eta_{\mathrm{v}} \equiv l_{vv}T^{-1}$ is the volume viscosity. The presence of a chemical reaction away from equilibrium does not violate the assumption of local thermodynamic equilibrium.

1.3.3 Tensorial Contribution

The tensorial contribution to the entropy production is

$$\sigma_{\mathrm{t}} = \boldsymbol{\Pi}^{\mathrm{st}} : \left(-\frac{(\boldsymbol{\nabla}\boldsymbol{v})^{\mathrm{st}}}{T}\right), \tag{1.30}$$

where the superscript st indicates the symmetric traceless part of the tensor. The resulting flux–force relation is

$$\begin{aligned}\Pi^{\mathrm{st}}_{\alpha\beta} &= -\,l_{ss}\frac{(\nabla v)^{\mathrm{st}}_{\alpha\beta}}{T} = -\frac{l_{ss}}{2T}\left(\frac{\partial v_\beta}{\partial x_\alpha} + \frac{\partial v_\alpha}{\partial x_\beta} - \frac{2}{3}\delta_{\alpha\beta}\sum_\gamma \frac{\partial v_\gamma}{\partial x_\gamma}\right)\\ &\equiv -\eta\left(\frac{\partial v_\beta}{\partial x_\alpha} + \frac{\partial v_\alpha}{\partial x_\beta} - \frac{2}{3}\delta_{\alpha\beta}\sum_\gamma \frac{\partial v_\gamma}{\partial x_\gamma}\right),\end{aligned} \tag{1.31}$$

where $\eta \equiv l_{ss}(2T)^{-1}$ is the shear viscosity.

1.3.4 Differential Equations

If we now substitute the flux–force relations into the balance equations, we obtain differential equations for the variables. The task is elaborate for multi-component systems. In a one-component system, the continuity eqn (1.17) for the density remains unaltered. In the absence of an electric field and a current density, the internal energy change is

$$\rho \frac{\mathrm{d}u}{\mathrm{d}t} = \lambda \Delta T - p\boldsymbol{\nabla}\cdot\boldsymbol{v} + 2\eta(\boldsymbol{\nabla}\boldsymbol{v})^{\mathrm{st}} : (\boldsymbol{\nabla}\boldsymbol{v})^{\mathrm{st}} + \eta_{\mathrm{v}}(\boldsymbol{\nabla}\cdot\boldsymbol{v})^2, \tag{1.32}$$

where $\Delta \equiv \frac{\partial^2}{\partial x^2} + \frac{\partial^2}{\partial y^2} + \frac{\partial^2}{\partial z^2}$ is the Laplace operator and η_v is the volume viscosity. By substituting eqn (1.29) and (1.31) into (1.11), we obtain the Navier–Stokes equation

$$\rho \frac{\mathrm{d}\boldsymbol{v}}{\mathrm{d}t} = -\boldsymbol{\nabla}p + \eta\Delta\boldsymbol{v} + \left(\frac{1}{3}\eta + \eta_v\right)\boldsymbol{\nabla}\boldsymbol{\nabla}\cdot\boldsymbol{v}, \tag{1.33}$$

Diffusion and electrical effects have been treated in the literature.[9,11,24,34,36,39]

1.4 The Entropy Production in a Heterogeneous System

1.4.1 The Surface as an Autonomous System

Systems that include surfaces or interfaces are heterogeneous, not homogeneous. A heterogeneous system consists of more than one phase. Real systems are heterogeneous, for instance, because a container is needed for the fluid of interest, or because a solid phase is in contact with the surroundings. Gibbs invented the chemical potential of a component in a homogeneous phase and used this variable to define equilibrium conditions in a heterogeneous system. Between the phases in a heterogeneous system there is a continuous variation, say in the density of components, *cf.*, Figure 1.1. There may even be an accumulation in the surface of one component, *cf.*, Figure 1.2.

The transition region between two homogeneous phases, where the densities differ from those of the adjacent phases, can be defined as the interface or the surface, see Figure 1.1. The region can also be regarded as a thermodynamic system. Gibbs described the surface in terms of excess variables, with respect to the mathematical construct, the dividing surface, see *e.g.* ref. 29. The position of the dividing surface is illustrated in Figure 1.3, and the determination of the excess variable with regard to this position is illustrated in Figure 1.2. It is well documented[29] that thermodynamic relations apply also in the surface, but here for the excess variables.

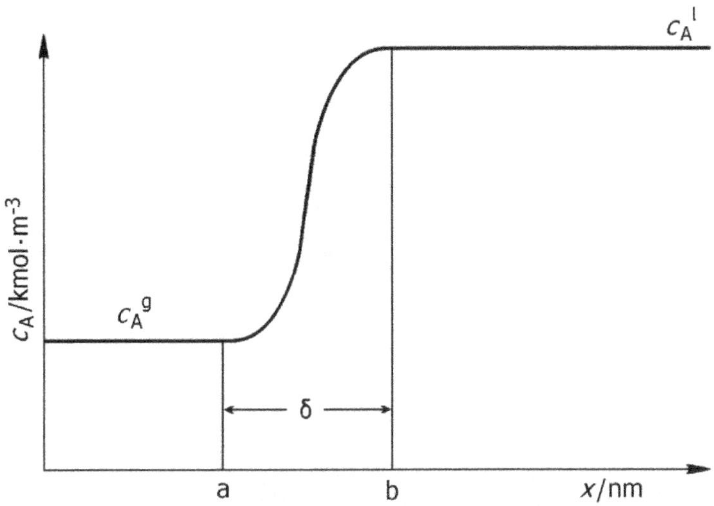

Figure 1.1 Variation in the concentration of a component A, c_A, going from the gas (g) to the liquid (l) state. The vertical lines at positions a and b give the extension of the surface, $\delta = b - a$. The scale of the x-axis is measured in nanometres.

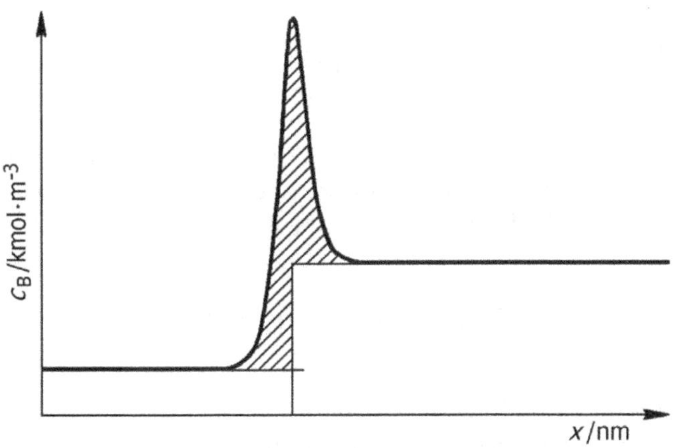

Figure 1.2 Variation in the concentration of a component B across the surface. The excess surface concentration of B is the integral under the curve in the sketch.

This autonomous nature of the surface, the fact that it can be regarded as an independent thermodynamic system away from equilibrium, has been questioned,[47] since the surface does not exist independent of the adjacent bulk phases. The surface has, however, unique properties, like the surface tension. The surface tension of a planar interface in a one-component system is furthermore a unique function of the surface temperature and the curvatures, serving therefore as an equation of state for the surface.

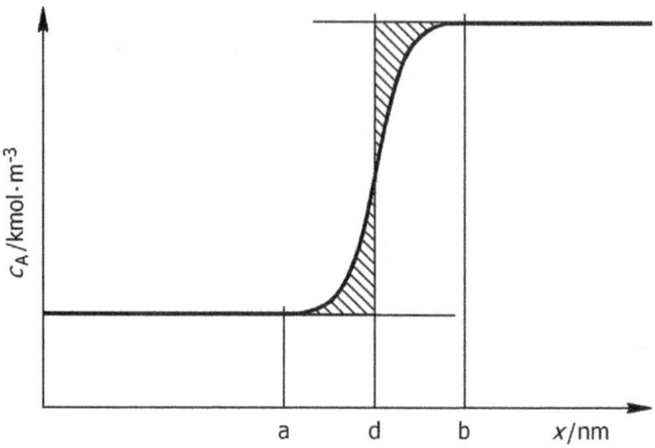

Figure 1.3 Determination of the position of the equimolar surface of component A. The vertical line at d is drawn so that the areas between the curve and the bulk phase concentrations are the same. The surface thickness is again given by $\delta = b - a$.

This supports the view that the surface is an autonomous thermodynamic system in the Gibbs description. The same can be said about a three-phase contact line.[29]

In the extension of classical non-equilibrium thermodynamics for homogeneous systems to heterogeneous systems, Kjelstrup and Bedeaux[29] treated the surface and the contact line as autonomous thermodynamic systems, not only at global equilibrium, but also away from equilibrium. Evidence in support for their hypothesis has been found, and will be discussed in Chapter 4. We refer to Chapter 17 for a discussion of the definition of excess densities.

1.4.2 The Entropy Production in a Surface and a Contact Line

The entropy production in a surface, or even at a contact line, can be derived following the same pattern as described above, by introducing the balance equations into the Gibbs equation. The Gibbs equation for the surface or the line refers then to the surface (line) excess variables defined, as pictured in Figure 1.3 following Gibbs.[29] Along the surface and the line there are now excess fluxes. The balance equations are written in a discrete form.

We shall not give the general expression for the entropy production in a typical system here. We hope it is more instructive to consider the examples that follow from this procedure, and are used in Chapters 8 to 12. These chapters give results for commonplace phenomena like evaporation, adsorption, crystallization, membrane transport, and electrode reactions. In all these cases, there is entropy production in the surface during heat, mass and charge transport, and chemical reactions. We shall see that the entropy production and the flux–force relations imposed by this quantity, can be significant and have a bearing on the whole system performance.

There is an important difference between transport properties of homogeneous phases, and surfaces or contact lines. The difference arises because of the lower symmetry level of surfaces and contact lines. Many homogeneous phases, like for instance fluids, are isotropic. They are symmetric for rotation, translation and parity. From the Curie principle it then follows that processes of a different tensorial character do not couple directly in the force–flux relations. Thus heat and diffusive fluxes, which are vectorial, do not couple to chemical reactions, which are scalar. A surface can only be isotropic for rotation, translation and parity in the plane of the surface. It follows that the normal components of heat and diffusive fluxes are scalar, and therefore *do couple* to chemical reactions in the interfacial region. This implies, for instance, that we can drive a chemical reaction at a membrane surface by temperature or chemical-potential differences across the surface. This is of great practical importance. Similarly, one can drive transport across the membrane by a reaction in the membrane. Similar properties are true for contact lines.

1.5 Scope

This book will review progress obtained the last decade on efforts to extend classical non-equilibrium thermodynamics. Much has been done to substantiate the basic assumptions, and this is reviewed in Chapters 2 to 4. The theory is now very well founded. The fluctuation-dissipation theorem is elucidated for equilibrium (Chapter 2). We shall see that the difference between local and global equilibrium can be found in the nature of the fluctuations. While the fluctuations are short-ranged in global equilibrium, they become long-ranged away from equilibrium. Progress on this issue is reviewed in Chapter 3. State-of-the-art knowledge on the validity of the hypothesis of local equilibrium is reviewed in Chapter 4.

We proceed to present studies of various bulk systems away from equilibrium (Chapter 5 to 7). Diffusion is frequently the rate-limiting step in a chemical process. A formula, derived using the Onsager relations, for the quantitative prediction of binary and ternary Fick diffusion coefficients is reported in Chapter 5. An important computational technique, which can be used to obtain transport coefficients, is non-equilibrium molecular dynamics simulations. The technique, which mimics an experimental situation, gives an essential tool, compared with other methods, *cf.*, Chapter 6. Biological or chemical structures can be maintained in a system by supply of energy and exchange of products with the outside. A chemical example of this non-equilibrium self-assembly is described in Chapter 7.

The entropy production can now be given also for surfaces and contact lines, following Gibbs' procedure using excess thermodynamic variables. For surfaces, the flux–force relations impose dynamic boundary conditions on phase transitions. This is described for the liquid–vapor transition in

Chapter 8, for the liquid–solid transition in Chapter 10, and for gas adsorption in Chapter 9. The porous media, discussed in Chapter 9, represent a challenge, with external as well as internal interfaces. It is concluded that the surface can be favourably dealt with as a separate thermodynamic system in all these cases with absence of equilibrium across the phase boundary. Non-equilibrium thermodynamics brings a new systematic tool to deal with these processes, adding new knowledge to them. Non-equilibrium thermodynamics for contact lines is still in its infancy, as reported in Chapter 8, and so are complex interfaces, see Chapter 17.

The book discusses three applications of the theory relevant to the industry. The coupling of transport processes becomes substantial at membrane surfaces, as discussed in Chapter 11. This is essential for membrane distillation processes. Chapter 12 explains how a detailed modelling of electrochemical cells can be done, including the coupled transport phenomena at the electrode surfaces.

The process industry may, in a not too distant future, have to give annual reports not only on the products that they produce, but also on their annual lost exergy or entropy production. The public sector can favour such a development, by giving benefits to those who limit their entropy production, or reduce the energy dissipation as heat. Chapter 13 on entropy production minimization with optimal control theory can be seen as an effort to create awareness on these issues, and inspire developments in neighbouring fields like control theory.[44] This book, as well as the references cited in this chapter, give guidelines that are needed to understand the nature of the entropy production and can therefore help to avoid it.

The linear regime of the theory covers many applications, but many central processes like chemical reactions, are non-linear. Non-equilibrium thermodynamics has been extended in recent years to include such processes. Non-linear flux–force relations are found by going first to a mesoscopic scale in time and space. This development is described in the end of the book in Chapter 14 to 16, first in general terms in Chapter 14, then for nucleation (Chapter 15), and finally for biological processes (Chapter 16). It is the hope that all these chapters can lay the foundations for further applications and for experiments.

Acknowledgements

ETH Zürich is thanked for guest professorships to S.K. and D.B. in 2014.

References

1. Thomson (Lord Kelvin), *Mathematical and Physical Papers. Collected from Different Scientific Periodicals from May 1841 and to the Present Time*, Cambridge University Press, London, 1884, vol. II.
2. L. Onsager, *Phys. Rev.*, 1931, **37**, 405.
3. L. Onsager, *Phys. Rev.*, 1931, **38**, 2265.

4. *The Collected Works of Lars Onsager*, ed. H. Hemmer, H. Holden and S. K. Ratkje, World Scientific, Singapore, 1996.
5. J. Meixner, *Ann. Phys.*, 1941, **39**, 333.
6. J. Meixner, *Ann. Phys.*, 1942, **41**, 409.
7. J. Meixner, *Ann. Phys.*, 1943, **43**, 244.
8. J. Meixner, *Z. Phys. Chem. B*, 1943, **53**, 235.
9. I. Prigogine, *Etude Thermodynamique des Phenomenes Irreversibles*, Desoer, Liège, 1947.
10. P. Mitchell, *Nature*, 1961, **191**, 144.
11. S. R. de Groot and P. Mazur, *Non-equilibrium Thermodynamics*, North-Holland, Amsterdam, 1962.
12. S. R. de Groot and P. Mazur, *Non-equilibrium Thermodynamics*, Dover, London, 1984.
13. R. Haase, *Thermodynamics of Irreversible Processes*, Addison-Wesley, Reading, MA, 1969.
14. R. Haase, *Thermodynamics of Irreversible Processes*, Dover, London, 1990.
15. A. Katchalsky and P. Curran, *Non-equilibrium Thermodynamics in Biophysics*, Harvard University Press, Cambridge, Massachusetts, 1975.
16. S. R. Caplan and A. Essig, *Bioenergetics and Linear Non-equilibrium Thermodynamics. The Steady State*, Harvard University Press, Cambridge, Massachusetts, 1983.
17. K. S. Førland, T. Førland and S. K. Ratkje, *Irreversible Thermodynamics. Theory and Application*, Wiley, Chichester, 1988.
18. K. S. Førland, T. Førland and S. Kjelstrup, *Irreversible Thermodynamics. Theory and Application*, Tapir, Trondheim, 3rd edn, 2001.
19. V. P. Carey, *Statistical Thermodynamics and Microsale Thermophysics*, Cambridge University Press, Cambridge, 1999.
20. D. Kondepudi and I. Prigogine, *Modern Thermodynamics. From Heat Engines to Dissipative Structures*, Wiley, Chichester, 1998.
21. D. Jou, J. Casas-Vásquez and G. Lebon, *Extended Irreversible Thermodynamics*, Springer, Berlin, 2nd edn, 1996.
22. H. C. Öttinger, *Beyond Equilibrium Thermodynamics*, Wiley-Interscience, Hoboken, 2005.
23. D. D. Fitts, *Non-equilibrium Thermodynamics*, MGraw-Hill, New York, 1962.
24. G. D. C. Kuiken, *Thermodynamics for Irreversible Processes*, Wiley, Chichester, 1994.
25. I. Pagonabarraga, A. Pérez-Madrid and J. M. Rubí, *Physica A*, 1997, **237**, 205.
26. I. Pagonabarraga and J. M. Rubí, *Physica A*, 1992, **188**, 553.
27. D. Reguera and J. M. Rubí, *J. Chem. Phys.*, 2001, **115**, 7100.
28. D. Bedeaux and P. Mazur, *Physica A*, 2001, **298**, 81–100.
29. S. Kjelstrup and D. Bedeaux, *Non-equilibrium Thermodynamics of Heterogeneous Systems*, Series on Advances in Statistical Mechanics, World Scientific, Singapore, 2008, vol. 16.
30. M. Doi, *Soft Matter Physics*, Oxford University Press, Oxford, 2013.

31. J. M. Ortiz de Zárate and J. V. Sengers, *Hydrodynamic Fluctuations in Fluids and Fluid Mixtures*, Elsevier, Amsterdam, 2006.
32. S. Kjelstrup, D. Bedeaux, I. Inzoli and J.-M. Simon, *Energy*, 2008, **33**, 1185.
33. K. G. Denbigh, *Chem. Eng. Sci.*, 1956, **6**, 1.
34. R. B. Bird, E. Stewart and E. N. Lightfoot, *Transport Phenomena*, Wiley, New York, 1960.
35. S. Kjelstrup, D. Bedeaux, E. Johannessen and J. Gross, *Non-equilibrium Thermodynamics for Engineers*, World Scientific, Singapore, 2010.
36. R. Taylor and R. Krishna, *Multicomponent Mass Transfer*, Wiley, New York, 1993.
37. E. L. Cussler, *Diffusion, Mass Transfer in Fluid Systems*, Cambridge University Press, Cambridge, 2nd edn, 1997.
38. Y. Demirel, *Non-equilibrium Thermodynamics*, Elsevier, Amsterdam, 2002, 2nd edn, 2007.
39. R. Krishna and J. A. Wesselingh, *Chem. Eng. Sci.*, 1997, **52**, 861.
40. A. Bejan, *J. Appl. Phys.*, 1996, **79**, 1191.
41. J. Szargut, D. R. Morris and F. R. Steward, *Exergy Analysis of Thermal, Chemical and Metallurgical Processes*, Hemisphere, New York, 1988.
42. R. S. Berry, V. Kazakov, S. Sieniutyz, Z. Szwastand and A. M. Tsirlin, *Thermodynamic Optimization of Finite Time Processes*, Wiley, Chichester, 2000.
43. K. H. Hoffmann, J. Burzler, A. Fischer, M. Schaller and S. Schubert, *J. Non-Equilib. Thermodyn.*, 2003, **28**, 233.
44. L. Luyben, B. Tyreus and M. L. Luyben, *Plantwide Process Control*, McGraw-Hill, New York, 1998.
45. J. Xu, S. Kjelstrup and D. Bedeaux, *Phys. Chem. Chem. Phys.*, 2006, **8**, 2017.
46. J. Ross and P. Mazur, *J. Chem. Phys.*, 1961, **35**, 19.
47. J. S. Rowlinson and B. Widom, *Molecular Theory of Capillarity*, Clarendon, Oxford, 1982.

CHAPTER 2

Fluctuating Hydrodynamics and Fluctuation–Dissipation Theorem in Equilibrium Systems

JOSÉ M. ORTIZ DE ZÁRATE*[a] AND JAN V. SENGERS[b]

[a] Departamento de Física Aplicada I, Facultad de Física, Universidad Complutense, 28040 Madrid, Spain; [b] Institute for Physical Science and Technology, University of Maryland, College Park MD 20742, USA
*Email: jmortizz@fis.ucm.es

2.1 Introduction

The method of fluctuating hydrodynamics for dealing with fluctuations in systems in thermodynamic equilibrium was originally developed by Landau and Lifshitz,[1–3] with relevant subsequent contributions by Fox and Uhlenbeck.[4,5] Fluctuating hydrodynamics was initially formulated for one-component fluid systems[3,4] and later extended, among other developments, to binary mixtures,[6] chemical reactions,[7] nematic liquid crystals[8] and, as further elucidated in Chapter 3, to systems out of equilibrium.

The central idea of fluctuating hydrodynamics is to consider as stochastic variables the fluxes that appear in the expression of the entropy production. As discussed in Chapters 1 and 4, entropy production (dissipation) in thermodynamic systems is related to the transfer of energy and/or momentum between degrees of freedom. Such a transfer is generically associated with molecular collisions or interactions, which are intrinsically random

Experimental Thermodynamics Volume X: Non-equilibrium Thermodynamics with Applications
Edited by Dick Bedeaux, Signe Kjelstrup and Jan V. Sengers

Published by the Royal Society of Chemistry, www.rsc.org

processes. Hence, it seems natural to assume that fluxes are, from a more fundamental perspective, stochastic variables reflecting the random nature of molecular motion. The next important ingredient in fluctuating hydrodynamics is the assumption that the linear flux–force relations (Section 1.3) continue to be valid, on average, in the presence of fluctuations. In practice, this means that eqn (1.2) for any flux J_i is modified into:

$$J_i = \sum_j L_{ij} X_j + \delta J_i, \tag{2.1}$$

where one distinguishes between systematic parts, that continue to be given by eqn (1.2) in terms of the thermodynamic forces X_j and the phenomenological coefficients L_{ij}, and random or fluctuating parts, δJ_i. In the more general case, $\delta J_i(t)$ are functions of time t so that, in the language of van Kampen,[9] the random fluxes are a set of stochastic processes. Hence, instead of considering a single value for a thermodynamic flux at each time, one treats $J_i(t)$ as consisting of a systematic part plus an (infinite) set of possible values of a random part. As further illustrated below, if instead of eqn (1.2), eqn (2.1) is substituted into the corresponding balance laws, one obtains, rather than regular differential equations, a set of stochastic differential equations that describe the evolution of the thermodynamic variables under the influence of random fluxes. The equations of non-equilibrium thermodynamics then transform into a Langevin-like set of stochastic differential equations, where the fluctuating fluxes play the role of random forces. The goal of fluctuating hydrodynamics is to deduce the statistical properties of the thermodynamic variables, such as their mean values or correlations from the statistical properties of the random fluxes. For a system in thermodynamic equilibrium, the first two moments of the Gaussian probability distribution of the random fluxes are:[2,3]

$$\begin{gathered}\langle \delta J_i(t) \rangle = 0, \\ \langle \delta J_i(t) \delta J_j(t') \rangle = k_B (L_{ij} + L_{ji})\, \delta(t - t') = 2 k_B L_{ij}\, \delta(t - t').\end{gathered} \tag{2.2}$$

where eqn (1.3) has been used. The first part of eqn (2.2) states that, by definition, the mean value of the random fluxes must be zero, so that the linear flux–force relations continue to be valid on average. The second part of eqn (2.2) is called the fluctuation–dissipation theorem (FDT). It basically states that the intensity (squared) of the fluctuations is determined by the same coefficients L_{ij} that control the entropy production (dissipation), with k_B being Boltzmann's constant. In the FDT (2.2), it is assumed that, as a consequence of the randomness of the molecular motion, the fluctuations in the fluxes at two different times are uncorrelated; in the language of stochastic-process theory,[9,10] the random fluxes, $\delta J_i(t)$, are a set of Markov processes.

In fluids, the systems considered by non-equilibrium thermodynamics are spatially extended. Then the fluxes (both the systematic and the random part) become fields $J_i(\boldsymbol{r},t)$ that depend also on the position $\boldsymbol{r}$. In addition, the phenomenological coefficients $L_{ij}(\boldsymbol{r})$ may also be dependent on the location. The FDT in these cases is formulated by assuming that random fluxes at different spatial points are uncorrelated, so that

$$\langle \delta J_i(\boldsymbol{r},t)\ \delta J_j(\boldsymbol{r}',t')\rangle = 2k_\mathrm{B}L_{ij}(\boldsymbol{r})\ \delta(\boldsymbol{r}-\boldsymbol{r}')\ \delta(t-t'). \tag{2.3}$$

One sees, either from eqn (2.2) or (2.3), that in a global-equilibrium state the values of the random fluxes have a multivariate spatiotemporal Gaussian probability distribution. Hence, eqn (2.2) or (2.3) completely specify the functional probability of the random fluxes, because one can obtain all moments of the distribution from the second moments. It should be noted that the FDT (*i.e.*, normal Gaussian distribution of fluctuating fluxes) has only been rigorously derived from statistical physics[11–13] for fluctuations in systems that are in equilibrium. Extension of the FDT to non-equilibrium steady states will be discussed in Chapter 3.

The FDT is traditionally rationalized on the following chain of arguments: First, the spatial correlations $\langle \delta J_i(\boldsymbol{r},t)\ \delta J_j(\boldsymbol{r}',t')\rangle$ in eqn (2.3) are expressed with arbitrary undetermined prefactors. Next, the fluctuations of the thermodynamic variables are calculated by substituting eqn (2.1) into the corresponding balance laws. Finally, the prefactors are determined by the requirement that the entropy must be a maximum, so that the equal-time (static) correlation functions among the fluctuating variables have entropy as a probability generating functional.[9,12] We shall return to this issue at the end of Section 2.4.1.

2.2 Fluctuating Hydrodynamics for a One-component Fluid

In the previous section we introduced the general formulation of fluctuating hydrodynamics for any thermodynamic system. To be more specific, we consider now a homogeneous phase of a one-component fluid. The relevant balance laws for this problem were presented in Chapter 1: balance of mass, eqn (1.7), balance of momentum, eqn (1.11), and balance of energy, eqn (1.12), in which we now neglect the term with the electric field (non-conducting and non-polarizable fluid). Fluctuating hydrodynamics in the presence of electric fields is a subject that still needs to be developed. Thus, we identify as the two thermodynamic fluxes associated with this problem, the heat flux $\boldsymbol{J}_q(\boldsymbol{r},t)$ and the viscous pressure tensor $\boldsymbol{\Pi}(\boldsymbol{r},t)$. Indeed in the entropy production (eqn (1.15)), these are the only two terms that remain when all the diffusion fluxes, the electric current and the reaction rate are set to zero. The (deterministic) linear flux–force relations for this problem were also presented in Chapter 1. Following eqn (2.1), we now have to set up the corresponding fluctuating versions. For the heat flux, the deterministic version is eqn (1.18), so that in terms of the experimentally preferred thermal conductivity λ, defined in (1.23), we now obtain:

$$J_q = \lambda T^2 \nabla \left(\frac{1}{T} \right) + \delta J_q, \tag{2.4}$$

where T is the temperature and $\delta J_q(r,t)$ the random heat flux. Since we consider in this section a one-component fluid, we note that there is no distinction between the heat flux and the measurable heat flux, like in eqn (1.13), and there are no diffusion fluxes (*i.e.*, the only one existing component velocity equals the centre of mass velocity, $v_1 = v$). We further notice that, as already mentioned in Chapter 1, in non-equilibrium thermodynamics the spatiotemporal derivatives are global, and it is not needed to indicate which variables are held constant; it is implicitly assumed that they are time and/or the other spatial coordinates.

Eqn (2.4) is the stochastic version of the classical Fourier law for heat conduction. Next, applying the generic FDT (2.3) to the particular case of the random heat flux, we have for an equilibrium system:[14,15]

$$\left\langle \delta J_{q,\alpha}(r,t)\, \delta J_{q,\beta}(r',t') \right\rangle = 2k_B T_0^2 \lambda\, \delta_{\alpha\beta}\, \delta(r - r')\, \delta(t - t'), \tag{2.5}$$

where the subscripts α,β span the three spatial coordinates: $\alpha,\beta \in \{x,y,z\}$, and where T_0 denotes the uniform temperature that characterizes an equilibrium system. As in Chapter 1, we assume the system to be isotropic, so that the thermal conductivity matrix is a constant λ times the unit tensor. For λ in eqn (2.5) the corresponding equilibrium value should be adopted.

The second flux of this problem, *i.e.*, the viscous pressure tensor, was separated into two parts $\Pi = 1/3(\mathrm{Tr}\Pi)\mathbf{1} + \Pi^{st}$ in the linear flux–force relations of Chapter 1. The corresponding linear relations were then eqn (1.29) for the trace and eqn (1.31) for the traceless part. In fluctuating hydrodynamics both are usually combined into a single equation:[3,15]

$$\Pi_{\alpha\beta} = -\eta_v\, \delta_{\alpha\beta}\, \nabla \cdot v - \eta \left(\frac{\partial v_\beta}{\partial x_\alpha} + \frac{\partial v_\alpha}{\partial x_\beta} - \frac{2}{3} \delta_{\alpha\beta}\, \nabla \cdot v \right) + \delta \Pi_{\alpha\beta}, \tag{2.6}$$

where η is the shear viscosity, η_v the bulk viscosity, $v(r,t)$ the fluid flow velocity, and where $\delta\Pi_{\alpha\beta}(r,t)$ are the components of the random viscous pressure tensor. In eqn (2.6) the notation x_α indicates the three spatial coordinates: $\{x_x = x,\ x_y = y,\ x_z = z\}$. As customary in non-equilibrium thermodynamics, it is not explicitly indicated that the other two spatial variables as well as the time are held constant in partial spatial derivatives. Eqn (2.6) is just the stochastic version of the classical Newton's viscosity law. Next, applying the generic FDT (2.3) to the particular case of the random viscous pressure, we have:[3,16]

$$\begin{aligned}\langle \delta\Pi_{\alpha\beta}(r,t)\, \delta\Pi_{\mu\nu}(r',t') \rangle = 2k_B T_0 \Big[&\eta \left(\delta_{\alpha\mu}\delta_{\beta\nu} + \delta_{\alpha\nu}\delta_{\beta\mu} \right) \\ &+ \left(\eta_v - \frac{2}{3}\eta \right) \delta_{\alpha\beta}\delta_{\mu\nu} \Big] \delta(r - r')\delta(t - t'),\end{aligned} \tag{2.7}$$

where the subscripts μ,ν also span the three spatial coordinates: $\mu,\nu \in \{x,y,z\}$.

In many practical applications the fluid flow is divergence free, $\nabla \cdot \boldsymbol{v} = 0$. In these cases the linear flux–force relation for the viscous stress matrix, eqn (2.6), becomes simpler and the associated FDT reduces to:[15]

$$\langle \delta \Pi_{\alpha\beta}(\boldsymbol{r},t)\, \delta \Pi_{\mu\nu}(\boldsymbol{r}',t') \rangle = 2k_\mathrm{B} T_0 \eta (\delta_{\alpha\mu}\delta_{\beta\nu} + \delta_{\alpha\nu}\delta_{\beta\mu})\, \delta(\boldsymbol{r}-\boldsymbol{r}')\delta(t-t'), \tag{2.8}$$

without any contribution from the bulk viscosity η_v.

Next, we substitute the stochastic linear relations, eqn (2.4) and (2.6), into the corresponding balance laws, eqn (1.11) for the momentum balance and eqn (1.12) for the energy balance. This procedure then yields the fluctuating hydrodynamics equations for a one-component fluid, namely:

$$\frac{\mathrm{d}\rho}{\mathrm{d}t} = -\rho\, \nabla \cdot \boldsymbol{v}, \tag{2.9a}$$

$$\rho \frac{\mathrm{d}\boldsymbol{v}}{\mathrm{d}t} = -\nabla p + \eta\, \nabla^2 \boldsymbol{v} + \left(\frac{1}{3}\eta + \eta_v\right) \nabla(\nabla \cdot \boldsymbol{v}) - \nabla \cdot \delta \boldsymbol{\Pi}, \tag{2.9b}$$

$$\rho \frac{\mathrm{d}u}{\mathrm{d}t} = \lambda\, \nabla^2 T - p\, \nabla \cdot \boldsymbol{v} - \nabla \cdot \delta \boldsymbol{J}_q, \tag{2.9c}$$

where, following the nomenclature of Chapter 1, $\rho(\boldsymbol{r},t)$ is the mass density, $p(\boldsymbol{r},t)$ the pressure and $u(\boldsymbol{r},t)$ the specific internal energy. In eqn (2.9) it is further assumed that the transport coefficients (thermal conductivity and viscosities) are constant, independent of position. Eqn (2.9a) simply reproduces the balance of mass, eqn (1.7). Notice that in eqn (2.9c) viscous heating is neglected in the energy balance (eqn (1.12)). This is the term with a double contraction of the viscous pressure matrix and the velocity derivatives matrix. This term is of second order in the fluxes, so that, in non-equilibrium thermodynamics at the linear level, it may be neglected. It is convenient to convert eqn (2.9c) for the specific energy into an equation for the temperature by using the thermodynamic relations

$$\mathrm{d}u = T\,\mathrm{d}s + p\frac{\mathrm{d}\rho}{\rho^2}, \qquad T\,\mathrm{d}s = c_\mathrm{p}\,\mathrm{d}T + \frac{\alpha T}{\rho}\mathrm{d}p,$$

where α is the thermal expansion coefficient, c_p the specific heat capacity at constant pressure, and $s(\boldsymbol{r},t)$ the specific entropy. Then one has:

$$\rho c_\mathrm{p} \frac{\mathrm{d}T}{\mathrm{d}t} + \alpha T \frac{\mathrm{d}p}{\mathrm{d}t} = \lambda \nabla^2 T - \nabla \cdot \delta \boldsymbol{J}_q. \tag{2.9d}$$

In addition, the thermodynamic equation of state $\rho = \rho(p,T)$ is needed. Hence, eqn (2.9a to d) represent a set of five coupled stochastic partial differential equations for five unknowns: the three velocity components and a pair of thermodynamic fields: $p(\boldsymbol{r},t)$ and $T(\boldsymbol{r},t)$, for instance. The two random fluxes act as forcing terms and, in the language of stochastic processes theory,[9,10] they are Gaussian white noise with intensity given by eqn (2.5) and (2.7).

2.3 Fluctuating Hydrodynamics for a Binary Mixture

In this section we present the fluctuating hydrodynamics equations for a fluid mixture of two nonreacting and nonconducting components so that, initially, there are two mass densities and two diffusion fluxes. However, the two diffusion fluxes, $\boldsymbol{J}_1$ and $\boldsymbol{J}_2$, are not independent since, as pointed out after eqn (1.7), $\boldsymbol{J}_1+\boldsymbol{J}_2=0$. Hence, eqn (1.16) for the entropy production simplifies to:

$$\sigma=\boldsymbol{J}'_q\cdot\boldsymbol{\nabla}\left(\frac{1}{T}\right)+\boldsymbol{J}_1\cdot\left[-\frac{1}{T}\boldsymbol{\nabla}_{\mathrm{T}}\mu\right]+\boldsymbol{\Pi}:\left(-\frac{\boldsymbol{\nabla v}}{T}\right), \tag{2.10}$$

where, in the absence of any electric current and of any chemical reaction, a single independent diffusion flux appears. For the sake of shortening notation, we use $\mu=\mu_1-\mu_2$ in eqn (2.10) where μ_1 and μ_2 are the specific chemical potentials of the two components, so that $\nabla_{\mathrm{T}}\mu$ represents the gradient at constant temperature of the chemical potential difference (see Chapter 1). One concludes from eqn (2.10) that in a binary mixture, in addition to the (measurable) heat flux and the viscous pressure tensor, we need to consider one extra thermodynamic flux: the diffusion $\boldsymbol{J}_1$ which is a vector flux. Since there is no coupling between the viscous pressure matrix and the two vector fluxes, the stochastic version of the pressure tensor is the same as for a one-component fluid, *i.e.*, eqn (2.6). Similarly, the FDT for the random viscous pressure matrix continues to be given by eqn (2.7) above in the general case, or by eqn (2.8) for divergence-free flows.

As elaborated in Section 1.3.1, diffusion and heat fluxes are coupled, which complicates the formulation of the stochastic version of the associated linear flux–force relations and the corresponding FDT. Additional complications arise from the fact that in Chapter 1 the linear flux–force relations were formulated in terms of velocities (no diffusion fluxes) and set up in the inverse way: forces as a function of fluxes, instead of fluxes as a function of forces. Anyway, setting $n=2$ and $\boldsymbol{j}=0$ in eqn (1.26), in the absence of any electrical resistances, we obtain in terms of Maxwell–Stefan diffusivities D_{12}:

$$\begin{aligned}
\boldsymbol{\nabla}T&=-\frac{1}{\lambda}\left[\boldsymbol{J}'_q-c_1q_1^*(\boldsymbol{v}_1-\boldsymbol{v}_2)\right],\\
\frac{\rho_1}{T}\boldsymbol{\nabla}_{\mathrm{T}}\mu_1&=-\frac{c_1q_1^*}{T^2}\boldsymbol{\nabla}T-cR\frac{x_1x_2}{D_{12}}(\boldsymbol{v}_1-\boldsymbol{v}_2),\\
\frac{\rho_2}{T}\boldsymbol{\nabla}_{\mathrm{T}}\mu_2&=-\frac{c_2q_2^*}{T^2}\boldsymbol{\nabla}T-cR\frac{x_1x_2}{D_{21}}(\boldsymbol{v}_2-\boldsymbol{v}_1).
\end{aligned}\tag{2.11}$$

In eqn (2.11) ρ_{k} is the mass density, x_{k} the mole fraction, $c_{\mathrm{k}}=\rho_{\mathrm{k}}/M_{\mathrm{k}}$ with M_{k} being the molar mass, q_{k}^* the measurable heat of transfer, and $\boldsymbol{v}_{\mathrm{k}}$ the velocity of component k. Of course, the three eqn (2.11) are not independent; since the second and the third add to zero, as required by the

Gibbs–Duhem eqn (1.19). By subtracting the second and the third eqn (2.11) we obtain,

$$\frac{\nabla_{\mathrm{T}}(\mu_1-\mu_2)}{T}=\frac{\nabla_{\mathrm{T}}\mu}{T}=-\frac{\rho c_1 q_1^*}{\rho_1\rho_2 T^2}\nabla T-cR\frac{\rho^2 x_1 x_2}{\rho_1^2\rho_2^2 D_{12}}\boldsymbol{J}_1, \tag{2.12a}$$

while we can rewrite the first as

$$\boldsymbol{J}'_q=-\lambda\nabla T+c_1 q_1^*\frac{\rho}{\rho_1\rho_2}\boldsymbol{J}_1. \tag{2.12b}$$

In eqn (2.12) we used $\boldsymbol{v}_1=\boldsymbol{v}+\boldsymbol{J}_1/\rho_1$, $\boldsymbol{v}_2=\boldsymbol{v}-\boldsymbol{J}_1/\rho_2$ and eqn (1.24). Eqn (2.12) are, for a binary mixture, the two independent linear flux–force relations. Solving for the two vector fluxes appearing in eqn (2.10) we arrive at

$$\begin{aligned}\boldsymbol{J}'_q&=-\lambda\left(1+\frac{(c_1q_1^*)^2}{\lambda T^2}\frac{D_{12}}{x_1x_2cR}\right)\nabla T-c_1q_1^*\frac{\rho_1\rho_2 D_{12}}{\rho c x_1 x_2}\frac{\nabla_{\mathrm{T}}\mu}{RT},\\ \boldsymbol{J}_1&=-c_1q_1^*\frac{\rho_1\rho_2D_{12}}{\rho c x_1x_2}\frac{\nabla T}{RT^2}-\frac{\rho_1^2\rho_2^2 D_{12}}{\rho^2 c x_1 x_2}\frac{\nabla_{\mathrm{T}}\mu}{RT}.\end{aligned} \tag{2.13}$$

Note how the Onsager relation (eqn (1.3)) also holds when inverting the linear flux laws of Chapter 1, so as to formulate it according to eqn (1.2). Instead of the chemical-potential gradient, in hydrodynamics one more commonly uses the concentration gradient. They are related by

$$\nabla_{\mathrm{T}}\mu=\left(\frac{\partial\mu}{\partial w}\right)_{\mathrm{T,p}}\nabla w.$$

We neglect baro-diffusion here, consistent with having neglected the pressure gradient term in the Gibbs–Duhem eqn (1.19). In the expression above, $w=\rho_1/\rho$ is the concentration of component 1 in weight fraction. Next, substitution of the above expression into eqn (2.12) yields, in terms of the common transport coefficients,[3] Fick diffusivity D (not to be confused with the Stefan–Maxwell diffusivity D_{12}) and thermal diffusion ratio k_{T},

$$\begin{aligned}\boldsymbol{J}'_q&=-\lambda\nabla T+k_{\mathrm{T}}\left(\frac{\partial\mu}{\partial w}\right)_{\mathrm{T,p}}\boldsymbol{J}_1,\\ \boldsymbol{J}_1&=-\rho D\left[\nabla w+\frac{k_{\mathrm{T}}}{T}\nabla T\right].\end{aligned} \tag{2.14}$$

Comparison of eqn (2.12) and (2.14) enables us to identify the relationship between the transport coefficients more commonly used in hydrodynamics and the transport coefficients of Chapter 1, namely

$$\frac{\rho c_1 q_1^*}{\rho_1\rho_2}=k_{\mathrm{T}}\left(\frac{\partial\mu}{\partial w}\right)_{\mathrm{T,p}},\qquad \rho DT=\frac{\rho_1^2\rho_2^2 D_{12}}{\rho^2 cRx_1x_2}\left(\frac{\partial\mu}{\partial w}\right)_{\mathrm{T,p}}. \tag{2.15}$$

From eqn (2.15),

$$\rho D\frac{k_T}{T}=\frac{c_1 q_1^*}{T^2}\frac{\rho_1\rho_2 D_{12}}{\rho c R x_1 x_2}.$$

Hence, eqn (2.12) and (2.14) are indeed consistent. We note that, making use of eqn (1.24), the first of eqn (2.15) may be rewritten in a more symmetric way, as the product of a thermodynamic factor times the difference of heats of transport:

$$k_T=\left[\frac{q_1^*}{M_1}-\frac{q_2^*}{M_2}\right]\left(\frac{\partial\mu}{\partial w}\right)_{T,p}^{-1},$$

which is the starting point of many thermodiffusion theories.[17,18]

Now, we can set up the stochastic version of the linear flux–force relations (2.13). Using eqn (2.14), we obtain

$$\boldsymbol{J}'_q=-\lambda\left[\left(1+\frac{D\varepsilon_D}{a}\right)\nabla T+\frac{DT}{k_T a}\varepsilon_D\nabla w\right]+\delta\boldsymbol{J}'_q, \quad (2.16a)$$

$$\boldsymbol{J}_1=-\rho D\left[\nabla w+\frac{k_T}{T}\nabla T\right]+\delta\boldsymbol{J}_1, \quad (2.16b)$$

where, following the general guidelines of Section 2.1, we introduce a random (measurable) heat flux $\delta\boldsymbol{J}'_q(\boldsymbol{r},t)$ and a random diffusion flux $\delta\boldsymbol{J}_1(\boldsymbol{r},t)$. Note that the random contribution $\delta\boldsymbol{J}_2(\boldsymbol{r},t)$ to the diffusion flux of the second component equals $-\delta\boldsymbol{J}_1(\boldsymbol{r},t)$, since the requirement that the two diffusion fluxes add to zero holds in general, not only for the average values. Furthermore, in eqn (2.16) we introduced the thermal diffusivity $a=\lambda/\rho c_p$, and the dimensionless Dufour effect ratio:

$$\varepsilon_D=\frac{k_T^2}{c_p T}\left(\frac{\partial\mu}{\partial w}\right)_{T,p}=\frac{\rho^2(c_1 q_1^*)^2}{c_p T\rho_1^2\rho_2^2}\left(\frac{\partial w}{\partial\mu}\right)_{p,T}=\frac{D_{12}}{D}\frac{(c_1 q_1^*)^2}{\rho c R c_p T^2 x_1 x_2}. \quad (2.17)$$

Next, following again the general guidelines from eqn (2.3), we obtain from eqn (2.13) the FDT for a binary mixture in equilibrium as[6,15,19]

$$\left\langle\delta J'_{q,\alpha}(\boldsymbol{r},t)\,\delta J'_{q,\beta}(\boldsymbol{r}',t')\right\rangle=2k_B T_0^2\lambda\left(1+\frac{D\varepsilon_D}{a}\right)\delta_{\alpha\beta}\,\delta(\boldsymbol{r}-\boldsymbol{r}')\,\delta(t-t'), \quad (2.18a)$$

$$\left\langle\delta J'_{q,\alpha}(\boldsymbol{r},t)\,\delta J_{1,\beta}(\boldsymbol{r}',t')\right\rangle=2k_B T_0\rho D k_T\,\delta_{\alpha\beta}\,\delta(\boldsymbol{r}-\boldsymbol{r}')\,\delta(t-t'), \quad (2.18b)$$

$$\langle\delta J_{1,\alpha}(\boldsymbol{r},t)\,\delta J_{1,\beta}(\boldsymbol{r}',t')\rangle=2k_B T_0\rho D\left(\frac{\partial w}{\partial\mu}\right)_{T,p}\delta_{\alpha\beta}\,\delta(\boldsymbol{r}-\boldsymbol{r}')\,\delta(t-t'), \quad (2.18c)$$

where all thermophysical properties are to be evaluated at their respective equilibrium values. The fluctuating hydrodynamics equations for a binary mixture are obtained by substitution of the stochastic linear flux–force relations into the balance laws. Regarding momentum balance, the only flux involved is the viscous pressure matrix, so that the fluctuating Navier–Stokes equation for a binary mixture is the same as for a one-component fluid, *i.e.*, eqn (2.9b).

For the mass balance, there are initially two equations in eqn (1.6), namely, for the two partial mass densities, ρ_1 and ρ_2. In practice, one often adopts as the independent variable only one of the partial densities, say ρ_1, and the total mass density ρ. The balance law for the total density ρ is the same as eqn (2.9a) for a one-component fluid. In many cases, the weight fraction $w = \rho_1/\rho$, whose balance law is given by eqn (1.10), is used as the concentration variable. In summary, for mass balance in fluctuating hydrodynamics of a binary mixture one has to consider, in addition to eqn (2.9a),

$$\frac{\mathrm{d}w}{\mathrm{d}t} = D\left[\nabla^2 w + \frac{k_\mathrm{T}}{T}\nabla^2 T\right] - \frac{1}{\rho}\boldsymbol{\nabla}\cdot\delta\boldsymbol{J}_1, \tag{2.19}$$

obtained upon substitution of the fluctuating linear relation, eqn (2.16b), into eqn (1.10).

The relevant energy balance law is eqn (1.12) which contains the total heat flux. Since the linear flux–force relations, eqn (2.12), are in terms of the measurable heat flow, we need first to express the right-hand side of eqn (1.12) in terms of $\boldsymbol{J}'_q$,

$$\rho\frac{\mathrm{d}u}{\mathrm{d}t} = -\boldsymbol{\nabla}\cdot\left(\boldsymbol{J}'_q - \sum_{j=1}^{2} h_j\boldsymbol{J}_j\right) - p\,\boldsymbol{\nabla}\cdot\boldsymbol{v},$$

where h_j are the specific partial enthalpies and we have again neglected viscous heating as was done for a one-component fluid. Next, we use eqn (1.5) to switch from internal energy to entropy as variable, so that:

$$\rho\left(T\frac{\mathrm{d}s}{\mathrm{d}t} + \frac{p}{\rho^2}\frac{\mathrm{d}\rho}{\mathrm{d}t} + \sum_{j=1}^{2}\mu_j\frac{\mathrm{d}w_j}{\mathrm{d}t}\right) = -\boldsymbol{\nabla}\cdot\left(\boldsymbol{J}'_q + \sum_{j=1}^{2} h_j\boldsymbol{J}_j\right) - p\boldsymbol{\nabla}\cdot\boldsymbol{v}.$$

By using total mass balance, eqn (1.7), and partial mass balances, eqn (1.10), the above expression simplifies to:

$$\rho T\frac{\mathrm{d}s}{\mathrm{d}t} = -\boldsymbol{\nabla}\cdot\boldsymbol{J}'_q + \sum_{j=1}^{2}(\mu_j - h_j)\boldsymbol{\nabla}\cdot\boldsymbol{J}_j - \sum_{j=1}^{2}\boldsymbol{J}_j\cdot\boldsymbol{\nabla}h_j,$$

which, when further transformed by casting it in the form of eqn (1.4), was indeed used to identify the entropy flux and the entropy production in eqn (1.14) and (1.15) (recall that $\mu_j = h_j - Ts_j$, with s_j the specific partial entropies). However, for our present purpose, we may neglect the last term in the right-hand side of the equation above, since it is, like viscous heating, of second order in the fluxes. Hence, we finally obtain:

$$\rho T \frac{\mathrm{d}s}{\mathrm{d}t} = -\boldsymbol{\nabla}\cdot\boldsymbol{J}'_q - T\sum s_j \boldsymbol{\nabla}\cdot\boldsymbol{J}_j = -\boldsymbol{\nabla}\cdot\boldsymbol{J}'_q + T\sum_{j=1}^{2}\left(\frac{\partial \mu_j}{\partial T}\right)_{\mathrm{p,w}} \boldsymbol{\nabla}\cdot\boldsymbol{J}_j,$$

which is the working equation for the energy balance in terms of the same fluxes used to set up the linear flux–force relations. Of course, for practical use, as in eqn (2.9d) for a one-component fluid, it is convenient to switch from entropy to measurable variables: temperature, pressure, *etc.*, by using thermodynamic relations. For a binary mixture, we have

$$T\mathrm{d}s = c_\mathrm{p}\,\mathrm{d}T + \frac{\alpha T}{\rho}\mathrm{d}p - T\left(\frac{\partial \mu}{\partial T}\right)_{\mathrm{p,w}} \mathrm{d}w.$$

Using the equation above in the energy balance, with eqn (1.10), we obtain

$$\rho c_\mathrm{p}\frac{\mathrm{d}T}{\mathrm{d}t} + \alpha T\frac{\mathrm{d}p}{\mathrm{d}t} = -\boldsymbol{\nabla}\cdot\boldsymbol{J}'_q, \tag{2.20}$$

which explains why $\boldsymbol{J}'_q$ is referred to as the *measurable heat* flux. Next, substituting the stochastic linear flux–force relation (2.16a) into the expression (2.20) above, we obtain the fluctuating hydrodynamics equation for the energy balance

$$\frac{\mathrm{d}T}{\mathrm{d}t} + \frac{\alpha T}{\rho c_\mathrm{p}}\frac{\mathrm{d}p}{\mathrm{d}t} = (a + D\varepsilon_\mathrm{D})\nabla^2 T + \frac{DT}{k_\mathrm{T}}\varepsilon_\mathrm{D}\,\nabla^2 w - \frac{1}{\rho c_\mathrm{p}}\,\boldsymbol{\nabla}\cdot\delta\boldsymbol{J}'_q. \tag{2.21}$$

Notice that, when $\varepsilon_\mathrm{D} = 0$, the stochastic heat equation for a binary mixture is the same as for a one-component fluid, eqn (2.9d). As anticipated, eqn (2.17) gives a quantitative measure of the relevance of the Dufour effect.

This completes the formulation of the fluctuating hydrodynamics equations for a binary mixture. In summary, compared with a one-component fluid, one has an extra thermodynamic variable, the concentration w, and an extra random diffusion flux $\delta\boldsymbol{J}_1$. The FDT couples the random diffusion flux and the random (measurable) heat flux, whose correlations are now given by eqn (2.18). The FDT for the random stress matrix is the same as for a one-component fluid, eqn (2.7) or (2.8). The fluctuating hydrodynamics equation for the total mass balance is eqn (2.9a) and for the momentum balance eqn (2.9b), which are the same as for a one-component fluid. In addition, one has eqn (2.19) for the partial density balance and eqn (2.21) for the energy balance. Because of the Dufour effect, the latter is slightly different from the fluctuating-energy balance for a one-component fluid.

2.4 Some Examples

2.4.1 Temperature Fluctuations at Constant Pressure

A deterministic stationary solution of the fluctuating hydrodynamics eqn (2.9) for a one-component fluid in equilibrium is: $\boldsymbol{v}(\boldsymbol{r},t) = 0$, $\rho(\boldsymbol{r},t) = \rho_0$,

$p(\boldsymbol{r},t) = p_0$ and $T(\boldsymbol{r},t) = T_0$, where ρ_0, p_0 and T_0 are the equilibrium density, pressure and temperature, independent of $\boldsymbol{r}$ and t, and related by an equation of state $\rho_0 = \rho(p_0, T_0)$. Due to random thermal forcing, spontaneous fluctuations around this equilibrium system will appear. The spatiotemporal evolution of these fluctuations can be studied by substituting $\rho(\boldsymbol{r},t) = \rho_0 + \delta\rho(\boldsymbol{r},t)$, $T(\boldsymbol{r},t) = T_0 + \delta T(\boldsymbol{r},t)$, $p(\boldsymbol{r},t) = p_0 + \delta p(\boldsymbol{r},t)$ and $\boldsymbol{v}(\boldsymbol{r},t) = 0 + \delta\boldsymbol{v}(\boldsymbol{r},t)$ into all the derivatives contained in the fluctuating hydrodynamics eqn (2.9), so as to obtain

$$\frac{\mathrm{d}(\delta\rho)}{\mathrm{d}t} = \frac{\partial(\delta\rho)}{\partial t} + \delta\boldsymbol{v} \cdot \boldsymbol{\nabla}(\delta\rho) = -\rho\, \boldsymbol{\nabla} \cdot \delta\boldsymbol{v}, \tag{2.22a}$$

$$\rho \frac{\mathrm{d}(\delta\boldsymbol{v})}{\mathrm{d}t} = -\boldsymbol{\nabla}(\delta p) + \eta\, \nabla^2(\delta\boldsymbol{v}) + \left(\eta_\mathrm{v} + \frac{1}{3}\eta\right) \boldsymbol{\nabla}(\boldsymbol{\nabla} \cdot \delta\boldsymbol{v}) - \boldsymbol{\nabla} \cdot \delta\boldsymbol{\Pi}, \tag{2.22b}$$

$$\rho c_\mathrm{p} \frac{\mathrm{d}(\delta T)}{\mathrm{d}t} + \alpha T \frac{\mathrm{d}(\delta p)}{\mathrm{d}t} = \lambda \nabla^2(\delta T) - \boldsymbol{\nabla} \cdot \delta\boldsymbol{J}_q. \tag{2.22c}$$

Eqn (2.22) represent the most general fluctuating hydrodynamics equations for a one-component fluid in equilibrium. Since fluctuations are small, eqn (2.22) can be linearized in the fluctuating thermodynamic variables, *i.e.*, replacing material time derivatives by partial time derivatives. They then can be readily solved by applying full spatiotemporal Fourier transforms, so as to obtain the spectrum of the fluctuations. The calculations are rather straightforward and have been presented in the literature,[4,15] leading to the well-known Rayleigh–Brillouin spectrum for the fluctuations.

For many applications one can use a Boussinesq approximation, namely assume that the equation of state can be approximated by $\rho = \rho_0 - \alpha\rho_0(T - T_0)$, or $\delta\rho = -\alpha\rho_0\delta T$, assuming that the isothermal compressibility is negligibly small, and fluctuations in pressure are thermodynamically independent of temperature or density fluctuations. This approximation is referred to as the Boussinesq approximation[20,21] in the context of fluids subjected to temperature gradients. The Boussinesq approximation is generally good for liquids that can sustain large temperature differences (fluctuations) with negligible small associated (through the equation of state) pressure differences. The Boussinesq approximation is not good for gases. A full rigorous discussion of the Boussinesq approximation[22] requires switching to dimensionless variables and a series expansion in some small parameters. As discussed in the relevant literature,[20,22] one obtains

$$0 = \boldsymbol{\nabla} \cdot \delta\boldsymbol{v}, \tag{2.23a}$$

$$\rho_0 \frac{\mathrm{d}(\delta\boldsymbol{v})}{\mathrm{d}t} = -\boldsymbol{\nabla}(\delta p) + \eta\, \nabla^2(\delta\boldsymbol{v}) - \boldsymbol{\nabla} \cdot \delta\boldsymbol{\Pi}, \tag{2.23b}$$

$$\rho_0 c_\mathrm{p} \frac{\mathrm{d}(\delta T)}{\mathrm{d}t} = \lambda \nabla^2(\delta T) - \boldsymbol{\nabla} \cdot \delta\boldsymbol{J}_q. \tag{2.23c}$$

In the Boussinesq approximation, the (fluctuating) flow is divergence free, eqn (2.23a). Furthermore, the temperature fluctuations are uncoupled from pressure fluctuations, and only nonlinearly coupled to velocity fluctuations through their appearance in material derivatives. However, since fluctuations are small, one can linearize in the fluctuating fields, in which case temperature fluctuations can be directly obtained from eqn (2.23c). Indeed, upon applying a full spatiotemporal Fourier transform to (linearized) eqn (2.23c) one obtains:

$$\delta T(\omega, \boldsymbol{q}) = \frac{-\mathrm{i} q_\alpha \delta J_{\mathrm{q},\alpha}(\omega, \boldsymbol{q})}{\rho_0 c_\mathrm{p}(\mathrm{i}\omega + a q^2)}, \tag{2.24}$$

where ω is the frequency and $\boldsymbol{q}$ the wave vector of the fluctuations, and summation over repeated indices is understood. To deduce the correlation function of the temperature fluctuations, one needs the correlation function of the random force appearing in the right-hand side of eqn (2.23c), which is obtained by applying Fourier transforms to the corresponding FDT, eqn (2.5). Introducing the notation $F_1(\omega,\boldsymbol{q}) = -\mathrm{i}(\rho_0 c_\mathrm{p})^{-1} q_\alpha \delta J_{q,\alpha}(\omega,\boldsymbol{q})$, one arrives at:[14–16]

$$\langle F_1^*(\omega, \boldsymbol{q})\, F_1(\omega', \boldsymbol{q}')\rangle = \frac{2k_\mathrm{B}\lambda T^2}{\rho^2 c_\mathrm{p}^2} q^2 (2\pi)^4 \delta(\omega - \omega')\, \delta(\boldsymbol{q} - \boldsymbol{q}'), \tag{2.25}$$

where the subscript 0 has been dropped from both the equilibrium density ρ_0 and the equilibrium temperature T_0 in the prefactor of the FDT, eqn (2.5). From eqn (2.25) one sees that the correlation function of the temperature fluctuations will be proportional to delta functions in frequency and wave vector, namely:

$$\langle \delta T^*(\omega,\boldsymbol{q})\ \delta T(\omega',\boldsymbol{q}')\rangle = S(\omega,q)(2\pi)^4\ \delta(\omega - \omega')\delta(\boldsymbol{q} - \boldsymbol{q}'), \tag{2.26}$$

where $S(\omega,q)$ is a structure factor representing the spectrum of the temperature fluctuations at a given wave vector. Fluctuations with different frequencies or wave vectors are uncorrelated. The structure factor depends only on the magnitude q of the wave vector $\boldsymbol{q}$, meaning that the spectrum is spatially isotropic. From eqn (2.24) and making use of eqn (2.25) and (2.26), one obtains the explicit expression:

$$S(\omega, \boldsymbol{q}) = \frac{k_\mathrm{B} T^2}{\rho c_\mathrm{p}} \frac{2aq^2}{\omega^2 + a^2 q^4}. \tag{2.27}$$

Eqn (2.27) represents a single Lorentzian centred at $\omega = 0$, known as Rayleigh line in the frequency spectrum. The Boussinesq approximation is applicable when the Rayleigh line and the Brillouin lines are well separated.

Next, applying double inverse Fourier transforms in the two frequencies to eqn (2.26) we obtain

$$\langle \delta T^*(\boldsymbol{q},t)\ \delta T(\boldsymbol{q}',t')\rangle = S(q,|t - t'|)(2\pi)^3\ \delta(\boldsymbol{q} - \boldsymbol{q}'), \tag{2.28}$$

with $S(q,\tau)$ being the time correlation function of temperature fluctuations. From eqn (2.27)

$$S(q,\tau) = \frac{k_B T^2}{\rho c_p} \exp(-a q^2 \tau), \tag{2.29}$$

and we see that temperature fluctuations in equilibrium decay diffusively with a single decay rate aq^2 that is proportional to the square of the wave number. Finally, applying inverse Fourier transforms to eqn (2.28) in the two wave vectors and setting $t = t'$, we obtain the real-space equal-time correlation of temperature fluctuations in equilibrium, namely:

$$\langle \delta T^*(\mathbf{r},t)\,\delta T(\mathbf{r}',t)\rangle = \frac{k_B T^2}{\rho c_p}\,\delta(\mathbf{r}-\mathbf{r}'), \tag{2.30}$$

that is proportional to a delta function, so that density fluctuations in equilibrium at two different spatial points are uncorrelated, and equal-time fluctuations are spatially short-ranged.

To conclude this section, it is interesting to consider fluctuations at the same spatial location. The value of the delta function when $\mathbf{r} = \mathbf{r}'$ is the inverse of the volume of one fluid element (or point) V^{-1}. With the Boussinesq-like incompressible assumption adopted in this section, fluctuations in the (internal) energy density are $\delta u = c_p \delta T$, and the fluctuations in the total energy U contained in an element of volume V are $\delta U = V\rho c_p \delta T$. Hence, setting $\mathbf{r} = \mathbf{r}'$ in eqn (2.30) we obtain

$$\langle (\delta U)^2 \rangle = V^2 \rho^2 c_p^2 \frac{k_B T^2}{\rho c_p} \frac{1}{V} = C_p\, k_B T^2,$$

where $C_p = \rho V c_p$ is the total heat capacity. Of course, one recognizes above (for the particular case of an incompressible system) the expression for the energy fluctuations in the canonical ensemble, as presented in any textbook on classical statistical physics.[2] This result confirms the consistency of fluctuating hydrodynamics and the fluctuation–dissipation theorem.

2.4.2 Concentration Fluctuations at Large Lewis Number

The fluctuating hydrodynamics equations for a binary mixture are eqn (2.9a,b), (2.19) and (2.21). Similarly to the case of a one-component fluid, a deterministic stationary solution is: $\mathbf{v}(\mathbf{r},t) = 0$, $\rho(\mathbf{r},t) = \rho_0$, $p(\mathbf{r},t) = p_0$, $w(\mathbf{r},t) = w_0$ and $T(\mathbf{r},t) = T_0$, where ρ_0, p_0 w_0 and T_0 are the equilibrium density, pressure, concentration and temperature, independent of $\mathbf{r}$ and t, and related by an equation of state $\rho_0 = \rho(p_0, T_0, w_0)$. Again, due to random thermal forcing, spontaneous fluctuations around these equilibrium values will appear. The spatiotemporal evolution of these fluctuations can be studied, in general, by substituting $\rho(\mathbf{r},t) = \rho_0 + \delta\rho(\mathbf{r},t)$, $T(\mathbf{r},t) = T_0 + \delta T(\mathbf{r},t)$, $p(\mathbf{r},t) = p_0 + \delta p(\mathbf{r},t)$, $w(\mathbf{r},t) = w_0 + \delta w(\mathbf{r},t)$ and $\mathbf{v}(\mathbf{r},t) = 0 + \delta \mathbf{v}(\mathbf{r},t)$ into all the derivatives contained in the fluctuating hydrodynamics equations for a binary mixture, so as to obtain:

$$\frac{\mathrm{d}(\delta\rho)}{\mathrm{d}t} = -\rho(\nabla\cdot\delta v), \tag{2.31a}$$

$$\rho\frac{\mathrm{d}(\delta\boldsymbol{v})}{\mathrm{d}t} = -\nabla(\delta p) + \eta\,\nabla^2(\delta\boldsymbol{v}) + \left(\eta_\mathrm{v} + \frac{1}{3}\eta\right)\nabla(\nabla\cdot\delta\boldsymbol{v}) - \nabla\cdot\delta\boldsymbol{\Pi}, \tag{2.31b}$$

$$\frac{\mathrm{d}(\delta w)}{\mathrm{d}t} = D\left[\nabla^2(\delta w) + \frac{k_\mathrm{T}}{T}\nabla^2(\delta T)\right] - \frac{1}{\rho}\nabla\cdot\delta\boldsymbol{J}_1, \tag{2.31c}$$

$$\frac{\mathrm{d}(\delta T)}{\mathrm{d}t} + \frac{\alpha T}{\rho c_\mathrm{p}}\frac{\mathrm{d}(\delta p)}{\mathrm{d}t} = (a + D\varepsilon_\mathrm{D})\nabla^2(\delta T) + \frac{DT}{k_\mathrm{T}}\varepsilon_\mathrm{D}\nabla^2(\delta w) - \frac{1}{\rho c_\mathrm{p}}\nabla\cdot\delta\boldsymbol{J}'_q. \tag{2.31d}$$

This is a system of six coupled partial stochastic differential equations. While upon linearization in the fluctuating fields and application of Fourier transforms eqn (2.31) can be analytically solved in general,[6,19] in many practical applications the following approximations apply:

1. A Boussinesq-like approximation,[20,21] similar to the one adopted in Section 2.4.1 for a one-component fluid. This approximation transforms the total mass balance (2.31a) into a divergence-free flow condition, and uncouples pressure from the heat eqn (2.31d).
2. Neglecting the Dufour effect: $\varepsilon_\mathrm{D} = 0$. Although Dufour effect is important for gases, because of the presence of a c_p^{-1} term in eqn (2.17), it can be neglected in most liquid mixtures.[23]
3. Considering the large Lewis-number limit.[24] The Lewis number is $\mathrm{Le} = a/D$, which is indeed large for many liquid mixtures. In the limit $\mathrm{Le} \to \infty$, any coupling between the concentration fluctuations and the temperature fluctuations vanishes.[24,25] Moreover, since a large Lewis number also implies that velocity fluctuations decay much faster than temperature and concentration fluctuations, the time derivative in the left-hand side of eqn (2.31b) can also been neglected. In some references,[26] this approximation is referred to as large-Schmidt-number approximation.

With the approximations listed above the fluctuating hydrodynamics equations for fluctuations around an equilibrium state of a binary mixture simplify to:

$$0 = \nabla\cdot\delta\boldsymbol{v}, \tag{2.32a}$$

$$0 = -\nabla(\delta p) + \eta\nabla^2(\delta\boldsymbol{v}) - \nabla\cdot\delta\boldsymbol{\Pi}, \tag{2.32b}$$

$$\frac{\mathrm{d}(\delta w)}{\mathrm{d}t} = \frac{\partial(\delta w)}{\partial t} + (\delta\boldsymbol{v}\cdot\nabla)\delta w = D\,\nabla^2(\delta w) - \frac{1}{\rho}\nabla\cdot\delta\boldsymbol{J}_1, \tag{2.32c}$$

$$\frac{\mathrm{d}(\delta T)}{\mathrm{d}t} = \frac{\partial(\delta T)}{\partial t} + (\delta \boldsymbol{v} \cdot \boldsymbol{\nabla})\delta T = a\,\nabla^2(\delta T) - \frac{1}{\rho c_\mathrm{p}} \boldsymbol{\nabla} \cdot \delta \boldsymbol{J}'_q. \tag{2.32d}$$

Notice that eqn (2.32d) for the temperature fluctuations is uncoupled from (2.32c) for the concentration fluctuations, while both are only nonlinearly coupled to eqn (2.32b) for the velocity fluctuations through the material time derivative. However, since fluctuations are small, nonlinear terms can be neglected, transforming eqn (2.32) into a set of linear stochastic partial differential equations, that can be readily solved.

For instance, we can consider (2.32c) for the concentration fluctuations, which has exactly the same structure as eqn (2.23c) for the temperature fluctuations in a one-component fluid. Hence, the correlation function of the concentration fluctuations in a binary mixture in equilibrium has a similar structure to that found in eqn (2.26) to (2.30) for the temperature fluctuations in a one-component fluid. In particular, one has

$$\langle \delta w^*(\omega,\boldsymbol{q})\ \delta w(\omega',\boldsymbol{q}')\rangle = C(\omega,q)(2\pi)^4\ \delta(\omega - \omega')\ \delta(\boldsymbol{q} - \boldsymbol{q}'), \tag{2.33}$$

where $C(\omega,q)$ is now the structure factor representing the spectrum of the concentration fluctuations at a given wave vector. The explicit expression for $C(\omega,q)$ can easily be obtained from the (linearized) eqn (2.32c) and the FDT for the random diffusion flux, eqn (2.18c):

$$C(\omega, q) = \frac{k_\mathrm{B}T}{\rho}\left(\frac{\partial w}{\partial \mu}\right)_{\mathrm{T,p}} \frac{2Dq^2}{\omega^2 + D^2 q^4}. \tag{2.34}$$

and the corresponding time-correlation function

$$C(q, \tau) = \frac{k_\mathrm{B}T}{\rho}\left(\frac{\partial w}{\partial \mu}\right)_{\mathrm{T,p}} \exp(-Dq^2\tau). \tag{2.35}$$

Hence, the concentration fluctuations decay diffusively with decay rate Dq^2. Finally, just as with the temperature fluctuations in a one-component fluid, the equal-time concentration fluctuations in a binary mixture are spatially short-ranged:

$$\langle \delta w^*(\boldsymbol{r}, t)\, \delta w(\boldsymbol{r}', t)\rangle = \frac{k_\mathrm{B}T}{\rho}\left(\frac{\partial w}{\partial \mu}\right)_{\mathrm{T,p}} \delta(\boldsymbol{r} - \boldsymbol{r}'). \tag{2.36}$$

To end this section, we note that eqn (2.34) to (2.36) reproduce what would be obtained from taking approximations 1 to 3 above in the full spectrum as calculated from eqn (2.31) and including all the couplings. For simplicity, we preferred to take the limit in the working equations before any detailed calculation.

2.5 Alternative Approaches

As reviewed by other authors,[27,28] there exists an alternative method to fluctuating hydrodynamics for dealing with thermal fluctuations in equilibrium systems, namely, the arbitrary-initial-condition method. This method was pioneered by Mountain *et al.*[29,30] and has been adopted in the well-known books of Berne and Pecora,[31] Boon and Yip,[32] and Hansen and McDonald.[33] It consists in solving deterministic (with no random contributions) linearized hydrodynamic equations with arbitrary initial conditions. For instance,[30] for concentration fluctuations in a binary mixture, one considers the deterministic version of eqn (2.31c), linearize it in the fluctuating fields and Fourier transform in space, so as to obtain:

$$\frac{\partial(\delta w)}{\partial t} = -Dq^2\,\delta w, \tag{2.37}$$

which can be readily solved for $\delta w(t)$ as a function of an arbitrary initial value $\delta w(0)$. One then multiplies the solution $\delta w(t)$ of eqn (2.37) by $\delta w(0)$, averages over all initial conditions, and obtains the dynamic $\langle\delta w(t)\delta w(0)\rangle$ correlation function as a function of the static $\langle\delta w^2(0)\rangle$ correlations, namely:

$$\langle\delta w(t)\delta w(0)\rangle = \langle\delta w^2(0)\rangle\ \exp(-Dq^2t). \tag{2.38}$$

Some other authors[34] multiply eqn (2.37) itself by an arbitrary initial condition, and then average over fluctuations, arriving at evolution equations for the dynamic correlations that are then solved, leading to the same eqn (2.38). Next, to complete the calculation by the Mountain[29] method one needs expressions for the static correlations $\langle\delta w^2(0)\rangle$, which are obtained from equilibrium statistical physics (thermodynamic fluctuation theory) and yield

$$\langle\delta w^2(0)\rangle = \frac{k_\mathrm{B}T}{\rho}\left(\frac{\partial w}{\partial\mu}\right)_{\mathrm{T,p}}. \tag{2.39}$$

The two approaches, fluctuating hydrodynamics and arbitrary initial conditions, are fully equivalent for fluctuations in equilibrium systems, and eqn (2.38) and (2.39) exactly reproduce eqn (2.35) above. However, unlike the Mountain approach, only fluctuating hydrodynamics can be extended to fluctuations in systems in non-equilibrium states, as will be shown in Chapter 3. In the Mountain method, eqn (2.39) is no longer valid outside equilibrium. The presence of gradients, due to mode-coupling phenomena at the linear level that are not present in equilibrium, leads to equal-time fluctuations that are substantially enhanced with respect to their equilibrium values. The arbitrary initial-condition approach[29] cannot be extended to non-equilibrium fluctuations because there is not, in general, an accepted expression for the "entropy", from which the proper alternative to eqn (2.39) can be deduced. However, as demonstrated in Chapter 3, adopting a local version of the FDT (2.18c) allows a natural extension of fluctuating hydrodynamics to deal with non-equilibrium systems.

The emphasis of this chapter has been on the basic theoretical background of fluctuating hydrodynamics and its connection with non-equilibrium thermodynamics, stressing the aspects more useful for the interpretation of light-scattering experiments.[31–33] Before concluding we should mention, however, that numerical simulation of fluctuating-hydrodynamics problems has become a very active field of research.[26,35–40] Numerical simulations are useful in understanding the nature of thermal fluctuations, particularly for complex systems,[26] where a full theory incorporating all physically relevant issues, like boundary conditions,[36] nonlinear terms,[37,40] spatial variability of thermophysical properties, *etc.*, becomes increasingly difficult.

Acknowledgements

Support from the UCM/Santander Research Grant PR6/13-18867 is gratefully acknowledged.

References

1. L. D. Landau and E. M. Lifshitz, *Zh. Eksp. Teor. Fiz.*, 1957, **32**, 618, English translation: *Sov. Phys. JETP*, 1957, **5**, 512.
2. L. D. Landau and E. M. Lifshitz, *Statistical Physics. Part I*, Pergamon, London, 1958.
3. L. D. Landau and E. M. Lifshitz, *Fluid Mechanics*, Pergamon, London, 1959.
4. R. F. Fox and G. E. Uhlenbeck, *Phys. Fluids*, 1970, **13**, 1893.
5. R. F. Fox, *Phys. Rep.*, 1978, **48**, 179.
6. J. Foch, *Phys. Fluids*, 1971, **14**, 893.
7. I. Pagonabarraga, A. Pérez-Madrid and J. M. Rubí, *Physica A*, 1997, **237**, 205.
8. M. Treiber and L. Kramer, *Phys. Rev. E: Stat. Phys., Plasmas, Fluids, Relat. Interdiscip. Top.*, 1994, **49**, 3184.
9. N. G. van Kampen, *Stochastic Processes in Physics and Chemistry*, North Holland, Dordrecht, 1982.
10. C. W. Gardiner, *Handbook of Stochastic Methods*, Springer, Heidelberg, 1985.
11. M. Bixon and R. Zwanzig, *Phys. Rev.*, 1969, **187**, 267.
12. R. F. Fox and G. E. Uhlenbeck, *Phys. Fluids*, 1970, **13**, 2881.
13. K. T. Mashiyama and H. Mori, *J. Stat. Phys.*, 1978, **18**, 385.
14. R. Schmitz and E. G. D. Cohen, *J. Stat. Phys.*, 1985, **40**, 431.
15. J. M. Ortiz de Zárate and J. V. Sengers, *Hydrodynamic Fluctuations in Fluids and Fluid Mixtures*, Elsevier, Amsterdam, 2006.
16. R. Schmitz and E. G. D. Cohen, *J. Stat. Phys.*, 1985, **39**, 285.
17. S. Hartmann, G. Wittko, W. Köhler, K. I. Morozov, K. Albers and G. Sadowski, *Phys. Rev. Lett.*, 2012, **109**, 065901.
18. S. Hartmann, G. Wittko, F. Schock, W. Groβ, F. Lindner, W. Köhler and K. I. Morozov, *J. Chem. Phys.*, 2014, **141**, 134503.
19. B. Law and J. C. Nieuwoudt, *Phys. Rev. A: At., Mol., Opt. Phys.*, 1989, **40**, 3880.

20. S. Chandrasekhar, *Hydrodynamic and Hydromagnetic Stability*, Oxford University Press, Oxford, 1961.
21. P. Manneville, *Dissipative Structures and Weak Turbulence*, Academic Press, London, 1990.
22. R. Pérez-Cordón and M. G. Velarde, *J. Phys.*, 1975, **36**, 591.
23. S. R. de Groot and P. Mazur, *Non-Equilibrium Thermodynamics*, North-Holland, Amsterdam, 1962.
24. M. G. Velarde and R. S. Schechter, *Phys. Fluids*, 1972, **15**, 1707.
25. J. M. Ortiz de Zárate, J. A. Fornés and J. V. Sengers, *Phys. Rev. E: Stat., Nonlinear, Soft Matter Phys.*, 2006, **74**, 046305.
26. K. Balakrishnan, A. L. García, A. Donev and J. B. Bell, *Phys. Rev. E: Stat., Nonlinear, Soft Matter Phys.*, 2014, **89**, 013017.
27. F. Vázquez and M. López de Haro, *J. Non-Equilib. Thermodyn.*, 2001, **26**, 279.
28. M. López de Haro, J. A. del Río and F. Vázquez, *Rev. Mex. Fís.*, 2002, **48**(Suppl. 1), 230.
29. R. D. Mountain, *Rev. Mod. Phys.*, 1966, **38**, 205.
30. R. D. Mountain and J. M. Deutch, *J. Chem. Phys.*, 1969, **50**, 1103.
31. B. J. Berne and R. Pecora, *Dynamic Light Scattering*, Wiley, New York, 1976.
32. J. P. Boon and S. Yip, *Molecular Hydrodynamics*, Mc-Graw-Hill, New York, 1980.
33. J. P. Hansen and I. R. McDonald, *Theory of Simple Liquids*, Academic Press, London, 1986.
34. C. Cohen, J. W. H. Sutherland and J. M. Deutch, *Phys. Chem. Liq.*, 1971, **2**, 213.
35. G. De Fabritiis, M. Serrano, R. Delgado-Buscalioni and P. V. Coveney, *Phys. Rev. E: Stat., Nonlinear, Soft Matter Phys.*, 2007, **75**, 026307.
36. R. Delgado-Buscalioni and A. Dejoan, *Phys. Rev. E: Stat., Nonlinear, Soft Matter Phys.*, 2008, **78**, 046708.
37. H. Spohn, *J. Stat. Phys.*, 2014, **154**, 1191.
38. N. Voulgarakis and J.-W. Chu, *J. Chem. Phys.*, 2009, **130**, 134111.
39. A. Donev, E. Vanden-Eijnden, A. L. Garcia and J. B. Bell, *Comm. App. Math. Comp. Sci.*, 2010, **5**, 149.
40. A. Donev, A. de la Fuente, J. B. Bell and A. L. Garcia, *Phys. Rev. Lett.*, 2011, **106**, 204501.

CHAPTER 3

Thermal Fluctuations in Non-equilibrium Steady States

JAN V. SENGERS,*[a] JOSÉ M. ORTIZ DE ZÁRATE[b] AND THEODORE R. KIRKPATRICK[a,c]

[a] Institute for Physical Science and Technology, University of Maryland, College Park MD 20742, USA; [b] Departamento de Física Aplicada I, Facultad de Física, Universidad Complutense, 28040 Madrid, Spain; [c] Department of Physics, University of Maryland, College Park MD 20742, USA

*Email: sengers@umd.edu

3.1 Introduction

This chapter will be concerned with thermal fluctuations in fluids and fluid mixtures that are in a quiescent (*i.e.*, non-convective and non-turbulent) steady non-equilibrium state. There is a qualitative difference between thermal fluctuations in fluids in equilibrium and out of equilibrium. Temperature and concentration fluctuations in fluids in thermodynamic equilibrium are generally uncorrelated at hydrodynamic length scales, as reviewed in Section 2.4 of Chapter 2. On the other hand, in non-equilibrium states these fluctuations exhibit an algebraic dependence on the distance variable r and become long ranged on hydrodynamic scales.

Non-equilibrium thermodynamics is based on a local-equilibrium assumption, which implies that at each point in space and time the local thermodynamic properties are related by the same thermodynamic relations as for a fluid in equilibrium. The evidence for the validity and applicability of this local-equilibrium assumption for the thermodynamic properties is

Experimental Thermodynamics Volume X: Non-equilibrium Thermodynamics with Applications
Edited by Dick Bedeaux, Signe Kjelstrup and Jan V. Sengers

Published by the Royal Society of Chemistry, www.rsc.org

reviewed in Chapter 4. While the local-equilibrium assumption remains valid for the fluctuation–dissipation theorem (FDT) in systems out of equilibrium, local equilibrium no longer holds for the non-equilibrium temperature and concentration correlation functions. This failure is obvious for fluids undergoing thermal convection or turbulence, but also occurs in quiescent non-equilibrium states far away from any hydrodynamic instability. That is, in the presence of a temperature gradient or a concentration gradient, the correlation functions always exhibit a non-equilibrium enhancement, which becomes long ranged on hydrodynamic scales. Hence, buoyancy and finite-size effects need to be included to specify the wave-number dependence of non-equilibrium correlation functions.

To elucidate the nature of non-equilibrium fluctuations we shall specifically consider in this chapter thermal fluctuations in fluids and fluid mixtures in the presence of a stationary temperature gradient, which is the most studied case so far. Reviews of the subject have been presented by two of us some time ago.[1,2] Some interesting subsequent developments will be included in the present chapter.

3.2 Non-equilibrium Enhancement of Thermal Fluctuations

We consider a one-component fluid located between two horizontal plates separated by a distance L and subject to a stationary temperature gradient of magnitude $\nabla T = (T_1 - T_2)/L$, where T_1 and T_2 are the temperatures of the upper and lower plates, respectively, as indicated schematically in Figure 3.1. We adopt a Cartesian coordinate system with the z-axis in the vertical direction, *i.e.*, opposite to the direction of the gravitational acceleration vector $\mathbf{g}$. The stability of this configuration, often referred to as the Rayleigh–Bénard problem, is controlled by the Rayleigh number

$$\mathrm{Ra} = \frac{\alpha L^4 \mathbf{g} \cdot \boldsymbol{\nabla} T}{\nu a}, \tag{3.1}$$

where α is the thermal expansion coefficient, a the thermal diffusivity, and ν the kinematic viscosity (related to the shear viscosity η and the density ρ by $\nu = \eta/\rho$). For normal fluids ($\alpha \geq 0$), the Rayleigh number is negative when the fluid layer is heated from above ($T_1 \geq T_2$) and positive when heated from

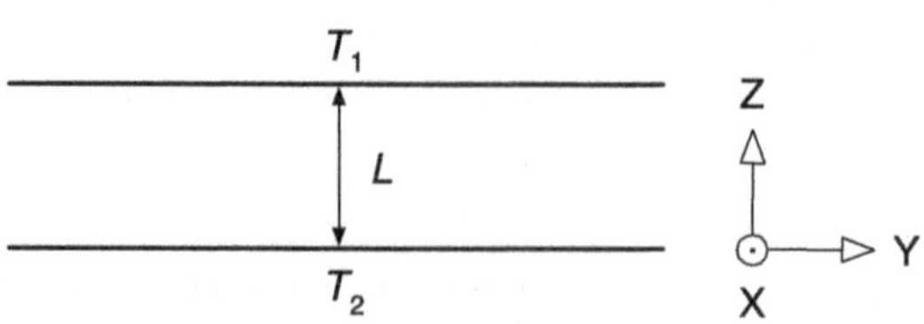

Figure 3.1 Schematic representation of a fluid layer in a temperature gradient $(T_1 - T_2)/L$.

below ($T_1 \leq T_2$). The fluid layer is in a quiescent state with zero fluid velocity provided that

$$\mathrm{Ra} \leq \mathrm{Ra}_c, \tag{3.2}$$

where Ra_c is the critical Rayleigh number associated with the onset of convection.[3] In this chapter we shall only consider quiescent states for which condition (3.2) is satisfied.

As in Chapter 2, we consider here temperature fluctuations associated with the Rayleigh line in the fluctuation spectrum, which can be observed experimentally with light scattering[4] or with shadowgraphy.[5,6] To obtain the relevant equations for the fluctuations, we need to consider $T(\boldsymbol{r},t) = T_0(\boldsymbol{r}) + \delta T(\boldsymbol{r},t)$, $\rho(\boldsymbol{r},t) = \rho_0(\boldsymbol{r}) + \delta\rho(\boldsymbol{r},t)$ and $\boldsymbol{v}(\boldsymbol{r},t) = 0 + \delta\boldsymbol{v}(\boldsymbol{r},t)$, where the fluctuating contributions $\delta T(\boldsymbol{r},t)$, $\delta\rho(\boldsymbol{r},t)$ and $\delta\boldsymbol{v}(\boldsymbol{r},t)$ depend on the location vector $\boldsymbol{r}$ and the time t. Unlike in equilibrium fluctuating hydrodynamics, the local equilibrium temperature $T_0(\boldsymbol{r})$ and the local equilibrium density $\rho_0(\boldsymbol{r})$ now also depend on the position $\boldsymbol{r}$. As in equilibrium, to derive the temperature fluctuations we may continue to use the Boussinesq approximation

$$\delta\rho = -\alpha\rho_0\delta T, \tag{3.3}$$

which gave eqn (2.23c) for the temperature fluctuations. Since $\nabla T_0 \neq 0$, the equation for the temperature fluctuations now becomes

$$\rho c_p \left[\frac{\partial \delta T}{\partial t} + \delta\boldsymbol{v} \cdot \boldsymbol{\nabla} T_0\right] = \lambda \boldsymbol{\nabla}^2 \delta T - \boldsymbol{\nabla} \cdot \delta\boldsymbol{J}_q, \tag{3.4}$$

where c_p is the isobaric specific heat capacity, λ the thermal conductivity and $\delta\boldsymbol{J}_q$ the fluctuating heat flux.

We note that the advective term in the material time derivative in eqn (2.23c) causes a coupling between the temperature fluctuations and the velocity fluctuations through the appearance of $\delta\boldsymbol{v} \cdot \nabla T_0$ in eqn (3.4). The general phenomenon that a gradient causes a coupling between two hydrodynamic modes, in this case between the heat mode and a viscous mode, is the fundamental reason why fluctuations in non-equilibrium states are qualitatively different from fluctuations in equilibrium states. Hence, to determine the magnitude and time dependence of the non-equilibrium (NE) temperature fluctuations, we must supplement eqn (3.4) with the corresponding equation for the velocity fluctuations, which is given in the Boussinesq approximation by eqn (2.23b):

$$\rho \frac{\partial \delta\boldsymbol{v}}{\partial t} = -\nabla \delta p + \eta \nabla^2 \delta\boldsymbol{v} - \boldsymbol{\nabla} \cdot \delta\boldsymbol{\Pi}, \tag{3.5}$$

where $\delta\boldsymbol{\Pi}$ is the fluctuating pressure tensor. Here we consider temperature fluctuations at constant pressure and we may substitute $\delta p = 0$ in eqn (3.5). Instead of temperature fluctuations at constant pressure, one may consider fluctuations of the entropy density s at constant pressure that are related to the temperature fluctuations by $\delta s = (c_p/T)\delta T$. We see from eqn (3.4) that the temperature fluctuations only couple with transverse velocity fluctuations δv_z in the direction of the temperature gradient: A convenient procedure in fluctuating hydrodynamics to obtain an equation for these wall-normal velocity fluctuations is to apply a double rotational to eqn (3.5):[1]

$$\rho \frac{\partial \nabla^2 v_z}{\partial t} = \eta \nabla^2 (\nabla^2 \delta v_z) - (\boldsymbol{\nabla} \cdot \boldsymbol{\nabla} \cdot \boldsymbol{\nabla} \cdot \delta \boldsymbol{\Pi})_z, \tag{3.6}$$

where $(\boldsymbol{\nabla} \cdot \boldsymbol{\nabla} \cdot \boldsymbol{\nabla} \cdot \delta \boldsymbol{\Pi})_z$ is the z component of $\boldsymbol{\nabla} \cdot \boldsymbol{\nabla} \cdot \boldsymbol{\nabla} \cdot \delta \boldsymbol{\Pi}$ and where the divergence-free flow condition (2.23a) has been used.

To solve eqn (3.4) and (3.6) we need the correlation functions for the fluctuating fluxes $\delta \boldsymbol{J}_q$ and $\delta \boldsymbol{\Pi}$. Since these are noise correlations produced by molecular collisions occurring at distances over which any spatial dependence of the average temperature T_0 is still negligibly small, we assume that these noise correlations are given by the local-equilibrium versions of the FDT. Hence, we assume that eqn (2.5) and (2.8) continue to be valid in nonequilibrium, but with the temperature T to be identified with the local temperature T_0:

$$\left\langle \delta J_{q,\alpha}(\boldsymbol{r}, t)\, \delta J_{q,\beta}(\boldsymbol{r}', t') \right\rangle = 2 k_B T_0^2 \lambda\, \delta(\boldsymbol{r} - \boldsymbol{r}') \delta(t - t'), \tag{3.7}$$

$$\langle \delta \Pi_{\alpha,\beta}(\boldsymbol{r},t)\, \delta \Pi_{\mu,\nu}(\boldsymbol{r}',t') \rangle = 2 k_B T_0 \eta (\delta_{\alpha,\mu} \delta_{\beta,\nu} + \delta_{\alpha,\nu} \delta_{\beta,\mu}) \delta(\boldsymbol{r} - \boldsymbol{r}') \delta(t - t'), \tag{3.8}$$

where k_B is Boltzmann's constant. In practice, the local equilibrium temperature T_0 and the transport coefficients λ and η in the prefactor in eqn (3.7) and (3.8) may be approximated by their average value in the fluid layer. In the subsequent equations, thermodynamic and transport properties without a subscript 0 are to be identified with their local-equilibrium values at the average local temperature, *i.e.*, at the center of the fluid layer.

If one applies a spatiotemporal Fourier transformation to eqn (3.4) and (3.6) and to the resulting correlation functions an inverse Fourier transformation, as was done in Section 2.4.1 for the equilibrium correlation functions, one readily obtains[1]

$$\langle \delta T(\boldsymbol{q}, \tau) \delta T(\boldsymbol{q}, 0) \rangle = \frac{k_B T^2}{\rho c_p} \left[(1 + A_T) \exp(-a q^2 \tau) - A_v \exp(-\nu q^2 \tau) \right], \tag{3.9}$$

where the NE amplitudes A_T and A_v are given by

$$A_T = \frac{c_p}{T(\nu^2 - a^2)} \frac{\nu}{a} \frac{(q_\parallel \nabla T_0)^2}{q^6}, \quad A_v = \frac{c_p}{T(\nu^2 - a^2)} \frac{(q_\parallel \nabla T_0)^2}{q^6}. \tag{3.10}$$

Here q is the magnitude of the wave vector $\boldsymbol{q}$ of the fluctuations and $q_\parallel$ the magnitude of its component $\boldsymbol{q}_\parallel$ in the horizontal x–y plane, *i.e.*, perpendicular to the temperature gradient. From eqn (3.9) we see that the NE temperature fluctuations contain a contribution from a heat mode decaying exponentially as a function of the correlation time τ with an exponential decay rate aq^2 and from a viscous mode with decay rate νq^2.

The amplitudes of the NE contributions in eqn (3.10) reach their maximum values for $\boldsymbol{q} = \boldsymbol{q}_{\parallel}$, *i.e.*, when the heat mode couples with the transverse viscous mode along the direction of the temperature gradient, so that

$$A_{\mathrm{T}} = \frac{c_{\mathrm{p}}}{T(\nu^2 - a^2)}\frac{\nu}{a}\frac{(\nabla T_0)^2}{q^4}, \quad A_{\mathrm{v}} = \frac{c_{\mathrm{p}}}{T(\nu^2 - a^2)}\frac{(\nabla T_0)^2}{q^4}. \tag{3.11}$$

Eqn (3.9) is the nonequilibrium extension of the time-dependent equilibrium correlation function given by eqn (2.29) in Chapter 2.

The equations above for the NE temperature fluctuations were originally discovered by Kirkpatrick *et al.*[7] Ronis and Procaccia[8] suggested that they could be obtained by extending Landau's fluctuating hydrodynamics to nonequilibrium states, a suggestion that was implemented by several investigators.[9–13] Light-scattering experiments of Sengers and collaborators have confirmed the validity of eqn (3.9) with considerable accuracy.[14–16] As an example, we show in Figure 3.2 the NE amplitudes A_{T} and A_{v} as a function of $(\nabla T_0)^2/q^4$. The symbols indicate the experimental amplitudes obtained for liquid toluene at an average temperature of 40 °C. The lines represent the amplitudes calculated from the available thermal-property data for toluene at 40 °C without any adjustable parameters. It was verified that the effect of any temperature dependence of the thermophysical properties over the height of the fluid layer was negligibly small, justifying identification of the properties with their average value at the center of the fluid layer.[15] Results from more recent shadowgraph experiments will be discussed in Section 3.3. All these experiments have confirmed the validity of extending the fluctuation–dissipation theorem to non-equilibrium states.

In this chapter we shall be especially interested in the total intensity of the NE fluctuations that follows from eqn (3.9) by taking $\tau = 0$:

$$\langle \delta T(\boldsymbol{q})\rangle^2 = \frac{k_{\mathrm{B}}T^2}{\rho c_{\mathrm{p}}}\left[1 + \frac{c_{\mathrm{p}}}{T\nu(\nu + a)}\frac{(q_{\parallel}\nabla T_0)^2}{q^6}\right]. \tag{3.12}$$

Thus, for any non-zero value of the temperature gradient ∇T_0, the temperature fluctuations will exhibit a NE enhancement that for small wave numbers will diverge as q^{-4}.

As discussed in Section 2.4.2 of Chapter 2, in binary mixtures one will encounter not only temperature fluctuations, but also concentration fluctuations. Extension of the theory to binary mixtures in the presence of a temperature gradient was first implemented by Law and Nieuwoudt.[17] In mixtures, not only the temperature fluctuations, but also the concentration fluctuations will exhibit pronounced NE enhancements. Both types of NE fluctuations have been observed experimentally with an intensity again diverging as q^{-4}.[16,18] In general there will also be a coupling between the concentration and temperature fluctuations.[16]

An important parameter is the Lewis number Le which is the ratio of the thermal diffusivity a governing the decay of the temperature fluctuations,

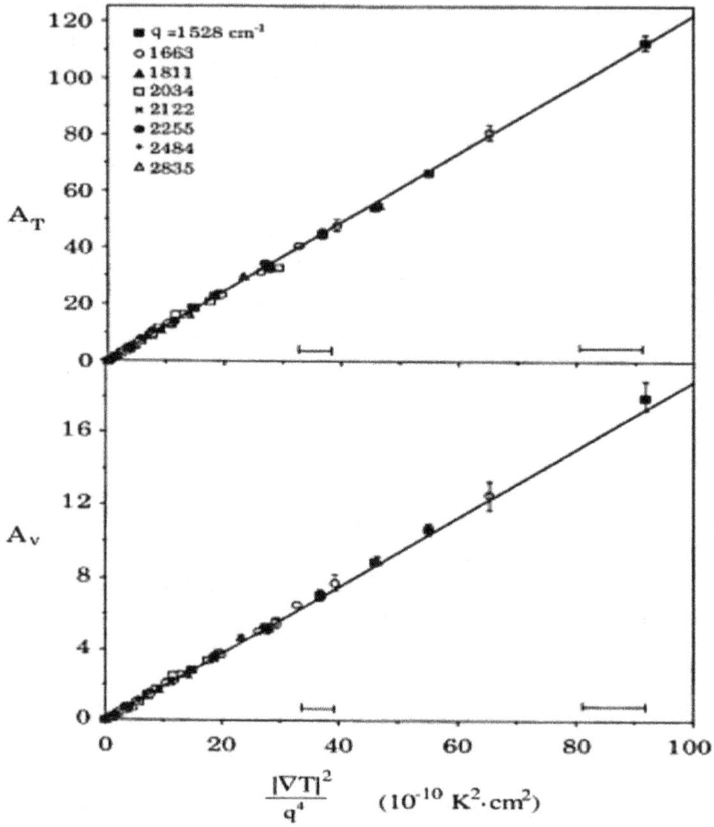

Figure 3.2 Nonequilibrium fluctuation amplitudes A_T and A_v as a function of $(\nabla T_0)^2/q^4$ for liquid toluene at 40 °C. The symbols indicate experimental values and the lines represent the theoretical prediction from fluctuating hydrodynamics.
Copyright (1992) by The American Physical Society.[15]

and the mass diffusion coefficient D governing the decay of the concentration fluctuations:

$$\mathrm{Le} = \frac{a}{D}. \tag{3.13}$$

A significant simplification occurs when the Lewis number is large, in which case the coupling between the heat and concentration modes and, hence, between temperature and concentration fluctuations, disappears and the concentration fluctuations become dominant.[19] As in Chapter 2, we write the concentration as $w(\boldsymbol{r},t) = w_0(\boldsymbol{r},t) + \delta w(\boldsymbol{r},t)$, where $w(\boldsymbol{r},t)$ is the weight fraction of one of the components and $w_0(\boldsymbol{r},t)$ its local equilibrium value but now with $\nabla w_0 \neq 0$. Then eqn (2.32c) for large values of Le becomes:

$$\rho\left[\frac{\partial \delta w}{\partial t} + \delta \boldsymbol{v} \cdot \nabla w_0\right] = \rho D \nabla^2 \delta w - \nabla \cdot \delta \boldsymbol{J}_1, \tag{3.14}$$

where $\delta \boldsymbol{J}_1$ is the fluctuating diffusion flux (see eqn (2.16b)). To solve eqn (3.14) we need the correlation function for the fluctuations $\delta \boldsymbol{J}_1$, for which we adopt again the local-equilibrium version of the FDT consistent with eqn (2.18c):

$$\left\langle \delta J_{1,\alpha}(\boldsymbol{r},t)\,\delta J_{1,\beta}(\boldsymbol{r}',t')\right\rangle = 2k_{\mathrm{B}}T\rho D\left(\frac{\partial w}{\partial \mu}\right)_{\mathrm{T,p}}\delta(\boldsymbol{r}-\boldsymbol{r}')\delta(t-t'), \tag{3.15}$$

where $(\partial w/\partial \mu)_{\mathrm{T,p}}$ is the osmotic compressibility. For the correlation function of the fluctuating pressure tensor we can continue using eqn (3.8). Just as the intensity of the NE temperature fluctuations eqn (3.10) follows from eqn (3.4) for the temperature fluctuations, we now obtain the intensity of the concentration fluctuations from eqn (3.14)

$$\langle \delta w(\boldsymbol{q})\rangle^2 = \frac{k_{\mathrm{B}}T}{\rho}\left(\frac{\partial w}{\partial \mu}\right)_{\mathrm{T,p}}\left[1+\frac{1}{\nu D}\left(\frac{\partial w}{\partial \mu}\right)^{-1}_{\mathrm{T,p}}\frac{(q_{\parallel}\,\nabla w_0)^2}{q^6}\right]. \tag{3.16}$$

A convenient experimental procedure for establishing a concentration gradient ∇w_0 is applying a temperature gradient ∇T_0 to the fluid mixture which induces a concentration gradient through the Soret effect:[16,18,20–22]

$$\nabla w_0 = -w_0(1-w_0)S_{\mathrm{T}}\nabla T_0, \tag{3.17}$$

where S_{T} is the appropriate Soret coefficient. The validity of eqn (3.16) has also been confirmed experimentally.[20,22] The theory has recently been extended to ternary mixtures by Ortiz de Zárate *et al.*[23]

Several investigators have also studied concentration fluctuations in the presence of transient concentration gradients in isothermal liquid mixtures induced by free diffusion.[24–27] The theory developed above pertains to NE fluctuations in stationary states. Nevertheless, these transient experiments have also confirmed that the NE concentration fluctuations vary as q^{-4}. The question of when spatial and temporal variations of the gradient will cause deviations of the stationary solution given by eqn (3.12) and (3.16) requires further research.[28]

It is interesting to compare the NE fluctuations with critical fluctuations in fluids. The intensity of the critical fluctuations is known to vary with the wave number q as[29–31] $q^{-2+\eta_c} \cong q^{-2}$ ($\eta_c = 0.033$). Hence, in real space the critical fluctuations will vary as[26] $r^{-1+\eta_c} \cong r^{-1}$. On the other hand, the dependence of the NE fluctuations on q^{-4} implies that in real space the NE fluctuations will depend linearly on r, so that the correlations extend over the entire system.[32] Hence, the NE fluctuations are having a major impact on further developments in nonequilibrium statistical physics. The NE fluctuations cannot increase indefinitely as $q \rightarrow 0$. At small wave numbers, the intensity of the NE will be suppressed by gravity and by the limited size of the system. In addition, the NE fluctuations are inducing NE forces that need to be relaxed. The remainder of this chapter will be devoted to a discussion of these remarkable features associated with NE fluctuations.

3.3 Gravity Effects

The fact that gravity will suppress the NE fluctuations at small wave numbers was originally predicted by Segrè *et al.*[33,34] Gravity induces a pressure gradient in the Navier–Stokes eqn (1.33)

$$\nabla p = \rho \boldsymbol{g} = \overline{\rho}_0 [1 - \alpha(T_0 - \overline{T}_0) g \hat{z}], \tag{3.18}$$

where $\overline{\rho}_0$ and $\overline{T}_0$ are the average density and temperature in the center of the fluid layer and where $\hat{z}$ is the unit vector in the z-direction. As a consequence, eqn (3.6) for the velocity fluctuations becomes[35]

$$\rho \frac{\partial \nabla^2 v_z}{\partial t} = \eta \nabla^2 (\nabla^2 \delta v_z) + \alpha g \left(\frac{\partial^2 \delta T}{\partial x^2} + \frac{\partial^2 \delta T}{\partial y^2} \right) - (\nabla \cdot \nabla \cdot \nabla \cdot \delta \boldsymbol{\Pi})_z. \tag{3.19}$$

In this section we consider the solution of eqn (3.4) for the temperature fluctuations in conjunction with eqn (3.19) for the velocity fluctuations for a fluid layer heated from above, *i.e.*, for negative values of the Rayleigh number Ra. If a small adiabatic contribution $(\alpha T/c_p)\boldsymbol{g}$ to the temperature gradient is neglected, one finds for $\mathrm{Ra} \leq 0$ that the intensity (3.12) of the temperature fluctuations in a one-component fluid changes into[1,33,35]

$$\langle \delta T(\boldsymbol{q}) \rangle^2 = \frac{k_B T^2}{\rho c_p} \left[1 + \frac{c_p (\nabla T_0)^2 L^4}{T(\nu + a)} \frac{\tilde{q}_\parallel^2}{(\tilde{q}^6 - \mathrm{Ra}\, \tilde{q}_\parallel^2)} \right], \tag{3.20}$$

where $\tilde{q} = qL$ and $\tilde{q}_\parallel = q_\parallel L$ are dimensionless wave numbers. When $\boldsymbol{q} = \boldsymbol{q}_\parallel$, eqn (3.20) reduces to

$$\langle \delta T(\boldsymbol{q}) \rangle^2 = \frac{k_B T^2}{\rho c_p} \left[1 + \frac{c_p (\nabla T_0)^2 L^4}{T(\nu + a)} \frac{1}{(\tilde{q}^4 - \mathrm{Ra})} \right]. \tag{3.21}$$

For large values of the wave number $\tilde{q}^4 \geq |\mathrm{Ra}|$ the NE enhancement of the fluctuations will indeed be proportional to $\tilde{q}^{-4}$ in agreement with eqn (3.12), but for $\tilde{q}^4 \leq |\mathrm{Ra}|$ the NE enhancement will reach a finite limit. The crossover will occur at a wave number[6,33]

$$q_{cr} = \left(\frac{\alpha g |\nabla T_0|}{\nu a} \right)^{1/4}. \tag{3.22}$$

As an example we show in Figure 3.3 the intensity $S(q) = \langle \delta T(\boldsymbol{q}) \rangle^2$, given by eqn (3.21), relative to the local equilibrium value $S_E(q) = k_B T^2 / \rho c_p$ as a function of the wave number $q = q_\parallel$, predicted for toluene with a temperature gradient of $\nabla T_0 = 200\ \mathrm{K \cdot cm^{-1}}$.[33] For large wave numbers the intensity increases as q^{-4} with decreasing wave number and then crosses over to a finite value as $q \to 0$. Note that the NE enhancements of the fluctuations in the fluid layer exceed values more than 10^6 times the local-equilibrium value for these fluctuations.

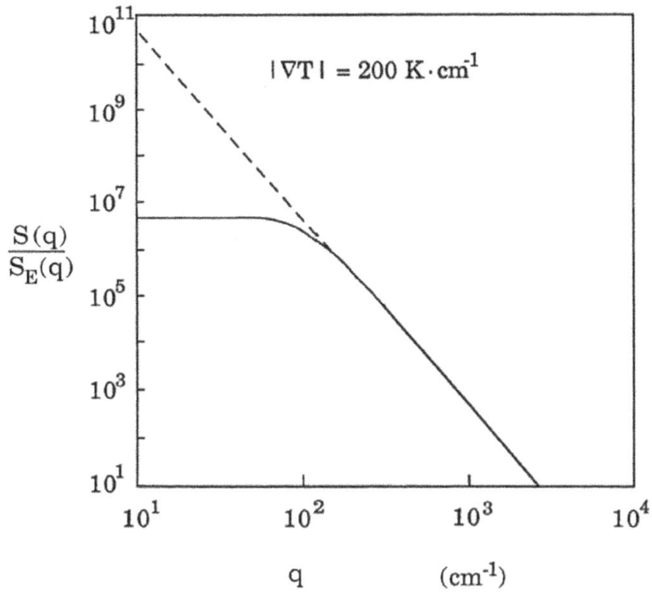

Figure 3.3 Nonequilibrium enhancement of the temperature fluctuations predicted for toluene in a temperature gradient of $\nabla T_0 = 200\ \mathrm{K \cdot cm^{-1}}$. The dashed line indicates the dependence on q^{-4} in the absence of gravity. The solid curve represents the actual wave number dependence in the presence of gravity.
Reprinted with permission from *Physica A*.[33]

For a binary mixture in the large Lewis-number approximation, eqn (3.16) changes (for $\boldsymbol{q} = \boldsymbol{q}_{\parallel}$) into[1,6,21,25,34,36,37]

$$\langle \delta w(\boldsymbol{q}) \rangle^2 = \frac{k_B T}{\rho} \left(\frac{\partial w}{\partial \mu} \right)_{T,p} \left[1 + \left(\frac{\partial w}{\partial \mu} \right)_{T,p}^{-1} \frac{(\nabla w_0)^2 L^4}{\nu D} \frac{\tilde{q}_{\parallel}^2}{(\tilde{q}^6 - \psi \mathrm{Ra\,Le}\, \tilde{q}_{\parallel}^2)} \right]. \quad (3.23)$$

In eqn (3.23) ψ is the separation ratio which is the ratio of the density gradient produced by the concentration gradient and the density gradient produced by the gradient: $\psi = \beta \nabla w_0 / \alpha \nabla T_0$, where β is the solutal expansion coefficient. The crossover from a $\tilde{q}^{-4}$ behavior for large wave numbers to a constant limiting value for small q now occurs at a wave number[6,34,37]

$$q_{cr} = \left(\frac{\beta g |\nabla w_0|}{\nu D} \right)^{1/4}. \quad (3.24)$$

The crossover from a q^{-4} behavior for large wave numbers to a limiting value at smaller wave numbers has been observed experimentally both with light scattering[20,21,25] and shadowgraphy.[24,26,27,38–40] In Figure 3.4 we show the results of early light-scattering measurements obtained by Vailati and Giglio[20] for a liquid mixture of aniline + cyclohexane. To get a large value of the Soret coefficient, the experiments were conducted at a temperature 10 K

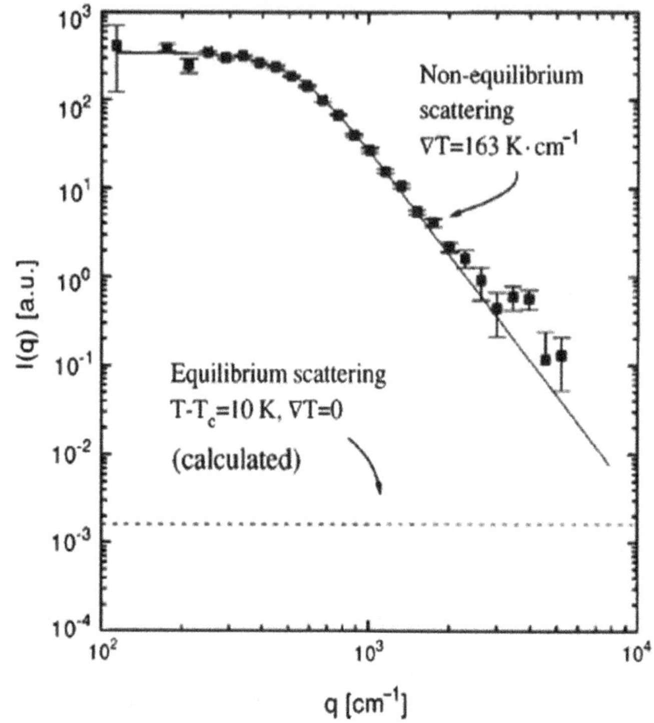

Figure 3.4 Intensity of light scattered by a mixture of aniline and cyclohexane as a function of the scattering wave number $q = q_{\parallel}$ reported by Vailati and Giglio.[20] Symbols indicate experimental intensities obtained with $\nabla T_0 = 163\ \mathrm{K \cdot cm^{-1}}$. The solid curve represents a fit to eqn (3.23) with ∇w_0 given by eqn (3.17).

above the critical phase-separation temperature T_c of the solution. The data do show the predicted crossover from an initial increase proportional to q^{-4} to a finite limiting value of the intensity of the fluctuations, which indeed becomes of the order of 10^6 times the local-equilibrium value. Not surprisingly, these huge NE fluctuations have been referred to as "giant" fluctuations in the literature.[24,38,41,42]

3.4 Finite-size Effects

As a consequence of their very long-ranged nature, the NE fluctuations extend over the entire spatial extent L of the fluid layer. Hence, to obtain a complete representation of the NE fluctuations one needs to obtain the solution of the equations for the fluctuations subject to the appropriate boundary conditions. For the temperature fluctuations one considers perfectly conducting walls, so that

$$\delta T = 0 \quad \text{at } z = 0, L. \tag{3.25}$$

For the velocity fluctuations one considers either stress-free or rigid-boundary conditions.[43] The advantage of stress-free boundary conditions is that one can obtain an exact analytic solution of the fluctuating-hydrodynamics equations. For an incompressible fluid, the stress-free boundary conditions become

$$\delta v_z = 0, \quad \frac{\mathrm{d}^2 v_z}{\mathrm{d}z^2} = 0 \quad \text{at } z = 0, L. \tag{3.26}$$

If we solve eqn (3.4) and (3.19) for the NE temperature fluctuations in a one-component fluid subject to the boundary conditions (3.25) and (3.26), we find[35]

$$\langle \delta T(\boldsymbol{q}) \rangle^2 = \frac{k_\mathrm{B} T^2}{\rho c_\mathrm{p}} \left[1 + \frac{c_\mathrm{p} (\nabla T_0)^2 L^4}{T(\nu + a)} \tilde{S}_\mathrm{NE}(\boldsymbol{q}) \right], \tag{3.27}$$

with

$$\tilde{S}_\mathrm{NE}(\boldsymbol{q}) = \frac{\tilde{q}_\parallel^2}{L} 2 \sum_{N=1}^{\infty} \frac{\sin(N\pi z/l) \sin(N\pi z'/L)}{\left(\tilde{q}_\parallel^2 + N^2\pi^2\right)^3 - \mathrm{Ra}\, \tilde{q}_\parallel^2}. \tag{3.28}$$

Here $\tilde{S}_\mathrm{NE}(\boldsymbol{q})$ is a normalized NE enhancement representing the modification of the NE enhancement of the fluctuations due to the combined gravity and finite-size effects. Eqn (3.27) is valid not only when the fluid layer is heated from above, but also when heated from below, since the finite-size effects stabilize the system as long as condition (3.2) is satisfied.

In Figure 3.5 we show this normalized $\tilde{S}_\mathrm{NE}$ for $\boldsymbol{q} = \boldsymbol{q}_\parallel$ as a function of the dimensionless wave number $\tilde{q} = qL$ for three different values of the Rayleigh number. For any value of the Rayleigh number, eqn (3.28) implies a crossover from a $\tilde{q}^{-4}$ behavior for large $\tilde{q}$ to a $\tilde{q}^2$ behavior for very small $\tilde{q}$. For negative Rayleigh numbers, *i.e.*, when the fluid layer is heated from above, the enhancement exhibits a crossover from a $\tilde{q}^{-4}$ behavior at large $\tilde{q}$ to a gravitationally induced plateau at intermediate wave numbers, as discussed in the preceding section, to ultimately a $\tilde{q}^2$ behavior for very small $\tilde{q}$. The curve for $\mathrm{Ra} = 0$ indicates the behavior to be expected in the absence of gravity ($g = 0$). For positive values of the Rayleigh number, gravity enhances the fluctuations causing a divergence of the NE fluctuations at a critical Rayleigh number $\mathrm{Ra_c}$ and a critical wave number q_c.[1,35,44]

For a comparison with experiments it is advisable to consider the more realistic rigid boundary condition obtained by replacing eqn (3.26) with[43]

$$\delta v_z = 0, \quad \frac{\mathrm{d} v_z}{\mathrm{d}z} = 0 \quad \text{at } z = 0, L. \tag{3.29}$$

In this case it is not possible to get an analytic solution, but a solution in a Galerkin approximation has been obtained by Ortiz de Zárate and Sengers.[45] The equivalent of eqn (3.27) and Figure 3.5 in this approximation can be

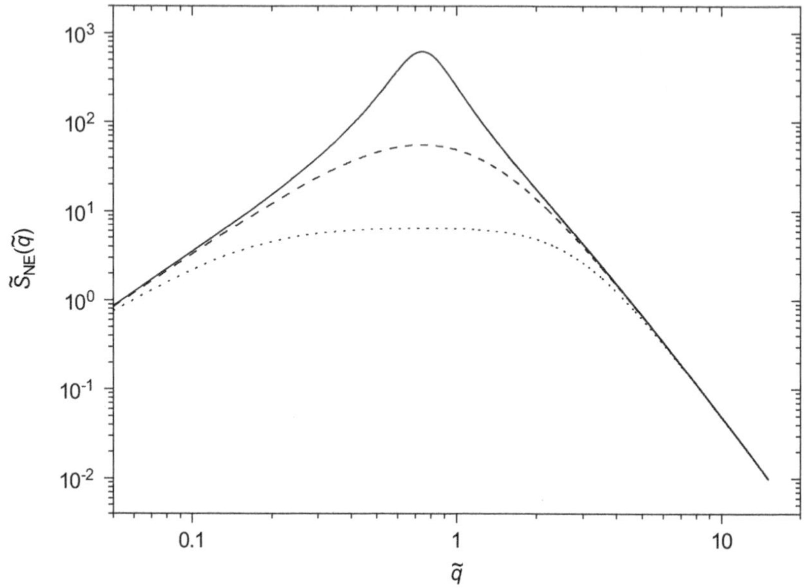

Figure 3.5 Normalized contribution $\tilde{S}_{NE}(\tilde{q})$ representing the modification of the NE temperature fluctuations due to gravity and finite-size effects as a function of the wave number calculated from eqn (3.28). The dotted curve corresponds to Ra = −5000 when the fluid layer is heated from below, the dashed curve corresponds to Ra = 0 (g = 0), and the solid curve corresponds to Ra = 600 close to the convective instability. Reprinted with permission from *Physica A*.[35]

found in ref. 45. The theory has been extended to binary fluid mixtures by Ortiz de Zárate *et al.*[46,47]

Early attempts to measure NE temperature fluctuations induced by a temperature gradient at larger wave numbers than accessible in traditional light-scattering experiments have been reported by Ahlers and collaborators.[48,49] They used a shadowgraph apparatus, originally developed to measure convection cells above the Rayleigh–Bénard instability,[50] but improved so as also to probe the NE fluctuations below the Rayleigh–Bénard instability.[5] A detailed quantitative comparison with the theoretical prediction was hampered by the fact that the experiments were performed in fluids close to the critical point, where the Rayleigh number becomes a rapidly varying function of temperature. In addition, corrections due to finite exposure times were needed. Nevertheless, the data are consistent qualitatively with the prediction from fluctuating hydrodynamics. In addition, they gave valuable information on the decay rate of the NE fluctuations close to the Rayleigh–Bénard instability.[49]

Subsequently, a beautiful new shadowgraph instrument was developed by Vailati *et al.*[6] as part of a gradient-driven fluctuations experiment (GRADFLEX) project of the European Space Agency in collaboration with the National Aeronautics and Space Administration of the USA.[6] With this instrument it became possible to investigate the NE fluctuations at negative Rayleigh numbers far away from any convective instability. Experiments were performed both on earth and at low gravity during a satellite mission.[51] The results obtained from the GRADFLEX experiment for liquid CS_2 are shown in Figure 3.6.[52] One sees that in the earth-based experiments the intensity of the NE fluctuations again exhibits the crossover from a q^{-4} behavior for large wave numbers to a plateau value at smaller wave numbers, just as the observations discussed in the previous section. The wave numbers covered in the earth-based experiments are not yet small enough to see the final limiting q^2 behavior at smaller q. However, at microgravity the NE fluctuations become so large that also the crossover to the limiting q^2 behavior at smaller q is clearly seen in the range of wave numbers covered in the experiment. These experimental results provide a dramatic confirmation of the effect of gravity and finite size on the NE fluctuations predicted by fluctuating hydrodynamics.

We note that gravity and finite-size effects will also modify the time dependence of the NE fluctuations, which no longer will be a simple sum of two exponentials as given in eqn (3.9).[1,26,47,49,53,54] This feature requires further research and is not addressed in the present chapter.

Fluctuating hydrodynamics also predicts an enhancement of the velocity fluctuations in fluids in the presence of a velocity gradient, *i.e.*, in laminar flow.[55,56] The NE velocity fluctuations also exhibit a crossover from a q^{-4} behavior for large q to a q^2 behavior as $q \rightarrow 0$.[57–61] However, the magnitude of

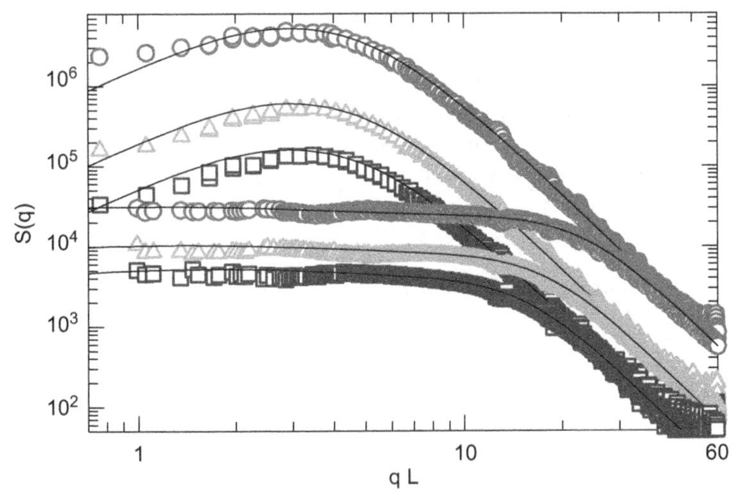

Figure 3.6 Log–log plot of the NE enhancement of the fluctuations in liquid CS_2 as a function of qL. The symbols indicate the experimental data obtained from the GRADFLEX instrument at $\nabla T_0 = 17.9\ \mathrm{K \cdot cm^{-1}}$ (squares), at $\nabla T_0 = 34.5\ \mathrm{K \cdot cm^{-1}}$ (triangles), and at $\nabla T_0 = 101\ \mathrm{K \cdot cm^{-1}}$, in microgravity (upper curves) and on earth. The curves represent the theoretical prediction from fluctuating hydrodynamics.[45]
Copyright (2011) by The American Physical Society.[52]

the NE velocity fluctuations is substantially smaller than those of the NE temperature and concentration fluctuations.[59]

3.5 Fluctuation-induced Nonequilibrium Forces

Generally, when thermal fluctuations are large and long ranged, they are expected to create fluctuation induced forces in the fluid.[62] They are frequently referred to as Casimir or Casimir-like forces, since they have some similarity with forces induced by electromagnetic fluctuations discovered by Casimir.[63] A common example of Casimir forces in fluids are those induced by critical fluctuations, which were originally predicted by Fisher and de Gennes[64] and which have been investigated extensively in the literature.[65–67] One finds a scale-dependent force per unit area, to be denoted as a critical Casimir pressure p_c, which is given by[68]

$$p_c = \frac{k_B T}{L^3}\Theta\left(\frac{L}{\xi}\right), \tag{3.30}$$

where Θ is a finite-size scaling function with ξ being the correlation length of the critical fluctuations.

As discussed in Section 3.2, the NE fluctuations are much larger and long ranged than critical fluctuations Hence, one should expect that the NE fluctuations will induce NE Casimir pressures even more significant than critical Casimir pressures, as recently pointed out by Kirkpatrick *et al.*[42,69] To elucidate the nature of the NE Casimir pressure, we consider the pressure as a function of the conserved thermodynamic quantities, the mass density ρ and the energy density e, to be distinguished from the specific energy u in Chapters 1 and 2. The fluctuating pressure $p(\rho+\delta\rho, e+\delta e)$ is then a function of the fluctuating mass density $\rho+\delta\rho$ and the fluctuating energy density $e+\delta e$. Upon expanding $p(\rho+\delta\rho, e+\delta e)$ in terms of $\delta\rho$ and δe we obtain

$$\begin{aligned} p(\rho+\delta\rho, e+\delta e) &= p(\rho, e) + \left(\frac{\partial p}{\partial \rho}\right)_e \delta\rho + \left(\frac{\partial p}{\partial e}\right)_\rho \delta e \\ &+ \frac{1}{2}\left[\left(\frac{\partial^2 p}{\partial \rho^2}\right)_e (\delta\rho)^2 + 2\left(\frac{\partial^2 p}{\partial \rho \partial e}\right)\delta\rho\delta e + \left(\frac{\partial^2 p}{\partial e^2}\right)_\rho (\delta e)^2\right]. \end{aligned} \tag{3.31}$$

As was discussed in Section 3.2, the equations for the NE temperature fluctuations in the preceding sections (and confirmed experimentally) were obtained in the Boussinesq approximation $\delta\rho = -\alpha\rho_0\delta T$ in accordance with eqn (3.3) and with zero linear pressure fluctuation

$$\delta p = \left(\frac{\partial p}{\partial \rho}\right)_e \delta\rho + \left(\frac{\partial p}{\partial e}\right)_\rho \delta e = 0. \tag{3.32}$$

From eqn (3.3) and (3.32) we deduce that

$$\delta e = -\left(\frac{\partial e}{\partial \rho}\right)_p \delta\rho = \left(\frac{\partial e}{\partial \rho}\right)_p \alpha\rho_0 \delta T. \tag{3.33}$$

If we substitute eqn (3.3) and (3.33) into the expansion (3.31) for the pressure and then take the average of the temperature fluctuations, we obtain a NE contribution to the pressure given by

$$p_{\rm NE}(\boldsymbol{r}) = -\frac{1}{2}\alpha^2\rho^2\left(\frac{\partial p}{\partial e}\right)_\rho\left(\frac{\partial^2 e}{\partial \rho^2}\right)_p \langle(\delta T(\boldsymbol{r}))^2\rangle_{\rm NE}. \tag{3.34}$$

In eqn (3.34) we only retain the NE contribution $\langle(\delta T(\boldsymbol{r}))^2\rangle_{\rm NE}$ to the temperature fluctuations, since the equilibrium contribution is already contained in the local equilibrium value of the pressure. With the help of some thermodynamic relations,[70] eqn (3.34) can be converted into

$$p_{\rm NE}(z) = \frac{\rho c_{\rm p}(\gamma-1)}{2T}\left[1 - \frac{1}{\alpha c_{\rm p}}\left(\frac{\partial c_{\rm p}}{\partial T}\right)_p + \frac{1}{\alpha^2}\left(\frac{\partial \alpha}{\partial T}\right)_p\right]\langle(\delta T(z))^2\rangle_{\rm NE}, \tag{3.35}$$

where γ is the ratio of the isochoric and isobaric specific heat capacities.[69] We note that the fluctuation induced NE pressure depends on the coordinate z in the direction of the temperature gradient.

The intensity $\langle(\delta T(z))^2\rangle_{\rm NE}$ of the NE temperature fluctuations can be readily obtained from eqn (3.27) by taking $z' = z$ in eqn (3.28) and integrating eqn (3.2) over the wave vector $\boldsymbol{q}_\parallel$. We thus obtain for the NE contribution to the temperature fluctuations

$$\langle(\delta T(z))^2\rangle_{\rm NE} = \frac{k_{\rm B}T^3}{48\pi\rho a(\nu + a)}F(z;\,{\rm Ra}) \tag{3.36}$$

with

$$F(z;\,{\rm Ra}) = 48\int d\tilde{q}_\parallel \sum_{N=1}^{\infty} \frac{\tilde{q}_\parallel^3 \sin^2(N\pi z/L)}{(\tilde{q}_\parallel^2 + N^2\pi^2)^3 - {\rm Ra}\,\tilde{q}_\parallel^2}. \tag{3.37}$$

The NE fluctuation induced pressure $p_{\rm NE}(z)$ is obtained by substituting eqn (3.36) into eqn (3.35):[69]

$$p_{\rm NE}(z) = \frac{c_{\rm p}k_{\rm B}T^2(\gamma-1)}{96\pi a(\nu + a)}\left[1 - \frac{1}{\alpha c_{\rm p}}\left(\frac{\partial c_{\rm p}}{\partial T}\right)_p + \frac{1}{\alpha^2}\left(\frac{\partial \alpha}{\partial T}\right)_p\right]F(z;\,{\rm Ra})L\left(\frac{\nabla T_0}{T}\right)^2. \tag{3.38}$$

We note that for a fixed value of the temperature gradient ∇T_0, the NE Casimir pressure *increases* with the distance L. Since $\nabla T = \Delta T/L$, for a fixed temperature across the fluid layer, the NE fluctuation induced pressure will vary as L^{-1}, still a much slower variation than the L^{-3} variation of the critical Casimir pressure given by eqn (3.30).

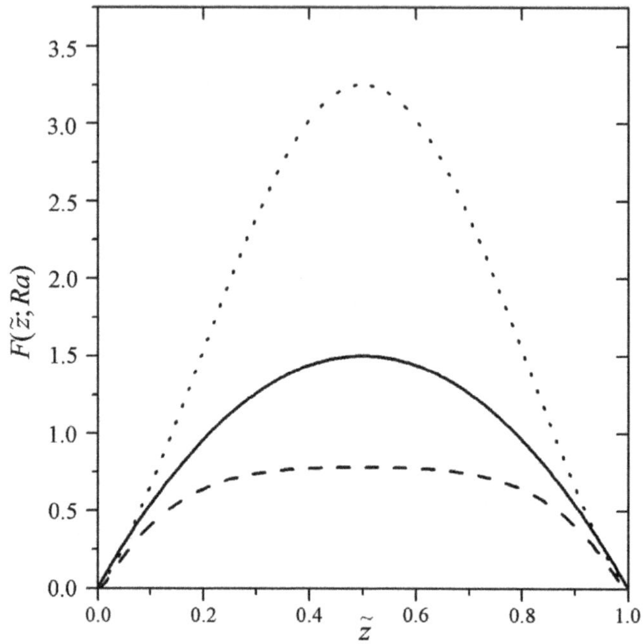

Figure 3.7 Amplitude $F(z;\ \mathrm{Ra})$ of the fluctuation induced NE pressure for free boundaries, given by eqn (3.36) as a function of $\tilde{z}=z/L$ for three values of the Rayleigh number. The solid curve is for $\mathrm{Ra}=0$ ($g=0$), dashed curve for negative $\mathrm{Ra}=-3000$, and dotted curve for positive $\mathrm{Ra}=570$ close to the convective instability.
Copyright (2014) by The American Physical Society.[69]

The fluctuation induced NE pressure depends on the coordinate z through the function $F(z;\ \mathrm{Ra})$, as shown in Figure 3.7. To estimate the order of magnitude of the NE Casimir pressures, we consider the effective value of the fluctuation induced NE pressures in the fluid layer by replacing $F(z;\mathrm{Ra})$ in eqn (3.38) by its average

$$\langle F(z;\ \mathrm{Ra})\rangle_z = \frac{1}{L}\int_0^L \mathrm{d}z\, F(z;\ \mathrm{Ra}). \tag{3.39}$$

The resulting values for this effective NE pressure $p_{\mathrm{NE}}(L)=\langle p_{\mathrm{NE}}(z)\rangle_z$ are shown in Figure 3.8 for n-heptane for temperature differences of $\Delta T=10$ K and $\Delta T=25$ K. The left panel shows the NE pressure when the fluid layer is heated from above and the right panel when heated from below. We see from this figure that the NE pressure $p_{\mathrm{NE}}(L)$ at $L=10^{-4}$ m is of the order of 10^{-3} Pa or larger, which may be compared with a critical Casimir pressure of the order of 10^{-9} Pa at this distance.[69] Thus, NE Casimir pressures are orders of magnitude larger than critical Casimir pressures.[42,69] When the fluid is heated from below, the average NE pressure first decreases with increasing

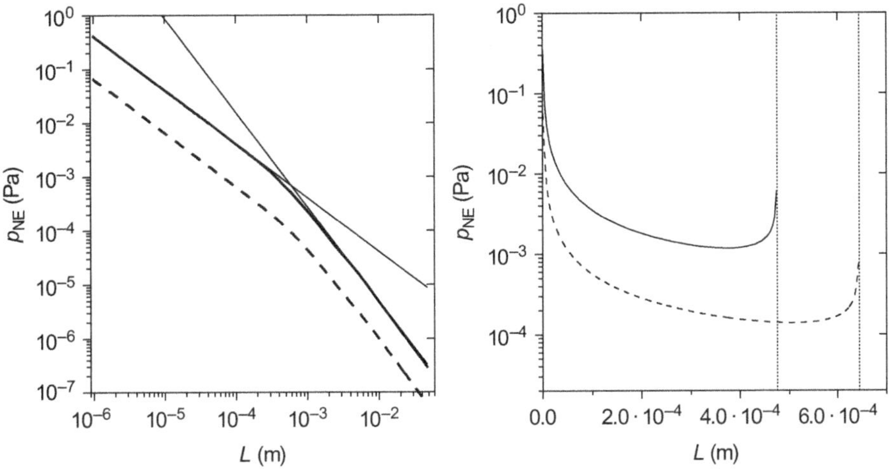

Figure 3.8 Average NE pressure $p_{\mathrm{NE}} = \langle p_{\mathrm{NE}}(z)\rangle_z$ as a function of the distance L between the plates for n-heptane at an average temperature of 25 °C at a temperature difference of $\Delta T = 10$ K (dashed curve) and $\Delta T = 25$ K (solid thick curve). The left panel shows the average NE pressure when the fluid layer is heated from above with the thin lines indicating the asymptotic behavior for small and large L. The right panel shows the average NE pressure when the fluid layer when heated from below. The vertical lines indicate the value of the plate separation L_c at which the critical Rayleigh number is reached for each ΔT.

plate separation L, but ultimately rises again so as to diverge as the critical Rayleigh number is approached.

As mentioned above, the actual NE Casmir pressures are functions of z, as shown in Figure 3.7, and thus induces a NE pressure gradient. Such an induced NE pressure gradient violates mechanical equilibrium and needs to be relaxed. In the absence of any particles in the fluid, this gradient will cause a rearrangement of the local-equilibrium (LE) density profile so that $\rho(z) = \rho_{\mathrm{LE}}(z) + \rho_{\mathrm{NE}}(z; L)$. However, the induced NE pressure gradient is sufficiently large so that it may affect the movement of microparticles in the fluid.[69] In principle, the presence of an induced NE pressure or NE density distribution is a violation of the assumption of local equilibrium. Since the NE pressures are of the order of Pa or less, they are still minuscule compared to normal fluid pressures of MPa. Hence, the assumption of local equilibrium remains valid for normal applications of nonequilibrium thermodynamics except, perhaps, very close to the convective instability. However, local equilibrium is no longer valid when considering forces on microparticles in the presence of a gradient.

Just as with the NE temperature fluctuations, also the NE concentration fluctuations in fluid mixtures will induce NE pressures. The fluctuation induced pressures in fluid mixtures may be even more interesting than those in one-component fluids. From eqn (3.12) and (3.16), we see that the ratio of the NE contribution to the fluctuations in fluids and fluid mixtures is essentially determined by $a(\nu+a)/\nu D \cong \mathrm{Le}$, which is large for most liquid solutions. Furthermore, by inducing a concentration gradient through the Soret effect in accordance with eqn (3.17), it should be possible to induce a substantial concentration gradient with much smaller temperature differences than needed to create similar Casimir pressures in a one-component fluid. This is a topic of continuing research.

Finally we note from nonequilibrium statistical physics that[69]

$$p_{\mathrm{NE}} = \kappa_{\mathrm{NL}}(\nabla T)^2, \tag{3.40}$$

where κ_{NL} is a non-linear Burnett coefficient.[71] The time-dependent correlation functions of the non-linear Burnett coefficients are known to diverge, so that κ_{NL} consists of a bare molecular contribution $\kappa_{\mathrm{NL}}^{(0)}$ and a divergent contribution $\kappa_{\mathrm{NL}}^{(1)}L$ proportional to L.[71–73] Thus, apart from a contribution at a molecular scale, we obtain

$$p_{\mathrm{NE}} = \kappa_{\mathrm{NL}}^{(1)}L(\nabla T)^2. \tag{3.41}$$

On comparing eqn (3.38) with eqn (3.41), we see that we have determined the contribution from the long-range NE fluctuations to a Burnett coefficient. Most interestingly, we conclude that a transport coefficient like κ_{NL} not only diverges, but also depends on gravity and finite-size effects. These features, though to a lesser extent, even apply to linear transport coefficients. Long-range hydrodynamic fluctuations are responsible for the so-called long-time tails[70,74] in the time-dependent correlation functions for the linear transport coefficients which also depend on gravity[75] and finite-size effects.[76] The ubiquitous presence of long-range hydrodynamics has important implication for the theory of transport phenomena.[41,77] As explained in Chapter 1, to obtain the hydrodynamic equations, one supplements the balance laws with linear relationships between fluxes and forces. The fact that the Onsager coefficients in these linear relations themselves become dependent on the solution of the hydrodynamic equations is sometimes referred to as renormalization of the transport coefficients.[78,79]

Attempts have been made to extend linear nonequilibrium thermodynamics by treating the entropy as a function of the classical variables and of the dissipative fluxes, but assuming that this dependence is still local in space.[80] Since the fluctuations extend over the entire system, the validity of such a locality assumption is questionable.

3.6 Conclusions

Fluctuating hydrodynamics has been confirmed experimentally, not only for equilibrium states, but also for nonequilibrium states.

Thermal fluctuations in nonequilibrium states always exhibit a strong NE enhancement.

Fluctuations in nonequilibrium are always long ranged, encompassing the entire system.

Fluctuations in nonequilibrium are affected by gravity.

Thermal fluctuations in nonequilibrium states are affected by the finite size of the system.

Fluctuations in nonequilibrium induce large NE Casimir forces.

Fluctuations cause a renormalization of the Onsager coefficients.

Acknowledgements

JMOZ acknowledges support from the UCM/Santander Research Grant PR6/13-18867. The research of TRK was supported by National Science Foundation Grant DMR-1401449.

References

1. J. M. Ortiz de Zárate and J. V. Sengers, *Hydrodynamic Fluctuations in Fluids and Fluid Mixtures*, Elsevier, Amsterdam, 2006.
2. J. V. Sengers and J. M. Ortiz de Zárate, *J. Non-Equilib. Thermodyn.*, 2007, **32**, 319.
3. P. Manneville, *Dissipative Structures and Weak Turbulence*, Academic Press, New York, 1990.
4. B. Chu, *Laser Light Scattering. Basic Principles and Practice*, Academic Press, Boston, 2nd edn, 1991.
5. S. P. Trainoff and D. S. Cannell, *Phys. Fluids*, 2002, **14**, 1340.
6. A. Vailati, R. Cerbino, S. Mazzoni, M. Giglio, G. Nikolaenko, C. J. Takacs, D. S. Cannell, W. V. Meyer and A. E. Smart, *Appl. Opt.*, 2006, **45**, 2155.
7. T. R. Kirkpatrick, J. R. Dorfman and E. G. D. Cohen, *Phys. Rev. A: At., Mol., Opt. Phys.*, 1982, **26**, 955.
8. D. Ronis and I. Procaccia, *Phys. Rev. A: At., Mol., Opt. Phys.*, 1982, **26**, 1812.
9. I. Heureux and I. Oppenheim, *Physica A*, 1988, **148**, 503.
10. R. Schmitz and E. G. D. Cohen, *J. Stat. Phys.*, 1985, **39**, 285.
11. R. Schmitz and E. G. D. Cohen, *J. Stat. Phys.*, 1985, **40**, 431.
12. R. Schmitz, *Phys. Rep.*, 1988, **171**, 1.
13. B. M. Law and J. V. Sengers, *J. Stat. Phys.*, 1989, **57**, 531.
14. B. M. Law, P. N. Segrè, R. W. Gammon and J. V. Sengers, *Phys. Rev. A: At., Mol., Opt. Phys.*, 1990, **41**, 816.
15. P. N. Segrè, R. W. Gammon, J. V. Sengers and B. M. Law, *Phys. Rev. A: At., Mol., Opt. Phys.*, 1992, **45**, 714.
16. W. B. Li, P. N. Segrè, R. W. Gammon and J. V. Sengers, *Physica A*, 1994, **204**, 399.
17. B. M. Law and J. C. Nieuwoudt, *Phys. Rev. A: At., Mol., Opt. Phys.*, 1989, **40**, 3880.
18. P. N. Segrè, R. W. Gammon and J. V. Sengers, *Phys. Rev. A: At., Mol., Opt. Phys.*, 1993, **47**, 1992.

19. M. G. Velarde and R. S. Schechter, *Phys. Fluids*, 1972, **15**, 1707.
20. A. Vailati and M. Giglio, *Phys. Rev. Lett.*, 1996, **77**, 1484.
21. A. Vailati and M. Giglio, *Prog. Colloid Polym. Sci.*, 1997, **104**, 76.
22. W. B. Li, K. J. Zhang, J. V. Sengers, R. W. Gammon and J. M. Ortiz de Zárate, *J. Chem. Phys.*, 2000, **112**, 9139.
23. J. M. Ortiz de Zárate, C. Giraudet, H. Bataller and F. Croccolo, *Eur. Phys. J. E: Soft Matter Biol. Phys.*, 2014, **37**, 77.
24. A. Vailati and M. Giglio, *Nature*, 1997, **390**, 262.
25. A. Vailati and M. Giglio, *Phys. Rev. E: Stat. Phys., Plasmas, Fluids, Relat. Interdiscip. Top.*, 1998, **58**, 4361.
26. F. Croccolo, D. Brogioli, A. Vailati, M. Giglio and D. S. Cannell, *Phys. Rev. E: Stat., Nonlinear, Soft Matter Phys.*, 2007, **76**, 041112.
27. A. Oprisan and A. L. Payne, *Opt. Commun.*, 2013, **290**, 100.
28. S. Delong, Y. Sun, B. E. Griffiths, E. Vanden-Eynden and A. Donev, *Phys. Rev. E: Stat., Nonlinear, Soft Matter Phys.*, 2014, **90**, 063312.
29. M. E. Fisher, *J. Math. Phys.*, 1964, **5**, 944.
30. A. Pelisso and E. Vicari, *Phys. Rep.*, 2002, **368**, 549.
31. J. V. Sengers and J. G. Shanks, *J. Stat. Phys.*, 2009, **137**, 857.
32. J. M. Ortiz de Zárate, R. Pérez Cordón and J. V. Sengers, *Physica A*, 2001, **291**, 113.
33. P. N. Segrè, R. Schmitz and J. V. Sengers, *Physica A*, 1993, **195**, 31.
34. P. N. Segrè and J. V. Sengers, *Physica A*, 1993, **198**, 46.
35. J. M. Ortiz de Zárate and J. V. Sengers, *Physica A*, 2001, **300**, 25.
36. J. V. Sengers and J. M. Ortiz de Zárate, in *Thermal Nonequilibrium Phenomena in Fluid Mixtures*, ed. W. Köhler and S. Wiegand, Springer, Berlin, 2002, pp. 121–145.
37. S. Mazzoni, R. Cerbino, A. Vailaiti and M. Giglio, *Ann. N. Y. Acad. Sci.*, 2006, **1077**, 351.
38. D. Brognioli, A. Vailati and M. Giglio, *J. Phys. Condens. Matter*, 2000, **12**, A39.
39. D. Brognioli, A. Vailati and M. Giglio, *Phys. Rev. E: Stat. Phys., Plasmas, Fluids, Relat. Interdiscip. Top.*, 2000, **61**, R1.
40. R. Cerbino and A. Vailati, *Curr. Opin. Colloid Interface Sci.*, 2009, **14**, 416.
41. A. Donev, T. G. Fai and E. Vanden-Eijnden, *J. Stat. Mech.: Theory Exp*, 2014, P04004.
42. T. R. Kirkpatrick, J. M. Ortiz de Zárate and J. V. Sengers, *Phys. Rev. Lett.*, 2013, **110**, 235902.
43. S. Chandrasekhar, *Hydrodynamic and Hydromagnetic Stability*, Oxford University Press, Oxford, 1961.
44. P. C. Hohenberg and J. Swift, *Phys. Rev. A: At., Mol., Opt. Phys.*, 1977, **46**, 4773.
45. J. M. Ortiz de Zárate and J. V. Sengers, *Phys. Rev. E: Stat., Nonlinear, Soft Matter Phys.*, 2002, **66**, 036305.
46. J. M. Ortiz de Zárate, F. Peluso and J. V. Sengers, *Eur. Phys. J. E: Soft Matter Biol. Phys.*, 2004, **15**, 310.

47. J. M. Ortiz de Zárate, J. A. Fornés and J. V. Sengers, *Phys. Rev. E: Stat., Nonlinear, Soft Matter Phys.*, 2006, **74**, 046305.
48. M. Wu, G. Ahlers and D. S. Cannell, *Phys. Rev. Lett.*, 1995, 1743.
49. J. Oh, J. M. Ortiz de Zárate, J. V. Sengers and G. Ahlers, *Phys. Rev. E: Stat., Nonlinear, Soft Matter Phys.*, 2004, **69**, 021106.
50. J. R. de Bruyn, E. Bodenswchatz, S. W. Morris, S. P. Trainoff, Y. Hu, D. S. Cannell and G. Ahlers, *Rev. Sci. Instrum.*, 1996, **67**, 2043.
51. A. Vailati, R. Cerbino, S. Mazzoni, C. J. Takacs, D. S. Cannell and M. Giglio, *Nat. Commun.*, 2011, **2**, 280.
52. C. J. Takacs, A. Vailati, R. Cerbino, S. Mazzoni, M. Giglio and D. S. Cannell, *Phys. Rev. Lett.*, 2011, **106**, 244502.
53. C. J. Takacs, G. Nikolaenko and D. S. Cannell, *Phys. Rev. Lett.*, 2008, **100**, 234502.
54. F. Croccolo, H. Bataller and F. Scheffold, *J. Chem. Phys.*, 2012, **137**, 234202.
55. J. Lutsko and J. W. Dufty, *Phys. Rev. A: At., Mol., Opt. Phys.*, 1985, **32**, 3040.
56. J. Lutsko and J. W. Dufty, *Phys. Rev. E: Stat., Nonlinear, Soft Matter Phys.*, 2002, **65**, 041206.
57. J. M. Ortiz de Zárate and J. V. Sengers, *Phys. Rev. E: Stat., Nonlinear, Soft Matter Phys.*, 2008, **77**, 026306.
58. J. M. Ortiz de Zárate and J. V. Sengers, *Phys. Rev. E: Stat., Nonlinear, Soft Matter Phys.*, 2009, **79**, 046308.
59. J. V. Sengers and J. M. Ortiz de Zárate, *J. Non-Newtonian Fluid Mech.*, 2010, **165**, 925.
60. J. M. Ortiz de Zárate and J. V. Sengers, *J. Stat. Phys.*, 2011, **144**, 774.
61. J. M. Ortiz de Zárate and J. V. Sengers, *J. Stat. Phys.*, 2013, **150**, 540.
62. M. Kardar and R. Golestanian, *Rev. Mod. Phys.*, 1999, **71**, 1233.
63. H. B. G. Casimir, *Proc. K. Ned. Akad. Wet.*, 1948, **51**, 793.
64. M. E. Fisher and P. G. de Gennes, *C. R. Acad. Sci. Paris B*, 1978, **287**, 207.
65. M. Krech, *The Casimir Effect in Critical Systems*, World Scientific, Singapore, 1994.
66. M. Krech, *J. Phys. Condens. Matter*, 1999, **11**, R391.
67. A. Gambassi, A. Maciolek, C. Hertlein, U. Nellen, L. Helden, C. Bechinger and S. Dietrich, *Phys. Rev. E: Stat., Nonlinear, Soft Matter Phys.*, 2009, **80**, 061143.
68. A. Gambassi, C. Hertlein, L. Helden, C. Bechinger and S. Dietrich, *Europhys. News*, 2009, **40**, 18.
69. T. R. Kirkpatrick, J. M. Ortiz de Zárate and J. V. Sengers, *Phys. Rev. E: Stat., Nonlinear, Soft Matter Phys.*, 2014, **89**, 022145.
70. M. H. Ernst, E. H. Hauge and J. M. J. van Leeuwen, *J. Stat. Phys.*, 1976, **15**, 7.
71. J. J. Brey, *J. Chem. Phys.*, 1983, **79**, 4585.
72. Y. Pomeau and P. Résibois, *Phys. Rep.*, 1975, **19**, 63.
73. M. H. Ernst and J. R. Dorfman, *J. Stat. Phys.*, 1975, **12**, 311.

74. J. R. Dorfman, T. R. Kirkpatrick and J. V. Sengers, *Annu. Rev. Phys. Chem.*, 1994, **45**, 213.
75. D. Brogioli and A. Vailati, *Phys. Rev. E: Stat. Phys., Plasmas, Fluids, Relat. Interdiscip. Top.*, 2000, **63**, 012105.
76. J. C. Nieuwoudt, T. R. Kirkpatrick and J. R. Dorfman, *J. Stat. Phys.*, 1984, **34**, 20.
77. A. Donev, J. B. Bell, A. de la Fuente and A. I. Garcia, *J. Stat. Mech.:Theory Exp.*, 2011, P06014.
78. D. Bedeaux and P. Mazur, *Physica*, 1974, **73**, 431.
79. D. Bedeaux and P. Mazur, *Physica*, 1974, **75**, 79.
80. D. Jou, J. Casas-Vásquez and G. Lebon, *Extended Irreversible Thermodynamics*, 2nd edn, Springer, Berlin, 1996.

CHAPTER 4

Local Equilibrium in Non-equilibrium Thermodynamics

SIGNE KJELSTRUP* AND DICK BEDEAUX

Department of Chemistry, Norwegian University of Science and Technology, 7491 Trondheim, Norway
*Email: signe.kjelstrup@ntnu.no

4.1 Introduction

Thermodynamic equations apply to systems in global equilibrium.[1–10] They apply evidently to homogeneous systems, but also to systems of lower dimensionality; surfaces and contact lines. The same equations hold between the variables everywhere throughout the volume, the area or the line. Most natural and industrial systems are, however, not in global equilibrium. In order to use thermodynamic equations away from global equilibrium, we need to assume that each volume element, area element or line element is in *local thermodynamic equilibrium* or just local equilibrium. Any classical thermodynamic modelling of systems away from global equilibrium will therefore use the *hypothesis of local equilibrium* (from now on also referred to as the hypothesis).

Assuming that the hypothesis is true, we can use normal thermodynamic relations between the variables for any volume, surface and line element. Away from global equilibrium the variables depend on time and position. The hypothesis of local equilibrium is explicitly stated as a basic assumption in classical non-equilibrium thermodynamics.[7–10] It is always used for

Experimental Thermodynamics Volume X: Non-equilibrium Thermodynamics with Applications
Edited by Dick Bedeaux, Signe Kjelstrup and Jan V. Sengers

Published by the Royal Society of Chemistry, www.rsc.org

thermodynamic modelling of irreversible processes.[1–6] For the homogeneous phases it forms the basis of the continuum hypothesis, which allows us to replace thermodynamic quantities by corresponding thermodynamic fields which are continuous functions of space and time. For heterogeneous systems, the fields are no longer continuous at surfaces and contact lines, and the hypothesis of local equilibrium must be formulated with Gibbs excess variables. In this Chapter we first review results from molecular dynamics simulations and from mesoscopic systems, which support the use of the hypothesis in homogeneous (Section 4.2.2) and heterogeneous systems (Section 4.2.3); work that has been done to strengthen the basis of classical non-equilibrium thermodynamics.[7–10]

Following the stated definitions, we can write, for instance, the Gibbs equation for systems of all dimensionalities. For a homogeneous (three-dimensional) system at location $\boldsymbol{r}=(x,y,z)$ and time t, we have

$$u(\boldsymbol{r},t)=T(\boldsymbol{r},t)s(\boldsymbol{r},t)-p(\boldsymbol{r},t)+\sum_j \mu_j(\boldsymbol{r},t)c_j(\boldsymbol{r},t). \tag{4.1}$$

The intensive variables are here the temperature T, the pressure p and the chemical potentials μ_j. The summation is carried out over all independent components. In order to be able to deal with all three types of system on the same footing, the densities of the internal energy u, entropy s or components c_j are given per unit of volume,[†] following ref. 7. In global equilibrium, the dependence on $\boldsymbol{r}$ and t disappears for homogeneous systems.

For a surface (a two-dimensional system) the corresponding Gibbs equation is

$$u^{\mathrm{s}}(\boldsymbol{r},t)=T^{\mathrm{s}}(\boldsymbol{r},t)s^{\mathrm{s}}(\boldsymbol{r},t)+\gamma(\boldsymbol{r},t)+\sum_j \mu^{\mathrm{s}}{}_j(\boldsymbol{r},t)\Gamma_j(\boldsymbol{r},t), \tag{4.2}$$

where $\boldsymbol{r}=(x,y)$ is now a position along the surface. The energy-, entropy- and particle-densities as well as the surface tension, γ, are excess variables per unit of surface area, defined according to Gibbs, as was explained in Chapter 1. Their values depend on the choice of the dividing surface. We return to that dependence in Section 4.3. Superscript s denotes a property particular to the surface. The equation indicates that the temperature and chemical potentials are unique to the surface and are not necessarily equal to the temperatures and chemical potentials near the surface in the adjacent phases. A surface, which obeys eqn (4.2), is therefore *autonomous*, *cf.* Section 1.4. The properties of the surface then do not depend on the temperatures and chemical potentials near the surface in the adjacent phases. This property has been questioned, *e.g.* in ref. 11, for surfaces away from equilibrium. We have confirmed the validity for surfaces, *cf.* Section 4.2.3, and will review this evidence below.

[†]According to IUPAC standard, the symbols u, s and c are reserved for specific quantities. Such quantities are not relevant here.

The corresponding equation for a contact line (a one-dimensional system) is

$$u^{\mathrm{c}}(r,t) = T^{\mathrm{c}}(r,t)s^{\mathrm{c}}(r,t) + \gamma^{\mathrm{c}}(r,t) + \sum_j \mu_j^{\mathrm{c}}(r,t)\Gamma_j^{\mathrm{c}}(r,t), \tag{4.3}$$

where $r = x$ is now a position along the line. Superscript c denotes the contact line between three coexisting phases. The line tension is γ^{c}. Excess densities are used, in units per m of line length. The systems described by eqn (4.1) to (4.3) differ in dimensionality, but they can all be regarded as macroscopic in the sense that variables enter with their thermodynamic limit value.

We review direct evidence and criteria for local equilibrium in the first part of Section 4.2.

In the second part of Section 4.2, we address descriptions on the mesoscopic level. These fall into three categories. Rubí and co-workers[12–14] have pioneered a method, called mesoscopic non-equilibrium thermodynamics, which introduces internal degrees of freedom, *cf.* Chapters 14 to 16. Small systems can also be handled with the ensemble construction of Hill,[15,16] but this work has not yet been carried to the non-equilibrium regime. Lebon and co-workers[3] defined an extension of the local-equilibrium hypothesis, by introducing fluxes as variables in the Gibbs equation. In their theory of extended irreversible thermodynamics,[2,3] intensive variables obtain new definitions.

The internal variables of mesoscopic non-equilibrium thermodynamics[7,12–14] cannot be controlled from the outside. A central example of such a variable is the probability density of a chemical reaction, *cf.* Chapters 14 to 16. Can we use the hypothesis of local equilibrium for this variable along the reaction coordinate? If so, in which form, and can it be defended? Results from the last two decades are used to substantiate the answers to these questions.

In order to understand why a property holds true, it is also often useful to construct and examine a model that violates the hypothesis. The square-gradient theory, a popular density functional theory, introduces density gradients as non-local properties of the thermodynamic model. It is remarkable that local equilibrium is valid for Gibbs excess variables that are found by integration over the density profiles.[18–20] We review the evidence for these cases in Section 4.3. We shall finally discuss in Section 4.4 how the hypothesis can be actively used to predict unknowns.

4.2 Validating the Hypothesis of Local Equilibrium

In this section, we review results which support the hypothesis of local equilibrium for homogeneous phases and surfaces.

4.2.1 Criteria and Tools

Some criteria have been formulated to test the validity of the hypothesis. For gases, a length criterion has been used; stating that the characteristic size of a volume element should be larger than the mean free path in all directions.[21] Such a comparable length is not available for liquids, where the correlation length away from the critical point is of the same order as the diameter of a molecule, but proposals for alternative length criteria[21] seem to be larger than necessary.[22,23] Independent of the aggregate state, a Maxwell velocity distribution is nevertheless expected.[7] It follows also that any of the velocity components can be used to compute the temperature of the volume element.

In this situation, when geometry and particle density are so important for an answer to the questions raised in the introduction, non-equilibrium molecular dynamics (NEMD) is indispensable. It provides a tool for investigation of the thermodynamic behaviour of particles. NEMD can be used to study bulk systems and interfaces. A special advantage of the method is that interfaces can be formed spontaneously in the system, by proper choice of boundary conditions, without prior knowledge of surface properties. This makes an unbiased investigation of interfaces possible. Volumes of different sizes can be compared with NEMD, and deviations from the thermodynamic limit can be determined. Simulation results can be compared to experimental results.[24] A source of error is often the choice of the interaction potential. The cut-off distance in this potential, for instance, is critical for reproduction of experimental results.

The diffuse interface model or the square-gradient model[18–20] is a tool for prediction of equilibrium properties of surfaces. This tool has now become useful in studies of systems far away from global equilibrium, *cf.* Section 4.4.

4.2.2 Homogeneous Phases

Ciccotti and co-workers[24,25] addressed the issue of local equilibrium already in the 1980s. They modelled liquid argon in temperature gradients amounting to $1.8 \cdot 10^9\ \mathrm{K \cdot m^{-1}}$, or with shear rates amounting to $1.14 \cdot 10^{11}\ \mathrm{s^{-1}}$. These are gradients which are difficult to realize in the laboratory. The volume element contained an average of about 32 Lennard-Jones-like particles. This volume can probably be counted as being macroscopic, in view of the results obtained. Particle fluctuations in such a volume should be within a few percent, and this was found. Fourier's law applied to the system when exposed to a temperature gradient. Systems under shear obeyed Newton's law. In both cases, the hypothesis of local equilibrium was verified by computing the same values for the density and potential energy in a volume element in a field as for the same element in global equilibrium.

Hafskjold and Ratkje[22] confirmed the findings of Tenenbaum *et al.*[25] for a Lennard-Jones-like system in a temperature gradient. The value of the enthalpy, pressure and potential energy were the same when the local volume element was in global equilibrium, and when it was exposed to a large gradient, *cf.* Figure 4.1. Linear laws and Onsager's reciprocal relations were established for transport of heat and mass. The local temperature fluctuation was of the same order of magnitude as the maximum temperature difference across a volume element. The velocity distribution in a control volume (around 32 particles) was Maxwellian within 1 % deviation.

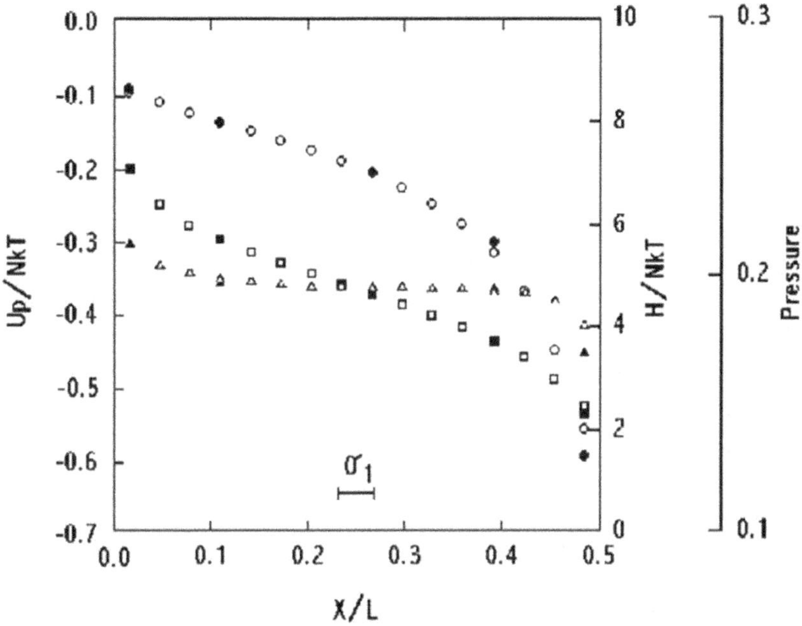

Figure 4.1 Potential energy (squares), enthalpy (circles) and pressure (triangles) in a two-component Lennard-Jones spline system exposed to a temperature gradient, (empty points) compared to corresponding values at global equilibrium (filled points). The particle diameter σ_1 is shown. B. Hafskjold and S. Kjelstrup Ratkje, *J. Stat. Phys.*, 1995, **78**, 463.[22] Figure 8. With kind permission from Springer Science and Business Media.

Maxwell distributions were found in good approximation for all components of Xu *et al.*[26] who studied the dissociation of fluorine to fluor atoms at chemical equilibrium in a temperature gradient. The peak of the velocity distribution shifted proportional to the thermal driving force, as proposed by de Groot and Mazur,[7] showing a net velocity, but no violation of local equilibrium. The distributions are shown in Figure 4.2.

Hafskjold and co-workers[27,28] gave further confirmations of the hypothesis for non-ideal Lennard-Jones particles,[27] and for ionic systems.[28] Bresme and co-workers[28–31] and Røsjorde *et al.*[32] confirmed linear relationships for fluxes and forces during polarization of water in a temperature gradient. Water was polarized by a temperature gradient of 10^8 K · m^{-1} in the liquid phase to create an electric field of 10^6 V · m^{-1}. Kjelstrup *et al.*[23] reviewed the efforts to elucidate these questions up to the year 2008, and concluded that the hypothesis of local equilibrium was obeyed in homogeneous phases when:

a. The volume element had more than 10 to 18 particles[23]
b. The system was exposed to thermal fields of up to about 10^8 K · m^{-1} in the liquid and 10^9 K · m^{-1} in the gas phase

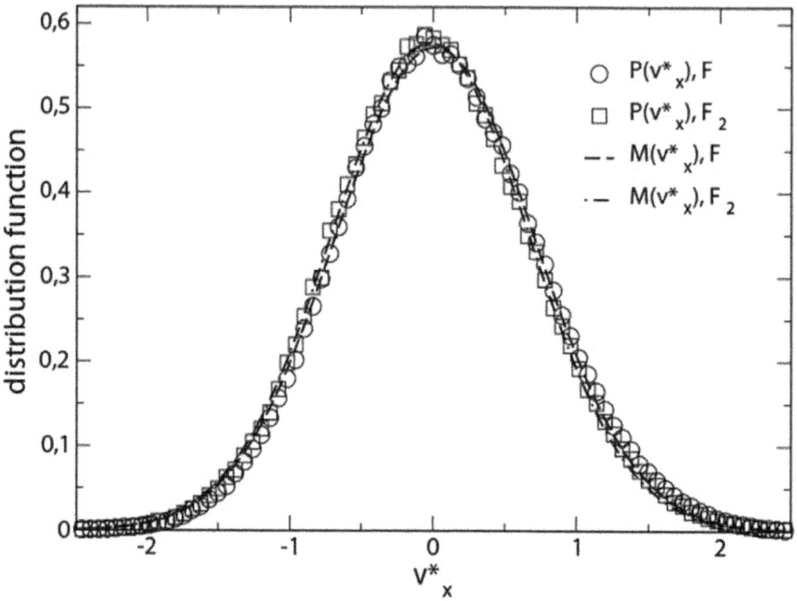

Figure 4.2 Velocity distributions of reactant (F_2) and product (F) of the dissociation reaction in a temperature gradient. The distributions are Maxwellian within 1 %, but slightly shifted to the left according to J. Xu, S. Kjelstrup and D. Bedeaux, *Phys. Chem. Chem. Phys.*, 2006, **8**, 2017.[25] Reproduced by permission of the PCCP Owner Societies.

c. The velocity distribution was Maxwellian with a peak shifted in response to the driving force[26]
d. Linear flux–force relations applied

We now know that when volume elements become smaller than those needed to produce data in the thermodynamic limit, we can also find a thermodynamic description from the method of Hill,[15] see Schnell and co-workers.[16,33–35] An extension of eqn (4.1) to (4.3) by the method of Hill could possibly also extend the validity of the hypothesis. Also the effect on the hypothesis of an extension of the linear flux relations to the non-linear domain, remains to be further investigated, see Section 4.2.4.

4.2.3 Surfaces

The first validation of the hypothesis of local equilibrium for surfaces was reported by Røsjorde *et al.*[36] The surface was described in terms of excess variables. For a one-component system, the surface in local equilibrium was fully characterised by the temperature dependence of the interfacial tension. The equation of state (4.4) was first verified by equilibrium simulations. In the computer experiment of Røsjorde *et al.*[36] with one-component

Lennard-Jones spline particles, the surfaces were allowed to form spontaneously in a large box under a temperature gradient up to $3 \cdot 10^8\ \mathrm{K \cdot m^{-1}}$. The interfacial tension and the temperature were computed in this situation and fitted to the function:

$$\gamma = \gamma_0 \left(\frac{T_c - T}{T_c} \right)^{2\nu}, \tag{4.4}$$

where $\nu = 0.63$ is a universal critical exponent, and c stands for critical, see Figure 4.3. The same relationship (4.4) was found also in non-equilibrium; confirming the hypothesis, see Figure 4.3. Ge *et al.*[37] confirmed the results for a Lennard-Jones interaction potential with a longer tail, changing the factor γ_0 and the critical temperature T_c of eqn (4.4). Similar results were obtained by Xu *et al.*[39] Simon *et al.*[38] verified the same result with a chain model of *n*-octane. The length of a molecule was five times larger than the layer thickness used in the sampling of data, and there were on average only 8 molecules in the volume element. The molecule was modelled with rotational and vibrational degrees of freedom. The temperature was calculated as usual in molecular dynamics from the kinetic energy of particles within the surface boundaries. The same surface temperature was found for in-plane velocity components of the flat surface as well as components

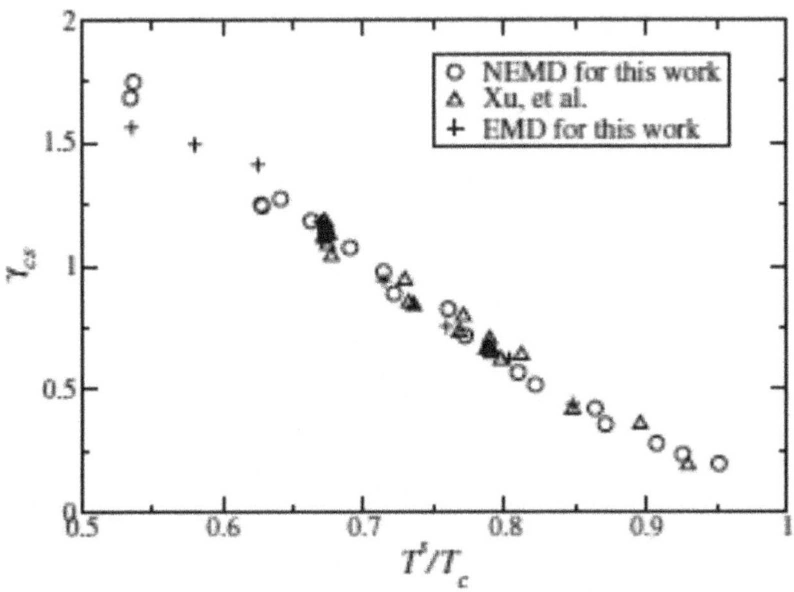

Figure 4.3 The interfacial tension as a function of temperature of a one-component Lennard-Jones spline system according to J. Ge, S. Kjelstrup, D. Bedeaux, J.-M. Simon and B. Rousseau,[37] *Phys. Rev. E*, 2007, 75, 061604, Figure 7. Copyright 2007 by The America Physical Society. The interfacial tension and temperature are measured relative to their critical values. We see that results from global equilibrium (EMD) coincide with results from non-equilibrium simulations (NEMD).

Table 4.1 Temperatures calculated from molecular dynamics simulations of one component in a thermal field.[39] The temperatures T_H and T_L, at the end and in the middle of the symmetric simulation box, were controlled. Results are reported for the absence and presence of a mass flux J for overall density c. The temperatures of the liquid and the vapour close to their interface are T^l and T^g, and T^s is the temperature of the surface.

$c/\text{mol}\cdot\text{m}^{-3}$	T_H/K	T_L/K	$J/\text{mol}\cdot\text{m}^{-2}\cdot\text{s}^{-1}$	T^l/K	T^g/K	T^s/K
8300	136.4	93.0	0	93.6	101.7	94.3
10 400	161.2	99.2	0	101.8	110.7	103.3
16 600	161.2	99.2	0	106.3	134.4	113.8
12 500	198.4	93.0	0	99.2	115.9	101.6
8300	136.4	93.0	6670	95.6	103.3	96.8
10 400	148.8	86.8	26 700	96.5	110.0	99.0
16 600	173.6	86.8	40 000	103.9	148.4	116.7
12 500	148.8	80.6	20 000	90.8	107.3	92.6

perpendicular to the surface in the interfacial region,[36] confirming the autonomous nature of the surface. The authors were able to distinguish the surface temperature from the adjacent liquid and vapour temperatures in the calculations.[39] Table 4.1 shows that the temperature can jump into the surface and across it by several degrees, even in the absence of mass transport. In the presence of mass transport, the jumps become even larger. Interestingly, the formulas for transfer coefficients from kinetic theory gave better fits to the results from the simulations when the surface temperature was used, rather than the gas temperature.[39] The table values give indirect support to the idea of the surface as an autonomous system and to the hypothesis of local equilibrium.

To create a stationary interface in a gradient using non-equilibrium molecular dynamics simulations is often computer-time consuming. Errors can arise in the analysis if the system is not given enough time to respond to the boundary conditions. This can become critical for obtaining good results with two or more components, where particles in the liquid phase have to interchange with the gas-phase particles to arrive at the stationary state. Inzoli *et al.*[40] studied two-component Lennard-Jones systems in this situation, and were able to confirm the validity of the hypothesis. Their results, shown in Figure 4.4, gave information on thermodynamic properties of the surface under temperature gradients, identical to the information obtainable at global equilibrium, thereby supporting the hypothesis. The concentration variation of the surface excess entropy was found. The heat capacity is equal to minus the second derivative of the interfacial tension. Given the analytic form of eqn (4.4), it is negative.

In studies of *n*-butane adsorption on and into zeolites, linear flux–force relations were confirmed for a wide range of conditions.[41] All studies combined have shown that the surface is in local equilibrium when

e. The area element has more than 10 atoms
f. The surface is exposed to thermal fields up to $10^9\ \text{K}\cdot\text{m}^{-1}$

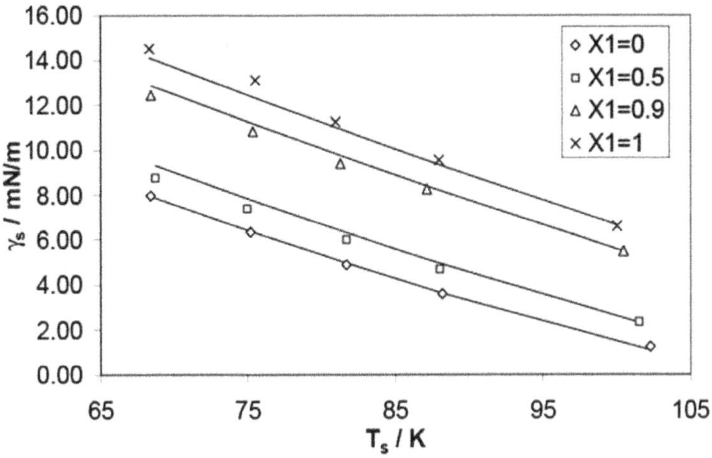

Figure 4.4 The interfacial tension of a two-component Lennard-Jones spline mixture as a function of temperature. Data are shown for mole fractions x_1 of the light component. Points represent calculations away from global equilibrium. I. Inzoli, S. Kjelstrup, D. Bedeaux and J.-M. Simon, *Chem. Eng. Sci.*, 2011, **66**, 4533, Figure 2.[40]
With kind permission from Elsevier.

g. The kinetic temperature calculated from one velocity component is the same for all velocity components *x,y,z*
h. Linear flux–force relations apply

The fact that the equation of state for the surface is the same, whether or not one applies a gradient to the system, makes it possible to circumvent extra equilibrium simulations. Non-equilibrium simulations can also give this information, see Section 4.4.

4.2.4 Mesoscopic Systems

The subject of mesoscopic equilibrium thermodynamics is concerned with equilibrium thermodynamics at intermediate length scales, *i.e.* scales between microscopic and macroscopic scales. Mesoscopic non-equilibrium thermodynamics can be understood as non-equilibrium thermodynamics at intermediate length scales. Hydrodynamic fluctuations are interesting phenomena at intermediate length scales *cf.* Chapter 3. These are described without introduction of new variables. In mesoscopic non-equilibrium thermodynamics we introduce internal degrees of freedom in order to describe the system on either the smaller length or time scales. We mentioned above the most prominent example of system with an internal variable, namely the chemical reaction. It has long been standard in chemical reaction kinetics to introduce a coordinate to measure the progress of the reaction. Along this coordinate, one pictures an activation energy barrier, leading to the concept of an activated (slow) process. The internal variable in this picture is the probability density along the coordinate axis to find the reacting mixture. The transport across the barrier, first proposed by Eyring and used by Kramers,[42] has later been adopted in many other processes. The macroscopic flux–force relations obtained for these systems are clearly nonlinear. From the results in Sections 4.2.2 to 4.2.3, it is therefore not obvious that the system is in local equilibrium.

The hypothesis has nevertheless been taken as a premise for mesoscopic non-equilibrium thermodynamics. For mesoscopic systems, the validity of the hypothesis of local equilibrium means that *we write the Gibbs entropy postulate for internal variables*, see Rubí and co-workers.[13,14,43–46] For an isolated system with one degree of freedom (a chemical reaction in an isolated box), a small change in the entropy density is due to a change in the probability density $c(\gamma)$ along the reaction coordinate γ[‡] by:

$$\delta s = -\frac{1}{T}\int_0^1 \mu(\gamma)\delta c(\gamma)\mathrm{d}\gamma. \tag{4.5}$$

At equilibrium, $\delta s = 0$. Given that $c(\gamma)$ is normalised, it follows that the chemical potential is independent of γ. Using $\partial c(\gamma)/\partial t = -\partial r(\gamma)/\partial \gamma$, we find for the entropy production along the γ-coordinate

$$\sigma(\gamma) = -r(\gamma)\frac{1}{T}\frac{\partial \mu(\gamma)}{\partial \gamma}, \tag{4.6}$$

where r is the rate of the chemical reaction. All variables along the γ-coordinate also depend on the time. The partial derivatives here and below imply that either γ or t are kept constant. The corresponding linear law is

$$r(\gamma) = -l(\gamma)\frac{1}{T}\frac{\partial \mu}{\partial \gamma}. \tag{4.7}$$

We recognize the bilinear form in the entropy production and the linear law, characteristic of classical non-equilibrium thermodynamics. The properties are the same as for classical thermodynamics, but do now apply to a sub-macro (meso) level.

The construction makes it clear that a direct test of eqn (4.5) to (4.7) is difficult, if not impossible. Unlike what we have described in Sections 4.2.2 and 4.2.3, only indirect proof of the hypothesis can therefore be obtained. Indirect proof is constituted by the many well-known formulas one can derive with mesoscopic non-equilibrium thermodynamics, like for instance nucleation theory,[14] the Butler–Volmer equation,[46] equations for molecular motors, *etc.*[47] The law of mass action in these cases[45–47] is always obtained from eqn (4.6) by integrating over the activation energy barrier, see also Chapters 14 to 16.

[‡]We use the symbol γ for the reaction coordinate, following ref. 7. It should not be confused with the same symbol for the surface tension.

4.2.5 Local *versus* Global Equilibrium: Molecular Fluctuations

We have seen above that the situation at local equilibrium is very much the same as in global equilibrium: The normal thermodynamic equations apply. A natural question is therefore: is there distinction between the two states? The answer is connected to molecular fluctuations. Fluctuations of densities and temperatures are very different in local and global equilibrium.

In global equilibrium, fluctuations in extensive variables have their equilibrium values, characterized by Gaussian white noise. Their correlations can be calculated from the probability of fluctuations around equilibrium, or from the fluctuation–dissipation theorem which regards random contributions to a fluctuating thermodynamic flux.

A local form of the fluctuation–dissipation theorem remains valid when the system is driven away from global equilibrium.[17] These parts of the fluctuations are Gaussian, as those in global equilibrium are. In addition, the system obtains long-range correlations, which are not present in global equilibrium.[17] *The long-range correlations will therefore characterise a state in local equilibrium* in a system far from global equilibrium. The correlations in such a system can be several orders of magnitude larger than in global equilibrium.[17,48] The systems described above in Section 4.2.1 showing Maxwell distributions[26] are too small for observation of long-range correlations. Ortiz de Zárate *et al.*[49] predicted that long-range correlations become observable for systems near 1 μm.

Correlations in local equilibrium are central to all energy converting processes. One may expect that they are particularly important for coupling of phenomena, and therefore for a possible decrease of energy losses. It is important to understand their origin and nature, and we refer the reader to Chapter 3 for more information.

4.3 Local Equilibrium and Density Functional Theory

Density functional theory is a sophisticated tool used in many fields. Its essence can be captured by van der Waals square-gradient theory, in which contributions proportional to the square of the density gradients are added to the Helmholtz energy density. From the equation of state, the Gibbs energy density and other thermodynamic properties are constructed. An equation of state for the pressure tensor can thus be constructed, relevant for a description of the liquid–vapour interface in terms of density gradient variables. All thermodynamic properties become non-local functions of the densities involved by these constructions.

With non-local variables involved, one may rightly question the meaning of local equilibrium in an interfacial region. In order to do so, the local entropy production and the resulting linear laws were first constructed.[18–20,50,51] In the next step, profiles were integrated across the interfacial region, to obtain Gibbs excess densities. These are the variables that enter eqn (4.2). The value of an excess density depends in principle on the choice made for the dividing surface. Therefore the square-gradient model was used to find several sets of thermodynamic properties for various choices of the dividing surface. The surface temperature is, however, independent of this choice, like the interfacial tension of a flat surface is. The temperature must be the same, wherever we choose to locate the dividing surface (it is gauge invariant). In order to be able to verify gauge invariance, the surface temperature was calculated for a one-component system in several ways, for instance:[9]

1. From the standard thermodynamic relation $T^s = (h^s - \mu^s)/s^s$ for an *equimolar surface*
2. From the same relation as in (1) but with variables for the *surface of tension*
3. From the interfacial tension, calculated from the surface equation of state $T^s = T^s(\gamma)$
4. From the definition of the interfacial tension using the van der Waals square-gradient model, using in the end $T^s = T^s(\gamma)$

Table 4.2 shows results for temperatures calculations, for vapour pressures deviating from the vapour pressure at saturation by (1 or ±2) % (cases PG98, PG99, and PG102 in Table 4.2) and for temperatures deviating from the equilibrium temperature of the vapour pressure by (1 or ±2) % (cases TL98, TL99, and TL102 in Table 4.1). More results for other conditions were also reported.[9,18,20] We see that the temperatures are the same within the accuracy of the numerical calculation, 0.01 K, see Table 4.2. This is a remarkable finding for a model which in the outset is non-local. The fact that local equilibrium holds true for the surface independent of the choice of the set of excess variables is a strong indication of its autonomous nature. It means that, provided we follow Gibbs, we can construct a two-dimensional thermodynamic system from a non-local model and expect it to be in local equilibrium, independent of the presence of an external field. The same was documented for a two-component system.[20] Magnanelli *et al.* confirmed the

Table 4.2 The surface temperature T^s (in K) as calculated by methods 1 to 4 (see text) for a surface exposed to three chemical potential differences PG98, P99 and P102 and three temperature differences TL98, TL 99 and TL102 (see text). The table was adapted from ref. 9, 18 and 50.

	T^s/K			
Method	1	2	3	4
PG98	560.6651	560.6650	560.6651	560.6651
PG99	561.1563	561.1563	561.1563	561.1563
PG102	562.6421	562.6420	562.6421	562.6421
TL98	553.1456	553.1451	553.0787	553.0804
TL99	557.3527	557.3525	557.3364	557.3369
TL102	570.4054	570.4043	570.3482	570.3541

hypothesis with temperature-dependent variables in the square-gradient model.[51]

We have seen from Table 4.2 that the analytical method is very precise.[18,50,51] Thanks to that precision, it was also possible to conclude that it is *not correct* to take the surface temperature equal to the liquid or the vapour temperature in the layer next to the surface (the liquid and vapour temperatures at the surface are the extrapolated temperatures from the bulk phase), see also Table 4.1. This supports the idea of eqn (4.2), that the surface has its own temperature, which can differ from that of the surroundings.

4.4 Predicting Properties from the Hypothesis of Local Equilibrium

From the collected evidence reviewed above, the hypothesis of local equilibrium appears true, even under very large gradients. It can then be an advantage to invert the situation. Rather than checking the validity, one may take it for granted and find new results.

This was done by Bedeaux and Ytrehus in 1990[52] in a description of sublimation and condensation of water vapour below 0 °C. By defining the entropy production in terms of surface excess properties, their flux–force relations were able to reproduce the results from kinetic theory. Since then, Römer *et al.*[53] have also actively used simulation results from non-equilibrium conditions to find equations of state for the system in question. This was done, for instance, for water as modelled with the SPC/E and TIP4P/2005 models.[53]

The autonomous nature of the surface enables predictions. Using the assumption of local equilibrium in the surface, it was predicted[9,18,50,51] that the surface temperature was accessible, away from global equilibrium, from knowledge of the molar density difference across the surface. The simple, useful formulae proposed for the concentration difference and the internal energy difference (both per unit of volume) were:

$$\begin{aligned} c^{\mathrm{l}} - c^{\mathrm{g}} &= c^{\mathrm{l}}_{\mathrm{eq}}(T^{\mathrm{s}}) - c^{\mathrm{g}}_{\mathrm{eq}}(T^{\mathrm{s}}), \\ u^{\mathrm{l}} - u^{\mathrm{g}} &= u^{\mathrm{l}}_{\mathrm{eq}}(T^{\mathrm{s}}) - u^{\mathrm{l}}_{\mathrm{eq}}(T^{\mathrm{s}}). \end{aligned} \tag{4.8}$$

Similar formulas can be written for other variables. The formulae states that a concentration, c, difference measured away from equilibrium can be used with information from equilibrium to find the actual temperature of the surface. All the relations of Gibbs for the surface can be derived by active use of the assumption of local equilibrium.

This statement was later generalized by Savin *et al.*[54] in the most far-reaching predictions made for surfaces so far. The authors pointed out first that a calculation of excess densities depends on the choice of the dividing interface, and this fact can be seen as an ambiguity because certain measurements, for instance of the interfacial tension of a flat surface, should not depend on such a choice. The authors went on to show that this gauge invariance implies local equilibrium. By characterizing different gauges in terms of the excess mass density, rather than the position of the dividing surface, a gauge transformation in a one-component system was written as an expansion around the equimolar surface ($c^s = 0$):

$$s^s(T^s, c^s) = s^s(T^s, 0) + \frac{\Delta s}{\Delta c} c^s. \tag{4.9}$$

For this gauge, we also know that

$$s^s(T^s, 0) = -\frac{d\gamma}{dT^s}. \tag{4.10}$$

On the other hand, the Clapeyron equation for global equilibrium says that

$$\frac{\Delta s}{\Delta c} = -\frac{d\mu}{dT}. \tag{4.11}$$

By comparing eqn (4.9) to (4.11) we see that the assumption of local equilibrium is consistent with the Clapeyron equation. This is remarkable, as the last equation is derived in classical thermodynamics for constant chemical potential throughout the system. It is now a consequence of the gauge invariance, stating that eqn (4.11) also applies when the surface is exposed to an external field, and with values of *variable jumps that arise from the field.*

This finding has several implications: The concept of "dynamic interfacial tension" becomes obsolete, as any interfacial tension must obey the equation of state. The surface *variables* (for instance, the temperature) can be time-dependent, however. The eqn (4.8) first given by Johannessen and Bedeaux,[9,50] have a parallel with other variables.[54]

Contact lines have not been investigated from this perspective so far. It is clear that the general conclusions above are also applicable to contact lines. The contact line in a non-equilibrium one-component system must have the triple temperature in order to obey local equilibrium. It would be a major achievement if this could be verified.

4.5 Conclusions and Perspectives

To summarize this chapter, let us answer again: how do we know that the hypothesis of local equilibrium is valid? In the first place, we know this because expected sensible results are produced from this hypothesis. Local equilibrium applies with all probability on the macroscopic level for homogeneous phases, interfaces, and contact lines. It is also likely on the mesoscopic level, because we can derive results from this hypothesis which agree with experiments. The law of mass action[45] and the Butler–Volmer equation[46] are examples of such results. In the second place, we can cite a series of results from non-equilibrium molecular dynamics simulations that agree with the hypothesis.[22–32] Even under extreme conditions with gradients larger than 10^8 K · m^{-1} in the liquid and 10^9 K · m^{-1} in the vapour phase, local equilibrium holds true.

The evidence for local equilibrium in a surface, obtained from work that uses density functional theory, is convincing, because the hypothesis is not made *a priori*, but is brought out by proper definitions of variables, and use of Gibbs excess densities.[20,50,51]

Thanks to these results, we are now in a position to actively use the hypothesis for predictive purposes.[53,54] With the hypothesis of local equilibrium on the meso-scale, one can predict new properties. The biological ion pump for Ca^{2+}-ions,[44] may for instance also be a heat pump. Equally interesting is the insight offered by the work[54] that variable jumps, like those in the well-known Clausius–Clapeyron formulae, can be seen as a consequence of local equilibrium in the interface, away from global equilibrium. Such knowledge can be useful in the modelling of interface transport.

Hill introduced an ensemble of replicas of the small system, and constructed thermodynamic relations for one replica from the ensemble.[15] Using his method, one can show that systems down to a few particles have definable thermodynamic properties. The concept of local equilibrium has not been studied in this context. But one consequence is seen already: it appears that a small system need not be macroscopic in order to be in local equilibrium. What is important is only to be able to account for finite-size effects.[16] The hypothesis of local equilibrium was stated for continuous descriptions, eqn (4.1) to (4.3). A continuous description is obtained by averaging the microscopic description over volume, area or line element. For a discussion of this, see for instance Section 2.1.2 in ref. 17, the magnitude of these elements is chosen so large that they can be considered to be macroscopic. Macroscopic thermodynamic relations can then be used, according to the hypothesis of local equilibrium. But also in the case of small systems[15,16] we expect the hypothesis of local equilibrium to be true. The thermodynamic relations will differ from the ones given for the thermodynamic limit, however.

Acknowledgements

ETH Zürich is thanked for guest professorships in 2014 to S.K. and D.B.

References

1. R. B. Bird, E. Stewart and E. N. Lightfoot, *Transport Phenomena*, Wiley, 2nd rev. edn, 2007.
2. D. Jou, J. Casas-Vázquez, and G. Lebon, *Extended Irreversible Thermodynamics*, Springer, Berlin, 2nd edn, 1996.
3. G. Lebon, D. Jou and J. Casas-Vázquez, *J. Phys. A: Math. Gen.*, 1980, **13**, 275.

4. H. C. Öttinger, *Beyond Equilibrium Thermodynamics*, Wiley-Interscience, Hoboken, 2005.
5. R. Taylor and R. Krishna, *Multicomponent Mass Transfer*, Wiley, New York, 1993.
6. E. L. Cussler, *Diffusion, Mass Transfer in Fluid Systems*, Cambridge, 2nd edn, 1997.
7. S. R. de Groot and P. Mazur, *Non-Equilibrium Thermodynamics*, Dover, London, 1984.
8. Y. Demirel, *Non-Equilibrium Thermodynamics*, Elsevier, Boston, 3rd edn, 2014.
9. S. Kjelstrup and D. Bedeaux, *Non-Equilibrium Thermodynamics of Heterogeneous Systems*, Series on Advances in Statistical Mechanics, World Scientific, Singapore, 2008, vol. 16.
10. S. Kjelstrup, D. Bedeaux, E. Johannessen and J. Gross, *Non-Equilibrium Thermodynamics for Engineers*, World Scientific, Singapore, 2010.
11. R. Defay, I. Prigogine, A. Bellemans and D. H. Everett, *Interfacial Tension and Adsorption*, Longmans, London, 1966.
12. A. Pérez-Madrid, I. Pagonabarraga and J. M. Rubí, *Physica. A*, 1997, **237**, 205.
13. I. Pagonabarraga and J. M. Rubí, *Physica. A*, 1992, **188**, 553.
14. D. Reguera and J. M. Rubí, *J. Chem. Phys.*, 2001, **115**, 7100.
15. T. L. Hill, Thermodynamics of Small Systems, *Part 1*, Benjamin, New York, 1963.
16. S. K. Schnell, T. J. H. Vlugt, J.-M. Simon, D. Bedeaux and S. Kjelstrup, *Chem. Phys. Lett.*, 2011, **504**, 199.
17. J. M. Ortiz de Zárate and J. V. Sengers, *Hydrodynamic Fluctuations in Fluids and Fluid Mixtures*, Elsevier, Amsterdam, 2006.
18. E. Johannessen and D. Bedeaux, *Physica. A*, 2003, **330**, 354.
19. D. Bedeaux, E. Johannessen and A. Røsjorde, *Physica. A*, 2003, **330**, 329.
20. K. S. Glavatskiy and D. Bedeaux, *Phys. Rev. E: Stat., Nonlinear, Soft Matter Phys.*, 2009, **79**, 031608.
21. H. J. Kreuzer, *Non-Equilibrium Thermodynamics and its Statistical Foundations*, Clarendon, Oxford, 1981.
22. B. Hafskjold and S. Kjelstrup Ratkje, *J. Stat. Phys.*, 1995, **78**, 463.
23. S. Kjelstrup, D. Bedeaux, I. Inzoli and J.-M. Simon, *Energy*, 2008, **33**, 1185.
24. C. Trozzi and G. Ciccotti, *Phys. Rev. A: At., Mol., Opt. Phys.*, 1984, **29**, 916.
25. A. Tenenbaum and G. Ciccotti, *Phys. Rev. A: At., Mol., Opt. Phys.*, 1982, **25**, 2778.
26. J. Xu, S. Kjelstrup and D. Bedeaux, *Phys. Chem. Chem. Phys.*, 2006, **8**, 2017.
27. B. Hafskjold, I. Fujihara and T. Ikeshoji, *Mol. Phys.*, 1997, **90**, 999.
28. F. Bresme, B. Hafskjold and I. Wold, *J. Phys. Chem.*, 1996, **100**, 1879.
29. F. Bresme, A. Lervik, D. Bedeaux and S. Kjelstrup, *Phys. Rev. Lett.*, 2008, **101**, 020602.
30. F. Römer, A. Lervik and F. Bresme, *J. Chem. Phys.*, 2012, **137**, 074503.

31. J. Muscatello, F. Römer, J. Sala and F. Bresme, *Phys. Chem. Chem. Phys.*, 2011, **13**, 19970.
32. A. Røsjorde, D. Bedeaux, S. Kjelstrup and B. Hafskjold, *J. Colloid Interface Sci.*, 2001, **240**, 355.
33. S. K. Schnell, X. Liu, J.-M. Simon, A. Bardow, D. Bedeaux, T. J. H. Vlugt and S. Kjelstrup, *J. Phys. Chem. B*, 2011, **115**, 10911.
34. S. K. Schnell, T. J. H. Vlugt, J.-M. Simon, D. Bedeaux and S. Kjelstrup, *Mol. Phys.*, 2011, **110**, 1069.
35. P. Krüger, S. K. Schnell, D. Bedeaux, S. Kjelstrup, T. H. J. Vlugt and J.-M. Simon, *J. Phys. Chem. Lett.*, 2013, **4**, 235.
36. A. Røsjorde, D. W. Fossmo, D. Bedeaux, S. Kjelstrup and B. Hafskjold, *J. Colloid Interface Sci.*, 2000, **232**, 178.
37. J. Ge, S. Kjelstrup, D. Bedeaux, J.-M. Simon and B. Rousseau, *Phys. Rev. E: Stat., Nonlinear, Soft Matter Phys.*, 2007, **75**, 06104.
38. J.-M. Simon, S. Kjelstrup, D. Bedeaux and B. Hafskjold, *J. Phys. Chem. B.*, 2004, **108**, 7186.
39. J. Xu, S. Kjelstrup, D. Bedeaux, A. Røsjorde and L. Rekvig, *J. Colloid Interface Sci.*, 2006, **299**, 452.
40. I. Inzoli, S. Kjelstrup, D. Bedeaux and J.-M. Simon, *Chem. Eng. Sci.*, 2011, **66**, 4533.
41. I. Inzoli, S. Kjelstrup, D. Bedeaux and J.-M. Simon, *Microporous Mesoporous Mater.*, 2009, **125**, 112.
42. H. A. Kramers, *Physica. A*, 1940, 7, 284.
43. J. M. Vilar and J. M. Rubí, *Proc. Natl. Acad. Sci. U. S. A.*, 2001, **98**, 11081.
44. A. Lervik, D. Bedeaux and S. Kjelstrup, *Eur. Biophys. J.*, 2012, **41**, 437.
45. J. M. Rubí, D. Bedeaux, S. Kjelstrup and I. Pagonabarraga, *Int. J. Thermophys.*, 2013, **34**, 1214.
46. J. M. Rubí and S. Kjelstrup, *J. Phys. Chem. B*, 2003, **107**, 13471.
47. D. Bedeaux and S. Kjelstrup, *Phys. Chem. Chem. Phys.*, 2008, **48**, 7304.
48. P. N. Segrè, R. Schmitz and J. V. Sengers, *Physica. A*, 1993, **196**, 31.
49. J. M. Ortiz de Zárate, J. V. Sengers, D. Bedeaux and S. Kjelstrup, *J. Chem. Phys.*, 2007, **107**, 034501.
50. E. Johannessen and D. Bedeaux, *Physica. A*, 2003, **333**, 252.
51. E. Magnanelli, Ø. Wilhelmsen, S. Kjelstrup and D. Bedeaux, *Phys. Rev. E: Stat., Nonlinear, Soft Matter Phys.*, 2014, **90**, 032402.
52. D. Bedeaux, L. F. J. Hermans and T. Ytrehus, *Physica. A*, 1990, **169**, 263.
53. F. Römer, Z. Wang, S. Wiegand and F. Bresme, *J. Phys. Chem. B*, 2013, **117**, 8209.
54. T. Savin, K. S. Glavatskiy, S. Kjelstrup, H. C. Öttinger and D. Bedeaux, *Eur. Phys. Lett.*, 2012, **97**, 40002.

CHAPTER 5

Diffusion in Liquids: Experiments, Molecular Dynamics, and Engineering Models

CHRISTINE PETERS,[a,†] LUDGER WOLFF,[a,†] THIJS J. H. VLUGT*[b] AND ANDRÉ BARDOW[a,b]

[a] RWTH Aachen University, Chair of Technical Thermodynamics, Schinkelstr. 8, 52062 Aachen, Germany; [b] Delft University of Technology, Process & Energy Laboratory, Leeghwaterstraat 39, 2628CB Delft, The Netherlands

*Email: t.j.h.vlugt@tudelft.nl

5.1 Introduction

Mass transfer plays an important role in many industrial and environmental processes. For engineers, the precise calculation of mass transfer is essential to design processes. Since mass transfer is usually limited by diffusion, the quantitative description of diffusion, especially in liquids, has been a challenge to scientists for decades. In recent years, our qualitative and quantitative understanding of diffusion in liquids has substantially improved. In this development, non-equilibrium thermodynamics has played a key role. In this chapter, we summarize the current state-of-the-art for studying

[†]Both authors contributed equally to this chapter.

Experimental Thermodynamics Volume X: Non-equilibrium Thermodynamics with Applications
Edited by Dick Bedeaux, Signe Kjelstrup and Jan V. Sengers

Published by the Royal Society of Chemistry, www.rsc.org

diffusion in liquids by experiments, molecular dynamics, and engineering models.

Physically, mass transfer is the movement of molecules in a mixture. Commonly, this movement is divided into a convective and a diffusive part: Convection describes the movement of the mixture as a whole, where each species moves with the same reference velocity. Diffusion refers to the relative movement to this reference velocity. Hence, the magnitude of the diffusive flux depends on the chosen velocity reference frame. Taylor and Krishna[1] provide a detailed overview on common reference frames and their transformations. Important reference frames in practice are the molar average velocity, the volume average velocity, and the barycentric reference velocity.

Different types of diffusion are distinguished:

- *Mutual diffusion* (also named inter-diffusion) describes the net transport of all molecules of species i relative to the reference velocity. Mutual diffusion is thus the quantity of interest in practice to describe mass transfer. Mutual diffusion is induced by chemical-potential gradients due to, *e.g.*, concentration, temperature, external force fields, and pressure.[2]
- *Self-diffusion* (also named intra-diffusion) describes the movement of individual molecules due to Brownian motion without any driving force. In binary mixtures, self- and mutual diffusion coefficients of species i coincide for the limit of zero concentration since the individual molecule is then equivalent to all molecules of species i.

In the modelling of diffusion, the relevant transport coefficient is the so-called diffusion coefficient which relates the driving forces to the diffusive flux. Diffusion coefficients of liquids are typically in the order of $1 \cdot 10^{-9}$ $\mathrm{m^2 \cdot s^{-1}}$. Diffusion coefficients depend on concentrations, pressure as well as temperature. In this chapter, we focus on the concentration dependence of diffusion in liquids. Poling *et al.*[2] provide an overview on the temperature dependence. Even today, the exact form of the dependence of diffusion coefficients on concentration and temperature is still unknown. This difficulty applies especially for multicomponent systems. Therefore, the prediction of diffusion coefficients is still a challenging task.

For the mathematical description of multicomponent mutual diffusion, two approaches are used: the phenomenological generalized Fick's law and the physically motivated Maxwell–Stefan (MS) equations.

In the molar reference frame, M, generalized Fick's law employs the mole fraction gradient $\boldsymbol{\nabla} x_k$ as driving force and relates it to the diffusive flux $\boldsymbol{J}_i^{\mathrm{M}}$ as follows:[1]

$$\boldsymbol{J}_i^{\mathrm{M}} = -c_{\mathrm{t}} \sum_{k=1}^{n-1} D_{ik}^{\mathrm{M}} \boldsymbol{\nabla} x_k, \quad i = 1, \ldots, n-1. \tag{5.1}$$

Here, c_{t} is the total molar concentration, n is the number of species in the system, and D_{ik}^{M} are the Fick diffusion coefficients. The diagonal elements D_{ii}^{M}

are the main diffusion coefficients describing the influence of the mole-fraction gradient ∇x_i of component i on its own diffusive flux. The off-diagonal elements D_{ij}^{M} are the cross diffusion coefficients describing the influence of the mole fraction gradient ∇x_j of component j on the diffusive flux of component i. The off-diagonal elements D_{ij}^{M} can also be negative. Usually, the main contribution to the diffusive flux is provided by the diagonal elements, since $D_{ii}^{\mathrm{M}} > D_{ij}^{\mathrm{M}}$. The entire, usually non-symmetric, matrix $\boldsymbol{D}^{\mathrm{M}}$ is always positive definite.

The MS equations[1,3] are physically motivated and can be derived from the thermodynamics of irreversible processes.[4] They are an example of an Onsager reciprocal relation. In the MS equations, the chemical-potential gradient as driving force is balanced by the friction forces between the moving molecules:

$$-\frac{1}{RT}\nabla\mu_i = \sum_{j=1, j\neq i}^{n} \frac{x_j(\boldsymbol{v}_i - \boldsymbol{v}_j)}{Đ_{ij}}. \tag{5.2}$$

In eqn (5.2), R is the gas constant, T the absolute temperature and μ_i the chemical potential of component i at pressure p and temperature T. $(\boldsymbol{v}_i - \boldsymbol{v}_j)$ is the difference between the average velocities of components i and j, leading to friction forces. In eqn (5.2), $Đ_{ij}$ are the MS diffusion coefficients, representing inverse friction coefficients. The MS diffusion coefficients are always symmetric: $Đ_{ij} = Đ_{ji}$.

The MS equations can be reformulated into a form analogous to generalized Fick's law in the molar reference frame:[5]

$$\boldsymbol{J}^{\mathrm{M}} = -c_{\mathrm{t}}\boldsymbol{B}^{-1}\boldsymbol{\Gamma}\nabla\boldsymbol{x}. \tag{5.3}$$

Here, $\boldsymbol{J}^{\mathrm{M}}$ and $\boldsymbol{x}$ are the conjunctions of the diffusive fluxes $\boldsymbol{J}_i^{\mathrm{M}}$ and the mole fractions x_i, respectively. The elements of the matrix $\boldsymbol{B}$ are given by

$$B_{ii} = \frac{x_i}{Đ_{in}} + \sum_{k=1, i\neq k}^{n} \frac{x_k}{Đ_{ik}} \quad \text{and} \quad B_{ij} = -x_i\left(\frac{1}{Đ_{ij}} - \frac{1}{Đ_{in}}\right). \tag{5.4}$$

The matrix $\boldsymbol{\Gamma}$ is the so-called matrix of *thermodynamic factors*. It contains thermodynamic information of the system and its elements are defined by

$$\Gamma_{ij} = \delta_{ij} + x_i\left(\frac{\partial \ln \gamma_i}{\partial x_j}\right)_{p,T,\Sigma}. \tag{5.5}$$

Here, δ_{ij} denotes the Kronecker delta and γ_i denotes the activity coefficient of component i. The differentiation of the activity coefficient is carried out at constant temperature and pressure, and at constant mole fraction of all other components except the n-th one, so that $\sum_{i=1}^{n} x_i = 1$ during the differentiation (indicated by the symbol Σ).[6,7] Usually, the thermodynamic factor $\boldsymbol{\Gamma}$ is not symmetric.

By comparing eqn (5.1) and (5.3), an expression for the relation between MS and Fick diffusion coefficients can be derived:[1]

$$\boldsymbol{D}^{\mathrm{M}} = \boldsymbol{B}^{-1}\boldsymbol{\Gamma}. \tag{5.6}$$

Although the MS equations provide a thermodynamically sound basis for the calculation of diffusion, their application is often limited in practice: on the one hand, the direct measurement of the chemical-potential gradient $\nabla\mu_i$ is practically impossible; on the other hand, the calculation of the thermodynamic factor with today's excess Gibbs energy (G^{E})-models and equations of state still introduces large uncertainties.[1] This dilemma has existed since the derivation of the MS equations 150 years ago. Recent approaches[8] to calculate the thermodynamic factor from Molecular Dynamics (MD) seem to be promising and are discussed in Section 5.3.

A comparison between the approach of Fick and MS shows the following important characteristics:

- While Fick diffusion coefficients depend on the velocity reference frame, MS diffusivities do not.
- The MS approach separates thermodynamic effects from binary molecular interactions, whereas Fick's approach lumps both effects into the diffusion coefficient (see eqn (5.6)).
- While the MS diffusivity matrix is symmetric, the Fick diffusivity matrix is not. Hence, $n(n-1)/2$ MS diffusivities are sufficient whereas $(n-1)^2$ Fick diffusivities are needed to describe an n-component mixture. Therefore, for $n > 2$, the Fick diffusivities are not independent.
- MS diffusivities are typically found to be less concentration-dependent than Fick diffusivities.[1] The increased concentration dependence of the Fick diffusivities results from the concentration dependence of the thermodynamic factor (see eqn (5.6)).
- Although the MS approach is physically based, the phenomenological approach of Fick is better suited to practical applications since only concentrations and not chemical potentials are directly measurable.

The reliable application of both the MS and the Fick approach depends on the knowledge of accurate values of their diffusion coefficients. In principle, three methods are available to obtain diffusion coefficients, which are discussed in the following sections of this chapter:

- Experimental investigations are the most significant source for diffusion coefficients. However, experiments are time-consuming and expensive. Several experiments are already required for a binary mixture due to the concentration (and temperature) dependence of the diffusivities. For multicomponent systems, the measurement effort scales with $(n-1)^2$. Therefore, only a limited number of multicomponent systems have been investigated experimentally. Selected experimental methods are discussed in Section 5.2.

- Molecular Dynamics (MD) simulations have been successfully applied to compute accurate diffusion coefficients. Still, MD simulations are time-consuming and require reliable force fields. An overview on MD simulations is given in Section 5.3.
- Engineering models allow the semi-empirical prediction of diffusion coefficients from other physical properties. Their aim is to reduce the required data to a minimal amount, *e.g.*, to the diffusion coefficients of the binary subsystems for a prediction of multicomponent diffusion coefficients. However, these models still contain large uncertainties. Nonetheless, recent research has led to considerable improvements in prediction accuracy. An overview of engineering models is given in Section 5.4.

5.2 Experimental Methods

Several methods exist for the measurement of mutual and self-diffusion coefficients. Table 5.1 provides an overview of current experimental methods. It is based on the compilation of Woolf *et al.*,[9] complemented by recent methods.

For each measurement method, Table 5.1 lists typical measurement temperatures and pressure ranges, as well as measurement durations. Additionally, the minimum number of independent measurements is provided which is required to determine the full ternary diffusion coefficient matrix in a ternary mixture. Since diffusion coefficients depend strongly on temperature, all techniques require careful temperature control. As is common for liquids, pressure plays a minor role in comparison to temperature.[10]

In the following, we present a brief introduction to the established measurement techniques. An excellent longer introduction has already been presented by Woolf *et al.*[9]

In the *diaphragm measurement technique*, two mixed phases of equal volume but different concentrations are separated by a porous membrane.[11] The concentration difference must fulfil two criteria: on the one hand, it should be small, such that the diffusion coefficient can be assumed to be constant. On the other hand, it has to be high enough to allow for a sufficient accuracy of the measurement of the concentration difference.[9] The diffusion coefficient is determined from the change of concentration in the two phases over time. The major disadvantage of this technique is the required calibration with a similar system of known diffusion coefficient due to the unknown system-dependent transport properties of the membrane. An advantage of this technique is the possibility to measure self- as well as mutual diffusion coefficients. This technique has already been shown to be even feasible for systems with more than three components.[12]

The *conductance method* is used to measure binary mutual diffusion coefficients, but it is restricted to electrolyte solutions.[9] Two methods exist for generating a concentration gradient between two solutions: either the solution of lower concentration is layered above the higher concentrated

Table 5.1 Measurement techniques for mutual and self-diffusion coefficients, expanded from Woolf *et al.*[9]

	Diffusion type						
	Self	Mutual	Accuracy	Temperature range	Pressure range	Experiment duration	Min. no. of experiments for ternary systems
Diaphragm cell	✓	✓	0.5 to 1 %	$<$400 K	$\leq$400 MPa	$>$2 days	2
Conductance		✓	0.2 %	$\approx$298 K	$\approx$0.1 MPa	$>$2 days	2
Taylor dispersion		✓	$\approx$1 %	$\leq$600 K	$\leq$100 MPa	$<$1 h	2
NMR	✓		$\approx$1 to 2 %	$\leq$700 K	$\leq$100 MPa	$<$1 h	—
Gouy		✓	$<$0.1 %	$\approx$298 K	$\approx$0.1 MPa	1 to 2 days	2
Rayleigh		✓	$<$0.1 %	$\approx$298 K	$\approx$0.1 MPa	1 to 2 days	2
DLS		✓	$\approx$0.8 to 10 %[22]	283 to 323 K[22]	$\approx$0.1 MPa	$\approx$1 h[83]	[a]
Raman		✓	$\approx$0.2 to 5 %[25,27]	$\leq$350 K[27]	$\leq$40 MPa[84]	$<$1 h	1
Microfluidic		✓	$\approx$ 1 to 2 %[31]	$\leq$400 K[34]	$\approx$ 0.1 MPa	$<$1 h	1 *in situ* 2 in-/offline

[a]Only eigenvalues of Fick diffusion matrix.[24]

solution or the concentration of the lower solution is increased by injecting additional electrolyte. Electrodes near the bottom and the top of the diffusion cell measure conductivities, which are first converted to concentrations and then to the diffusion coefficient. Though the measurement of ternary diffusion coefficients is possible, multiple measurements are required.[13]

The *Taylor dispersion* technique features a simple experimental setup, short measurement times, and comparatively low sample consumption. The diffusion takes place in a capillary, which ensures laminar flow. A sample is injected into a solution of different concentration flowing through the capillary. The interaction of axial convection and radial diffusion transforms the injected step profile into a Gaussian profile. At the outlet of the capillary, a detector measures the time-dependent concentration profile. From the variance of the Gaussian profile, the mutual diffusion coefficient is calculated. Detailed guidelines for the measurement of ternary diffusion coefficients have been presented in ref. 14–16.

NMR spectroscopy is primarily used for the measurement of self-diffusion coefficients. A small sample is exposed to a magnetic-field, eventually with a gradient. In this setup, a radio frequency pulse or a series of pulses is directed on the sample and the echo is analysed. Due to transverse relaxation and absorption by responding diffusing nuclei, the amplitude of the spin-echo declines. The diffusion coefficient can be determined from the amplitude of the echo.[9] While the accuracy of conventional NMR measurements is moderate, it was recently further improved.[17] NMR measurements are feasible for multicomponent mixtures, as long as the frequencies of different species are distinguishable.[18]

In the setup of the *Gouy interferometric method*, a vertical diffusion cell, layering two liquids above each other, is used. To observe concentration changes, monochromatic light is directed to the diffusion zone at different positions in axial direction.[9,19] The developing fringe patterns behind the diffusion cell are recorded and transferred to spatially resolved refractive indices. In binary mixtures, the spatially resolved refractive indices are converted to concentration profiles from which the mutual diffusion coefficient is determined. For the measurement of ternary diffusion coefficients, at least two independent measurements are required.[20] The *Rayleigh interferometric method* is similar to the Gouy interferometric method. However, the monochromatic light is directed to the measurement cell as well as to a reference cell.[9]

In *dynamic light scattering* (*DLS*) measurements of mutual diffusion coefficients, a laser is directed to the diffusion cell and the scattered light is detected, usually under a 90° angle.[21,22] DLS analyzes microscopic fluctuations and can thus be conducted in homogeneous samples, *i.e.*, without a macroscopic concentration gradient. DLS is especially well suited for macromolecules due to good scattering properties.[23] With regard to the data reduction, an autocorrelation function is fitted to the time-dependent intensity of the scattered light. For ternary systems, only the eigenvalues of the diffusion coefficient matrix are determined.[24]

5.2.1 1D Raman Spectroscopy Diffusion Measurement

In 1D Raman spectroscopy diffusion measurements, two solutions of different concentrations containing all components are layered on top of each other.[25,26] Even though the data analysis does not require a sharp initial interface, the resulting diffusion coefficient improves, if the concentration gradient at the interphase is sharpened.[27] The diffusion process is observed *in situ* by spatially resolved Raman spectroscopy: a laser is used as a monochromatic light source and traverses the diffusion cell in the direction of diffusion, while the Raman scattering is observed perpendicular to the direction of diffusion with a spectrometer and a CCD-chip (see Figure 5.1

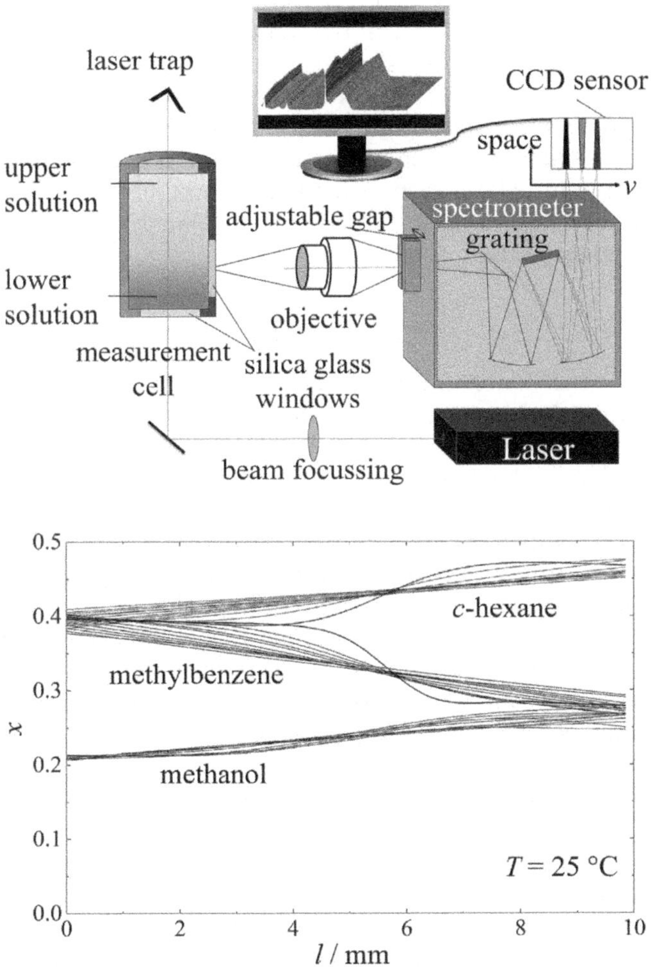

Figure 5.1 (Top) Schematic set-up of 1D Raman spectroscopy diffusion measurement. (Bottom) Typical profiles of the mole fractions x along the cell height l over time.

(top)).[25] This technique is suitable for organic as well as aqueous solutions with Raman active species.

The CCD-chip collects spatially resolved Raman spectra over time. From this information, spatially and temporally resolved concentration profiles (see Figure 5.1 (bottom)) can be obtained *via* the Beer–Lambert law.[28] This results in a high information content per experiment; multi-component diffusion coefficients can be obtained from only one experiment, since spectroscopy resolves the individual concentration profile of each species.[29]

5.2.2 Microfluidic Diffusion Measurement

Advantages of microfluidics are, in general, small dimensions accompanied by small volumes and operation in the laminar regime.[30] Shorter diffusion lengths offer the advantage of shorter diffusion times.[31] Kamholz *et al.*[32] presented fundamental work on diffusion in microfluidics. This measurement technique is often based on two liquid phases co-flowing in a microfluidic device: see Figure 5.2 (top). Diffusion occurs perpendicular to the flow and is negligible in axial direction for Peclet numbers $\mathrm{Pe} > 100$. The mass transfer has been described in more detail by Häusler *et al.*[31] and can be modelled for binary mixtures as

$$\frac{\partial \bar{c}}{\partial \mathrm{Fo}} = p(\bar{z}) \frac{\partial^2 \bar{c}}{\partial \bar{z}^2}. \tag{5.7}$$

with dimensionless concentration $\bar{c}$, Fourier number $\mathrm{Fo} = Dt \cdot L^{-2}$ with D as Fick diffusion coefficient, t as mean retention time at each axial position, and L as channel half width. $\bar{z}$ is the distance in the direction of diffusion, scaled by the channel half width, $p(\bar{z})$ is the axial velocity distribution, scaled by its mean velocity.

The diffusion coefficients are determined from the concentration profiles. Each concentration profile depends on its corresponding contact time (see Figure 5.2 (bottom)). At zero contact time, no change of concentration occurs, whereas after a very large time, concentrations are almost homogenous. Hence, both cases provide no information on the diffusion coefficient. Therefore, an optimal contact time exists. Häusler *et al.*[31] showed that the optimal contact time, represented by the Fourier number, is in the range $\mathrm{Fo} = 0.3$ to $\mathrm{Fo} = 0.4$.

Three basic concepts exist for the analysis of these concentrations profiles:

- Concentrations are measured *in situ* with spectroscopic methods, *e.g.*, fluorescence[33] or Raman spectroscopy.[34] This offers the potential for the evaluation of concentration-dependent diffusion coefficients.
- A detection unit measures the constant concentrations inline at the outlet, *e.g.*, a conductivity meter or an UV-VIS spectrometer.[31] This set-up produces less information than spatially resolved measurements due to the spatially integrated concentration measurements.

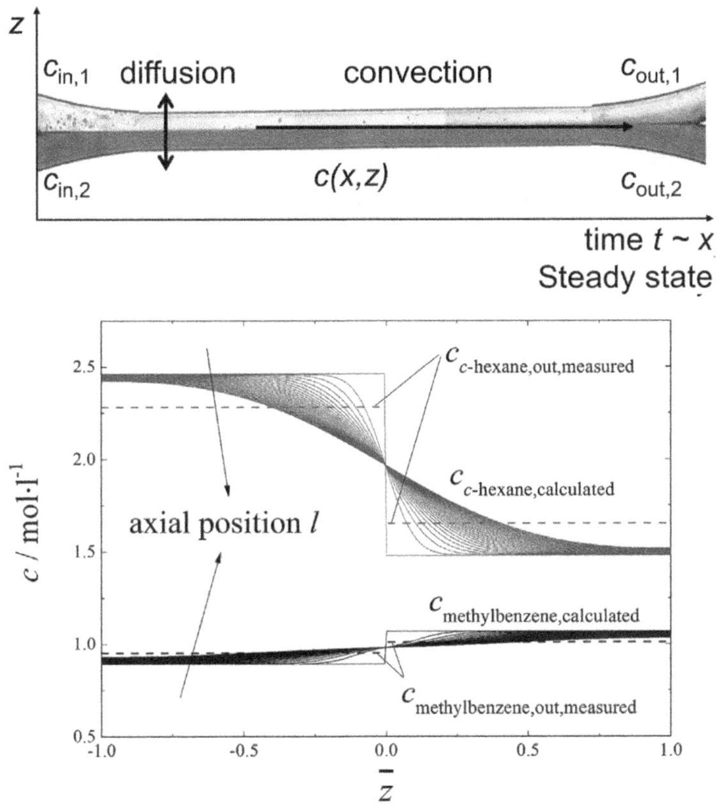

Figure 5.2 (Top) Directions of mass transport in a microfluidic diffusion experiment. Change of concentration is indicated by fading gray colour. (Bottom) Change of concentration c in the microfluidic diffusion experiment with axial position l for the system c-hexane + methylbenzene + methanol at 25 °C.

- The two outlets are sampled and offline analysis allows for the concentration determination with any analytical system, *e.g.*, UV/VIS.[35] This concept is feasible for all types of concentrations and components due to the unlimited choice of analytical systems.

The feasibility of multicomponent-diffusion measurements has been shown.[35] While offline analyses require multiple independent measurements to obtain multicomponent or concentration-dependent diffusion coefficients, only one *in situ* Raman spectroscopy experiment would be sufficient.

5.3 The Use of Molecular Dynamics to Compute Diffusivities

In this section, we provide an overview of the use of Molecular Dynamics (MD) to compute self- and mutual diffusion coefficients in liquid mixtures.

We focus on equilibrium MD and briefly discuss the use of non-equilibrium MD. This section concludes with a discussion on how to compute the matrix of thermodynamic factors from MD simulations.

5.3.1 Molecular Dynamics

Molecular Dynamics (MD) is a computational technique to study properties of a collection (ensemble) of many particles (atoms, molecules). The central idea is that atoms/molecules interact with each other; these interactions either follow from quantum mechanics or a parameterized functional form (the so-called force field). The forces on atoms/molecules that follow from these interactions are used to integrate Newton's second law (force equals mass times acceleration) numerically. In this way, a trajectory of all the atoms and molecules in the system is obtained. Essentially, the result of an MD simulation is a "movie" that shows how all atoms and molecules move around as a function of time. From such a "movie", average properties of the system can be computed: (1) static (thermodynamic) properties, *e.g.*, the average total energy, average total pressure of the system; and (2) dynamic properties (properties related to the time evolution of the system), *e.g.*, diffusivity, heat conductivity, and viscosity. MD simulations thus produce time averages, in sharp contrast to Monte Carlo simulations that provide configurational averages for a system of interacting particles.[36]

There are many excellent textbooks on the MD technique,[36–39] and several open-source software packages for MD are available online.[40–43] Instead of providing a detailed description of the MD technique, below we briefly outline some of the characteristics of MD that are needed in order to understand how MD can be used to compute diffusivities of typical liquid systems:

- The use of a parameterized functional form for interactions between atoms and molecules (force field) is necessary to access the time-scale required to study diffusion in liquids. Typical force fields consist of both intramolecular interactions (interactions inside a molecule such as bond-stretching and bond-bending) and intermolecular interactions (interactions between molecules, such as Lennard-Jones and electrostatic interactions). In many cases, the intermolecular interactions can be described by pair potentials (so that the total energy can be expressed as a summation over all particle pairs).
- The system of interacting particles is contained inside the so-called simulation box which has a certain volume. To minimize the effect of the boundaries of the system, usually periodic boundary conditions are applied in combination with the nearest image convention. This is illustrated in Figure 5.3. Making the system larger reduces finite-size effects, but increases the required CPU time (the computational cost usually scales linearly with the total number of particles in the system).

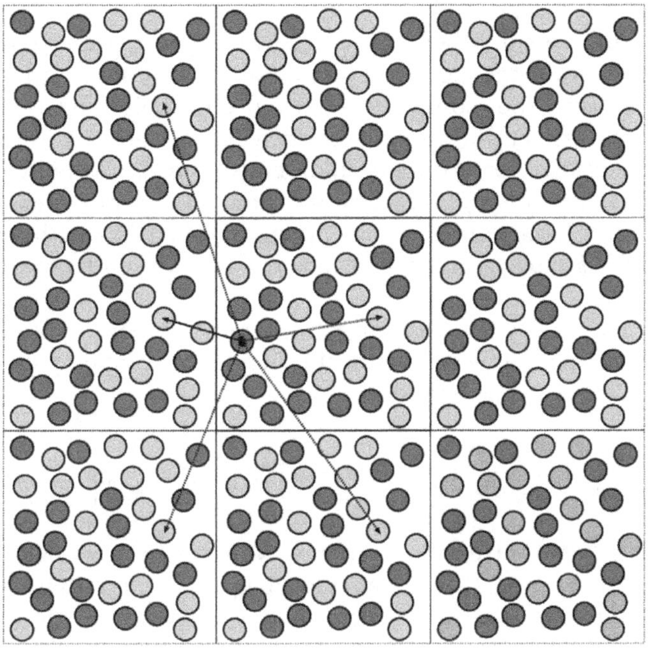

Figure 5.3 Schematic representation of periodic boundary conditions. A system is surrounded by copies of itself so there are no artificial walls in the system. If a particle leaves the central simulation box, it enters on the other side. According to the nearest image convention (assuming pairwise interactions), only the nearest images of two particles interact (as an example: the solid line shows the nearest neighbours of a particle pair, while the dotted lines show the same particle pair in several other periodic images, not being nearest neighbours.

Typically, in MD simulations the number of particles varies between 10^2 and 10^7, depending on the complexity of the used force field and the time scale that needs to be accessed.

- The time step Δt to integrate the equations of motion for the system numerically is set by the steepness of the interaction potential. Typically, $\Delta t = 10^{-15}$ s $= 10^{-3}$ ps. Due to this small time step, timescales of typically 1 μs are accessible with MD. This is possible by using modern computers, but the macroscopic timescale (seconds) is clearly beyond reach of today's technology.
- If no thermostat is used, MD conserves the total energy E which is the sum of kinetic and potential (interaction) energy. The temperature of the system is related to the kinetic energy of the system *via* the equipartition theorem: the kinetic energy equals $k_BT/2$ per degree of freedom in the system (T is the absolute temperature and k_B is the Boltzmann constant). As the volume of the simulation box and the number of particles are also constant, computed averages correspond

to the microcanonical ensemble. It is also possible to perform MD simulations in other ensembles by using a thermostat or barostat, *e.g.*, the canonical (constant temperature, volume, and number of particles) and the isobaric–isothermal ensemble (constant temperature, pressure, and number of particles). The latter is especially convenient, as in experiments temperature and pressure are often constant too. The so-called grand-canonical ensemble (constant temperature, volume, and chemical potential, but a fluctuating number of particles) is much harder to realize in MD. We return to this issue in Section 5.3.4.

- Differences in Gibbs or Helmholtz energies can be computed from MD trajectories as well. The chemical potential of component i follows *via* Widom's test particle method

$$\mu_i = \left(\frac{\partial F}{\partial N_i}\right)_{T,V,N_{j\neq i}} = -k_B T \ln\left\langle \exp\left[-\frac{\Delta U_i}{k_B T}\right]_{N,V,T}\right\rangle \tag{5.8}$$

in which F is the Helmholtz energy, N_i is the number of molecules of component i, and ΔU_i is the energy change if an additional molecule of species i would be inserted at a random position in the system. The brackets $\langle\cdots\rangle_{N,V,T}$ denote an average in the canonical ensemble and an average over all random positions of the test particle. For dense liquids, this method often fails as ΔU_i is often very large due to overlaps with other molecules in the systems. Improvements of Widom's test particle method use a more gradual insertion of the test particle in the system thereby avoiding overlaps. Examples of this are methods where an additional coupling parameter switches on/off interactions between the test particle and the surrounding molecules.[44]

5.3.2 Computing Diffusivities from Equilibrium Molecular Dynamics

At equilibrium, the self-diffusivity of component i in a three-dimensional isotropic system follows from the mean square displacement of particles of this component[36]

$$D_{i,\text{self}} = \frac{1}{6N_i}\lim_{m\to\infty}\frac{1}{m\cdot\Delta t}\left\langle\sum_{l=1}^{N_i}\left|\boldsymbol{r}_{l,i}(t+m\cdot\Delta t)-\boldsymbol{r}_{l,i}(t)\right|^2\right\rangle \tag{5.9}$$

in which N_i is the number of molecules of component i, Δt is the time-step used in the MD simulation to integrate the equations of motion, and $\boldsymbol{r}_{l,i}(t)$ is the position of the l-th molecule of component i in the system at time t. The brackets denote an average over t. It is important to note that eqn (5.9) considers an average of displacements of individual molecules, *i.e.*, correlation between the motion of different molecules are irrelevant when calculating the self-diffusivity of a component. One takes advantage of the fact that on average, the mean square displacements of all molecules of the same species are identical. Alternatively, one can integrate the velocity autocorrelation function to obtain $D_{i,\text{self}}$:

$$D_{i,\text{self}} = \frac{1}{3N_i}\int_0^\infty \mathrm{d}t' \left\langle \sum_{l=1}^{N_i} (\boldsymbol{v}_{l,i}(t) \cdot \boldsymbol{v}_{l,i}(t+t')) \right\rangle \tag{5.10}$$

in which $\boldsymbol{v}_{l,i}(t)$ is the velocity of the l-th molecule of component i in the system at time t. For large t', the velocity auto-correlation function usually decays to zero very slowly and therefore it is often difficult to set the integration limit in eqn (5.10). Therefore, we prefer the use of eqn (5.9). In practice, one plots the logarithm of the average mean square displacement $\left\langle \sum_{l=1}^{N_i} \left|\boldsymbol{r}_{l,i}(t+m\cdot\Delta t) - \boldsymbol{r}_{l,i}(t)\right|^2 \right\rangle \Big/ N_i$ *versus* $\ln(m)$ and identifies the linear regime, from which $D_{i,\text{self}}$ can be computed. It is also important that in this regime the mean square displacement is sufficiently large, *e.g.*, of the order of the size of the simulation box squared. The method in principle requires the storage of the full trajectory of all the molecules in the system. The use of coarse graining significantly reduces the computation time and the amount of data that needs to be stored, and therefore it is often used in MD simulations.[36,45]

Calculating mutual diffusion coefficients from equilibrium MD simulations is a bit more complicated. Expressions for the Maxwell–Stefan diffusivities can be obtained by applying the so-called linear response theory.[46] The central idea of this theory is that a disturbance created in a system by a weak external perturbation decays in the same way as a disturbance caused by a spontaneous fluctuation in the system caused by the thermal motion of the molecules. Therefore, one can compute transport coefficients from the decay of spontaneous fluctuations of observable microscopic properties of the system, while the system is at equilibrium (*i.e.*, no gradients in temperature, concentration *etc.* in the system). The resulting transport equations are called Green–Kubo relations and they can be derived for many transport properties, *e.g.*, mutual diffusion coefficients, electric and thermal conductivities, shear viscosity, and bulk viscosity. For a detailed derivation of the transport equations for fluids, we refer the reader to the excellent textbook of Morriss and Evans.[46] For deriving the equations for mutual diffusion, we would also refer the reader to ref. 47 and 48. As the thermodynamic driving force for isothermal mass transfer is a chemical-potential gradient, the mutual diffusion coefficients computed from equilibrium MD simulations are Maxwell–Stefan (MS) diffusivities. There are no expressions for Fick diffusivities from the molecular motion of the molecules without additional terms regarding information on how a concentration gradient can be converted to a gradient in chemical potential.

In practice, one first computes the so-called Onsager coefficients Λ_{ij} from the trajectory of the molecules,[49,50]

$$\Lambda_{ij} = \frac{1}{6N}\lim_{m\to\infty}\frac{1}{m\cdot\Delta t}\cdot\left\langle\left(\sum_{l=1}^{N_i}(\boldsymbol{r}_{l,i}(t+m\cdot\Delta t)-\boldsymbol{r}_{l,i}(t))\right)\cdot\left(\sum_{k=1}^{N_j}(\boldsymbol{r}_{k,j}(t+m\cdot\Delta t)-\boldsymbol{r}_{k,j}(t))\right)\right\rangle \tag{5.11}$$

in which i and j refer to the different species in the liquid mixture. Analogous to eqn (5.10), this equation can also be formulated in term of velocity correlation functions:

$$\Lambda_{ij} = \frac{1}{3N}\int_0^{\infty}\mathrm{d}t'\left\langle\sum_{l=1}^{N_i}\boldsymbol{v}_{l,i}(t)\cdot\sum_{k=1}^{N_j}\boldsymbol{v}_{k,j}(t+t')\right\rangle. \tag{5.12}$$

The key difference between eqn (5.11) and (5.9) is that eqn (5.11) considers the motion of the centre of mass of all molecules of a certain species, rather than the motion of individual molecules. Therefore, to compute Onsager coefficients is much harder than to compute self-diffusivities, as one cannot average over the trajectories of individual molecules. Note that the integration of eqn (5.12) is difficult due to long-time tails of the correlation function. Therefore, it is preferred to use eqn (5.11) instead.[46,50] In practice, MD simulations of at least 100 ns are needed to compute Λ_{ij}, while usually a few nanoseconds is sufficient to compute self-diffusivities.[51–53] The matrix Λ is symmetric, *i.e.*, $\Lambda_{ij}=\Lambda_{ji}$ and the Onsager coefficients are constrained by $\sum_i M_i\Lambda_{ij}=0$, in which M_i is the molar mass of component i.[49] The MS diffusivities directly follow from the Onsager coefficients Λ_{ij} and Λ_{ii}. In binary systems, the MS diffusivity $Đ_{12}$ is related to the Onsager coefficients by[49]

$$Đ_{12} = \frac{x_2}{x_1}\Lambda_{11} + \frac{x_1}{x_2}\Lambda_{22} - 2\Lambda_{12}. \tag{5.13}$$

For ternary systems and systems with more than three components, the resulting expressions are more complex and we refer the reader to Krishna and van Baten, and Liu *et al.*[49,54]

5.3.3 Non-equilibrium Molecular Dynamics to Compute Mutual Diffusion Coefficients

Besides the Green–Kubo approach that requires equilibrium MD simulations, it is also possible to perform MD simulation out of equilibrium and compute transport coefficients from the fluxes and driving forces (that can both be measured in the MD simulation). These so-called non-equilibrium MD simulations (NEMD) are often very similar in spirit to actual physical experiments.[46,55–63] For computing transport diffusivities, there are roughly two types of approaches to compute transport diffusivities using NEMD:

- Boundary-driven NEMD. One can construct the MD simulation in such a way that there is a concentration gradient present in the system. One can either study how this concentration gradient vanishes as a function of time, or it can be maintained by exchanging molecules of different species between various locations in the simulation box, or by coupling of various locations of the system to grand-canonical reservoirs. In the latter methods, one can calculate the fluxes and concentration gradients, from which Fick and/or Maxwell–Stefan diffusivities follow.
- External-field-driven NEMD. One can artificially create a chemical-potential gradient by adding a force to molecules of a certain species. As a consequence, the average velocities of different species will differ. The Maxwell–Stefan diffusivities can be calculated from these velocity differences.[46]

For both types of approaches, there are many variations possible on the algorithm, and there are variants available for other transport properties like viscosity and heat conductivity. For a detailed overview, we refer the reader to ref. 37, 39, 46 and 50 and references to the original articles in these. A more detailed description of NEMD algorithms is also available in Chapter 6 of this book. When performing NEMD simulations, one should always take care of the following:

- As MD simulations deal with systems that are very small compared to macroscopic systems, gradients in NEMD are much larger than in experiments. Therefore, one should always extrapolate the computed transport coefficient to zero driving force.
- As in NEMD simulations the system is not in equilibrium, heat will be dissipated as a consequence. In order for the system to be in a steady state, this heat needs to be removed by a thermostat. It is important to verify that the thermostat does not cause any artifacts. An example of this is the formation of "traffic lanes" of different species in external field driven NEMD that can either be enhanced or suppressed by the choice of the thermostat.[46,60]

For various systems and transport properties it has been reported in the literature that equilibrium MD and NEMD yield very similar values of the transport coefficients. It is not clear *a priori* which approach is computationally more efficient.

5.3.4 Thermodynamic Factor from Simulations

The thermodynamic factor (eqn (5.5)) requires the knowledge of the activity coefficient γ_i, which is directly related to the chemical potential μ_i by

$$\mu_i = \left(\frac{\partial G}{\partial N_i}\right)_{T,p,N_{j\neq i}} = \mu_i^0(T,p) + RT\ln(x_i\gamma_i). \tag{5.14}$$

in which $\mu_i^0(T,p)$ is the chemical potential of the pure component i at T,p. Calculation of γ_i from μ_i requires $\mu_i^0(T,p)$, and both can be computed from Gibbs energy calculations, see for example the work by Sadowski and co-workers.[64] The chemical potential μ_i is exactly the same as the partial molar Gibbs energy g_i. Therefore, the elements of the matrix of thermodynamic factors Γ_{ij} correspond to a second derivative of the Gibbs energy G with respect to the composition of the system. In principle, one could use any approach to compute the Helmholtz or Gibbs energy (*e.g.*, Widom's test particle method) and take the derivative of μ_i with respect to composition to obtain Γ_{ij}. Such an indirect approach has been used by Vrabec and co-workers for systems of water and various alcohols.[65] One could avoid errors introduced by numerical differentiation of $\gamma_i(x_i)$ by using an activity coefficient model to fit $\gamma_i(x_i)$.[1,7]

It is also possible to use a more direct approach to obtain Γ_{ij}, without using a differentiation of numerically obtained results. As Γ_{ij} involves the second derivative of the Gibbs energy with respect to the number of particles, in terms of Widom's test particle method this would correspond to the simultaneous insertion of two test particles. Balaji and co-workers have developed the so-called permuted Widom test particle method that uses combinatorics to compute these second derivatives of G directly from a single simulation.[66,67]

Alternatively, the elements of the matrix of thermodynamic factors can be computed from the Kirkwood–Buff (KB) approach.[48,52,53,68–70] This approach relates thermodynamic properties to fluctuations in the number of particles in the grand-canonical ensemble, *i.e.*, an ensemble where the temperature, volume, and chemical potential of each species is constant and the number of particles of each component fluctuates. For a system of volume V, in this ensemble, the so-called KB coefficients G_{ij}^V are defined as

$$G_{ij}^V = V\frac{\langle N_iN_j\rangle - \langle N_i\rangle\langle N_j\rangle}{\langle N_i\rangle\langle N_j\rangle} - \frac{V\delta_{ij}}{\langle N_i\rangle} \tag{5.15}$$

in which N_i is the number of molecules of species i inside V, the brackets $\langle\ldots\rangle$ denote ensemble averages in the grand-canonical ensemble, and δ_{ij} is the Kronecker delta. The thermodynamic factors directly follow from the KB coefficients G_{ij}, see ref. 52, 53, 69 and 70 for the resulting expressions. The grand-canonical ensemble can be realized in Monte Carlo simulations,[36] but its efficiency critically relies on the insertion and removal of molecules to and from the system. For large molecules and dense liquid systems, these insertions and removals are often not very efficient as nearly always there is an overlap between an inserted molecule and molecules already present in the system, resulting in poor statistics. Specialized algorithms are in principle available to facilitate molecule transfers[36,71–73] but still these approaches are computationally expensive.

An ensemble in which T and p are constant and particle numbers fluctuate can also be realized in MD simulations by considering a small subsystem of volume V inside a large simulation box (which contains a constant number of particles), see Figure 5.4. The fluctuations in the number of particles inside the small subsystem result from the thermal motion of the molecules inside the system,[8] instead of changing the total number of molecules inside the simulation box (as is done in Monte Carlo simulations in the grand-canonical ensemble). It is crucial however to take into account the smallness of the subsystem: for a subsystem of volume V, one can show that the difference between G_{ij}^{V} and its value in the thermodynamic limit (denoted by G_{ij}^{∞}) scales as $1/V^{(1/d)}$ in which d is the dimensionality of the system.[8]

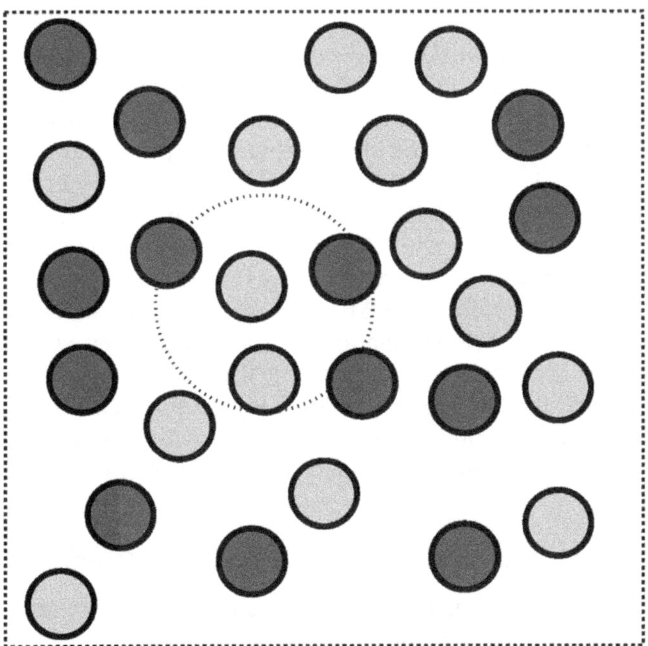

Figure 5.4 Schematic illustration of the small subsystem approach. A small subsystem of the volume (circle) is embedded in the simulation box (square). As the small subsystem can exchange particles and energy with the surroundings, it can be considered as a grand-canonical reservoir. KB coefficients can be computed from particle fluctuations inside according to eqn (5.15). An identical approach is to integrate the radial distribution function (eqn (5.17)). As explained in the main text, macroscopic properties can be obtained by extrapolating to the thermodynamic limit.

Alternatively, one can compute G_{ij}^V of a finite system by integration of the radial distribution function $g_{ij}(r)$ over volume. In the 1950s, Kirkwood and Buff derived the resulting expression for an infinitely large system[68]

$$G_{ij}^{\infty} = \lim_{V\to\infty} G_{ij}^V = 4\pi \int_0^{\infty} [g_{ij}(r') - 1] r'^2 \, \mathrm{d}r'. \tag{5.16}$$

This expression is only valid for infinite systems.[74] This aspect has been overlooked in many MD studies, even though by definition MD deals with systems that are not infinitely large. Simon, Krüger, and co-workers have shown that for $d=3$ the correct KB integral for a finite system equals

$$G_{ij}^V = 4\pi \int_0^{2r} [g_{ij}(r') - 1] \left(1 - \frac{3r'}{4r} + \frac{r'^3}{16r^3}\right) r'^2 \, \mathrm{d}r' \tag{5.17}$$

in which r is the radius of a sphere of volume V. It is important to note that this expression is identical to eqn (5.15). In the limit of $V\to\infty$, eqn (5.17) reduces to eqn (5.16). A simple truncation of the integration of eqn (5.16) does not result in a well-defined physical property. Extrapolation of G_{ij}^V to $1/r\to 0$ results in G_{ij} in the thermodynamic limit.

Vlugt, Bardow, and co-workers have used the small-subsystem approach to compute thermodynamic factors of binary and ternary liquid systems of acetone, methanol, tetrachloromethane, and chloroform at room temperature and pressure.[50,52–54] In addition, the thermodynamic factors were used to compute Fick diffusivities from MS diffusivities obtained by equilibrium MD simulations. In the vast majority of cases, results for Γ_{ij} and Fick diffusivities were in excellent agreement with experimental data and the small-subsystem approach turned out to be a convenient way for determining thermodynamic factors directly from equilibrium MD. The thermodynamic factor for these systems obtained from MD simulations is in better agreement with the one derived from experimental VLE data than the thermodynamic factor computed by using the COSMO-SAC method.[70] This is not surprising, as COSMO approaches do not take the precise details of the long-range liquid structure into account explicitly, while this is included (by definition) in MD simulations.

5.4 Engineering Models for Predicting Diffusivities

To save time and costs, it is desirable to predict diffusion coefficients from simple molecular information or physical properties. However, several challenges have to be addressed:

- The prediction of Fick diffusion coefficients is difficult, due to their empirical character. Therefore, only few approaches exist.[2,75]
- The prediction of MS diffusion coefficients is more common, since the MS approach separates thermodynamic effects from binary molecular interactions.[7] However, additional, often significant, uncertainties are introduced in the calculation of the thermodynamic factor (see eqn (5.6)).

A common approach to predict diffusion coefficients in concentrated solutions is to apply mixing rules: typically, these mixing rules combine diffusion coefficients at infinite dilution which are much easier to predict.[7,76,77] Thus, only these diffusion coefficients at infinite dilution have to be determined experimentally, by MD simulations, or again by prediction.

In this section, first, predictive models for diffusion coefficients at infinite dilution are presented, before predictive models for diffusion coefficients in concentrated solutions are introduced.

5.4.1 Prediction of Binary Diffusion Coefficients at Infinite Dilution

One of the most fundamental predictive equations for diffusion coefficients at infinite dilution is the *Stokes–Einstein equation*:

$$D_{AB}^{x_A \to 0} = \frac{k_B T}{6\pi \eta_B r_A}. \tag{5.18}$$

Here, the binary diffusion coefficient $D_{AB}^{x_A \to 0}$ of component A infinitely diluted in the solvent B is computed from the viscosity of the solvent η_B and the radius of the solute molecules r_A. The Stokes–Einstein equation can be derived from hydrodynamic theory and is therefore physically based.[10] It is strictly applicable only for macroscopic systems in which diluted, spherical macromolecules are solved in smaller solvent molecules.[2]

Nevertheless, the Stokes–Einstein equation serves as a framework for several models with the same structural concept: $D \propto T/\eta$. The most prominent example for such a model is the *Wilke–Chang equation*:[2]

$$D_{AB}^{x_A \to 0} = \frac{7.4 \cdot 10^{-8} \left[\Phi (M_B/\mathrm{g \cdot mol^{-1}})\right]^{0.5} (T/K)}{(\eta_B/\mathrm{mPa \cdot s})(V_A/\mathrm{cm^3 \cdot mol^{-1}})^{0.6}}. \tag{5.19}$$

Here, Φ is an association factor of the solvent B (which equals 1 if the molecules are not associated), M_B is the molar mass of the solvent B, T is the temperature, η_B is the viscosity of the solvent B, and V_A is the molar volume of the solute A.

Further modifications of the Wilke–Chang equation and additional models for the prediction of diffusion coefficients at infinite dilution are reported by Poling and Prausnitz.[2]

5.4.2 Prediction of Concentration-dependent Binary Diffusion Coefficients

The diffusion coefficients at infinite dilution are usually interpolated to determine concentration-dependent binary diffusion coefficients. The most prominent examples for the prediction of MS diffusion coefficients of concentrated binary mixtures are the equations by Vignes[78] and Darken[79].

The *Vignes equation* interpolates the diffusion coefficients at infinite dilution:

$$Đ_{ij} = \left(Đ_{ij}^{x_i \to 0}\right)^{x_j} \left(Đ_{ji}^{x_j \to 0}\right)^{x_i}. \tag{5.20}$$

While the equation has been derived empirically, it applies very well for ideal or nearly ideal mixtures.[2]

The *Darken equation*

$$Đ_{ij} = x_i D_{j,\text{self}} + x_j D_{i,\text{self}} \tag{5.21}$$

combines the self-diffusion coefficients $D_{i,\text{self}}$ and $D_{j,\text{self}}$ in the concentrated solution for the estimation of the mutual MS diffusion coefficient. This equation can be derived by assuming that velocity correlations between different molecules are small.[80] Since self-diffusion coefficients in concentrated solutions are scarcely available, Krishna and van Baten[49] suggest a mixing rule using more common diffusion coefficients at infinite dilution instead:

$$D_{i,\text{self}} = \sum_{j=1}^{n} w_j D_{i,\text{self}}^{w_j \to 0}. \tag{5.22}$$

There are more models that can predict concentration-dependent diffusion coefficients using information on the viscosity of the mixture (see ref. 2). However, this requirement severely limits their applicability.

5.4.3 Multicomponent Extensions of the Darken and Vignes Equations

Wesselingh and Krishna[76] proposed a *generalized Vignes equation* for multicomponent mixtures:

$$Đ_{ij} = \left(Đ_{ij}^{x_i \to 0}\right)^{x_j} \left(Đ_{ji}^{x_j \to 0}\right)^{x_i} \prod_{k=1, k \neq i,j}^{n} \left(Đ_{ij}^{x_k \to 1}\right)^{x_k}. \tag{5.23}$$

Here, the diffusion coefficients $Đ_{ij}^{x_k \to 1}$ describe the friction between components i and j when both are diluted in component k. Since these diffusion coefficients are not available experimentally, they have to be estimated.[81] Recently, Liu, Bardow, and Vlugt[51] derived a physically motivated model for this diffusion coefficient:

$$Đ_{ij}^{x_k \to 1} = \frac{D_{i,\text{self}}^{x_k \to 1} D_{j,\text{self}}^{x_k \to 1}}{D_{k,\text{self}}^{x_k \to 1}}. \tag{5.24}$$

The derivation is based on the assumption that velocity correlations between different molecules are small. The required self-diffusion coefficient $Đ_{k,\text{self}}^{x_k \to 1}$

can be calculated by predictive models such as the Wilke–Chang equation.[49] Therefore, the prediction of $Ð_{ij}$ based on pure component and binary mixture data at infinite dilution is possible. The combination of eqn (5.23) and (5.24) is called the *Vignes–LBV equation*.[51]

Liu, Bardow, and Vlugt[51] derived a multicomponent Darken-like equation from statistical mechanics. Assuming again negligible velocity cross-correlations, the following relation results:

$$Ð_{ij} = \frac{D_{i,\text{self}} D_{j,\text{self}}}{D_{\text{mix}}} \tag{5.25}$$

$$\text{with } \frac{1}{D_{\text{mix}}} = \sum_{i=1}^{n} \frac{x_i}{D_{i,\text{self}}}. \tag{5.26}$$

The equation reduces to the binary Darken equation for two component mixtures and satisfies well-defined limits for multicomponent mixtures. Again, the MS self-diffusion coefficients in the concentrated solution $D_{i,\text{self}}$ have to be estimated. Motivated by eqn (5.26), Liu, Bardow, and Vlugt propose the following mixing rule to estimate the $D_{i,\text{self}}$ from the self-diffusion coefficients at infinite dilution $D_{i,\text{self}}^{x_j \to 1}$:

$$\frac{1}{D_{i,\text{self}}} = \sum_{j=1}^{n} \frac{x_j}{D_{i,\text{self}}^{x_j \to 1}}. \tag{5.27}$$

The combination of eqn (5.25) and (5.26) is called the *predictive Darken–LBV equation*. For more details about the derivation we refer the reader to ref. 54.

The equations have been shown to hold well for fluids described by a Weeks–Chandler–Anderson (WCA) potential.[82] For systems with strong interactions (with no WCA potential), additional cross-correlations have to be considered for the evaluation of the diffusion coefficients.

Figure 5.5 shows literature values as well as computational results of diffusion coefficients calculated with the Vignes- and the Darken–LBV equations for the binary system benzene + cyclohexane. Both calculations are based only on the self-diffusion coefficients at the boundaries of the system $x \to 0$ and $x \to 1$, respectively. Although the Vignes equation reproduces the correct curvature of the concentration dependence of the diffusion coefficients, it lacks accuracy. In contrast, the predictive Darken–LBV equation is able to reproduce the literature data quite well.

Table 5.2 shows mean square deviations of the diffusion coefficients between literature data and calculations with the Vignes–LBV and the predictive Darken–LBV equations, respectively, for the ternary system methylbenzene + chlorobenzene + bromobenzene. Again, the superior performance of the predictive Darken–LBV equation is apparent.

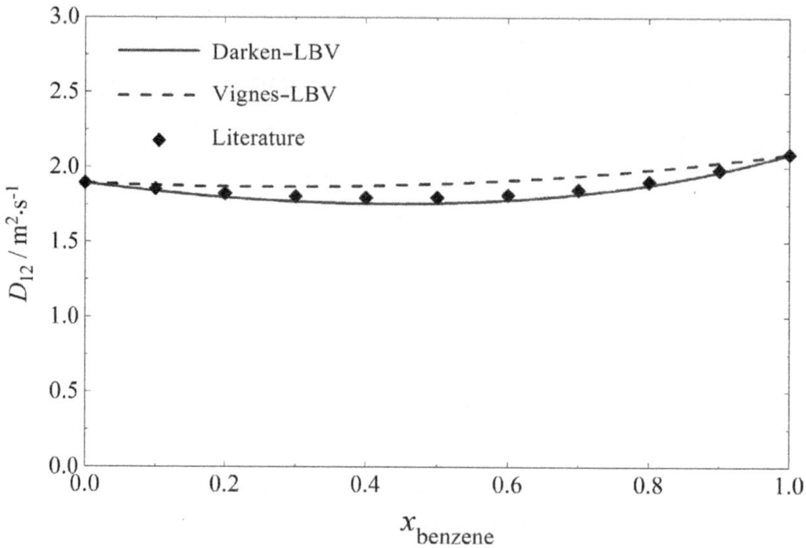

Figure 5.5 Comparison of literature values and predictions of the Vignes– and the predictive Darken–LBV equations. Absolute values of the diffusion coefficients for the binary system benzene + cyclohexane.

Table 5.2 Comparison of literature values and predictions of the Vignes– and the predictive Darken–LBV equations. Relative mean square deviations of the diffusion coefficients for the ternary system methylbenzene + chlorobenzene + bromobenzene.

Relative Mean Square Deviation	Vignes–LBV	Predictive Darken–LBV
D_{11}	0.0463	0.0110
D_{12}	0.2475	0.0729
D_{21}	0.2094	0.0580
D_{22}	0.0535	0.0029

5.5 Conclusions

Wherever mass transport occurs, diffusion is important. For the determination of diffusion coefficients, we highlighted three methods and their recent developments:

- Experimental methods provide accurate values for diffusion coefficients. While established methods require high experimental effort, recent improvements allow for reduced costs and time.
- Equilibrium Molecular Dynamics show good results for MS diffusion coefficients as well as for the thermodynamic factor. This offers the possibility for a fully predictive determination of Fick diffusion coefficients from force-field models.

- Engineering models are preferred for the rapid calculation of diffusion coefficients. The newly introduced models by Liu, Bardow and Vlugt have a physically sound basis and improve prediction accuracies of MS diffusion coefficients.

For all three methods, the full potential of these recent developments has not been fully explored and we expect further progress towards a quantitative understanding of diffusion in liquids.

Acknowledgements

The work of T.J.H. Vlugt was sponsored by the Stichting Nationale Computerfaciliteiten (National Computing Facilities Foundation, NCF) for the use of supercomputing facilities, with financial support from NWO-EW. The work of C. Peters, L. Wolff and A. Bardow was performed as part of the Cluster of Excellence "Tailor-Made Fuels from Biomass", which is funded by the Excellence Initiative by the German federal and state governments to promote science and research at German universities.

References

1. R. Taylor and R. Krishna, *Multicomponent Mass Transfer*, Wiley, New York, 1993.
2. B. E. Poling, J. M. Prausnitz and J. P. O'Connell, *The Properties of Gases and Liquids*, 5th edn, Digital Engineering Library @ McGraw-Hill, 2004.
3. R. Krishna and J. A. Wesselingh, *Chem. Eng. Sci.*, 1997, **52**, 861.
4. G. D. Kuiken, *Thermodynamics of Irreversible Processes*, Wiley, New York, 1994.
5. A. Bardow, Model-based Experimental Analysis of Multicomponent Diffusion, dissertation, RWTH Aachen University, 2004.
6. R. Taylor and H. A. Kooijman, *Chem. Eng. Commun.*, 1991, **102**, 87.
7. H. A. Kooijman and R. Taylor, *Ind. Eng. Chem. Res.*, 1991, **30**, 1217.
8. S. K. Schnell, T. J. H. Vlugt, J.-M. Simon, D. Bedeaux and S. Kjelstrup, *Chem. Phys. Lett.*, 2011, **504**, 199.
9. L. A. Woolf, R. Mills, D. Leaist, C. Erkey, A. Akgerman, A. J. Easteal, D. Miller, J. G. Albright, S. F. Y. Li and W. Wakeham, *Measurement of the Transport Properties of Fluids*, ed. W. A. Wakeham, A. Nagashima and J. V. Sengers, Blackwell Science Publications, Oxford, 1991, pp. 228–320.
10. R. B. Bird, W. E. Stewart and E. N. Lightfoot, *Transport Phenomena*, Wiley, New York, 2nd edn, 2002.
11. E. L. Cussler, *Diffusion – Mass Transfer in Fluid Systems*, Cambridge University Press, Cambridge, 3rd edn, 2007.
12. A. Sethy and H. T. Cullinan, *AIChE J.*, 1975, **21**, 571.
13. D. G. Leaist, *Can. J. Chem.*, 1985, **63**, 2933.
14. D. G. Leaist, *J. Phys. Chem.*, 1990, **94**, 5180.
15. W. E. Price, *J. Chem. Soc., Farad. Trans. 1*, 1988, **84**, 2431.

16. I. Van de Ven-Lucassen, M. Kemmere and P. Kerkhof, *J. Solution Chem.*, 1997, **26**, 1145.
17. C. D'Agostino, J. A. Stephens, J. D. Parkinson, M. D. Mantle, L. F. Gladden and G. D. Moggridge, *Chem. Eng. Sci.*, 2013, **95**, 43.
18. P. Stilbs, *Prog. Nucl. Magn. Reson. Spectrosc.*, 1987, **19**, 1.
19. D. G. Miller, R. Sartorio, L. Paduano, J. A. Rard and J. G. Albright, *J. Solution Chem.*, 1996, **25**, 1185.
20. D. G. Miller, *J. Phys. Chem.*, 1988, **92**, 4222.
21. D. A. Ivanov and J. Winkelmann, *Int. J. Thermophys.*, 2008, **29**, 1921.
22. M. H. Rausch, L. Hopf, A. Heller, A. Leipertz and A. P. Fröba, *J. Phys. Chem. B*, 2013, **117**, 2429.
23. O. Annunziata, D. Buzatu and J. G. Albright, *Langmuir*, 2005, **21**, 12085.
24. A. Bardow, *Fluid Phase Equilibr.*, 2007, **251**, 121.
25. A. Bardow, W. Marquardt, V. Göke, H.-J. Koß and K. Lucas, *AIChE J.*, 2003, **49**, 323.
26. R. W. Berg, S. Brunsgaard Hansen, A. A. Shapiro and E. H. Stenby, *Appl. Spectrosc.*, 2007, **61**, 367.
27. C. Blesinger, P. Beumers, F. Buttler, C. Pauls and A. Bardow, *J. Solution Chem.*, 2013, **43**, 144.
28. F. Alsmeyer, H.-J. Koß and W. Marquardt, *Appl. Spectrosc.*, 2004, **58**, 975.
29. A. Bardow, V. Göke, H.-J. Koß, K. Lucas and W. Marquardt, *Fluid Phase Equilibr.*, 2005, **228**, 357.
30. G. M. Whitesides, *Nature*, 2006, **442**, 368.
31. E. Häusler, P. Domagalski, M. Ottens and A. Bardow, *Chem. Eng. Sci.*, 2012, **72**, 45.
32. A. E. Kamholz, E. A. Schilling and P. Yager, *Biophys. J.*, 2001, **80**, 1967.
33. A. E. Kamholz and P. Yager, *Biophys. J.*, 2001, **80**, 155.
34. Y. Lin, X. Yu, Z. Wang, S.-T. Tu and Z. Wang, *Anal. Chim. Acta*, 2010, **667**, 103.
35. C. Blesinger, S. E. Yalcin, C. Pauls and A. Bardow, *μFlu-12 Proceedings*, 3rd European Conference on Microfluidics, Heidelberg, 2012.
36. D. Frenkel and B. Smit, *Understanding Molecular Simulation: from Algorithms to Applications*, Academic Press, San Diego, 2002.
37. M. P. Allen and D. J. Tildesley, *Computer Simulation of Liquids*, Clarendon Press, Oxford, 1987.
38. M. E. Tuckerman, *Statistical Mechanics: Theory and Molecular Simulation*, Oxford University Press, Oxford, 2010.
39. D. C. Rapaport, *The Art of Molecular Dynamics Simulation*, Cambridge University Press, Cambridge, 2004.
40. http://lammps.sandia.gov, 2014.
41. S. Plimpton, *J. Comput. Phys.*, 1995, **117**, 1.
42. http://www.gromacs.org, 2014.
43. B. Hess, C. Kutzner, D. van der Spoel and E. Lindahl, *J. Chem. Theory Comput.*, 2008, **4**, 435.
44. A. Torres-Knoop, S. P. Balaji, T. J. H. Vlugt and D. Dubbeldam, *J. Chem. Theory Comput.*, 2014, **10**, 942.

45. D. Dubbeldam, D. C. Ford, D. E. Ellis and R. Q. Snurr, *Mol. Simul.*, 2009, **35**, 1084.
46. D. J. Evans and G. P. Morriss, *Statistical Mechanics of Nonequilibrium Liquids*, Cambridge University Press, Cambridge, 2008.
47. D. R. Wheeler and J. Newman, *J. Phys. Chem. B*, 2004, **108**, 18353.
48. M. Schoen and C. Hoheisel, *Mol. Phys.*, 1984, **53**, 1367.
49. R. Krishna and J. M. van Baten, *Ind. Eng. Chem. Res.*, 2005, **44**, 6939.
50. X. Liu, S. K. Schnell, J.-M. Simon, P. Krüger, D. Bedeaux, S. Kjelstrup, A. Bardow and T. J. H. Vlugt, *Int. J. Thermophys.*, 2013, **34**, 1169.
51. X. Liu, T. J. H. Vlugt and A. Bardow, *Ind. Eng. Chem. Res.*, 2011, **50**, 10350.
52. X. Liu, S. K. Schnell, J.-M. Simon, D. Bedeaux, S. Kjelstrup, A. Bardow and T. J. H. Vlugt, *J. Phys. Chem. B*, 2011, **115**, 12921.
53. X. Liu, S. K. Schnell, J.-M. Simon, D. Bedeaux, S. Kjelstrup, A. Bardow and T. J. H. Vlugt, *J. Phys. Chem. B*, 2012, **116**, 6070.
54. X. Liu, A. Bardow and T. J. H. Vlugt, *Ind. Eng. Chem. Res.*, 2011, **50**, 4776.
55. A. P. Thompson, D. M. Ford and G. S. Heffelfinger, *J. Chem. Phys.*, 1998, **109**, 6406.
56. A. P. Thompson and G. S. Heffelfinger, *J. Chem. Phys.*, 1999, **110**, 10693.
57. E. J. Maginn, A. T. Bell and D. N. Theodorou, *J. Phys. Chem.*, 1993, **97**, 4173.
58. D. MacGowan and D. J. Evans, *Phys. Rev. A: At., Mol., Opt. Phys.*, 1984, **34**, 2133.
59. S. Sarman and D. J. Evans, *Phys. Rev. A: At., Mol., Opt. Phys.*, 1992, **45**, 2370.
60. D. R. Wheeler and J. Newman, *J. Phys. Chem. B*, 2004, **108**, 18362.
61. T. Ikeshoji and B. Hafskjold, *Mol. Phys.*, 1994, **81**, 251.
62. B. Hafskjold, T. Ikeshoji and S. K. Ratkje, *Mol. Phys.*, 1993, **80**, 1389.
63. B. Hafskjold and S. K. Ratkje, *J. Stat. Phys.*, 1995, **78**, 463.
64. S. Hempel, J. Fischer, D. Paschek and G. Sadowski, *Soft Mater.*, 2012, **10**, 26.
65. S. Parez, G. Guevara-Carrion, H. Hasse and J. Vrabec, *Phys. Chem. Chem. Phys.*, 2013, **15**, 3985.
66. S. P. Balaji, S. K. Schnell, E. S. McGarrity and T. J. H. Vlugt, *Mol. Phys.*, 2013, **111**, 285.
67. S. P. Balaji, S. K. Schnell and T. J. H. Vlugt, *Theor. Chem. Acc.*, 2013, **132**, 1333.
68. J. G. Kirkwood and F. P. Buff, *J. Chem. Phys.*, 1951, **19**, 774.
69. A. Ben-Naim, *Molecular Theory of Solutions*, Oxford University Press, Oxford, 2006.
70. X. Liu, A. Martín-Calvo, E. McGarrity, S. K. Schnell, S. Calero, J.-M. Simon, D. Bedeaux, S. Kjelstrup, A. Bardow and T. J. H. Vlugt, *Ind. Eng. Chem. Res.*, 2012, **51**, 10247.
71. W. Shi and E. J. Maginn, *J. Chem. Theory Comput.*, 2007, **3**, 1451.
72. W. Shi and E. J. Maginn, *J. Comput. Chem.*, 2008, **29**, 2520.
73. A. Torres-Knoop, S. P. Balaji, T. J. H. Vlugt and D. Dubbeldam, *J. Chem. Theory Comput.*, 2014, **10**, 942.

74. P. Krüger, S. K. Schnell, D. Bedeaux, S. Kjelstrup, T. J. H. Vlugt and J.-M. Simon, *J. Phys. Chem. Lett.*, 2013, **4**, 235.
75. Y.-D. Hsu, M. Tang and Y.-P. Chen, *Fluid Phase Equilibr.*, 2000, **173**, 1.
76. J. Wesselingh and R. Krishna, *Elements of Mass Transfer*, Faculty of Chemical Engeneering and Materials Science, TU Delft, 1989.
77. S. Rehfeldt and J. Stichlmair, *Fluid Phase Equilibr.*, 2007, **256**, 99.
78. A. Vignes, *Ind. Eng. Chem. Fundam.*, 1966, **5**, 189.
79. L. S. Darken, *Trans. AIME*, 1948, **175**, 184.
80. H. J. V. Tyrell and K. R. Harris, *Diffusion in Liquids, A Theoretical and Experimental Study*, Butterworth, London, 1984.
81. A. Bardow, E. Kriesten, M. A. Voda, F. Casanova, B. Blümich and W. Marquardt, *Fluid Phase Equilibr.*, 2009, **278**, 27.
82. J. D. Weeks, D. Chandler and H. C. Andersen, *J. Chem. Phys.*, 1971, **54**, 5237.
83. W. Krahn, G. Schwelger and K. Lucas, *J. Phys. Chem.*, 1983, **87**, 4515.
84. H. Guo, Y. Chen, W. Lu, L. Li and M. Wang, *Fluid Phase Equilibr.*, 2013, **360**, 274.

CHAPTER 6

Non-equilibrium Molecular Dynamics

FERNANDO BRESME,*[a,b] ANDERS LERVIK[a,b] AND JEFF ARMSTRONG[a]

[a] Department of Chemistry, Imperial College London, SW7 2AZ, London, United Kingdom; [b] Department of Chemistry, Norwegian University of Science and Technology, 7491 Trondheim, Norway
*Email: f.bresme@imperial.ac.uk

6.1 Introduction

Non-equilibrium molecular dynamics (NEMD) simulations provide a powerful tool to investigate non-equilibrium phenomena from a microscopic perspective by taking into account the position and momenta of individual particles. External or fictitious fields can be employed to induce energy or mass fluxes and the concomitant temperature and chemical-potential and density gradients. In this chapter we discuss NEMD methods to drive mass, momentum and energy fluxes, and displacement currents. We also review the application of NEMD algorithms to compute thermophysical properties, rationalize transport mechanisms and uncover novel non-equilibrium coupling phenomena.

Several NEMD methods were developed in the 1970's. These developments have been discussed in excellent reviews and monographs.[1–3] Non-equilibrium simulations have traditionally complemented the equilibrium computations of transport coefficients based on the time-correlation formalism and the Green–Kubo (GK) equations.[4] The latter rely on the analysis

Experimental Thermodynamics Volume X: Non-equilibrium Thermodynamics with Applications
Edited by Dick Bedeaux, Signe Kjelstrup and Jan V. Sengers

Published by the Royal Society of Chemistry, www.rsc.org

of spontaneous fluctuations. The signal-to-noise ratio associated with the GK approach can decrease significantly in computations of collective properties, such as the thermal conductivity or viscosity. This has been a problem in the past, as it was difficult to compute long correlation times with enough statistics. NEMD provides an alternative to reduce the signal-to-noise ratio through the use of external or fictitious fields. Further, it has been used to investigate stationary states and coupling phenomena, such as thermo-diffusion. Non-equilibrium thermodynamics (NET) provides the formalism to describe these coupling phenomena and to understand how they modify the transport properties.[5] In turn, NEMD simulations, which do not make assumptions about coupling equations, can be used to validate the NET hypotheses.[6] Moreover, by varying the force field details (*e.g.*, the mass ratio in a binary mixture or the intermolecular interactions) it is possible to establish a direct connection between the non-equilibrium coupling effects and the atomic and molecular properties.[7–9]

NEMD has become an ideal tool to investigate problems where the application of macroscopic approaches such as NET or kinetic theory is limited. This is the case for nanoscopic systems, activated processes or systems very far from equilibrium. Shockwaves and crack propagation provide good examples of problems that challenge traditional non-equilibrium and kinetic theories. These issues were recognized shortly after the development of the first NEMD algorithms.[1]

Current NEMD simulation methods can be employed to investigate transient and stationary states. The former involves the simulation of the system relaxation towards equilibrium, *e.g.*, the relaxation of a hot molecule with time. The relaxation can be fitted to continuum equations (*e.g.*, the heat diffusion equation), enabling a route to estimate transport coefficients.[10,11] Early simulations of stationary states relied on the use of *boundary* regions, either to modify the energy of the particles in contact with the regions or to impart momentum.[12] It was recognized in early works that the use of boundary regions located at the edges of the simulation box is not compatible with the use of standard periodic boundary-conditions in the direction of the flux. An alternative setup for the boundary-conditions has been introduced more recently that results in a system that is fully periodic, and hence compatible with the boundary-conditions used in most computer simulations.[13,14] Moreover, the boundary-condition problem can be circumvented by using fictitious fields in the equations of motion (synthetic algorithms).[2] The latter approach provides a route to perform NEMD preserving the homogeneity/periodicity of the system of interest. In contrast, Boundary-Driven methods use boundary regions operated by inducing explicit gradients (temperature, density, concentration, *etc.*) inside the simulation box. This feature can be exploited to investigate a whole range of thermodynamic states in a single simulation.[15]

Non-equilibrium algorithms are not general, in the sense that different strategies are employed to compute specific transport properties: thermal transport, mass transport, *etc.* Often non-equilibrium thermodynamics and the GK approach play an important role in the design of the algorithms or in the interpretation of the results.

This chapter is structured as follows. We start by providing a background on computer simulations and linear response theory. A detailed discussion of the available NEMD algorithms follows *viz.* synthetic, Boundary-Driven and transient. We devote the final part of the chapter to discuss the applicability of NEMD to quantify thermophysical properties and to investigate coupled phenomena. We focus our discussion on fluids and fluid mixtures. However, we note that many of the computational approaches discussed here are applicable to solids too.

6.2 Background

In this section, we provide a succinct review of the main concepts related to equilibrium molecular dynamics and transport-coefficients calculation within the linear response theory. The reader is referred to excellent monographs on computer simulations for a full discussion of molecular dynamics simulations.[16–18]

6.2.1 Molecular Dynamics

Molecular dynamics (MD) simulations were first performed by Alder and Wainwright.[19] They rely on the solution of Newton's equation of motion of N particles. The potential energy, U, of the particles can be expressed as a sum of pair, u_{ij}, triplet, u_{ijk}, and higher-order terms,

$$U(\boldsymbol{r}_1, \boldsymbol{r}_2, \ldots \boldsymbol{r}_N) = \frac{1}{2}\sum_{i,j} u_{ij}(\boldsymbol{r}_i, \boldsymbol{r}_j) + \frac{1}{6}\sum_{i,j,k} u_{ijk}(\boldsymbol{r}_i, \boldsymbol{r}_j, \boldsymbol{r}_k) + \cdots, \tag{6.1}$$

where $\boldsymbol{r}_i$ is the vector corresponding to the position of particle i. Often, the interactions are approximated at the pair level by using effective potentials, such as the truncated Lennard-Jones potential (LJ), which is zero for distances $r_{ij} = |\boldsymbol{r}_i - \boldsymbol{r}_j|$ greater than some prescribed cut-off and otherwise equal to

$$u_{ij}(\boldsymbol{r}_i, \boldsymbol{r}_j) = u_{ij}(r_{ij}) = 4\varepsilon_{ij}\left[\left(\frac{\sigma_{ij}}{r_{ij}}\right)^{12} - \left(\frac{\sigma_{ij}}{r_{ij}}\right)^{6}\right], \tag{6.2}$$

where ε_{ij} and σ_{ij} are the interaction strength and particle diameter, respectively. The LJ potential provides a good representation of simple liquids, and it is widely used to model dispersion interactions of complex fluids and materials. To avoid a discontinuity in the force at the cut-off, the force can be smoothed to zero as described in ref. 16.

The particles' positions are evolved using the $N(N-1)/2$ independent forces that can be derived from the pair potential, $\boldsymbol{f}_{ij} = \partial u_{ij}/\partial \boldsymbol{r}_i$. In order to minimize the impact that the small system sizes investigated in MD ($N \sim 10$ to 10^5) have on structural and thermodynamic properties, the simulations are performed by using periodic boundary-conditions, whereby a central box is surrounded by an infinite array of replicas (see Figure 6.1). The use of boundary-conditions has to be combined with simulations of systems that are not too small. Lateral box lengths smaller than 10σ, where σ is the diameter of the largest atom in the simulation can result in the "periodic error". In such small boxes the thermodynamic properties, *e.g.*, pressure or surface tension, can differ significantly from the thermodynamic limit values.[20,21]

Typically, the centre of mass of the simulation box is kept constant so that the barycentric frame of reference is a natural choice for molecular dynamics simulation methods. However, this is not always the case, especially when a net mass flow or current is present.

The equations of motion in MD simulations are usually integrated numerically by using the Verlet algorithm or its variants. This algorithm can also be employed in conjunction with Boundary-Driven NEMD, while more sophisticated equations of motion have been developed to drive, *e.g.*, a heat flux by using synthetic methods (see Section 6.3.1).

6.2.2 Transport Coefficients and Linear Response

Linear response theory[22] provides a powerful approach to quantify mechanical and thermal transport coefficients and to design non-equilibrium algorithms.[2] The GK approach furnishes a link between the linear non-equilibrium response and the time-dependent equilibrium fluctuations.

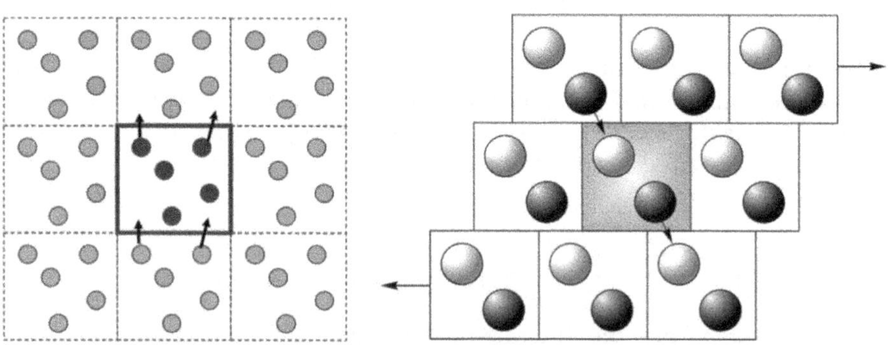

Figure 6.1 (Left) Illustration of periodic boundary-conditions in a two-dimensional system. The central simulation box has been highlighted. The arrows indicate the displacement of particles outside the central box and how they re-enter the box through the surrounding images boxes. (Right) Illustration of Lees–Edwards boundary-conditions employed in computations of planar Couette flow. The horizontal arrows pointing in opposite directions indicate the direction of the velocity in the upper and lower parts of the simulation box. The x and y axes are defined along the horizontal and vertical directions, respectively.

To exemplify the GK approach, we consider a system of N_c components in which there is transfer of mass and heat. The flux–force relations are then (as described in the previous chapters)

$$\begin{aligned} \boldsymbol{J}_i &= -L_{iq}\frac{\nabla T}{T^2} - \sum_{k=1}^{N_c-1} L_{ik}\frac{\nabla_T(\mu_k - \mu_{N_c})}{T}, \quad i=1 \ldots N_c - 1 \\ \boldsymbol{J}'_q &= -L_{qq}\frac{\nabla T}{T^2} - \sum_{k=1}^{N_c-1} L_{qk}\frac{\nabla_T(\mu_k - \mu_{N_c})}{T}, \end{aligned} \tag{6.3}$$

where $\boldsymbol{J}_i$ is the mass flux of component i in the barycentric frame of reference $\left(\sum_{j=1}^{N_c} \boldsymbol{J}_j = 0\right)$, T is the temperature, μ_i the chemical potential of component i, $\boldsymbol{J}'_q$ the measurable heat flux in the barycentric frame of reference, and $L_{ij} = L_{ji}$ are the Onsager coefficients. The subscript T in the gradient (∇_T) indicates that the gradient is taken at constant temperature. The GK method relies on the computation of autocorrelation functions, for instance, the Onsager coefficients related to the mutual diffusion of components i and j can be obtained by,[23,24]

$$L_{ij} = \frac{1}{3k_B V}\int_0^{\infty}\left\langle \boldsymbol{J}_i(t)\cdot\boldsymbol{J}_j(0)\right\rangle \mathrm{d}t = \frac{1}{3k_B V}\int_0^{\infty}\left\langle m_i \sum_{k=1}^{N_i} \boldsymbol{v}_{k,i}(t) \cdot m_j \sum_{l=1}^{N_j} \boldsymbol{v}_{l,j}(0) \right\rangle \mathrm{d}t, \tag{6.4}$$

where V is the volume of the simulation box, k_B is the Boltzmann constant, t is the time, m_j is the mass of component j and $\boldsymbol{v}_{l,j}$ is the velocity of molecule l of component j in the barycentric frame of reference. The factor of 3 applies to three-dimensional systems. The total number of molecules of component j is N_j. These Onsager coefficients can be related to the Maxwell–Stefan or Fick diffusivities.[25] For a thorough discussion on how the different diffusion coefficients are related, we refer to the previous chapter. We would like to point out that the Fick diffusion coefficients will depend on the choice of the frame of reference, and they are also non-symmetric. The Maxwell–Stefan diffusivities are, on the other hand, both independent on the frame of reference and symmetric.

One can also access the Fick diffusion coefficients by considering the generalized Fick's law of diffusion (in the average molar frame of reference)[25] for a constant temperature

$$\boldsymbol{J}_i = -c\sum_{j=1}^{N_c-1} D_{ij}\nabla x_j, \tag{6.5}$$

where $\boldsymbol{J}_i$ is the molar flux of component i, D_{ij} is the Fick diffusivity, c is the total molar concentration and x_j is the molar fraction of component j. Non-equilibrium simulation approaches for inter-diffusion make use of eqn (6.5) by computing the mass fluxes of the different components, arising from an imposed molar fraction gradient.

The GK approach to calculate the thermal conductivity, λ, relies on the autocorrelation function of the heat flux,

$$\lambda = \frac{L_{qq}}{T^2} = \frac{V}{3k_BT^2}\int_0^{\infty}\left\langle \boldsymbol{J}'_q(t)\cdot\boldsymbol{J}'_q(0)\right\rangle \mathrm{d}t, \tag{6.6}$$

where, $\boldsymbol{J}'_q$, in the notation of de Groot and Mazur,[5] is the reduced (or measurable) heat flux, and L_{qq}, the Onsager coefficient. The factor of 3 applies again to three-dimensional systems. We note that alternative GK equations can be obtained using other definitions of the heat-flux, which do include enthalpic terms.

In simulations, the heat flux is often obtained from the microscopic equation for the flux of internal energy, $\boldsymbol{J}_U$, which was derived by Irving and Kirkwood.[26] For zero barycentric velocity, pair wise interactions, and a local volume V, we have,

$$\boldsymbol{J}_U = \frac{1}{V}\sum_{i=1}^{N\in V}\left[\frac{1}{2}m_i v_i^2\,\boldsymbol{v}_i + u_i\,\boldsymbol{v}_i + \frac{1}{2}\sum_{j\neq i}^{N}(\boldsymbol{v}_i\cdot\boldsymbol{f}_{ij})\,\boldsymbol{r}_{ij}\right], \tag{6.7}$$

where $\boldsymbol{v}_i$, is the velocity of particle i, $\boldsymbol{f}_{ij}$ is the force of particle j on i, $\boldsymbol{r}_{ij} = \boldsymbol{r}_i - \boldsymbol{r}_j$ and $u_i = \frac{1}{2}\left(\sum_{j\neq i=1}^{N} u_{ij}(r_{ij})\right)$ is the potential energy of particle i. The heat flux is related to the internal energy flux through, $\boldsymbol{J}'_q = \boldsymbol{J}_U - \sum_{i=1}^{N_c-1}(h_i - h_{N_c})\boldsymbol{J}_i$, where h_i is the partial specific enthalpy of component i. For a stationary state $J_i = 0$ and therefore $\boldsymbol{J}'_q = \boldsymbol{J}_U$. Several approaches have been developed to compute the internal-energy flux given in eqn (6.7). The method of planes is particularly suitable for inhomogeneous fluids.[27] Eqn (6.7) has also been extended to perform simulations of molecules by using rigid-body dynamics algorithms,[28,29] and to take into account long-range interactions arising in charged and polar systems.[30–32]

When more than one component is present in the system of interest, the computation of transport coefficients requires special attention, as cross effects arising from the coupling of different fluxes are possible. One of such effects was observed by Ludwig in 1856[33] and later investigated by Soret.[34] The Ludwig–Soret effect can be predicted and understood by using NET and the linear flux–force (LFF) relations. For a binary mixture,[5]

$$\boldsymbol{J}_1 = -L_{1q}\frac{\nabla T}{T^2} - L_{11}\frac{\nabla_T(\mu_1 - \mu_2)}{T} \tag{6.8}$$

$$\boldsymbol{J}'_q = -L_{qq}\frac{\nabla T}{T^2} - L_{q1}\frac{\nabla_T(\mu_1 - \mu_2)}{T}, \tag{6.9}$$

Applying the Gibbs–Duhem equation to the gradient of chemical potentials we get, $\nabla_T(\mu_1 - \mu_2) = w_2^{-1}(\partial\mu_1/\partial w_1)_{T,P}\nabla w_1$, where w_1 and w_2 are the mass fractions of component 1 and 2, respectively. The mass flux can then be rewritten in terms of the thermal diffusion coefficient, $D_T = L_{1q}(\rho w_1 w_2 T^2)^{-1}$ and the diffusion coefficient, $D = L_{11}(\rho w_2 T)^{-1}(\partial\mu_1/\partial w_1)_{T,P}$, to get, $\boldsymbol{J}_1 = -\rho w_1 w_2 D_T \nabla T - \rho D \nabla w_1$, where ρ is the density. The Soret coefficient, S_T, is defined when $\boldsymbol{J}_1 = 0$ and given by $S_T = D_T D^{-1} = -(w_1 w_2)^{-1}(\nabla w_1/\nabla T)\big|_{J_1=0}$.

The coefficients in the LFF relations eqn (6.8) are connected to transport coefficients. For example (as in eqn (6.6)), $L_{qq} = \lambda T^2$ is connected to the thermal conductivity, while the cross-coefficients, $L_{1q} = L_{q1}$, are related to the Soret and Dufour coefficients.[5] These coefficients can be computed with appropriate GK equations (see, *e.g.*, ref. 6). It is important to note that the heat flux, $\boldsymbol{J}_q'$, does not contain the partial enthalpies. These enthalpic terms must be considered in order to compute the measurable heat flux correlation functions needed in the GK approach (see eqn (6.6) and ref. 6). Neglecting the enthalpic contributions can lead to large overestimations of the thermal conductivity. The partial enthalpies can be computed by using equilibrium[35] and non-equilibrium approaches.[36]

The GK approach can be used to compute the shear viscosity, η,[37]

$$\eta = \frac{V}{k_B T}\int_0^{\infty} \langle P_{xy}(t) \cdot P_{xy}(0) \rangle \mathrm{d}t, \tag{6.10}$$

via the computation of the off-diagonal components, P_{xy}, of the pressure tensor, $\boldsymbol{P}$,

$$\boldsymbol{P}V = \sum_{i=1}^{N} \frac{\boldsymbol{p}_i \boldsymbol{p}_i}{m_i} - \frac{1}{2}\sum_{i \neq j=1}^{N} \boldsymbol{r}_{ij}\boldsymbol{f}_{ij}, \tag{6.11}$$

where $\boldsymbol{p}_i$ is the momentum of particle i. The pressure tensor is needed to define the Hamiltonian employed in fictitious non-equilibrium methods for viscous-flow computations.[38,39]

6.3 Non-equilibrium Molecular Dynamics Simulations

Non-equilibrium molecular dynamics simulations offer a direct approach to quantify transport coefficients and coupled phenomena. NEMD approaches can be subdivided in synthetic, Boundary-Driven and transient methods. The first two approaches drive the system to the stationary state while in the latter the transport coefficients are computed by inspecting the relaxation of the system towards equilibrium. These different methods are discussed below.

6.3.1 Synthetic NEMD

Synthetic NEMD (S-NEMD) is a direct extension of the ideas contained in the linear response theory. S-NEMD relies on the introduction of a fictitious field into the equations of motion and, hence, it has little relation to an experimental setup. The equations of motion are consistent with the system's homogeneity and with the use of periodic boundary-conditions, which represents an advantage of the method.

The background of the S-NEMD approach is as follows.[2,40] For a general flux, $\boldsymbol{J}_i$, of a conserved quantity (energy, mass or momentum) the linear transport coefficient, L_{ij}, can be defined by, $\boldsymbol{J}_i = \sum_j L_{ij}\boldsymbol{X}_j$, which follows the linear phenomenological equations defined in the NET theory, where $\boldsymbol{X}_j$ is the thermodynamic force, *e.g.*, a thermal gradient. S-NEMD operates by first identifying the GK relation defining the transport coefficient of interest (see Section 6.2.2). A fictitious external field $\boldsymbol{F}_e$ is then "invented" and coupled to the system of interest. The S-NEMD method relies on the calculation of the average response of a phase space current, $\boldsymbol{B}$, under the influence of the fictitious field. The linear response theory states the average of $\boldsymbol{B}$ is defined by,[2]

$$\langle \boldsymbol{B}(t)\rangle = -\frac{V}{k_B T}\int_0^t \mathrm{d}t'\, \phi(t-t')\cdot \boldsymbol{F}_e(t'), \tag{6.12}$$

where $\phi(t) = \langle \boldsymbol{B}(t)\cdot \boldsymbol{J}(0)\rangle$ is the equilibrium time correlation function of $\boldsymbol{B}$ and $\boldsymbol{J}$, the field $\boldsymbol{F}_e$ acts over time t, and $\boldsymbol{J}$ is one of the currents appearing in the GK equations (*e.g.*, the mass flux or the measurable heat flux). Eqn (6.12) can be used to obtain the non-equilibrium steady-state response of the current $\boldsymbol{B}$ (again this could be the mass flux for diffusion or the heat flux for the thermal conductivity), while the equilibrium result for the transport coefficient is recovered in the limit of zero field $\boldsymbol{F}_e$, $L_{ii} = \lim_{F_e\to 0}\lim_{t\to\infty}\langle J_\alpha(t)\rangle/F_e$ where J_α is the flux (*e.g.*, the heat flux) in the direction α of the field $\boldsymbol{F}_e$ whose magnitude is given by F_e. For a system subjected to the external perturbation, the non-Hamiltonian equations of motion can be written as,[2] $\dot{\boldsymbol{r}}_i = \boldsymbol{p}_i/m_i + \boldsymbol{C}_i\cdot \boldsymbol{F}_e(t)$ and $\dot{\boldsymbol{p}}_i = \boldsymbol{f}_i + \boldsymbol{D}_i\cdot \boldsymbol{F}_e(t)$, where $\boldsymbol{r}_i$, $\boldsymbol{p}_i$, $\boldsymbol{f}_i$ and m_i are the position, momentum of, force on, and mass of particle i, and the dot denotes a time derivative. $\boldsymbol{F}_e(t)$ is the fictitious field and $\boldsymbol{C}_i$ and $\boldsymbol{D}_i$ are tensors that describe the coupling of the field and the system. The dissipative flux, $\boldsymbol{J}$, is defined by,

$$\dot{H}_0 = \sum_{i=1}^{N}\left[\boldsymbol{D}_i\cdot\frac{\boldsymbol{p}_i}{m_i} - \boldsymbol{C}_i\cdot \boldsymbol{f}_i\right]\boldsymbol{F}_e(t) = -V\boldsymbol{J}\cdot \boldsymbol{F}_e(t), \tag{6.13}$$

where $\dot{H}_0$ is the rate of change of internal energy due to the field. The $\boldsymbol{C}_i$ and $\boldsymbol{D}_i$ are chosen (see ref. 2) to ensure that $\boldsymbol{B}$ corresponds to the flux of interest, *e.g.*, the mass fluxes if the diffusion coefficient is required. The transport coefficient, L_{ii} (see above), which corresponds to the limit of a vanishing external force can then be obtained from a linear fitting using a few simulations at varying, small, fields. The reason why the S-NEMD method is more efficient at calculating transport coefficients than the equilibrium GK is that it produces a higher signal-to-noise ratio.

The S-NEMD method has been implemented to compute diffusion coefficients, shear viscosities and thermal conductivities. Because the self-diffusion coefficient is a single-particle property, equilibrium GK computations are preferred over S-NEMD ones. The latter can be advantageous to compute mutual diffusion coefficients. The computation of shear viscosities otherwise on the use of special boundary-conditions; the so-called Lees–Edwards (LE) boundary-conditions (see Figure 6.1),[41] which have been widely employed to investigate Couette flow. We have shaded grey the LE unit cell (see Figure 6.1). The periodic images located above and below the central layer move with velocities v_x and $-v_x$, respectively. These velocities define the strain rate $\gamma = \partial v_x/\partial y$. If a particle leaves the unit cell (dark particle in Figure 6.1) it will be replaced by its periodic image (light particle in Figure 6.1). However, unlike in a regular MD simulation, the image will in general have a different velocity and position relative to its cell's origin. This discontinuity of particle velocities and positions drives the system into a linear streaming situation, which can be exploited in Boundary-Driven NEMD simulations. The advantage of using S-NEMD is that it circumvents the problem associated with the time delay between the boundaries' motion and the shearing of the fluid, which is determined by the speed of sound of the fluid and, hence, precludes the investigation of time-dependent flows. By introducing the DOLLS algorithm, a solution to this problem was devised by Hoover *et al.*[38] Later Evans and Morris introduced a modification for the equations of motion of the momenta, this is the so called SLLOD algorithm.[2] The equations of motion for the latter are given by: $\dot{\boldsymbol{q}}_i = \boldsymbol{p}_i/m_i + \boldsymbol{q}_i \cdot \nabla \boldsymbol{u}$ and $\dot{\boldsymbol{p}}_i = \boldsymbol{f}_i - \boldsymbol{p}_i \cdot \nabla \boldsymbol{u}$, where $\boldsymbol{u}$ is the streaming velocity. The corresponding linear response of the pressure tensor is given by, $\langle \boldsymbol{P}(t) \rangle = -(V/k_B T)\int_0^t dt' \langle \boldsymbol{P}(t-t') \cdot \boldsymbol{P}(t') \rangle : \nabla \boldsymbol{u}(t')$, and the shear viscosity (eqn (6.10)) follows from, $\eta = \lim_{t\to\infty} \lim_{\gamma \to 0} \left[-\langle P_{xy}(t) \rangle / \gamma \right]$.

The S-NEMD computation of the thermal conductivities requires the definition of a fictitious field whereby hot particles are driven with the field, while cold particles are driven against the field.[2] This method predicts thermal conductivities consistent with those obtained by using Boundary-Driven methods, although instabilities in the S-NEMD heat-flow algorithm have been found for relatively large systems ($N > 900$ particles) and large fictitious fields.[42,43]

6.3.2 Boundary-Driven NEMD

Boundary-Driven NEMD (BD-NEMD) methods emulate a real experiment. An external perturbation is imposed to the system, by adding energy or applying an external force in a region of the simulation box. It was noted in the first BD-NEMD implementations of shear viscosities that the existence of unidirectional fluxes is incompatible with the periodic boundary-conditions.[12] This issue was resolved by setting up reflecting walls,[12,44] such that the particles undergo elastic collisions. It is also possible to simulate an atomistic wall in direct contact with the fluid.[45] The existence of an explicit wall may be a desirable feature in computer simulations of friction or in computations of thermal resistances, where the focus might be the quantification of interfacial transport coefficients. However, the wall might be an undesirable feature if bulk transport coefficients are required. This problem can be solved with appropriate boundary-conditions, which preserve the periodicity of the simulation box and eliminate the need for confining walls.[13,14] Figure 6.2 shows a typical setup used in BD-NEMD simulations of heat transport.[6,46–51]

6.3.2.1 BD-NEMD Algorithms for the Computation of Heat Transport

Many BD-NEMD simulations rely on the setup represented in Figure 6.2. A stationary heat flux can be achieved by adding and withdrawing kinetic energy from specific regions defined in the middle and the edges of the simulation box. Because this process is performed without modifying the configurational energy, the change in kinetic energy is equivalent to a change in the system's internal energy. This was the initial simulation setup by Hafskjold *et al.*[13,52] in their Heat-Exchange algorithm (HEX). Particles in "hot" and "cold" regions, R, are thermostatted by adding/withdrawing a specific amount of internal energy, ΔU. Because the potential energy remains constant after the thermostatting process (no change in particle coordinates), the change in internal energy, ΔU, is equal to the change in

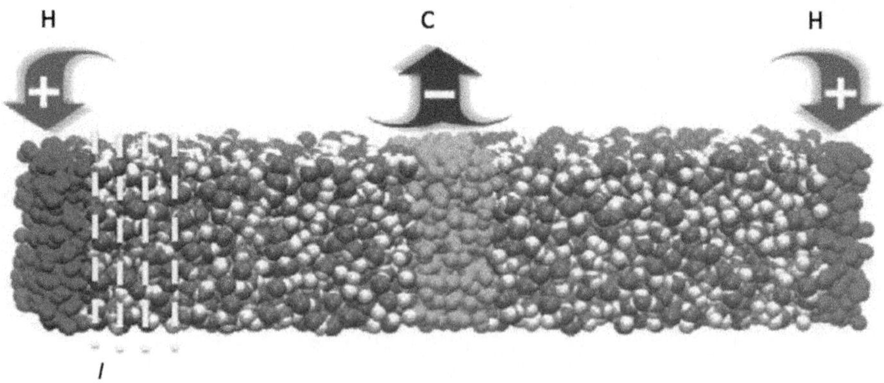

Figure 6.2 (Left) Snapshot of a simulation box containing water that illustrates the boundary-conditions employed in BD-NEMD simulations of heat transport. The regions in the centre and edges of the box are employed to add/withdraw energy or thermostat the molecules at specific temperatures. The dashed lines represent the position of slabs used to compute the local properties and l denotes a specific slab in the cell.

kinetic energy, ΔE_k The velocities of the particles, $\boldsymbol{v}_{i,b}$, are rescaled at regular time intervals according to, $\boldsymbol{v}_{i,a}=(1+\alpha)\,\boldsymbol{v}_{i,b}+\boldsymbol{\beta}$, where $\boldsymbol{\beta}$ is defined by

$$\boldsymbol{\beta}=-\left(\alpha\sum_{i=1}^{N}m_i\boldsymbol{v}_{i,b}\right)\Big/\sum_{i=1}^{N}m_i, \tag{6.14}$$

and α and $\boldsymbol{\beta}$ are connected to ΔU through $\Delta U=\sum_{i\in R}\left(m_i[(1+\alpha)\boldsymbol{v}_{i,b}+\boldsymbol{\beta}]^2-m_i\boldsymbol{v}_{i,b}^2\right)/2$. This defines a quadratic equation for α, with the sum running over all the atoms lying in the hot or cold regions, R, and the subscripts "b" and "a" refer to the velocities before and after the rescaling. The HEX algorithm does not work when the region R contains less than 2 particles, hence limiting the applicability of the algorithm to high densities. For simple fluids, molten salts and simple molecular fluids, the HEX algorithm generates the stationary state (constant heat flux) in a few hundred picoseconds. The stationary heat flux can then be calculated by using the microscopic eqn (6.7) and the continuity equation,

$$\boldsymbol{J}_U=\left\{\pm\frac{\langle\Delta U\rangle}{2\,\delta t A},0,0\right\}, \tag{6.15}$$

where angular brackets denote an ensemble average, δt is the time interval between rescaling events, A the cross sectional area of the simulation box and the thermal gradient is applied along the x direction. The factor of 2 takes into account the existence of two fluxes inside the simulation box, and the positive and negative signs refer to the average change in internal energy in the hot and cold regions, respectively. This and other NEMD methods discussed below generate thermal gradients preserving mechanical equilibrium, $\nabla\boldsymbol{P}=0$. This makes it possible to calculate equations of state using a single simulation by calculating the local temperature and density in the simulation cell.[15]

Reverse Non-equilibrium Molecular Dynamics (RNEMD) is a modification of the HEX algorithm that uses the same boundary-conditions discussed above.[46] The thermal gradient results from the exchange of the velocity vector of atoms lying in cold and hot regions. The exchange involves atoms with the lowest and highest speeds. The resulting heat flux can be quantified with eqn (6.15). The RNEMD has been extended to simulate molecular fluids,[53] non-periodic systems,[54] and mesoscopic models.[49]

In many heat-transfer problems it is convenient to set the temperature of the thermostatting regions to predefined values. In this approach the velocities of the atoms/molecules inside the hot and cold regions are rescaled to obtain the desired temperatures.[56] A simple rescaling by a factor, $\alpha=\sqrt{K_t/K}$, can be employed, where K_t is the target kinetic energy and K the instantaneous kinetic energy. The linear momentum must be reset after each rescaling. The velocity-rescaling algorithm[55] provides a good approach to control the temperature and momentum. The target kinetic energy, K_t, is drawn from the canonical distribution function,

$P(K_t)\,dK_t \propto K_t^{N_f/2-1} e^{-\beta K_t} dK_t$, where N_f, is the number of degrees of freedom, f, of the thermostatted molecules and $\beta = 1/(k_B T)$. The corresponding heat flux in the stationary state can then be obtained from eqn (6.15) and,

$$\langle \Delta U \rangle = \left\langle \frac{1}{2}\left[\left(\sum_{i=1}^{N\in \mathrm{TL}} m_i \mathbf{v}_i^2\right)_a - \left(\sum_{i=1}^{N\in \mathrm{TL}} m_i \mathbf{v}_i^2\right)_b\right]\right\rangle, \tag{6.16}$$

where the sums run over all the particles lying in either the hot or the cold Thermostatting Layers (TL) and "a" and "b" refer to the kinetic energy before and after the velocity rescaling. This thermostatting approach has been used to investigate both atomic and molecular fluids.[6,15,56]

One common feature of all the algorithms discussed above is that the dynamics of the atoms/molecules outside the thermostatting regions is Newtonian. Also, once a thermal gradient has been set up, the local temperature in a particular slab l (see Figure 6.2) of the simulation box can be calculated by using the equipartition principle, $N_f k_B T_l = \sum_{i=1}^{N\in l} m_i \mathbf{v}_i^2$, where N_f is the number of degrees of freedom of the particles lying in the slab l. Figure 6.3 shows a representative temperature profile for water obtained with BD-NEMD using the thermostatting algorithm. The profile is well defined and appears to be linear. However, we note that deviations from linearity are *always* present, since the thermal conductivity depends on the local temperatures and densities.[15] We show in Figure 6.3 one example of the internal-energy-flux computation in the stationary state, where we compare results from the microscopic and continuity approaches, eqn (6.7) and (6.15), respectively. The results in Figure 6.3 show that the microscopic

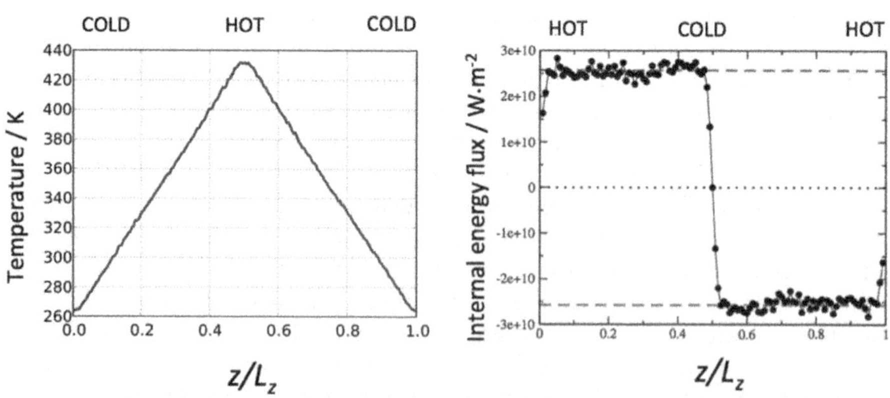

Figure 6.3 (Left) Temperature profile of the SPC/E water model obtained with BD-NEMD using the thermostatting NEMD algorithm. (Right) Heat-flux of the MCFM water model.[136] Symbols represent computations with the heat-flux microscopic expression (eqn (6.7)) and the horizontal dashed lines show the continuity equation result (eqn (6.15)). L_z is the box length in the direction of the thermal gradient.

eqn (6.7) is accurate for the density gradients obtained in representative BD-NEMD simulations of dense fluids.

6.3.2.2 BD-NEMD Algorithms for the Computation of Mass Transport

The BD-NEMD algorithms for the simulation of a mass flux follow an approach similar to that explained for heat flow. In the mass-exchange (MEX) algorithm,[57] particles are exchanged between reservoirs inside the simulation cell in order to create a concentration gradient and a mass flux. Alternatively, in the Dual Control Volume Grand Canonical Molecular Dynamics (DCV-GCMD) approach the chemical potential is fixed to specific values in different regions of the simulation cell.[58] The algorithms are discussed in more detail below.

The MEX algorithm allows the investigations of heat and mass coupled transport. High- and low-concentration regions are defined in the simulation cell and particles of different components are swapped between these regions, hence generating mass fluxes through the system. The swapping is performed ensuring that both energy and momentum are conserved. When a particle, labelled s, with velocity $\boldsymbol{v}'_s$ and mass m'_s is swapped, the velocity after swapping ($\boldsymbol{v}_s$) of that particle is set to,

$$\boldsymbol{v}_s = (1+\alpha)\left(\frac{m'_s}{m_s}\right)^{1/2}\boldsymbol{v}'_s + \boldsymbol{\beta}, \tag{6.17}$$

where m_s is the mass after swapping. The velocities of all the other particles $i \neq s$ in each region, are rescaled according to, $\boldsymbol{v}_i = (1+\alpha)\boldsymbol{v}'_i + \boldsymbol{\beta}$, where the velocity shift, $\boldsymbol{\beta}$, is now given by,

$$\boldsymbol{\beta} = \frac{-\alpha(\boldsymbol{p}+\Delta\boldsymbol{p})+\Delta\boldsymbol{p}}{M}, \tag{6.18}$$

where $M = \sum m'_i$ is the total mass in the region under consideration, $\boldsymbol{p} = \sum m'_i \boldsymbol{v}'_i$ the momentum and $\Delta\boldsymbol{p} = ((m_s m'_s)^{1/2} - m'_s)\boldsymbol{v}'_s$. The energy conservation in each region leads to a quadratic equation for α,

$$\alpha(2+\alpha)E_K + (1+\alpha)\boldsymbol{\beta}\cdot(\boldsymbol{p}+\Delta\boldsymbol{p}) + \frac{1}{2}\boldsymbol{\beta}^2 M + \Delta\phi_s = 0, \tag{6.19}$$

where $E_K = \left(\sum m'_i \boldsymbol{v}_i^2\right)/2$ and $\Delta\phi_s$ is the change in potential energy associated with the particle swapping. The stationary mass flux for component k can be obtained from (assuming transport takes place along the x-direction),

$$\boldsymbol{J}_k = \left\{\frac{\Delta m_k}{2\,\Delta t A}, 0, 0\right\}, \tag{6.20}$$

where Δt is the time between two particle swaps, A the cross-sectional area perpendicular to the x-direction, and Δm_k is the change in mass of species k due to the swapping ($\Delta m_k = m_k$ for regions where particles of type k are inserted and $\Delta m_k = -m_k$ where they are removed).

The DCV-GCMD method[58] is a hybrid MD/Monte Carlo (MC) approach. The boundaries of the system are fixed at different chemical potentials, hence inducing a chemical-potential gradient. Each time-step, a specified number of MC steps are performed to insert or delete particles according to a grand-canonical Monte Carlo scheme. Particles of species i are inserted into a control volume with chemical potential μ_i^c and the insertion is accepted if,

$$N_i^c \exp\left(-\frac{\mu_i^c}{k_B T} - \ln\frac{V^c}{\Lambda_i^3} - \frac{\Delta U_{p,i}}{k_B T}\right) \geq \xi, \tag{6.21}$$

where V^c is the volume of the control volume, N_i^c is the number of particles of species i in the control volume, Λ_i the de Broglie wavelength, $\Delta U_{p,i}$ the change in potential energy associated with the creation of a particle of type i, and ξ is a random number drawn from a uniform distribution (0,1). Created particles are then given velocities drawn from a Maxwell–Boltzmann distribution. A similar criterion is employed to accept particle deletion. The resulting mass flux is obtained by computing the net movement of particles in the box, or from the difference in the number of insertions/deletions in the two control volumes. The number of MC steps performed per MD step, typically (10 to 50), is a critical parameter to maintain the correct steady state.[58,59] Several authors[59] have discussed the advantages of Equilibrium Molecular Dynamics and Synthetic-NEMD approaches over the DCV-GCMD to study diffusion in micropores.[60]

6.3.2.3 BD-NEMD Algorithms for the Computation of Viscosity

In synthetic methods such as SLLOD, a shear rate is imposed on the system, and the resulting steady-state stress is measured. Alternatively, the stress or momentum flux can be imposed and the resulting shear rate computed.[61–63] This approach belongs to the general RNEMD technique, and therefore avoids the computation of the momentum flux. Following the general Boundary-Driven approach, the simulation box is divided into two regions, which are used to exchange momenta. At pre-defined time intervals, particles are swapped between these layers, hence generating a momentum flux. The swapping involves particles with the smallest and largest momentum components. When the steady state is reached, the rate of momentum transfer equals the momentum flowing back through the fluid by friction. By summing the total momentum, P_x, transferred up to time t in the simulation, the momentum flux, j_z, can be calculated, $j_z = P_x/(2tA)$, where A is the cross-sectional area of the simulation box. The velocity profile, v_x, can be calculated directly, and the viscosity obtained from $\eta = -j_z(\partial v_x/\partial z)^{-1}$. High momentum fluxes lead to non-linear velocity profiles and violations of the linear equations for the viscosity. Infrequent swapping may lead to non-stationary velocity profiles. Tenney and Maginn[64] have solved some of these issues.

6.3.3 Thermophoretic Forces and Soret Coefficient

The Soret coefficient of binary mixtures consisting of particles of similar size can be obtained from BD-NEMD simulations by computing the derivative, $\mathrm{d}w_1/\mathrm{d}T$ that appears in the Soret coefficient (see Section 6.2.2). When the weight fraction of the solute is significantly lower than that of the solvent, (*e.g.*, in aerosol, dilute colloidal suspensions) the computation of concentration profiles is affected by statistical accuracy, and the direct computation of the thermophoretic-force offers a more precise alternative.[65] Under the influence of a thermal gradient, a particle will feel a thermophoretic force, $\boldsymbol{F}_\mathrm{T}$, which induces a drift velocity. For a single particle, the drift velocity can be approximated by[66] $\boldsymbol{v}_\mathrm{T} = -D_\mathrm{T}\nabla T$, where D_T is the particle thermal diffusion coefficient. When the inertial effects can be neglected, the drag force is approximated by $\boldsymbol{F}_\mathrm{D} = \zeta\boldsymbol{v}_\mathrm{T}$, where ζ is the friction coefficient. The balance of the thermophoretic force and drag forces, $\boldsymbol{F}_\mathrm{D} = -\boldsymbol{F}_\mathrm{T}$, along with the Stokes–Einstein equation for the self-diffusion coefficient, can be used to obtain an equation for the thermal diffusion factor of a single particle,[65]

$$\alpha_\mathrm{T} = \frac{-F_\mathrm{T}}{k_\mathrm{B}\left(\dfrac{\mathrm{d}T}{\mathrm{d}z}\right)}, \tag{6.22}$$

where F_T is the force on the particle along the vector defining the direction of the thermal gradient setup in the z-direction. Eqn (6.22) connects the thermophoretic force, with the thermal diffusion factor, α_T, and the Soret coefficient, $S_\mathrm{T} = \alpha_\mathrm{T}/T$.[5] Eqn (6.22) represents a good approximation in suspensions involving solutes much heavier/larger than the solvent. The thermophoretic-force approach can also be employed to investigate binary fluids, but in this case the forces of both components must be included to quantify the Soret coefficient.[67]

The thermophoretic-force method discussed above is technically very simple to implement and exploits the same Boundary-Driven NEMD simulation setup shown in Figure 6.2. Two solute particles can be simulated at the same time. Each solute is tethered in space to its initial equilibrium position, *via* a harmonic restraint with a predefined force constant. Typically the initial position will be at a central point between the hot and cold thermostats. Once the system has reached the stationary state, the ensemble average of the solute position can be computed, and the thermophoretic force, F_T, extracted from Hooke's law, $F_\mathrm{T} = -K\Delta z$, where K is the force constant and Δz the displacement from the equilibrium position. This method has been used to investigate nanoparticles in solution,[65] binary mixture models,[67] and biomolecules.[68]

6.3.4 Transient Non-equilibrium Molecular Dynamics

One of the first implementations of Transient NEMD (TNEMD) focused on modelling cooling experiments of proteins.[69] The TNEMD method has been extended to compute the thermal conductivity and conductance of nanoparticles and nanoparticle–fluid interfaces.[11,70] Figure 6.4 illustrates how the TNEMD method works. The fluid or material of interest is initially heated to a target temperature, T_1, and equilibrated for a short time, typically (50 to 100) ps, while the surrounding solvent is thermostatted at a different temperature, T_2. The relaxation time required to evolve from T_1 to T_2 is determined by the thermal diffusivity, $a = \lambda/\rho c_\mathrm{p}$, where ρ and c_p and λ are the mass density, the isobaric heat capacity and thermal conductivity of the solute, respectively. The relaxation curve can be fitted to the heat-diffusion equation by using different boundary-conditions that assume the cooling is controlled by the thermal conductance (G) only (model A in Figure 6.4), or the thermal conductivity of the solute only (model B). More sophisticated models have been explored,[11,70] that take into account both the thermal conductance (G) of the interface and the thermal conductivity (λ) of the solute (model C). For nanoscale solutes, both G and λ contribute towards the temperature relaxation as can be assessed through the Biot number, $Bi = GR/\lambda$ where R defines the curvature of the solute of interest. For nanoscale solutes, $Bi \sim 1$, and heat-diffusion models that incorporate both G and λ must be considered.

The TNEMD approach has also been employed to investigate the relaxation of bulk fluids and solids using the BD-NEMD boundary-conditions (see Figure 6.2),[32] as well as two regions spanning the whole simulation cell, which are maintained at different temperatures.[71] These simulations have provided reasonable estimates of the thermal conductivity. One advantage of the TNEMD approach is that the analysis requires a short simulation time, (50 to 500) ps, and therefore it can be used in very large systems. Often λ and G are assumed to be independent of the temperature, hence the transport properties obtained from this approach must be interpreted as an average over the temperature range sampled during the relaxation process.

Transient approaches have also been developed to quantify the diffusivity of molecules in pores.[23] This approach is based on the use of Fick's second law,

$$\frac{\partial c}{\partial t} = D\frac{\partial^2 c}{\partial y^2}, \tag{6.23}$$

where $c(y,t)$ is the concentration as a function of time and position, D the diffusion coefficient, and y is the spatial direction where the concentration varies. By using appropriate boundary-conditions for the concentration, numerical or analytical equations for $c(y,t)$ can be obtained, making possible the estimation of D. A disadvantage of this approach, and in general of all the TNEMD methods, is that several simulations must be performed to obtain acceptable statistics. Furthermore, the simulations are not always in the linear regime, making the application of Fick's law questionable.[23] Finally, D, is assumed to be a constant; hence the diffusion coefficient must be interpreted as an effective coefficient.

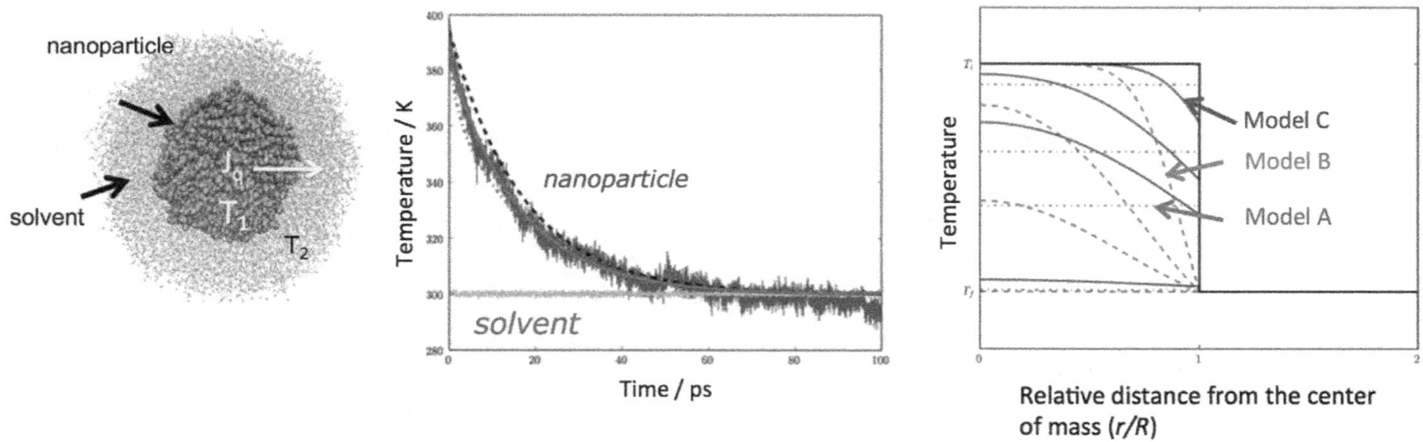

Figure 6.4 (Left) Snapshot of an alkane nanoparticle in water, illustrating the initial conditions for the TNEMD method, where the nanoparticle and solvent are initially equilibrated at temperatures T_1 and T_2. (Middle) Temperature relaxation of the nanoparticle as a function of time. Dashed line (model A), dotted line (model B) and dash-dotted line model C. The instantaneous temperature computed during the simulation is represented by the fluctuating line (Right) Sketch illustrating the different relaxation models, A, B, and C employed to fit the temperature relaxation of the nanoparticle.[11]

Arya *et al.*[72] have introduced a transient approach for the viscosity, which monitors the decay of a Maxwellian velocity profile introduced in an equilibrated system. This approach exploits the solution of the Navier–Stokes equation for the decay of an initial Gaussian velocity profile. The TNEMD provides satisfactory results for the viscosity of argon and *n*-butane.

6.4 NEMD Applications

NEMD simulations open a route to quantify the transport coefficients of pure fluids and mixtures and to test the theoretical basis of NET, particularly the validity of the Onsager reciprocal relations and the hypothesis of local equilibrium. We discuss in the following the applicability of the NEMD technique to address these questions.

6.4.1 Verification of Local-equilibrium and Onsager Reciprocal Relations (ORR)

NEMD has been employed to verify the linearity of the flux–force relations in systems involving transport of heat and mass. The Onsager reciprocal relations (ORR), $L_{ij} = L_{ji}$, and the local-equilibrium hypothesis have also been examined. These relations are fundamental in the development of the NET theory, hence their verification is an important objective. The ORR have been found to hold in cross thermo-transport problems. Simulations of heat and mass transport in the Ar/Kr mixture,[73] showed that the cross coefficients quantifying the Soret and the Dufour effects are equal within the uncertainty of the computations. Similar conclusions have been achieved by using synthetic and BD-NEMD simulations,[57,74] and GK computations.[75]

The local-equilibrium hypothesis in Non-equilibrium Thermodynamics is discussed in Chapter 4. This hypothesis states that the Gibbs relation for the entropy is valid for small volume elements in the system of interest. Hence, the equations of state and transport coefficients obtained under non-equilibrium conditions should agree with the equilibrium ones. This idea has been successfully verified in NEMD,[15,74] despite the very large thermal gradients employed in BD-NEMD $\sim 10^{10}$ K · m^{-1}, thus showing the local response remains in the linear regime.[15] This fact is not so surprising. In the local-equilibrium state the gradients of the thermodynamic quantities should be small. For thermal transport the inequality, $|\nabla T(r)|\ T^{-1}(r)\ L \ll 1$, should be fulfilled,[76] where L is the mean free path. For dense fluids L is of the order of an atom diameter. Considering gradients of 10^{10} K · m^{-1}, $T \sim 10^2$ K and $L \sim 10^{-10}$ m it is clear that most NEMD simulations fulfil the inequality above, and therefore conform to the local-equilibrium state. This makes it possible to compute the equation of state or the thermal conductivity along an isobar by using a single BD-NEMD simulation (see Figure 6.5).[15,30,56]

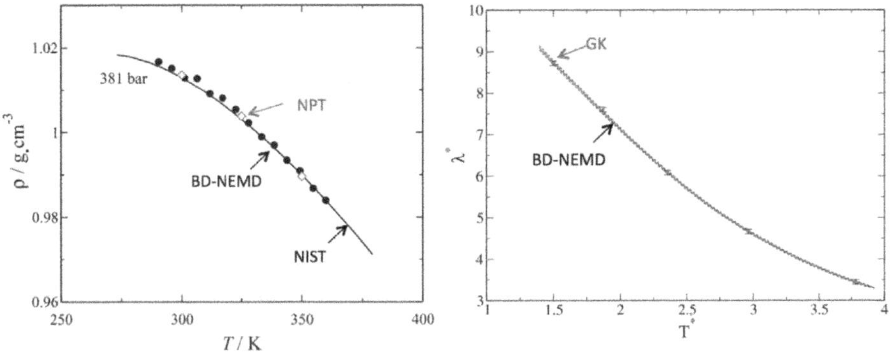

Figure 6.5 (Left) Equation of state of TIP4P/2005 water obtained from BD-NEMD (filled symbols), equilibrium simulations (diamonds)[56] and experimental data for water from the National Institute of Standards and Technology (NIST).[142] (Right) Local thermal conductivity of a supercritical fluid of diatomic molecules obtained from BD-NEMD (line) and GK equilibrium computations (symbols).[15] The temperature, T^*, and thermal conductivity, λ^* are given in Lennard-Jones reduced units.[16]

6.4.2 NEMD Computation of Transport Coefficients in Bulk Fluids and Interfaces

6.4.2.1 Viscosity

It was mentioned in the introduction how the low signal-to-noise ratio inherent in GK methods means very lengthy simulations are required to achieve well converged integrals for the Onsager coefficients. The shear viscosity coefficient has the added problem that the time scales of the fluctuations associated to its microscopic fluxes are very large, and so the correlation times that must to be considered can be much longer than those of other transport coefficients (*e.g.*, diffusion coefficient or thermal conductivity).

As one would expect, the first NEMD simulations of shear viscosity were performed for simple liquids. The shear viscosity of more complex molecular fluids has been calculated *via* both equilibrium and NEMD methods and compared to experiment over a wide range of thermodynamic conditions.[77–79] Initial discrepancies between NEMD and equilibrium methods near the triple point were related to the errors associated with the long-time tails in GK method.[80] Recently, the shear-viscosity calculations have been performed in a wide range of complex fluids; from relatively simple low-viscosity liquids such as water,[81–83] to complex liquids (nano-composite polymer melts).[84] The agreement between simulations and experiments is very satisfactory for fluids involving fairly different chemical compositions (see Figure 6.6-Top).

6.4.2.2 Thermal Conductivities

NEMD simulations have been applied to quantify the thermal conductivity (TC) of a wide range of fluids: atomic, molecular, polar and charged, as well

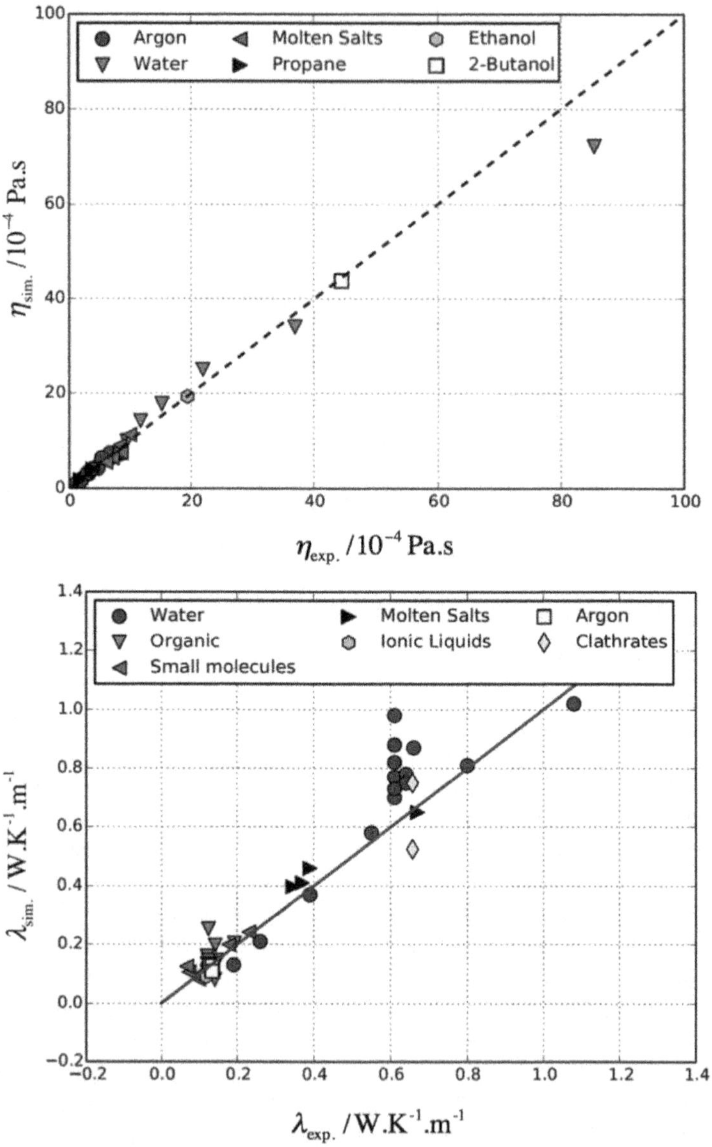

Figure 6.6 Correlation plots of the simulated (NEMD and GK computations) and experimental viscosities (Top) and thermal conductivities (Bottom), for simple and molecular fluids; molten salts (NaCl, KCl), small molecules (CO_2, HCl, CO), organic molecules (butane, hexane, cyclohexane, benzene, octane), room temperature ionic liquids (see text for details), water and water clathrates. Representative data taken from the references cited in the text are shown.

as fluid mixtures. Early computations focused on the hard-sphere (HS) model,[85] by using equilibrium simulations. The simulation results showed excellent agreement with the Enskog-theory predictions. These works highlighted the importance of collective effects in determining the decay of the autocorrelation functions. The thermal conductivity of soft-repulsive models has also been investigated by NEMD.[86] The Lennard-Jones potential has been the subject of intense investigation both in the liquid phase and near supercritical conditions.[6,46,51,74,75,87–93] Generally, the TC of simple liquids is well predicted with this model and the NEMD approach.

The computation of thermal conduction has been extended to charged fluids by using a variety of forcefields.[30,94,95] These studies are very difficult to perform experimentally given the high temperatures required in the molten salt regime. The equivalence of the NEMD and GK routes has been established[96] and the relevance of thermoelectric effects in determining the thermal conductivity has been highlighted.[97]

The simulation of molecular fluids, particularly organic molecules, is important in the oil industry. In addition to the investigation of small molecules (CO_2,[98,99] HCl,[28] CO[100]) the TC of organic molecules and their mixtures has been investigated *via* NEMD or GK methods.[48,53,101–105] Some of these works have highlighted the relevance of the rotational contribution to the thermal conductivity of molecular fluids.[98,106,107] Ionic liquids of different compositions have also been computed by using RNEMD. The force-fields tend to underestimate the thermal conductivities.[107]

The computation of the thermal conductivity of water deserves special attention. There have been a large number of investigations at standard conditions,[31,53,87,102,105,108–111] as well as at extreme pressures in the GPa range.[112] The TC of water features an anomalous increase with temperature. This behaviour was first reproduced in NEMD simulations of the MCFM water model,[109] and more recently in an extensive NEMD investigation of the SPC/E and TIP4P/2005 water models, which also confirmed the existence of maxima in the TC.[56] The TC is well reproduced in the liquid region but deviates significantly at near ambient conditions as well as low densities. The origin of this overestimation, which extends into the supercooled regime,[108] is still unclear. One important aspect in the computations of the TC of water is the treatment of the electrostatic interactions. It has been noted that the use of cut-off based techniques gives TCs $\sim$10 % lower than those obtained by using the Ewald method.[32]

The TC of aqueous solutions, containing acetone,[113] DMSO,[114] alkali halide solutions,[115] and clathrates[116,117] has also been considered. The simulation models account well for the TC decrease with increasing salt concentration, although the TCs are over predicted, a feature that is connected to the over prediction of the TC of pure water.

We show in Figure 6.6-Bottom a correlation plot that compiles simulation and experimental TCs. Current force fields predict well the TC of simple liquids, molten salts and small molecules. Deviations between simulation and experiments are observed in organic molecules, while current classical force fields significantly overestimate the TC of water at near ambient ($\lambda_{exp} = 0.61\ W \cdot K^{-1} \cdot m^{-1}$) and super-cooled (not shown)[108] conditions.

6.4.2.3 *Interfacial Transport*

NEMD simulations are playing an important role in quantifying the thermal conductance, *G*, of fluid–material interfaces, a problem of relevance in heat-management problems. The RNEMD method has been employed to compute the thermal conductance of hydrophobic–water interfaces,[102] finding in general good agreement with experiments.[118] Extension of these investigations to curved interfaces with nm curvatures has provided new physical insights. It was first shown in ref. 11 that *G* can decrease with interfacial curvature. This observation has been confirmed in TNEMD simulations of oxide nanoparticles[119] and protein–water interfaces[70] and by theoretical analyses.[120,121] BD-NEMD has also been employed to quantify the thermal resistance of nanoparticle–solvent interfaces,[54,122] and liquid–vapour interfaces, providing quantitative data to test the accuracy of the kinetic theory.[123]

NEMD approaches have also been used to investigate transport in nanopores[124] by combining simulations with NET. Studies on model nanopores range from simulations of model slit pores[125] to more complex systems, such as polymer-grafted[126] and charged nanopores.[127] NEMD techniques have been employed to investigate electric field-driven DNA translocation through nanopores,[128] and to quantify the efficiency of mass separation in thermophoretic devices.[45]

6.4.3 Coupling Phenomena

The existence of concurrent fluxes in a system, *e.g.*, heat and mass, leads to coupled phenomena. The Soret (thermodiffusion and thermophoresis) and Dufour and Peltier effects (charge transport) are well known, and are actively being investigated in energy-conversion problems. We discuss below simulation studies of coupling effects.

6.4.3.1 *Thermophoresis and Thermoelectric Coupling*

Multicomponent systems in the presence of a temperature gradient tend to exhibit cross coupling effects between the heat flux and the relative mass flux of the various components. This is the Ludwig–Soret effect. Mass/heat flux coupling has been extensively investigated for a wide range of systems. The first works focused on simple binary systems (Ar/Kr mixture).[73,74,129] The Soret coefficients were computed both *via* NEMD and GK methods. Good agreement was found with experimental data.[73,74,90,129] In a recent NEMD simulation, the impact of size and mass asymmetry on the TC was analysed systematically.[6] The study validated previous ideas that suggested that the cross coupling modifies the thermal conductivity by at most a few percent.[5]

Over the years there have been numerous simulations examining the key variables determining the Soret coefficient in simple binary mixtures. The Soret coefficient may feature a sign change as a result of changing thermodynamic conditions. This sign inversion has been observed both experimentally and in simulations. Simulations of simple Lennard-Jones systems indicate that the key variables influencing the Soret coefficient are the molar fraction, thermodynamic state, mass and size ratio of the two species.[6,7,130] These studies have highlighted the relevance of cross interactions in determining the sign inversion of the Soret effect too. Cross-coupling simulations have been extended to complex molecular mixtures.[48,107]

NET predicts the existence of coupling effects in charged fluids, such as molten salts.[5] It has been shown that Coulomb interactions inhibit the thermal separation of ions, although it is possible to polarize a charged fluid if the Coulombic interactions are sufficiently screened.[30] On the practical side, the equation for the measurable heat flux must be used (see Section 6.2.2), and cross coefficients must be considered in the computation of TCs. The coupling between the thermal gradient and the electric-charge fluctuations add non-negligible contributions to the TC. It has been found that neglecting coupling effects can lead to large overestimations of the TC of MgO and Mg_2SiO_4 molten salts.[97] Although such coupling effects seem to play a smaller role in other molten salts, such as NaCl.[96]

6.4.3.2 Thermal Orientation and Thermo-mechanical Coupling

Shortly after the discovery of liquid crystals, Lehman[131] found that the director of a cholesteric liquid crystal undergoes rotation under the application of a thermal gradient parallel to the cholesteric axis. This type of thermo-mechanical coupling effect has been rationalized by using flux–force linear phenomenological relations.[132] The coupling between the thermal gradient and the angular velocity of the liquid crystal director is determined by the Leslie coefficient, which has been quantified by using GK and NEMD simulations.[133,134] The coefficient is found to be two orders of magnitude smaller than the heat conductivity and the twist viscosity,[135] hence very long simulations are required to obtain accurate results.

Recently the response of water to a thermal gradient was investigated using BD-NEMD and NET.[136] It was found that the application of a thermal gradient could induce polarization in water. This physical effect, thermo-molecular orientation (TMO), can be rationalized by using NET. It arises from the coupling of polarization and heat fluxes,[136] which results in a polarization field ($\boldsymbol{E}$) that depends linearly the thermal gradient,

$$\boldsymbol{E} = \left(1 - \frac{1}{\varepsilon_r}\right)\frac{L_{pq}}{L_{pp}}\frac{\nabla T}{T}, \tag{6.24}$$

where ε_r is the relative permittivity.

It follows from this equation that the strength of the TMO effect is determined by the relative permittivity, the thermal gradient, the temperature and the ratio of coefficients, L_{pq}/L_{pp}, where the coefficient in the numerator controls the degree of coupling of the polarization and the heat flux, and the coefficient in the denominator is connected to the Debye relaxation time.[137]

NEMD simulations have played an instrumental role in determining the magnitude of this ratio. Fields of the order of (10^3 to 10^5) $V \cdot m^{-1}$ can be generated with large thermal gradients of the order of (10^6 to 10^8) $K \cdot m^{-1}$,[110,137–138] which are achievable using micro and nanoheaters.[139,140] The TMO effect has been observed in NEMD simulations of non-polar fluids consisting of anisotropic molecules, providing a link with the Soret effect in fluid mixtures.[100,141]

6.5 Conclusions

We have discussed in this chapter the diverse approaches available to perform NEMD simulations and the application of NEMD to quantify transport properties (diffusion, thermal conductivity and viscosity) of simple and complex fluids as well as their interfaces. NEMD provides in general a higher signal-to-noise ratio than GK and therefore enables the computation of coefficients with higher accuracy, especially in those cases where the correlation functions feature long-time tails and poor convergence. Otherwise, many simulations have shown the equivalence of NEMD and GK methods. However, the comparison of these two approaches requires careful consideration in multicomponent systems, since the existence of coupling effects might contribute to the transport properties such as the thermal conductivity (TC). While BD-NEMD methods provide direct access to the TC including the coupling effects, the GK approach requires the computation of several auto/cross correlation functions, which must be carefully defined. NET provides the necessary expressions to combine the different correlations and quantify the transport properties. Computer simulations of transport coefficients for a wide variety of fluids show in general good agreement with the available experiments, while significant deviations have been reported for the TC of water. The origin of these deviations, either forcefield, algorithmic or a combination of the two, is still unclear. Further work to improve our ability to compute the transport properties of polar and ionic fluids is needed.

Considerable effort has been devoted to study the consistency of the different NEMD approaches. The equivalence of homogeneous (synthetic) and BD-NEMD methods is now well established. Homogeneous approaches that use fictitious fields in "non-Newtonian" equations of motion avoid the use of walls, and, hence, are fully consistent with the use of periodic boundary-conditions. These methods provide a direct connection to linear response theory, and require previous knowledge of the correlation functions defining the transport property of interest. BD-NEMD methods on the other hand, emulate real experimental situations, the particle dynamics away from the boundaries is Newtonian, and they do not rely on the use of correlation functions. By construction, BD-NEMD approaches operate with explicit velocity, temperature and/or density gradients. These gradients can be analysed locally making it possible the computation of transport coefficients over a wide range of thermodynamic states within a single simulation, provided that the system response is in the linear regime. This represents a definite advantage of Boundary-Driven methods. We have further discussed Transient NEMD methods, which rely on a hybrid molecular-continuum approach, since the estimation of the transport coefficients is performed by fitting the simulated relaxation response to continuum diffusion equations. These methods have been shown to be suitable for the computation of interfacial properties (*e.g.*, interfacial conductance).

One important aspect of NEMD methods is the possibility of testing fundamental questions, such as the Onsager reciprocal relations or the local equilibrium hypothesis, which are two of the pillars of the NET theory. Simulations generally support the validity of these, and therefore NET. NEMD has also been employed to uncover novel physical effects, such as thermal orientation, a coupling phenomenon that can be studied in detail taking advantage of stationary NEMD simulations.

It is clear from our discussion that NEMD algorithms need to be specifically developed to quantify transport properties or investigate specific coupling phenomena. It is expected that NEMD will play an increasing role in the investigation of coupled effects, working alongside experiments to rationalize experimental results and uncover new non-equilibrium phenomena that can potentially be used for novel energy-conversion approaches.

Acknowledgements

We acknowledge support from EPSRC UK grant EP/J003859/1 and The Research Council of Norway (Project 221675). FB would like to acknowledge EPSRC for the award of a Leadership Fellowship.

References

1. W. G. Hoover and W. T. Ashurst, in *Theoretical Chemistry: Advances and Perspectives*, ed. H. Eyring and D. Henderson, Academic Press, New York, 1975, vol. 1 , p. 1.
2. D. J. Evans and G. P. Morris, *Statistical Mechanics of Non-equilibrium Liquids*, Academic Press, New York, 1990.
3. W. H. Hoover, *Ann. Rev. Phys. Chem.,*, 1983, **34**, 103.
4. D. A. McQuarrie, *Statistical Mechanics*, Harper & Row, New York, 1976.
5. S. R. de Groot and P. Mazur, *Non-Equilibrium Thermodynamics*, North-Holland, Amsterdam, 1962.
6. J. Armstrong and F. Bresme, *Phys. Chem. Chem. Phys.*, 2014, **16**, 12307.
7. D. Reith and F. Müller-Plathe, *J. Chem. Phys.*, 2000, **112**, 2436.
8. P. A. Artola and B. Rousseau, *Phys. Rev. Lett.*, 2007, **98**, 2.
9. B. Hafskjold, in *Thermal Nonequilibrium Phenomena in Fluid Mixtures*, Lecture Notes in Physics, ed. W. Köhler and S. Wiegand, Springer, 2001, p. 3.
10. M. Tesch and K. Schulten, *Chem. Phys. Lett.*, 1990, **169**, 97.
11. A. Lervik, F. Bresme and S. Kjelstrup, *Soft Matter*, 2009, **5**, 2407.
12. W. T. Ashurst and W. G. Hoover, *Phys. Rev. A: At., Mol., Opt. Phys.*, 1975, **11**, 658.

13. T. Ikeshoji and B. Hafskjold, *Mol. Phys.*, 1994, **81**, 251.
14. A. Tenenbaum, *Phys. Rev. A: At., Mol., Opt. Phys.*, 1983, **28**, 3132.
15. F. Bresme and J. Armstrong, *J. Chem. Phys.*, 2014, **140**, 016102.
16. M. P. Allen and D. J. Tildesley, *Computer Simulation of Liquids*, Clarendon Press, Oxford 1987.
17. M. E. Tuckerman, *Statistical Mechanics: Theory and Molecular Simulation*, Oxford University Press, Oxford, 2010.
18. D. Frenkel and B. Smit, *Understanding Molecular Simulation*, Academic Press, London, 2002.
19. B. J. Alder and T. E. Wainwright, *J. Chem. Phys.*, 1957, **27**, 1208.
20. M. González-Melchor, F. Bresme and J. Alejandre, *J. Chem. Phys.*, 2005, **122**, 104710.
21. M. González-Melchor, P. Orea, K. López-Lemus, F. Bresme and J. Alejandre, *J. Chem. Phys.*, 2005, **122**, 094503.
22. R. Zwanzig, *Non-Equilibrium Statistical Mechanics*, Oxford University Press, New York, 2001.
23. Y. Zhou and G. H. Miller, *J. Phys. Chem.*, 1996, **100**, 5516.
24. S. Viscardy, J. Servantie and P. Gaspard, *J. Chem. Phys.*, 2007, **126**, 184513.
25. R. Taylor and R. Krishna, *Multicomponent Mass Transfer*, Wiley, New York, 1993.
26. J. H. Irving and J. G. Kirkwood, *J. Chem. Phys.*, 1950, **18**, 817.
27. B. D. Todd, P. J. Davis and D. J. Evans, *Phys. Rev. E: Stat. Phys., Plasmas, Fluids, Relat. Interdiscip. Top.*, 1995, **51**, 4362.
28. D. J. Evans and S. Murad, *Mol. Phys.*, 1989, **68**, 1219.
29. R. Inoue, H. Tanaka and K. Nakanishi, *J. Chem. Phys.*, 1996, **104**, 9569.
30. F. Bresme, I. Wold and B. Hafskjold, *J. Phys. Chem.*, 1996, **100**, 1879.
31. T. W. Sirk, S. Moore and E. F. Brown, *J. Chem. Phys.*, 2013, **138**, 064505.
32. J. Muscatello and F. Bresme, *J. Chem. Phys.*, 2011, **135**, 234111.
33. C. Ludwig, Sitzungsber, *Akad. Wiss. Wien Math.-Naturwiss. Kl*, 1856, **20**, 1539.
34. C. Soret, *Arch. Génève*, 1879, **3**, 48.
35. P. Sindzingre, C. Massobrio, G. Ciccotti and D. Frenkel, Chem. Phys., 1989, **129**, 213.
36. B. Hafskjold and T. Ikeshoji, *Fluid Phase Equilib.*, 1995, **104**, 173.
37. J.-P. Hansen and I. R. McDonald, *Theory of Simple Liquids*, Academic Press, London, 2006.
38. W. G. Hoover, D. J. Evans, R. B. Hickman, A. J. Ladd, W. T. Ashurst and B. Moran, *Phys. Rev. Ser. A*, 1980, **22**, 1690.
39. D. J. Evans and G. P. Morris, *Comput. Phys. Rep.*, 1984, **1**, 300.
40. D. Evans, *Phys. Rev. A: At., Mol., Opt. Phys.*, 1981, **23**, 2622.
41. A. W. Lees and S. F. Edwards, *J. Phys. C: Solid State Phys.*, 1972, **5**, 1921.
42. D. J. Evans and H. J. M. Hanley, *Mol. Phys.*, 1989, **68**, 97.
43. D. P. Hansen and D. J. Evans, *Mol. Phys.*, 1994, **81**, 767.
44. J. M. Kincaid, X. Li and B. Hafskjold, *Fluid Phase Equilib.*, 1992, **76**, 113.
45. A. Lervik and F. Bresme, *Phys. Chem. Chem. Phys.*, 2014, **16**, 13279.

46. F. Müller-Plathe, *J. Chem. Phys.*, 1997, **106**, 6082.
47. W. Evans, J. Fish and P. Keblinski, *J. Chem. Phys.*, 2007, **126**, 154504.
48. A. Perronace, C. Leppla, F. Leroy, B. Rousseau and S. Wiegand, *J. Chem. Phys.*, 2002, **116**, 3718.
49. D. Lüsebrink and M. Ripoll, *J. Chem. Phys.*, 2012, **136**, 084106.
50. P. K. Schelling, S. R. Phillpot and P. Keblinski, *Phys. Rev. B*, 2002, **65**, 144306.
51. G. Galliero and C. Boned, *Phys. Rev. E: Stat., Nonlinear, Soft Matter Phys.*, 2009, **80**, 061202.
52. B. Hafskjold, T. Ikeshoji and S. Kjelstrup Ratkje, *Mol. Phys.*, 1993, **80**, 1389.
53. M. Zhang, E. Lussetti, L. E. S. de Souza and F. Müller-Plathe, *J. Phys. Chem. B*, 2005, **109**, 15060.
54. K. M. Stocker and J. D. Gezelter, *J. Chem. Theory Comput.*, 2014, **10**, 1878.
55. G. Bussi, D. Doandio and M. Parrinello, *J. Chem. Phys.*, 2007, **126**, 014101.
56. F. Römer, A. Lervik and F. Bresme, *J. Chem. Phys.*, 2012, **137**, 074503.
57. B. Hafskjold and S. K. Ratkje, *J. Stat. Phys.*, 1995, **78**, 463.
58. G. S. Heffelfinger and F. van Swol, *J. Chem. Phys.*, 1994, **10**, 7548.
59. G. Arya, H.-C. Chang and E. J. Maginn, *J. Chem. Phys.*, 2001, **115**, 8112.
60. S. Chempath, R. Krishna and R. Q. Snurr, *J. Phys. Chem. B*, 2004, **108**, 13481.
61. F. Müller-Plathe, *Phys. Rev. E: Stat. Phys., Plasmas, Fluids, Relat. Interdiscip. Top.*, 1999, **59**, 4894.
62. P. Bordat and F. Müller-Plathe, *J. Chem. Phys.*, 2002, **116**, 3362.
63. S. Butler and F. Müller-Plathe, *J. Mol. Liq.*, 2014, **192**, 114.
64. C. M. Tenney and E. J. Maginn, *J. Chem. Phys.*, 2010, **132**, 014103.
65. G. Galliero and S. Volz, *J. Chem. Phys.*, 2008, **128**, 064505.
66. E. Bringuier, *Philos. Mag.*, 2007, **87**, 873.
67. M. Yang and M. Ripoll, *J. Phys.: Condens. Matter*, 2012, **24**, 195101.
68. M. Belkin, S.-H. Chao, G. Giannetti and A. Aksimentiev, *J. Comput. Electron.*, 2014, **13**, 826.
69. M. Tesch and K. Schulten, *Chem. Phys. Lett.*, 1990, **169**, 97.
70. A. Lervik, F. Bresme, S. Kjelstrup, D. Bedeaux and J. M. Rubí, *Phys. Chem. Chem. Phys.*, 2010, **12**, 1610.
71. C. Melis, R. Dettori, S. Vandermeulen and L. Colombo, *Eur. Phys. J. B*, 2014, **87**, 96.
72. G. Arya, E. Maginn and H. C. Chang, *J. Chem. Phys.*, 2000, **113**, 2079.
73. D. MacGowan and D. J. Evans, *Phys. Rev. A: At., Mol., Opt. Phys.*, 1986, **34**, 2133.
74. G. V. Paolini and G. Ciccotti, *Phys. Rev. A: At., Mol., Opt. Phys.*, 1987, **35**, 5156.
75. N. A. Miller, P. J. Daivis, I. K. Snook and B. D. Todd, *J. Chem. Phys.*, 2013, **139**, 144504.
76. E. G. D. Cohen and R. L. Merlino, *J. Comput. Theory Trans.*, 2014, **43**, 3.
77. D. Heyes, *J. Chem Soc., Faraday Trans. 2*, 1983, **79**, 1741.

78. W. T. Ashurst and W. G. Hoover, *Phys. Rev. Lett.*, 1973, **31**, 4.
79. G. A. Fernandez, J. Vrabec and H. Hasse, *Fluid Phase Equilib.*, 2004, **221**, 157.
80. B. Holian and D. Evans, *J. Chem. Phys.*, 1983, **78**, 5147.
81. H. Zhang and J. F. Ely, *Fluid Phase Equilib.*, 2004, **217**, 111.
82. S. H. Lee, *Bull. Korean, Chem. Soc.*, 2014, **35**, 2.
83. N. Galamba and C. A. Nieto de Castro, *J. Chem. Phys.*, 2005, **122**, 224501.
84. P. T. Cummings and D. J. Evans, *Ind. Eng. Chem. Res.*, 1992, **31**, 1237.
85. B. J. Alder, D. M. Gass and T. E. Wainwright, *J. Chem. Phys.*, 1970, **53**, 3813.
86. D. M. Heyes, *J. Chem. Soc. Faraday Trans.*, 1984, **80**, 1363.
87. S. Kuang and J. D. Gezelter, *J. Chem. Phys.*, 2010, **133**, 164101.
88. S. Sarman and D. J. Evans, *Phys. Rev. A: At., Mol., Opt. Phys.*, 1992, **46**, 1960.
89. R. Vogelsang and C. Hoheisel, *Phys. Rev. A: At., Mol., Opt. Phys.*, 1988, **38**, 6296.
90. R. Vogelsang, C. Hoheisel, G. V. Paolini and G. Ciccotti, *Phys. Rev. A: At., Mol., Opt. Phys.*, 1987, **36**, 3964.
91. P. J. Gardner, D. M. Heyes and S. R. Preston, *Mol. Phys.*, 1991, **73**, 141.
92. R. Vogelsang and C. Hoheisel, *Phys. Rev. A: At., Mol., Opt. Phys.*, 1987, **35**, 3487.
93. R. Vogelsang, C. Hoheisel and G. Ciccotti, *J. Chem. Phys.*, 1987, **86**, 6371.
94. Y. Ishii, K. Sato, M. Salanne, P. A. Madden and N. Ohtori, *J. Phys. Chem. B*, 2014, **118**, 3385.
95. P. Sindzingre and M. J. Gillan, *J. Phys.: Condens. Matter*, 1990, **2**, 7033.
96. N. Galamba, C. A. N. de Castro and J. F. Ely, *J. Chem. Phys.*, 2007, **126**, 204511.
97. M. Salanne, D. Marrocchelli, C. Merlet, N. Ohtori and P. A. Madden, *J. Phys.: Condens. Matter*, 2011, **23**, 102101.
98. B. Y. Wang, P. T. Cummings and D. J. Evans, *Mol. Phys.*, 1992, **75**, 1345.
99. B. Y. Wang and P. T. Cummings, *Fluid Phase Equilib.*, 1989, **53**, 191.
100. F. Römer and F. Bresme, *Mol. Simul.*, 2012, **38**, 1198.
101. J. M. Simon, D. K. Dysthe, A. H. Fuchs and B. Rousseau, *Fluid Phase Equilib.*, 1998, **150**, 151.
102. H. Patel, S. Garde and P. Keblinski, *Nano Lett.*, 2005, **5**, 2225.
103. D. Bedrov, G. D. Smith and T. D. Sewell, *Chem. Phys. Lett.*, 2000, **324**, 64.
104. H. Luo and C. Hoheisel, *J. Chem. Phys.*, 1992, **96**, 3173.
105. D. Bedrov and G. D. Smith, *J. Chem. Phys.*, 2000, **113**, 8080.
106. H. Liu, E. Magin, A. E. Visser, N. J. Bridges and E. B. Fox, *Ind. Eng. Chem. Res.*, 2012, **51**, 7242.
107. H. M. Schaink, H. Luo and C. Hoheisel, *J. Chem. Phys.*, 1993, **99**, 9912.
108. F. Bresme, J. W. Biddle, J. V. Sengers and M. A. Anisimov, *J. Chem. Phys.*, 2014, **140**, 161104.
109. F. Bresme, *J. Chem. Phys.*, 2001, **115**, 7564.
110. J. Muscatello, F. Römer, J. Sala and F. Bresme, *Phys. Chem. Chem. Phys.*, 2011, **13**, 19970.

111. P. Kumar and H. E. Stanley, *J. Phys. Chem. B*, 2011, **115**, 14269.
112. F. Bresme and F. Römer, *J. Mol. Liq.*, 2013, **185**, 1.
113. C. Nieto-Draghi and J. Bonet-Avalos, *Mol. Phys.*, 2003, **101**, 2303.
114. C. Nieto-Draghi, J. Bonet-Avalos and B. Rousseau, *J. Chem. Phys.*, 2003, **119**, 4782.
115. F. Römer, Z. Wang, S. Wiegand and F. Bresme, *J. Phys. Chem. B*, 2013, **117**, 8209.
116. R. Inoue, H. Tanaka and K. Nakanishi, *J. Chem. Phys.*, 1996, **104**, 9569.
117. E. J. Rosenbaum, N. J. English, J. K. Johnson, D. W. Shaw and R. P. Warzinski, *J. Phys. Chem. B*, 2007, **111**, 13194.
118. Z. Ge, D. G. Cahill and P. V. Braun, *Phys. Rev. Lett.*, 2006, **96**, 186101.
119. M. Hu, D. Poulikakos, C. P. Grigoropoulos and H. Pan, *J. Chem. Phys.*, 2010, **132**, 164504.
120. K. S. Glabatskiy and D. Bedeaux, *J. Chem. Phys.*, 2014, **140**, 104708.
121. Ø. Wilhemsen, D. Bedeaux and S. Kjelstrup, *Phys. Chem. Chem. Phys.*, 2014, **16**, 10573.
122. S. Merabia, S. Shenogin, L. Joly, P. Keblinski and J. L. Barrat, *Proc. Natl. Acad. Sci. U. S. A.*, 2009, **106**, 15113.
123. S. Kjelstrup and D. Bedeaux, *Non-Equilibrium Thermodynamics of Heterogeneous Systems*, World Scientific, Singapore, 2008.
124. I. Inzoli, S. Kjelstrup, B. Bedeaux and J. M. Simon, *Microporous and Mesoporous Mater.*, 2009, **125**, 112.
125. K. Travis and K. Gubbins, *Langmuir*, 1999, **15**, 6050.
126. S. Adiga and D. Brenner, *Nano Lett.*, 2005, **5**, 2509.
127. A. P. Thompson, *J. Chem. Phys.*, 2003, **119**, 7503.
128. C. Sathe, X. Zou, J.-P. Leburton and K. Schulten, *ACS Nano*, 2011, **5**, 8842.
129. D. J. Evans and D. MacGowan, *Phys. Rev. A: At., Mol., Opt. Phys.*, 1987, **36**, 948.
130. P. A. Artola and B. Rousseau, *Mol. Phys.*, 2013, **111**, 22.
131. O. Lehmann, *Ann. Phys.*, 1900, **2**, 649.
132. F. M. Leslie, *Symp. Faraday Soc.*, 1971, **5**, 33.
133. S. Sarman and A. Laaksonen, *Phys. Chem. Chem. Phys.*, 2013, **15**, 3442.
134. S. Sarman, *J. Chem. Phys.*, 1994, **101**, 480.
135. S. Chandrasekhar, *Liquid Crystals*, Cambridge University Press, Cambridge, 1994.
136. F. Bresme, A. Lervik, D. Bedeaux and S. Kjelstrup, *Phys. Rev. Lett.*, 2008, **101**, 020602.
137. J. Armstrong, A. Lervik and F. Bresme, *J. Phys. Chem. B*, 2013, **117**, 14817.
138. J. Armstrong and F. Bresme, *J. Chem. Phys.*, 2013, **139**, 014504.
139. S. Duhr and D. Braun, *Proc. Natl. Acad. Sci. U. S. A.*, 2006, **103**, 19678.
140. A. Govorov, W. Zhang, T. Skeini, H. Richardson, J. Lee and N. Kotov, *Nanoscale Res. Lett.*, 2006, **1**, 84.
141. F. Römer, F. Bresme, J. Muscatello, D. Bedeaux and J. M. Rubí, *Phys. Rev. Lett.*, 2012, **108**, 105901.
142. NIST Chemistry Webbook, http://webbook.nist.gov/chemistry/.

CHAPTER 7

Non-equilibrium Molecular Self-assembly

GER KOPER* AND WOUTER HENDRIKSEN

Department of Chemical Engineering, Delft University of Technology, The Netherlands
*Email: G.J.M.Koper@tudelft.nl

7.1 Introduction

Structure formation in fluids is often too complex to allow for a full-fledged thermodynamic description. The classical approach to solution thermodynamics would be to capture such deviations from ideality in terms of activity coefficients or osmotic virial coefficients.[1] These coefficients in principle can be rationalized by considering the molecular interactions between the solute as modulated by the solvent as is done by the McMillan–Mayer theory.[2] Strong non-idealities leading to large values for these coefficients are usually the result of more interesting chemical dissociation and association processes. For instance, activities in the case of ionic dissociation are almost completely described by what is known as the Van't Hoff factor and the degree of dissociation, and to a minor extent by the electrostatic interactions between the ions.[3] Similarly, this is true for the association processes. This chapter will focus on the weak processes that give rise to aggregation while the molecules keep their identity and interact, for instance through Van der Waals forces or hydrogen bonding. Strong processes such as polymerization involving covalent bonds are not considered here.

Experimental Thermodynamics Volume X: Non-equilibrium Thermodynamics with Applications
Edited by Dick Bedeaux, Signe Kjelstrup and Jan V. Sengers

Published by the Royal Society of Chemistry, www.rsc.org

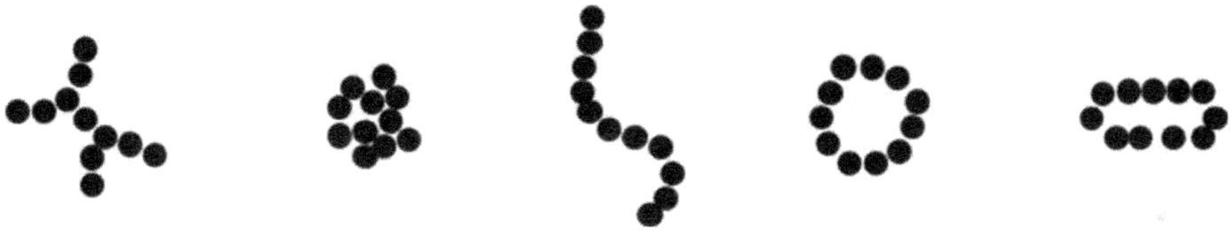

Figure 7.1 Various aggregate structures, from left to right: ramified, compact, linear, spherical micellar and wormlike micellar.

Weak association occurs in almost all mixtures, even in so-called simple solutions. In actual fact, weak association is often the precursor of phase separation which becomes more prominent the closer the mixture approaches the limiting temperature or concentration (pressure is generally not a relevant variable). It usually gives rise to "soft" clusters of solute molecules, as can be detected by light-scattering techniques,[4] or crystals when a solid phase is near. The structure of the clusters highly depends on the nature of the solute molecules as well as on the solvent, and can either be ramified or compact with ill-defined aggregation numbers, or they can exhibit explicit structures such as spherical micelles or threadlike micelles, see Figure 7.1. Linear aggregates are the typical intermediates between a random structure yet with a clear organization. The formation of relatively well-defined clusters is what is presently called *molecular self-assembly*, the creation of soft matter in complex liquids.[5]

Even if self-assembly itself is a kinetic process, and hence involves non-equilibrium processes, it is generally believed that there are striking differences depending on whether self-assembly occurs under equilibrium or non-equilibrium conditions. The spontaneously formed morphologies for equilibrium self-assembly show different dynamics and response than those for non-equilibrium self-assembly, sustained through the continuous pumping of energy into the system. An example of the first kind can be found in the use of chemically designed systems to provide specific functionalities that allow for a rich variety of thermodynamic phases as a route to tune their self-assembly and material properties.[6,7] An example of the second kind from Nature is the microtubule network, which is found in all eukaryotic cells that are involved in mitosis, cell motility, intracellular transport, and the maintenance of cell shape. This network is a non-covalent polymer network and assembles upon consumption of guanosine triphosphate (GTP), but depolymerises fast when the bound guanosine diphosphate (GDP) is lost. Inspired by processes in Nature, such as transient self-assembly, dynamic instability and ability to reconfigure, new systems have been developed.[8–10] The self-assembling structures are generated through chemical reactions by supply of energy to these uphill conversions, determining the temporal properties of these structures. Alternatively, driving fields, such as light, can be used to convert energy into molecular motion.[11]

The emergence of new, so-called "dissipative structures", can be understood theoretically and can be traced back to the pioneering work of Prigogine on dissipative self-sustained structures, see for instance the book by Nicolis and Prigogine.[12] However, despite the work carried out relating these structures to dynamic systems, chaos theory, non-equilibrium phase transitions, *etc.*, a fundamental, general theoretical framework to understand non-equilibrium molecular self-assembly is still missing.

In equilibrium self-assembly, the structures and morphologies that can be obtained are constrained by the underlying thermodynamics of the system. Therefore, the possibility to design or tailor particular structures, relies essentially on the proper design of the system. One needs to modify the physical constituents to obtain desired morphologies. On the other hand, in non-equilibrium self-assembly, the external input might tune the properties of the emerging, self-assembled domains.[13] As a result, this scenario in principle offers more flexibility to generate different kinds of structures. Moreover, these structures, sustained in steady state by the external supply of energy, can exhibit adaptive behaviour that offers the possibility to develop smart, highly responsive materials.

In general, the self-assembled systems out of equilibrium have stronger adaptability than their equilibrium counterparts, because they can change in response to modifications of the applied external fields that sustain them. Under certain conditions they can self-heal, such as when an external force returns to its initial state, the corresponding self-assembled structures react by returning to their previous configurations. However, since the response is dynamic, different scenarios can be explored, providing further malleability to self-assembly. In some cases, the driving force could lead to two or more competing structures. Under those conditions, if variations with respect to a given driving force are small, the system will relax to the same type of structures, while under a large additional forcing it can lead the system to accommodate the second, allowed configuration. This will be discussed in Section 7.4. The energy consumption of these dissipative structures also provides a natural connection with the ability of living systems to self-replicate. However, the control of self-assembly to produce structures that self-replicate requires careful tuning and control of ambient conditions.[14]

In this chapter we will discuss the fundamentals of molecular self-assembly in Section 7.2, and explain how this corresponds to out-of-equilibrium fuelled self-assembly in Section 7.3. The important effect of bi-stability and autocatalysis are discussed and related to the dynamic behaviour for natural out-of-equilibrium systems in Sections 7.4 and 7.5. Finally, the role of non-equilibrium thermodynamics is discussed in relation to self-assembly and future applications of the presented approach in Section 7.6.

7.2 General Principles

7.2.1 Equilibrium

From a formal point of view, molecular self-assembly can be regarded as an infinite chain of successive association equilibria as

$$\begin{cases} S_1 + S_1 \rightleftarrows S_2 \\ S_2 + S_1 \rightleftarrows S_3 \\ \vdots \\ S_{n-1} + S_1 \rightleftarrows S_n \\ S_n + S_1 \rightleftarrows S_{n+1} \\ \vdots \end{cases}, \tag{7.1}$$

where S_n denotes an aggregate of n species. The set of equations implicitly contains the more general equilibria $S_n + S_m \rightleftarrows S_{n+m}$ $(n, m = 1, 2, \ldots)$. When all these species are in chemical equilibrium, their chemical potentials (per molecule) are identical and for sufficiently low mole fractions one may write

$$\mu_n = \mu_n^{\ominus} + \frac{k_B T}{n} \ln \frac{x_n}{n} = \mu_1 \quad (n = 1, 2, \ldots), \tag{7.2}$$

in which k_B is Boltzmann's constant and T is temperature. The mole fractions x_n for molecules in aggregate S_n are constrained by the total amount of species x_t as

$$\sum_{n=1}^{\infty} x_n = x_t. \tag{7.3}$$

Once the standard chemical potentials $\mu_n^{\ominus}$ are specified, the set of eqn (7.1) to (7.3) can be solved, albeit most of the time only numerically. For instance, for linear aggregates with $n-1$ bonds in a chain of n molecules, the standard chemical potentials are defined as

$$\mu_n^{\ominus} = \mu_1^{\ominus} + \left(1 - \frac{1}{n}\right) B \quad (n = 1, 2, \ldots), \tag{7.4}$$

with B the bond Gibbs energy.[15] For many common aggregates, such as sphere-like micelles, wormlike micelles and vesicles, approximate forms for the chemical potentials have been derived by Nagarajan and Ruckenstein.[16,17]

Many types of aggregates, in particular the globular ones such as micelles, have a rather well defined aggregation number N, typical values are 50 to 100, so that for these structures the chain of equilibria in eqn (7.1) reduces to

$$N S_1 \overset{K}{\rightleftarrows} S_N, \tag{7.5}$$

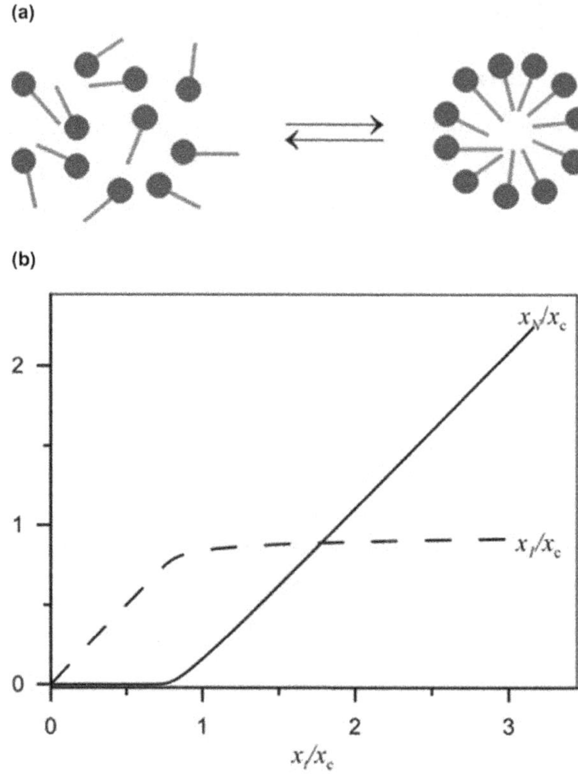

Figure 7.2 Schematic representation of self-assembly of monomers into aggregates (a) graphical example of monomer x_1, and (b) aggregate mole fraction x_N as a function of total mole fraction x_t; x_c is the critical mole fraction of the self-assembly process, $N=30$.

see Figure 7.2a. The behaviour of this model is determined by two parameters only, the aggregation number, N, and the equilibrium constant,† K, that describe the equilibrium, eqn (7.5), between monomers and aggregates as

$$x_N = N(Kx_1)^N. \tag{7.6}$$

Together with mass conservation, this model can be numerically solved, see Figure 7.2b for an example.[15]

As long as the total mole fraction x_t remains below the critical aggregation concentration (cac), no aggregates are formed and the monomer mole fraction follows the total mole fraction. Beyond this cac, the monomer mole fraction remains virtually constant at the cac value and the remainder of the total mole fraction of monomers is aggregated. The cac is related to the

†The conventional definition of association equilibrium, adopted here, deviates from the IUPAC convention. We follow literature convention[15] here to avoid confusion.

chemical potential of the free monomers and of the monomers in aggregates by

$$x_c = K^{-N/(N-1)} \sim K^{-1} \text{ for } N \gg 1, \tag{7.7a}$$

where the equilibrium constant is given by

$$K = \exp\{-(\mu_N^{\ominus} - \mu_1^{\ominus})/(k_B T)\} \tag{7.7b}$$

in terms of the standard chemical potentials introduced by eqn (7.2).

7.2.2 Gibbs Energy of Formation

In the hypothetical initial, non-self-assembled state, the chemical potential of the monomers (considered dilute) is given by

$$\mu_t = \mu_1^{\ominus} + k_B T \ln x_t. \tag{7.8}$$

Hence, the molar Gibbs energy of formation[1] $\Delta_f G$ of the self-assembled state with respect to the non-self-assembled state (the methodology to actually evaluate the Gibbs energies of formation is relatively unknown, see the discussion by Gerhartl[18] for an example) is given by

$$\Delta_f G = x_1\mu_1 + x_N\mu_N - x_t\mu_t - k_B T \ln\left\{1 - x_N\left(1 - \frac{1}{N}\right)\right\}. \tag{7.9}$$

Here the last term accounts for the change in the solvent entropy due to the change in mole fractions of monomers and aggregates; when considered at low monomer concentrations, it can be absorbed in the aggregate chemical potential. In Figure 7.3 the Gibbs energy of formation is plotted *versus* the conversion of monomers for a specific value of the total monomer mole fraction x_t. The conversion ξ is defined such that $x_1 = x_t(1 - \xi)$ and $x_N = x_t\xi$, and as such gives the fraction of monomers that have been converted into aggregates.

7.2.3 Gibbs Energy of Reaction, the Thermodynamic Driving Force

The Gibbs energy of reaction $\Delta_r G$ is the slope of the Gibbs energy of formation *versus* conversion, see Figure 7.3.

$$\Delta_r G = N(\mu_N - \mu_1) = k_B T \ln\left(\frac{x_N}{N(Kx_1)^N}\right). \tag{7.10}$$

It vanishes at the point ξ_{eq} where equilibrium is reached, see Figure 7.4. In addition, it provides for the driving force towards equilibrium when the system is away from it. We have expressed the Gibbs energy of reaction per aggregate rather than per monomer.

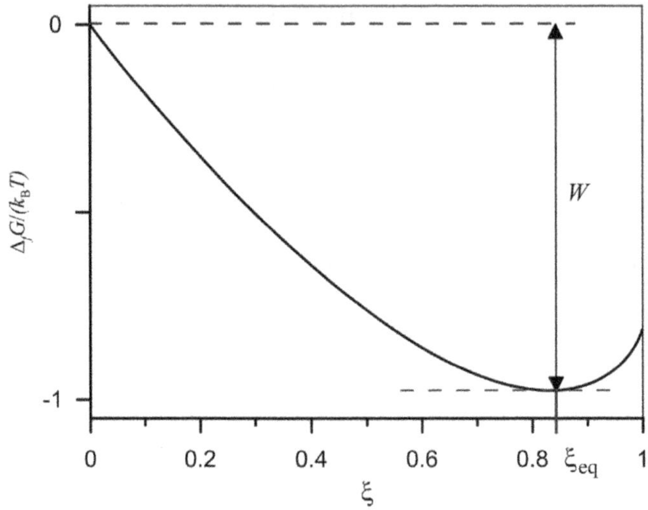

Figure 7.3 Graphical example of Gibbs energy of formation for the self-assembled state as a function of the conversion ξ. The equilibrium conversion ξ_{eq} where the slope, the reaction Gibbs energy, vanishes is indicated. The formation Gibbs energy difference between the between the initial state and the equilibrium state is the maximum work W that is available by the self-assembly process.

7.2.4 Conversion Rate

The set of kinetic equations belonging to the chain of successive equilibria, given by eqn (7.1), forms a set of Becker–Döring equations[19] as have been formulated by Aniansson and Wall[20]

$$\left.\begin{array}{c} S_1 + S_1 \underset{\tilde{k}'_2}{\overset{\tilde{k}_2}{\rightleftharpoons}} S_2 \\ S_2 + S_1 \underset{\tilde{k}'_3}{\overset{\tilde{k}_3}{\rightleftharpoons}} S_3 \\ \vdots \\ S_{n-1} + S_1 \underset{\tilde{k}'_n}{\overset{\tilde{k}_n}{\rightleftharpoons}} S_n \\ \vdots \end{array}\right\} \Rightarrow \left\{\begin{array}{l} \dfrac{\mathrm{d}}{\mathrm{d}t}x_1 = -2\tilde{k}_2 x_1^2 + 2\tilde{k}'_2 x_2 \\ \qquad\qquad + \displaystyle\sum_{m=3}^{\infty} \left(-\tilde{k}_m x_{m-1} x_1 + \tilde{k}'_m x_m\right) \\ \qquad\qquad\qquad \vdots \\ \dfrac{\mathrm{d}}{\mathrm{d}t}x_n = \tilde{k}_n x_1 x_{n-1} - \tilde{k}'_n x_n \\ \qquad\qquad - \tilde{k}_{n+1} x_n x_1 + \tilde{k}'_{n+1} x_{n+1} \\ \qquad\qquad\qquad \vdots \end{array}\right. , \tag{7.11}$$

with for each aggregate size $n = 2, 3, \ldots$ forward and backward rate coefficients $\tilde{k}_n$ and $\tilde{k}'_n$ respectively; t represents time. From experimental observations, as well as from an analysis of the equilibrium-size distribution,[21] it is known that there is a significant amount of monomer still present and coexisting with a large amount of aggregated material. Typically, it is quite sharply peaked around the aggregation number, N, as discussed above. There is little material in aggregated form with aggregation numbers between about 5 and N or much larger than N. Thus, the range of aggregate sizes for which significant amounts of monomer are present divides naturally into two regions: firstly a region of small aggregates, dominated by the monomer form, but including a small number of dimers, trimers, and so on; and secondly a region centred on the most probable aggregate size, at which the size distribution has its maximum. These two regions are separated by a region of extremely low concentration of aggregates. The kinetics determining how such a system approaches equilibrium itself falls into two stages. Initially, the matter in each of these two regions self-equilibrates and then, over a much longer timescale, molecules are transferred from one region to the other, until the two regions eventually reach global equilibrium. This latter process has been modelled by a contraction of the full Becker–Döring scheme, eqn (7.11), into the simple rate equation[22] of monomers into aggregates that for the present situation reads

$$r \equiv r_1 - r_N = k_1 x_1^N - k_N \frac{x_N}{N}, \tag{7.12}$$

with forward and backward rate constants k_1 and k_N, respectively (rate constants differ from those in eqn (7.11) because of the contraction procedure). By solving eqn (7.12) for stationarity, one easily verifies that the rate constants satisfy

$$\frac{k_1}{k_N} = K^N. \tag{7.13}$$

With the definitions of the chemical potentials in eqn (7.2), the conversion rate defined by eqn (7.12) can be written in terms of the driving force given by eqn (7.10) as

$$r = k_1 x_1^N \left[1 - \exp\left\{-\frac{N(\mu_N - \mu_1)}{k_\mathrm{B} T}\right\}\right] = k_1 x_1^N \left[1 - \exp\left\{-\frac{\Delta_\mathrm{r} G}{k_\mathrm{B} T}\right\}\right]. \tag{7.14}$$

The final exponential form of reaction rate in terms of the driving force is what is found typically for chemical reactions and can be found in many text books.[23,24] It has been derived from Mesoscopic Non-equilibrium Thermodynamics (MNET),[25] see also Chapter 14.

In Figure 7.4 the relationship of the formation rate as a function of the driving thermodynamic force is presented. It exhibits – as expected – a striking nonlinearity not unlike that of an electrical diode or electrode.

7.2.5 Work: Available and Lost

The available work from the self-assembly reaction is given by the Gibbs energy of reaction, see also Figure 7.3. The rate at which the conversion takes place determines the rate at which this work can be delivered, so that

$$\dot{W} = -r \Delta_\mathrm{r} G. \tag{7.15}$$

Under conditions where there is no coupling, this work is lost to the environment as there are no means to use it. It then is dissipated in the form of

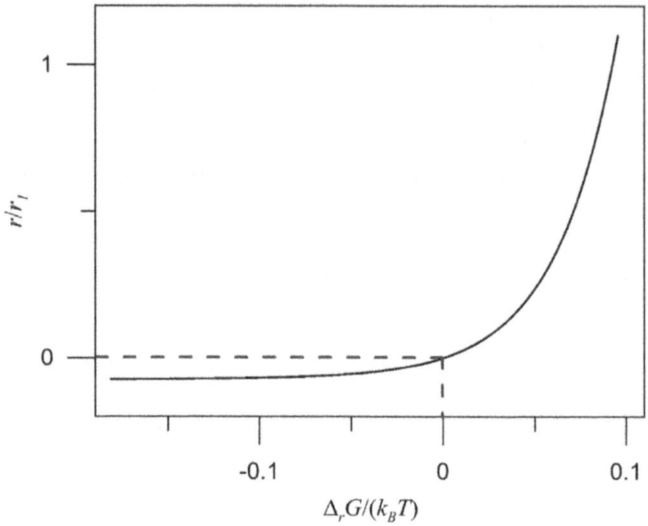

Figure 7.4 Graphical example of the relationship between formation rate and thermodynamic force for self-assembling systems.

heat. With a coupled electrochemical reaction, however, (some of) this work can be useful, see Chapter 12.

As mentioned above, see eqn (7.14), the theoretical framework provided by MNET, see Chapter 14, has been used to describe the dynamics of self-assembly[25] and to identify the relevant entropy production; it can be used, for example, to understand the energetics associated to the self-assembly process under generic non-equilibrium conditions. The change in entropy during the self-assembly process is closely linked to the reduction in phase space.[26] Therefore, there is a clear connection between MNET and the complementary theoretical treatments that focus on quantifying the rate at which this configuration space decreases during self-assembly. The latter does not identify the relevant fluxes and forces controlling self-assembly, while the expression for the entropy production provided[25] by MNET can be used to quantify such shrinkage, hence connecting more clearly the intrinsic dissipative process with the theory of dynamical systems. Moreover, MNET does not only allow a proper understanding of self-assembly kinetics; its thermodynamic basis provides also a natural basis to describe consistently thermal fluctuations and correlations in such kinetic processes.[27] It is also required to understand coupling with other processes, such as an electrochemical process that uses the work produced by the self-assembly or thermal driving forces as we shall elaborate more on in Section 7.6.

7.3 Fuelled Self-assembly

The simple self-assembly process described in the previous section has only one stationary state, which is the equilibrium state. Its behaviour is fully described in terms of the "equilibrium constant" K that is related to the rate coefficients defined by eqn (7.13) and the aggregation number N. Driving such a system out of equilibrium by a constant supply of monomers might reveal more interesting dynamics. The aim is to reach a steady state other than the equilibrium state and, hence, monomers have to be taken out of the system as well.

In Figure 7.5 an out-of-equilibrium self-assembly experiment is sketched, where at a given constant feed rate, r_t, monomers are formed from pre-monomers, *i.e.* parts that form a monomer upon association. Subsequently, there are two decay channels from monomers back to pre-monomers: directly, with rate r_{L1}, and indirectly through the self-assembly step with rate r_{LN}. The direct rate depends on the population of the monomer state x_1 and is given by

$$r_{L1} = k' x_1, \tag{7.16}$$

whereas the indirect rate depends on the population of the self-assembled state x_N as

$$r_{LN} = k x_N \tag{7.17}$$

with rate constants k and k'.

The indirect channel is coupled to the self-assembly process, for which the rate is given by eqn (7.12). The continuous formation of new monomers drives the self-assembly out of equilibrium. The monomer formation reaction and the subsequent destruction may be assumed to be controlled by external reactions, hence the name fuelled self-assembly. At low feed rates,

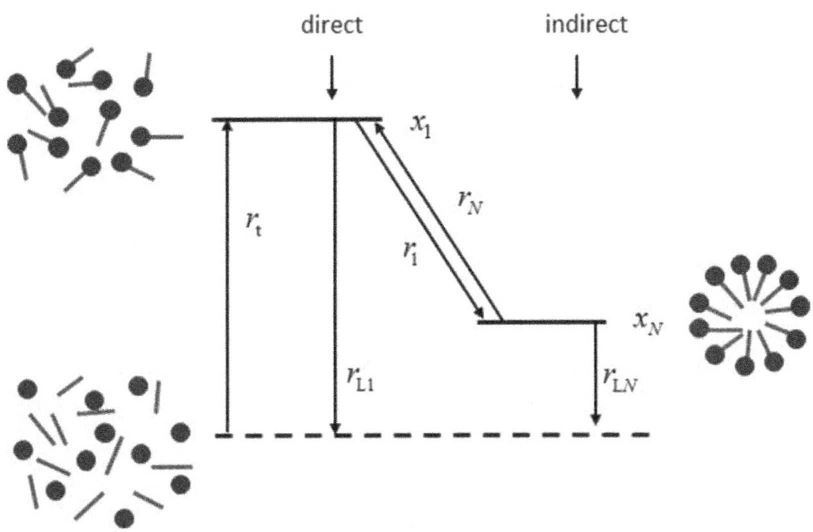

Figure 7.5 Schematic representation of a dual channel out-of-equilibrium self-assembly process. See the text for further details.

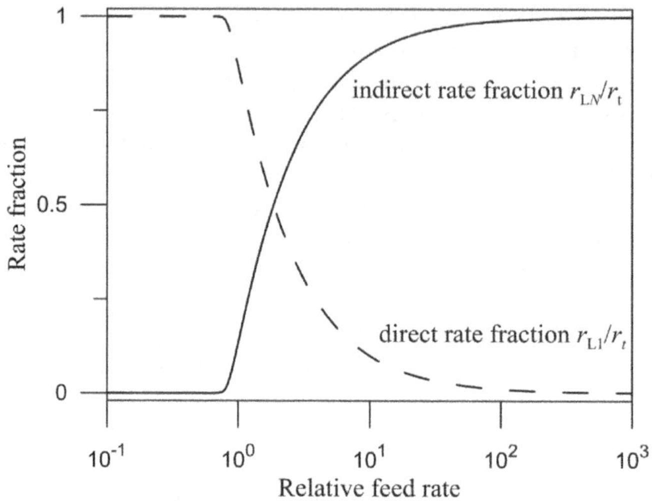

Figure 7.6 Direct and indirect rate fractions in the dual channel dissipative self-assembly process as a function of the relative feed rate $r_t/(k_1 x_c^N)$.

the monomer mole fraction x_1 remains low, below the cac, and, hence, there is no population of the self-assembled state. Therefore, the decay rate is solely *via* the direct channel as visualized in Figure 7.6. For larger feed rates, the self-assembled state is fully occupied and only a minority of the monomers – with mole fraction close to the cac – are still in the free monomer state. As a consequence, the larger part of the decay is through the indirect channel and the fraction that follows the direct channel is negligible, see Figure 7.6. The actual cross-over takes place around the critical aggregation concentration and is controlled by the relative rate factor

$$K_r = r_{L1}^{(c)} \left(\frac{1}{r_{LN}^{(c)}} + \frac{1}{r_1^{(c)}} \right) \tag{7.18}$$

where $r_{L1}^{(c)}$ is the direct rate, see eqn (7.16), $r_{LN}^{(c)}$ the indirect rate, eqn (7.17), and $r_1^{(c)}$ the forward aggregation rate, all at the cac as indicated by the superscript (c).

The result, as sketched in Figure 7.6, albeit reminiscent of the equilibrium behaviour illustrated by Figures 7.3 and 7.4, is only achieved when the system is out of equilibrium. The true equilibrium state of the model is where there are only pre-monomers and no monomers or aggregates at all. This state is attained at vanishing feed rates. With finite feed rates, a non-equilibrium stationary state sets in, of which the characteristics are described above. Importantly, the chemical potentials have values that differ significantly from those in equilibrium. In actual fact, work is delivered at a rate exactly given by eqn (7.15) with a non-vanishing Gibbs energy of reaction; it will vanish in equilibrium only.

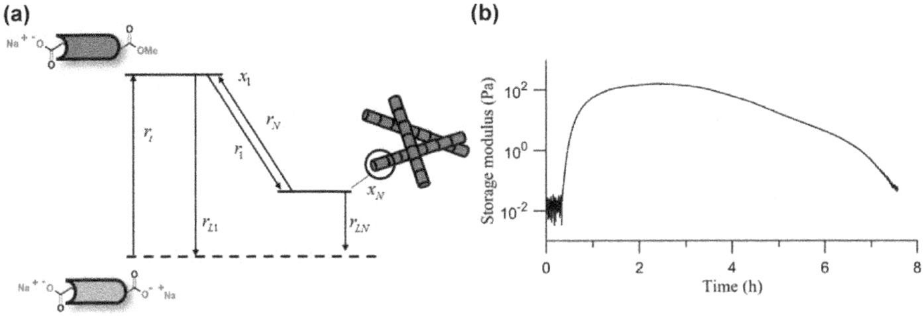

Figure 7.7 (a) Chemically fuelled transient self-assembly. In a reaction cycle, pre-monomers DBC react with methylating agent DMS to give monomers that subsequently self-assemble into fibrous aggregates. The formed monomers can hydrolyse both in the assembled and free state to revert to the original pre-monomer. Overall, one full cycle produces methanol and monomethylsulfate as waste products. (b) Elastic behaviour, as measured by variation of the storage modulus, of the reaction fluid that clearly indicates the development of a gel upon aggregation of monomers into fibres.

Recently, we have shown the out-of-equilibrium formation of active materials[8,28] that exhibits a behaviour that is very reminiscent of what is described above, see Figure 7.7. In this system, the pre-monomer is a low molecular weight gelator *N,N′*-dibenzoyl-(L)-cystine (DBC) containing two carboxylate groups per molecule that reacts with alkylating fuels to form a neutral ester. The monomer is formed by removing the charge of the carboxylate, by forming the ester, and hence allows for the molecular self-assembly process and eventually gelation. The leakage reaction from monomer back to pre-monomer takes place when the ester undergoes spontaneous hydrolysis in aqueous environments leading to the formation of a charged carboxylate and an alcohol waste product. The hydrolysis of the ester group is dependent on the pH, and occurs faster at more basic conditions. During a reaction cycle, the pre-monomer reacts with the commercially available strong methylating agent dimethylsulfate (DMS, $(CH_3)_2SO_4$) under basic conditions. With the batch-wise addition of DMS to buffered solutions, the pre-monomer results in its transient methylation, yielding increasing monomer levels.

The monomers self-assemble into fibres with a monodisperse 8 nm diameter and multiple micrometres in length. Typically, 20 minutes after addition of the fuel, the monomer concentration will reach the Critical Gelation Concentration (CGC), where the formation of fibres leads to macroscopic, semi-transparent gel materials. Over time, the monomers hydrolyse back to pre-monomers, leading to the breakdown of fibres and the dissolution of these gels. The viscoelastic behaviour of the gel as measured during the reaction cycle, see Figure 7.7b, reveals dynamics that are reminiscent of what is displayed in Figure 7.6. Importantly, in the experiment the decay rate from the self-assembled state to the pre-monomer state is much less than what results from the model described above. This is largely due to the fact that in linear aggregates the terminating monomers are less tightly bound to the aggregates than those in between.

7.4 Bi-stability

Micellar systems that catalyse their own formation have attracted considerable scientific interest.[22,29–32] In particular, the example reported by Bachmann, Luisi, and Lang in 1992[29] attracted quite some attention largely because of their hypothesis of "autopoietic self-replication" of micelles and their discussion with respect to the basic chemical mechanisms at the origin of Life.[29] It was claimed that the biphasic alkaline hydrolysis of an ester shows highly nonlinear kinetics that could be attributed to the effect of micellar autocatalysis. In this reaction, ethyl caprylate (EC), which is practically immiscible with water, undergoes alkaline hydrolysis when placed in contact with an underlying aqueous solution. The reaction yields amphiphilic sodium caprylate, which is known to form anionic micelles in aqueous media. Many attempts have been made to model this reaction until in 1997 Buhse *et al.* revealed that the behaviour was not so much due to autocatalysis, but rather arose from a transport phenomenon of EC-swollen micelles to the aqueous phase which sped up the hydrolysis reaction significantly.[32]

Nevertheless, there are many reports in the literature where bistable behaviour is claimed for associating molecular systems, see for instance the short overview in the paper by Ball and Haymet.[30] A relatively clear-cut experiment was reported by these authors[30] in which the charge-transfer interactions of drug molecules were studied. The conductivity of an aqueous iodine solution was measured while a solution of the amphiphilic drug amiodarone was added. In the reverse experiment, the conductivity of the amiodarone solution was followed as iodine solution was added. The conductivity curves provide clear evidence of bi-stability: the forward and reverse curves do not coincide. Similar results were obtained for the alkaline hydrolysis of C-4 to C-8 ethyl alkanoates.[33] Another, relatively recent, example involves the Krafft temperature, *i.e.* the temperature at which a micellar solution of given concentration forms micelles.[34] Such a transition can be followed by conductivity measurements, and also here the forward and reverse scans to find the temperature do not provide overlapping curves. In these experiments, significant time intervals of 30 minutes or more were built in to equilibrate the system. Unfortunately, no dependence on equilibration time was reported for these systems. Such information would aid in assessing the non-equilibrium character of these solutions.

A kinetic bi-stability can be demonstrated in a Continuously Stirred Tank Reservoir (CSTR), as shown by Buhse *et al.* for the above discussed biphasic surfactant system[31] and further discussed by Ball and Haymet.[30] We shall here give a more intuitive presentation based on graphical analysis. Let us first review the CSTR, an idealized vessel where it is assumed that the contents are ideally mixed at all times. A solution of monomers is fed at a given volumetric rate Q and in order to maintain a constant volume V of liquid in the vessel, the content is discarded at the same rate. The important characteristic of a CSTR is its residence time of which the inverse is the refreshment rate

$$k_0 = \frac{Q}{V}. \tag{7.19}$$

Stationary states are reached after waiting out a few times the residence time.

For the bi-stability experiment, let us assume a CSTR that initially is solely filled with solvent and that its feed consists of a solution of monomer mole fraction x_0. As long as the reactor content has not reached the cac, the actual monomer mole fraction x_1 in the reactor increases with time t according to

$$\frac{\mathrm{d}x_1}{\mathrm{d}t} = k_0(x_0 - x_1) \quad \text{for } x_0 \ll x_c, \tag{7.20}$$

where the first term accounts for the feed and the second term for the discharge. Once the monomer concentration is sufficient, the self-assembly reaction proceeds. Typically, there is a slow and non-catalysed process that runs whether aggregates are already present or not; for simplicity its rate is modelled as being first order, k_1x. A second, faster process does involve autocatalysis and, hence, must involve aggregates. The backward reaction, where aggregates disassemble into monomers is neglected here under the assumption that its time scale is too large compared to the residence time. A simple model rate would be $k_3x(x_0 - x)^2$ which involves the presence of two aggregates and one additional monomer. Other forms of rate are equally permissible, though. The overall reaction rate equation now takes the form

$$\frac{\mathrm{d}x_1}{\mathrm{d}t} = k_0(x_0 - x_1) - k_1x_1 - k_3x_1(x_0 - x_1)^2 \quad \text{for } k_0t \gg 1. \tag{7.21}$$

Graphically, the rate can be represented as in Figure 7.8a, where the feed rate is represented by the straight line and the chemical reaction is represented by the other line with a clear maximum and a small but nonzero rate when the monomer mole fraction in the reactor equals that of the feed. Crossings of these two lines indicate stationary states, *i.e.*, it is where the overall reaction rate vanishes. The outer two crossings are stable stationary states and the middle one is not stable as one may easily verify by considering the mole fraction dependence of the rate. The bi-stability is best visualized in Figure 7.8b, where the stationary-state mole fraction of the reactor is given as a function of the feed rate. There are two branches: one at low feed rates, where the monomer mole fraction is low and aggregation takes place, and one at high rates, where there is no significant aggregation. The region in-between exhibits hysteresis as is typical for bi-stability.

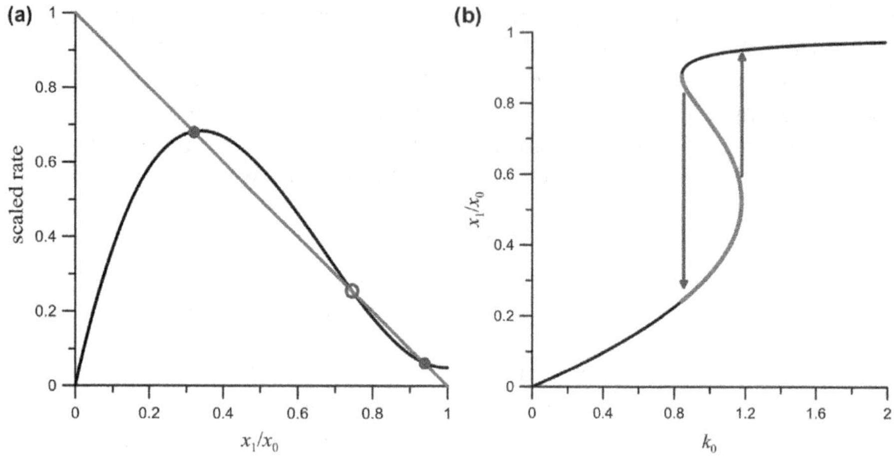

Figure 7.8 Graphical example of bi-stability in a continuously stirred tank reactor. (a) Analysis of rate equation in linear part (red) and nonlinear part lpar;black) where the closed circles indicate stable stationary points and the open circle an unstable one; rates scaled by k_0x_0. (b) Relative stationary monomer mole fraction as a function of feed rate. Hysteresis between two stable modes is indicated by arrows. Parameters chosen are $k_1 = 0.05\ s^{-1}$ and $k_3 = 4.5\ s^{-1}$; see eqn (7.21).

7.5 Autocatalysis

The bi-stability example with the CSTR discussed above, Section 7.4, clearly illustrates that multi-stability, *i.e.*, the occurrence of more than one stable stationary state, can be achieved with self-assembling molecular systems provided that the formation reaction is autocatalytic. The situation was different for the fuelled self-assembly, Section 7.3, which involved a regular aggregate formation reaction and did not exhibit multi-stability. Despite the obvious non-linearity of the aggregate formation rate there is only one stationary state. It is clear, however, that if the aggregate formation rate equation for this example was autocatalytic this system would also show multi-stability. As discussed above, bi-stability has been found in self-assembling reactions, but so far these have not been related to autocatalysis. The bi-stability in the ethyl caprylate example discussed above was due to a phase transfer step and for the other two examples, which involved hysteresis in concentration and temperature respectively, no molecular mechanism was provided.

The question then arises whether molecular self-assembly itself can be autocatalytic. So far, the starting point for a theoretical analysis has been the set of Becker–Döring (BD) equations,[19] which assumes the formation of aggregates to pass through a sequence of step-wise monomer additions or removals. The model has recently acquired some renewed interest for the archetypal molecular self-assembly system of micellar solutions.[21] The result of a detailed analysis is that – given the presence of one stationary state identified as the equilibrium state – the kinetics as described by the BD-equations is not capable of properly describing the kinetics of experimental systems. Already long before the seminal work of Aniansson and Wall on the same model,[20] it was known that the dynamics of micelle formation involves two characteristic time scales: a short time scale associated with the exchange of free monomers and monomers in aggregates and a long time scale associated with the break-up and formation of a whole aggregate. The problem that has been identified recently is that the model predictions for the ratio of the long and the short time scale appears to be too large.[21] In a subsequent analysis, the same authors discuss an extension of the model where the condition of step-wise monomer addition is relaxed. In particular, the break-up of aggregates in two halves as well as the formation of super-aggregates consisting of twice the average amount of monomers is allowed. It is the formation of super-aggregates that brings the separation of the short and long-time scale within the experimentally accessible range.[35] The authors conclude by the statement: "Implicit in our model is that the formation of micelles in a supersaturated solution proceeds by accretion of monomers to a proper micelle to form a super-micelle followed by fission of the super-micelle to two proper micelles; the formation of micelles is catalysed by micelles."

Fusion and fission have been experimentally observed in micellar systems, for instance, involving the non-ionic surfactant Triton X-100.[36,37] It is deployed in systems where various reactants for a chemical reaction are transported by swollen micelles to react upon fusion. Also theoretically, fusion and fission have been assessed[35,38] and, apart from accelerating the formation kinetics, no information is available on autocatalytic effects. A simple effective rate equation, as obtained by contraction of the Becker–Döring equations, such as eqn (7.12), is hence not known for such a case. The suggestion for a catalytic rate equation as made by writing eqn (7.21) for the demonstration of kinetic bi-stability was in actual fact inspired by the idea that two aggregates and monomers would coalesce into one super-aggregate that upon further growth would split into three proper aggregates. A similar approach can be found in the modelling of the dynamic instability for microtubules.[39] For these systems, more is known about the actually occurring phenomena[40,41] in relation to their dynamic behaviour, albeit that many more factors contribute that are difficult to control. More experimental work is required to uncover the actual mechanisms responsible for autocatalytic behaviour in natural and synthetic systems.

7.6 Role of Non-equilibrium Thermodynamics for Self-assembly

The Second Law of Thermodynamics provides a general framework for the assessment of the efficiency of a process. For the examples discussed here, it would state that the entropy produced during the process in a stationary state can be related to the difference in entropies of the input stream and the output stream. For the example of the fuelled self-assembly in Section 7.3, an earlier analysis has led to the conclusion that the dissipation is largely due to the reactions driving the self-assembly out of equilibrium.[25] A similar conclusion is expected for the case where the reaction is carried out in a CSTR, see Section 7.4, as friction-loss values are typically orders of magnitude less than those connected to chemical energy changes.

The formalism of non-equilibrium thermodynamics allows for the identification of the various contributions to the entropy production. The examples here feature a set of coupled reactions[25] and the more general expression for the available and – in the absence of utilization – lost work reads

$$\dot{W} = -\sum_k r_k \Delta_r G_k, \tag{7.22}$$

where the sum includes all reactions. For the fuelled self-assembly, the summation would include a reaction driving the self-assembly, the self-assembly process itself and the leakage reaction. The reactions in general will be coupled and it is important to realize that for any conjugate set of reaction rates and Gibbs energies of reactions, the expression (7.22) will hold.[24] As for chemical reactions, the relation between self-assembly rate and the Gibbs energy of reaction can be non-linear, see eqn (7.14). As shown explicitly for chemical reactions, this does not impose any limitation to the validity of the non-equilibrium thermodynamics formalism.[42] In particular, we showed the validity of eqn (7.14).[25] The work needed to maintain the non-equilibrium structure is drawn from the coupling of reactions, at the cost of some of the energy supply.

The coupling between fluxes and forces that contribute to entropy production becomes even more interesting when heat and mass transfer as well as electrochemical reactions are involved, see Chapter 12. As surmised before, the work lost in the self-assembly process could actually be put to some use in an electrochemical device. As an example, the hydrolysis occurring in the fuelled self-assembly process is readily performed electrochemically. This would open the route to store and retrieve energy from self-assembling systems. The associated coupling of reactions will be non-linear in character and as such only very few examples are worked out and none in detail.[42–44]

7.7 Conclusion

In this chapter we have discussed the principles of molecular self-assembly processes and some systematic approaches to run these processes out of equilibrium. It was found with regular aggregate-formation kinetics that a single stationary state is obtained that is reminiscent of the equilibrium steady state, albeit that entropy is produced. Multiple stationary states are possible when aggregate formation is partially autocatalytic. So far, very few, if any, experimental model systems have been reported that exhibit this kind of behaviour although it is to be expected that some well-known systems showing fusion and fission of aggregates might demonstrate multi-stability.

The quantification of the lost work in terms of a non-linear relation between reaction rate and thermodynamic driving forces is one of the main contributions of this chapter regarding the non-equilibrium thermodynamic analysis. Further possibilities for thermodynamic analysis have been discussed for more complete systems involving self-assembly.

Acknowledgements

The authors are indebted to Ignacio Pagonabarraga, and J. Miguel Rubí for their contributions regarding MNET and to S. Kjelstrup and D. Bedeaux and to J. Boekhoven, R. Eelkema and J. H. van Esch for discussions. This work was supported by the Netherlands Organization for Scientific Research (NWO) through a Complexity program grant. The Chemistry Department of Cambridge University is thanked for hosting G.K. during the summer of 2014.

References

1. J. M. Smith, H. C. Van Ness and M. M. Abbott, *Introduction to Chemical Engineering Thermodynamics*, McGraw-Hill, New York, 2001.
2. T. L. Hill, *An Introduction to Statistical Thermodynamics*, Dover Publications, New York, 1960.
3. E. J. King, *Acid-base Equilibria*, Macmillan, London, 1965.
4. M. Soos, M. Lattuada and J. Sefcik, *J. Phys. Chem. B*, 2009, **113**, 14962.
5. B. W. Ninham and P. L. Nostro, Molecular Forces and Self Assembly, in *Colloid, Nano Sciences and Biology*, Cambridge University Press, Cambridge, 2010.
6. T. Aida, E. W. Meijer and S. I. Stupp, *Science*, 2012, **335**, 813.
7. P. W. K. Rothemund, *Nature*, 2006, **440**, 297.
8. J. Boekhoven, A. M. Brizard, K. N. K. Kowlgi, G. J. M. Koper, R. Eelkema and J. H. van Esch, *Angew. Chem., Int. Ed.*, 2010, **49**, 4825.
9. S. Debnath, S. Roy and R. V. Ulijn, *J. Am. Chem. Soc.*, 2013, **135**, 16789.
10. X. M. Miao, W. Cao, W. T. Zheng, J. Y. Wang, X. L. Zhang, J. Gao, C. B. Yang, D. L. Kong, H. P. Xu, L. Wang and Z. M. Yang, *Angew. Chem., Int. Ed.*, 2013, **52**, 7781.
11. R. Eelkema, M. M. Pollard, J. Vicario, N. Katsonis, B. S. Ramon, C. W. M. Bastiaansen, D. J. Broer and B. L. Feringa, *Nature*, 2006, **440**, 163.
12. G. Nicolis and I. Prigogine, *Self-Organization in Nonequilibrium Systems: From Dissipative Structures to Order Through Fluctuations*, Wiley, New York, 1977.
13. M. Fialkowski, K. J. M. Bishop, R. Klajn, S. K. Smoukov, C. J. Campbell and B. A. Grzybowski, *J. Phys. Chem. B*, 2006, **110**, 2482.
14. P. L. Luisi and F. J. Varela, *Origins Life. Evol. Biospheres*, 1989, **19**, 633.

15. J. N. Israelachvili, *Intermolecular and Surface Forces*, Elsevier, Amsterdam, 2011.
16. R. Nagarajan and E. Ruckenstein, *Langmuir*, 1991, **7**, 2934.
17. R. Nagarajan, One Hundred Years of Micelles: Evolution of the Theory of Micellization, in: *Surfactant Science and Technology: Retrospects and Prospects*, ed. L. Romsted, Taylor and Francis, New York, 2014.
18. F. J. Gerhartl, *J. Chem. Educ.*, 1994, **71**, 539.
19. R. Becker and W. Döring, *Ann. Phys.*, 1935, **24**, 719.
20. E. A. G. Aniansson, S. N. Wall, M. Almgren, H. Hoffmann, I. Kielmann, W. Ulbricht, R. Zana, J. Lang and C. Tondre, *J. Phys. Chem.*, 1976, **80**, 905.
21. I. M. Griffiths, C. D. Bain, C. J. W. Breward, D. M. Colegate, P. D. Howell and S. L. Waters, *J. Colloid Interface Sci.*, 2011, **360**, 662.
22. P. V. Coveney and J. A. D. Wattis, *Proc. R. Soc. London, Ser. A*, 1996, **452**, 2079.
23. S. R. de Groot and P. Mazur, *Nonequilibrium Thermodynamics*, Dover publications, New York, 1984.
24. D. Kondepudi and I. Prigogine, Modern Thermodynamics, *From Heat Engines to Dissipative Structures*, Wiley, New York, 1998.
25. G. J. M. Koper, J. Boekhoven, W. E. Hendriksen, J. H. van Esch, R. Eelkema, I. Pagonabarraga, J. M. Rubí and D. Bedeaux, *Int. J. Thermophys.*, 2013, **34**, 1229.
26. R. Zwanzig, *Nonequilibrium Statistical Mechanics*, Oxford University Press, USA, 2001.
27. I. Pagonabarraga, A. Perez Madrid and J. M. Rubí, *Phys. A*, 1997, **237**, 205.
28. J. Boekhoven, *Multicomponent and Dissipative Self-Assembly Approaches*, Thesis, Delft University of Technology, Delft, the Netherlands, 2012.
29. P. A. Bachmann, P. L. Luisi and J. Lang, *Nature*, 1992, **357**, 57.
30. R. Ball and A. D. J. Haymet, *Phys. Chem. Chem. Phys.*, 2001, **3**, 4753.
31. T. Buhse, V. Pimienta, D. Lavabre and J. C. Micheau, *J. Phys. Chem. A*, 1997, **101**, 5215.
32. T. Buhse, R. Nagarajan, D. Lavabre and J. C. Micheau, *J. Phys. Chem. A*, 1997, **101**, 3910.
33. C. Roque, V. Pimienta, D. Lavabre and J. C. Micheau, *J. Phys. Chem. A*, 2001, **105**, 5877.
34. J. Z. Manojlovic, *Therm. Sci.*, 2012, **16**, S631.
35. I. M. Griffiths, C. J. W. Breward, D. M. Colegate, P. J. Dellar, P. D. Howell and C. D. Bain, *Soft Matter*, 2013, **9**, 853.
36. Y. Rharbi, M. A. Winnik and K. G. Hahn, *Langmuir*, 1999, **15**, 4697.
37. Y. Rharbi, M. Li, M. A. Winnik and K. G. Hahn, *J. Am. Chem. Soc.*, 2000, **122**, 6242.
38. A. K. Shchekin, M. S. Kshevetskiy and O. S. Pelevina, *Colloid J.*, 2011, **73**, 406.
39. R. Padinhateeri, A. B. Kolomeisky and D. Lacoste, *Biophys. J.*, 2012, **102**, 1274.
40. A. Desai and T. J. Mitchison, *Annu. Rev. Cell Dev. Biol.*, 1997, **13**, 83.

41. P. Maddox, A. Straight, P. Coughlin, T. J. Mitchison and E. D. Salmon, *J. Cell Biol.*, 2003, **162**, 377.
42. J. M. Rubí, D. Bedeaux, S. Kjelstrup and I. Pagonabarraga, *Int. J. Thermophys.*, 2013, **34**, 1214.
43. T. L. Hill, *Free Energy Transduction and Biochemical Cycle Kinetics*, Springer-Verlag, New York, 1989.
44. H. Qian and D. A. Beard, *Biophys. Chem.*, 2005, **114**, 213.

CHAPTER 8

Non-equilibrium Thermodynamics for Evaporation and Condensation

DICK BEDEAUX*[a] AND SIGNE KJELSTRUP[a,b]

[a] Department of Chemistry, Norwegian University of Science and Technology, 7491 Trondheim, Norway; [b] Process and Energy Laboratory, Delft University of Technology, Leeghwaterstraat 39, 2628CB Delft, The Netherlands
*Email: dick.bedeaux@chem.ntnu.no

8.1 Non-equilibrium Theories for Phase Transitions

The common phase transitions that take place everywhere, in nature as well as in industry, are well described and studied under equilibrium conditions. In practice, they take place under non-equilibrium conditions. It is well known that the interface can pose an additional resistance to transport. The Kapitza resistance for heat transfer[1] is one such example. Phase transitions have nevertheless been modelled in the engineering literature[2,3] with the assumption of continuity in the temperature and chemical potentials, *i.e.*, equilibrium, at the interface. To understand and describe phase transitions under realistic conditions is of central importance, since the phenomenon is so commonplace. In this chapter we review efforts that use non-equilibrium thermodynamics, kinetic theory, experiments, molecular dynamics simulations as well as the van der Waals square gradient theory. This is done for pure fluids and binary mixtures with focus on the liquid–vapour transition.

Experimental Thermodynamics Volume X: Non-equilibrium Thermodynamics with Applications
Edited by Dick Bedeaux, Signe Kjelstrup and Jan V. Sengers

Published by the Royal Society of Chemistry, www.rsc.org

Whenever mass is transferred from the liquid to the gas phase, or *vice versa*, there is an accompanying measurable heat flux arising from the difference in enthalpy between the two phases. The total heat flux, $\boldsymbol{J}_q$, and the mass flux, $\boldsymbol{J}$, may be controlled from the outside *e.g.* through temperature differences and by changing the pressure. In a stationary state of pure fluid evaporation, for example, the total heat flux is composed of the measurable heat flux $\boldsymbol{J}'^{\mathrm{g}}_q$ (or $\boldsymbol{J}'^{\mathrm{l}}_q$) and the latent heat H^{g} (or H^{l}) carried by the mass flux. Superscripts g and l refer to vapour (gas) and liquid, respectively. When heat and mass flow from the vapour to the liquid in a stationary state, the components of the total heat and mass fluxes normal to the surface, are continuous. It then follows from energy conservation in a pure fluid[4] that:

$$\boldsymbol{J}_q = \boldsymbol{J}'^{\mathrm{g}}_q + H^{\mathrm{g}}\boldsymbol{J} = \boldsymbol{J}'^{\mathrm{l}}_q + H^{\mathrm{l}}\boldsymbol{J}. \tag{8.1}$$

But this is not the only equation that links the measurable heat fluxes to the mass flux at the interface. The second law requires that the measurable heat fluxes and the mass flux through and into a surface are also *coupled* in the sense that one flux depends on the main driving force of the other flux, and *vice versa*. This coupling, described by non-equilibrium thermodynamics[3–6] arises from microscopic reversibility of molecular fluctuations, a property that is reflected in the Onsager relations. Such coupling takes place at interfaces *also* in pure (one-component) fluids, unlike the situation in homogeneous phases, and must not be violated.

In the rest of this chapter we will only consider transport of heat and mass through a flat surface. All fluxes and gradients will be assumed to be normal to this surface. The description therefore contains only these normal components, which are scalar. This is the reason why we will further use non-bold italic symbols for the fluxes.

Actual values of transfer resistivities are needed in order to compute concentration and temperature profiles at the interface during phase transformations. Beyond the famous results from the kinetic theory of gases,[5–9] not much is known about such coefficients. But serious experimental efforts have documented temperature jumps at the interface,[10–15] and transfer resistivities, as defined by non-equilibrium thermodynamics, have been calculated from these.[16–18] Non-equilibrium molecular dynamics simulations (*cf.,* also Chapter 6) can also supply such information, and have been used in studies of Lennard-Jones particles.[19–24] Studies of molecules with realistic particle interaction potentials are few, however.[21]

Square gradient theory is the first approximation to density functional theory, first formulated for single-component systems by van der Waals[25] and extended to mixtures by Cahn and Hilliard.[26] It was extended to the non-equilibrium domain by Bedeaux and co-workers for single-component systems,[27–29] and by Glavatskiy and Bedeaux for mixtures.[30–34] The theory was used to study the effect of surface curvature[35,36] on the resistivities, and extended to temperature-dependent influence parameters.[37] The square gradient theory offers rapid access to transfer resistivities *via* the so-called integral relations. These relations, derived by Bedeaux and co-workers for single and multicomponent fluids,[30,38,39] can be used to compute the transfer resistivities, once the local resistivities, the concentration and the partial-enthalpy profiles through the surface at equilibrium are known. The local resistivities, the concentration and the partial-enthalpy profiles can be found from molecular dynamics simulations[21,24] or modelled in the context of the square gradient theory.[29,33,35,36] An extension of the square gradient analysis to calculate the interfacial transfer resistivities for a one-component fluid with classical density functional theory was found.[40]

In this chapter we review the experimental and computational findings for the non-equilibrium liquid–vapour transition of one or two components. We discuss first the entropy production in the heterogeneous systems, before we define the transfer resistivities. We shall see that non-equilibrium thermodynamics can provide information about the resistivity that is special for this theory, namely the coupling resistivity. The sign and the order of magnitude of the coupling coefficient are related to the enthalpy change of the phase transition and to local thermal resistivities.

At the end we comment on a new area of application, transport phenomena in and around the three-phase contact line.

8.2 Heterogeneous Systems. Entropy Production at Surfaces

A heterogeneous system is made up by several homogeneous phases separated by interfaces or surfaces. We use the names surface and interface interchangeably. The thermodynamic description of an interface in equilibrium was given by Gibbs.[41] He introduced the surface *excess variables*, which are obtained by first extrapolating the corresponding variable in the adjacent phases until the dividing surface and by next integrating the difference over the surface thickness (the so-called surface of discontinuity). Chapter 17 and ref. 4 give a precise definition. This mathematical construct describes the interface as a two-dimensional system. For surfaces which are not in equilibrium, one can introduce not only excess densities, but also excess fluxes along the surface. For an extensive discussion, see ref. 42 and 43. This allows for an extension of classical non-equilibrium thermodynamics from homogeneous to heterogeneous systems, in other words, an extension to surfaces.[4] In the construction, we assume that the surface itself is always in *local* equilibrium, *cf.*, Chapter 4 for definitions and supporting evidence, and ref. 44 for implications. The surface will in general not be in equilibrium with the adjacent homogeneous phases, however. This means, for instance, that the values of the temperature and chemical potentials of the surface can differ from the values next to the surface. The last values can also differ from one another.[4] The assumption of local equilibrium and the two-dimensional description enable us to obtain a discrete formulation of the excess entropy production in the surface, and to obtain correct boundary conditions in the modelling of transport phenomena. In order to find the entropy production in the surface we follow the prescription outlined in Chapter 1.

Consider first the transport of heat and mass in a one-component fluid through an interface, *e.g.*, the evaporation of pure water. The evaporation can, for instance, take place from a funnel as pictured in Figure 8.1. Fluid is supplied at a constant rate to the bottom of the funnel, while vapour is withdrawn at the top at the same rate. The pressure is constant in the container. The flux of fluid through the interface, J, is constant in stationary evaporation or condensation. The fluid flux is relative to the position of the surface. The temperature can be measured in the direction perpendicular to the surface both in the vapour and in the liquid.

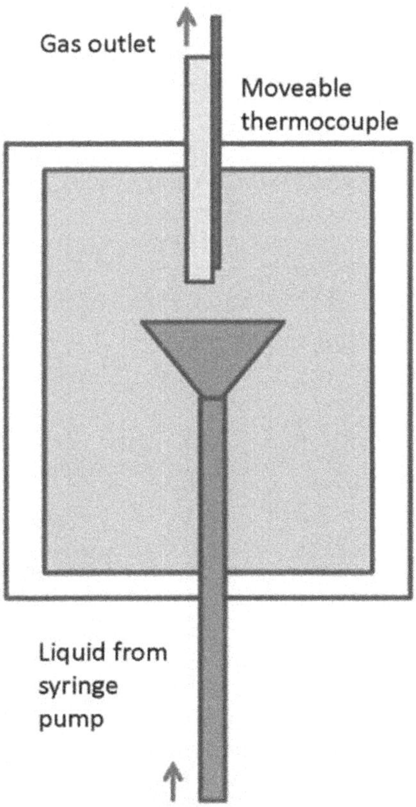

Figure 8.1 Sketch of experimental setup to measure temperature profiles during evaporation. Vapour is withdrawn from the cell at the same rate as liquid is supplied. The thermocouple can be lowered to measure temperatures near the surface of the evaporating liquid. There is a curvature of the meniscus in the funnel container for the liquid, but the surface can be regarded as flat on a molecular scale.

The general expression for the entropy production has two terms for heat and two terms for mass transport into and out of the interface:[4]

$$\sigma^{\mathrm{s}} = J'^{\mathrm{l}}_{q}\left[\frac{1}{T^{\mathrm{s}}}-\frac{1}{T^{\mathrm{l}}}\right] + J'^{\mathrm{g}}_{q}\left[\frac{1}{T^{\mathrm{g}}}-\frac{1}{T^{\mathrm{s}}}\right] - J^{\mathrm{l}}\left[\frac{\mu^{\mathrm{s}}(T^{\mathrm{s}})-\mu^{\mathrm{l}}(T^{\mathrm{s}})}{T^{\mathrm{s}}}\right] - J^{\mathrm{g}}\left[\frac{\mu^{\mathrm{g}}(T^{\mathrm{s}})-\mu^{\mathrm{s}}(T^{\mathrm{s}})}{T^{\mathrm{s}}}\right]. \tag{8.2}$$

Here T is the temperature and μ is the chemical potential. The superscript s denotes the surface. The entropy production is a product sum of each variable flux and its conjugate driving force. A practical frame of reference for the mass flux is the equimolar surface.[4]

In a stationary state $J^{\mathrm{l}} = J^{\mathrm{g}} = J$, and the last two terms can be contracted. We can also eliminate one of the heat fluxes using eqn (8.1). For this purpose, we use the Gibbs–Helmholtz relation

$$\left.\frac{\partial(\mu/T)}{\partial(1/T)}\right|_{p} = H, \tag{8.3}$$

for the liquid as well as the vapour, and obtain to linear order in the chemical driving forces

$$\begin{aligned}\frac{1}{T^{\mathrm{s}}}\left[\mu^{\mathrm{g}}(T^{\mathrm{s}})-\mu^{\mathrm{s}}(T^{\mathrm{s}})\right] &= \frac{1}{T^{\mathrm{l}}}\left[\mu^{\mathrm{g}}(T^{\mathrm{l}})-\mu^{\mathrm{s}}(T^{\mathrm{l}})\right] \\ &\quad + \left\{\frac{\partial}{\partial(1/T)}\left[\frac{\mu^{\mathrm{g}}(T)-\mu^{\mathrm{s}}(T)}{T}\right]_{p}\right\}_{T=T^{\mathrm{l}}}\left(\frac{1}{T^{\mathrm{s}}}-\frac{1}{T^{\mathrm{l}}}\right) \\ &= \left[\frac{\mu^{\mathrm{g}}(T^{\mathrm{l}})-\mu^{\mathrm{s}}(T^{\mathrm{l}})}{T^{\mathrm{l}}}\right] + (H^{\mathrm{g}}-H^{\mathrm{l}})\left(\frac{1}{T^{\mathrm{s}}}-\frac{1}{T^{\mathrm{l}}}\right).\end{aligned} \tag{8.4}$$

Here p is the pressure. With these relations the entropy production reduces to two terms:

$$\sigma^{\mathrm{s}} = J'^{\mathrm{g}}_{q}\left[\frac{1}{T^{\mathrm{g}}}-\frac{1}{T^{\mathrm{l}}}\right] - J\left[\frac{\mu^{\mathrm{g}}(T^{\mathrm{l}})-\mu^{\mathrm{l}}(T^{\mathrm{l}})}{T^{\mathrm{l}}}\right] \equiv J'^{\mathrm{g}}_{q}\Delta_{\mathrm{l,g}}\left(\frac{1}{T}\right) - J\left[\frac{\Delta_{\mathrm{l,g}}\mu(T^{\mathrm{l}})}{T^{\mathrm{l}}}\right], \tag{8.5}$$

when we eliminate the measurable heat flux in the liquid. The symbol Δ indicates a difference and is defined in both cases by the second equality. The chemical-potential difference across the interface must be evaluated at the temperature of the adjacent liquid. This is required to keep the entropy production invariant. The surface temperature and chemical potential have disappeared as variables in eqn (8.5). This can be explained by the stationary-state condition. In this state there is no net exchange of surface energy and mass with the adjacent liquid or the vapour (the surface Gibbs energy and the excess mass are constant). As a consequence, the surface temperature and chemical potential are not needed in the description of stationary evaporation or condensation. Both variables can differ from the values in the adjacent phases when the system is not in equilibrium!

We can alternatively choose to eliminate the heat flux in the vapour phase. This gives:

$$\sigma^{s} = J'^{l}_{q}\left[\frac{1}{T^{g}} - \frac{1}{T^{l}}\right] - J\left[\frac{\mu^{g}(T^{g}) - \mu^{l}(T^{g})}{T^{g}}\right] \equiv J'^{l}_{q}\Delta_{l,g}\left(\frac{1}{T}\right) - J\left[\frac{\Delta_{l,g}\mu(T^{g})}{T^{g}}\right]. \quad (8.6)$$

The standard state for the liquid is pure liquid at pressure $p = p_0 = 10^5$ Pa. The chemical potential of the pure liquid in equilibrium with an ideal vapour at the liquid temperature is

$$\mu^{l}(T) = \mu^{g,0}(T) + RT\ln\frac{p^{*}(T)}{p_0}. \quad (8.7)$$

where $p^{*}(T)$ is the pressure of the vapour in equilibrium with the liquid at temperature T and R is the gas constant. This pressure is referred to as the vapour pressure of the liquid or the saturation pressure. The chemical potential of an ideal vapour at the temperature T and at the real pressure, p, is:

$$\mu^{g}(T) = \mu^{g,0}(T) + RT\ln\frac{p}{p_0}. \quad (8.8)$$

By subtracting eqn (8.8) from eqn (8.7), we find:

$$\frac{\Delta_{l,g}\mu(T)}{T} = R\ln\frac{p}{p^{*}(T)}. \quad (8.9)$$

The temperature to use in eqn (8.5) is T^{l}, and in eqn (8.6) we use T^{g}. Eqn (8.9) assumes the gas to be ideal. For a non-ideal gas, one must replace the pressure by the fugacity. The curvature of the meniscus in Figure 8.1 is too small to have an impact on the chemical potential. By introducing these relations into the entropy production, eqn (8.5) and (8.6), we obtain two equivalent forms:

$$\sigma^{s} = J'^{g}_{q}\Delta_{l,g}\left(\frac{1}{T}\right) - J\left[R\ln\frac{p}{p^{*}(T^{l})}\right] = J'^{l}_{q}\Delta_{l,g}\left(\frac{1}{T}\right) - J\left[R\ln\frac{p}{p^{*}(T^{g})}\right]. \quad (8.10)$$

The liquid evaporates at constant temperature if the vapour pressure of the liquid is larger than the vapour pressure of the vapour. When it is lower, there is condensation. A temperature difference may also give rise to evaporation or condensation. Some temperature conditions may give rise to a vapour which is supersaturated with respect to the liquid. The interaction of these phenomena is described by the flux equations and the dynamic boundary conditions.

8.3 Symmetry Rules

Transports at interfaces (or contact lines) obey different symmetry rules than transport processes in homogeneous phases. The Curie principle says that tensors of different order do not couple. For transport of heat and mass through a surface, only the components normal to the surface are relevant. This leads immediately to special possibilities for coupling.[4]

The coupling between heat and mass fluxes at interfaces[4] differs from the coupling in the homogeneous phases. Unlike in the homogeneous phases, the coupling coefficients for heat and mass transport are large for phase transitions.[4] This will be brought out in the remaining part of the chapter. Also, there are new possibilities for coupling which we will bring out in detail in Chapter 12. The fluxes in the expressions for the entropy production above are the scalar normal components of vectors. These scalars can couple to other scalars. The scalar driving force of a chemical reaction does not couple to vectorial fluxes in the homogeneous phase, but a scalar chemical driving force in a surface can couple to the scalar normal component of vectorial fluxes to the surface. This applies to electrochemical reactions as well, also outside the range of linearity in the flux–force relations.[45] The jump in the electric potential and its coupling to transport of heat and mass are discussed in Chapter 12. In this chapter, we describe coupling between scalar components of the mass and heat fluxes.

The flux equations for the surface predict jumps in intensive variables across or into the surface, and define in this manner dynamic boundary conditions, essential for integration of variables through the surface or for the modelling of phase transitions.

8.4 Evaporation or Condensation in a Pure Fluid

8.4.1 Interfacial Transfer Resistivities

Once the entropy production is obtained, we can write constitutive relations. We choose the force–flux representation, and obtain from eqn (8.5), where the measurable heat flux in the vapour is used:

$$\begin{aligned} \Delta_{\mathrm{l,g}}\left(\frac{1}{T}\right) &= r_{qq}^{\mathrm{s,g}} J'^{\mathrm{g}}_q + r_{q\mu}^{\mathrm{s,g}} J, \\ -\frac{\Delta_{\mathrm{l,g}}\mu(T^{\mathrm{l}})}{T^{\mathrm{l}}} &= r_{\mu q}^{\mathrm{s,g}} J'^{\mathrm{g}}_q + r_{\mu\mu}^{\mathrm{s,g}} J. \end{aligned} \tag{8.11}$$

The coefficients $r_{ij}^{\mathrm{s,g}}$ are interface resistivities of the surface or *transfer resistivities*. As stated above, the component flux is relative to the position of the dividing surface. Equivalently we obtain from eqn (8.6), where the measurable heat flux in the liquid is used:

$$\begin{aligned} \Delta_{\mathrm{l,g}}\left(\frac{1}{T}\right) &= r_{qq}^{\mathrm{s,l}} J'^{\mathrm{l}}_q + r_{q\mu}^{\mathrm{s,l}} J, \\ -\frac{\Delta_{\mathrm{l,g}}\mu(T^{\mathrm{g}})}{T^{\mathrm{g}}} &= r_{\mu q}^{\mathrm{s,l}} J'^{\mathrm{l}}_q + r_{\mu\mu}^{\mathrm{s,l}} J. \end{aligned} \tag{8.12}$$

Again the coefficients $r_{ij}^{\mathrm{s,l}}$ are interface resistivities of the surface or *transfer resistivities*. The extra superscript of the transfer resistivities refers to the measurable heat flux used in their definition. Onsager relations apply, so $r_{q\mu}^{s,g}=r_{\mu q}^{s,g}$ and $r_{q\mu}^{s,l}=r_{\mu q}^{s,l}$. Both sets apply to the whole surface. By using the relation between the measurable heat fluxes given in eqn (8.1), the resistivities in eqn (8.12) can be found from the resistivities in eqn (8.11), see eqn (8.13). In the stationary state, there are therefore only three independent transfer resistivities for the description of heat and mass transport *across* a surface.†

It is common[1–3] in the analysis of evaporation and condensation to set the temperatures and chemical potentials equal on both sides of the surface (and in the surface). The liquid and the vapour phases are then in equilibrium at the surface, and Clausius–Clapeyron's equation applies. To assume equilibrium across the surface *during* evaporation or condensation means that the resistivities in eqn (8.11) and (8.12) are negligible, however see also Chapter 4. When the interface resistivities are sizable, jumps in temperatures and chemical potentials will develop at the surface.

Since the descriptions are alternative representations of the entropy production, the coefficients are related. From the energy balance, eqn (8.1), and the Onsager relations we obtain[4]

$$\begin{aligned} r_{qq}^{s,l}&=r_{qq}^{s,g}=r_{qq}^{s},\\ r_{q\mu}^{s,l}&=r_{\mu q}^{s,l}=r_{q\mu}^{s,g}-\Delta_{vap}Hr_{qq}^{s}=r_{\mu q}^{s,g}-\Delta_{vap}Hr_{qq}^{s},\\ r_{\mu\mu}^{s,l}&=r_{\mu\mu}^{s,g}-2\Delta_{vap}Hr_{\mu q}^{s,g}-(\Delta_{vap}H)^2r_{qq}^{s}. \end{aligned} \tag{8.13}$$

Here $\Delta_{vap}H\equiv H^g-H^l$ is the enthalpy of evaporation. The temperature, the pressure and the mass flux can be measured. In order to relate the equations to measurements, it is convenient to write the flux–force relations in the following form:

$$\begin{aligned} \Delta_{l,g}T&=-\frac{1}{\lambda^{s,g}}\left(J_q'^g-q^{*s,g}J\right),\\ -\frac{\Delta_{l,g}\mu(T^l)}{T^l}&=q^{*s,g}\frac{\Delta_{l,g}T}{T^lT^g}-\left(r_{\mu\mu}^{s,g}-r_{qq}^{s,g}(q^{*s,g})^2\right)J. \end{aligned} \tag{8.14}$$

The thermal conductivity at zero mass flux and the heat of transfer at the vapour side of the surface are defined by

$$\lambda^{s,g}\equiv-\left[\frac{J_q'^g}{\Delta_{l,g}T}\right]_{J=0}=\frac{1}{T^lT^gr_{qq}^{s,g}}\quad\text{and}\quad q^{*s,g}\equiv\left(\frac{J_q'^g}{J}\right)_{\Delta_{l,g}T=0}=-\frac{r_{q\mu}^{s,g}}{r_{qq}^{s,g}}. \tag{8.15}$$

†In ref. 4–6 and 16 we absorbed a common factor T^l in these interface resistivities.

The symbol q^* is commonly used for a heat of transfer. In the absence of a mass flux the heat of transfer can also be found from the known temperature and chemical potential differences:

$$q^{*\mathrm{s,g}} = -\left[\frac{T^{\mathrm{g}}\Delta_{\mathrm{l,g}}\mu(T^{\mathrm{l}})}{\Delta_{\mathrm{l,g}}T}\right]_{J=0}, \tag{8.16}$$

where we used the Onsager relation. Using the measurable heat flux on the liquid side one may alternatively write the flux–force relations in the following form

$$\begin{aligned} \Delta_{\mathrm{l,g}}T &= -\frac{1}{\lambda^{\mathrm{s,l}}}\left(J'^{\mathrm{l}}_q - q^{*\mathrm{s,l}}J\right), \\ -\frac{\Delta_{\mathrm{l,g}}\mu(T^{\mathrm{g}})}{T^{\mathrm{g}}} &= q^{*\mathrm{s,l}}\frac{\Delta_{\mathrm{l,g}}T}{T^{\mathrm{l}}T^{\mathrm{g}}} - \left(r^{\mathrm{s,l}}_{\mu\mu} - r^{\mathrm{s,l}}_{qq}\left(q^{*\mathrm{s,l}}\right)^2\right)J. \end{aligned} \tag{8.17}$$

where the thermal conductivity at zero mass flux and the heat of transfer at the liquid side of the surface are defined by

$$\lambda^{\mathrm{s,l}} \equiv -\left[\frac{J'^{\mathrm{l}}_q}{\Delta_{\mathrm{l,g}}T}\right]_{J=0} = \frac{1}{T^{\mathrm{l}}T^{\mathrm{g}}r^{\mathrm{s,l}}_{qq}} \quad \text{and} \quad q^{*\mathrm{s,l}} \equiv \left(\frac{J'^{\mathrm{l}}_q}{J}\right)_{\Delta_{\mathrm{l,g}}T=0} = -\frac{r^{\mathrm{s,l}}_{q\mu}}{r^{\mathrm{s,l}}_{qq}}. \tag{8.18}$$

In the absence of a mass flux, the heat of transfer can alternatively be found from the known temperatures and chemical potentials:

$$q^{*\mathrm{s,l}} = -\left[\frac{T^{\mathrm{l}}\Delta_{\mathrm{l,g}}\mu(T^{\mathrm{g}})}{\Delta_{\mathrm{l,g}}T}\right]_{J=0}, \tag{8.19}$$

where we used the Onsager relation. When the chemical and the thermal forces are such that $J=0$. In this situation, which applies to eqn (8.16) as well, the surface is in *Soret equilibrium*. In a homogeneous phase of a one-component fluid, the entropy production is equal to the product of the measurable heat flux and the gradient of the inverse temperature. In this case, with only one flux–force term, there is no coupling coefficient between the measurable heat flux and the mass flux. The situation changes at the surface, where the measurable heat flux and the component flux are independent variables. The fluxes across the interface relative to the position of the surface are then coupled.

Bedeaux *et al.*,[5,6] Johannessen *et al.*[39] and Kjelstrup *et al.*[46] used these equations to discuss the possibility of an inverted temperature profile in front of the surface of an evaporating liquid.

8.4.2 Interfacial Transfer Resistivities from Kinetic Theory

Considerable work has been done on evaporation and condensation, using the kinetic theory of gases, *cf.*, ref. 4–9 and references therein. The transfer resistivities of eqn (8.11) are, according to kinetic theory:

$$\begin{aligned} r_{qq}^{s,g} &= \frac{\sqrt{\pi}}{4c_{eq}^{g}(T^{s})R(T^{s})^{2}v_{mp}(T^{s})}\left(1+\frac{104}{25\pi}\right), \\ r_{q\mu}^{s,g} = r_{\mu q}^{s,g} &= \frac{\sqrt{\pi}}{8T^{s}c_{eq}^{g}(T^{s})v_{mp}(T^{s})}\left(1+\frac{16}{5\pi}\right), \\ r_{\mu\mu}^{s,g} &= \frac{2R\sqrt{\pi}}{c_{eq}^{g}(T^{s})v_{mp}(T^{s})}\left(\frac{1}{\sigma}+\frac{1}{\pi}-\frac{23}{32}\right). \end{aligned} \tag{8.20}$$

Here $v_{mp}(T^{s}) \equiv \sqrt{2RT^{s}/M}$ is the most probable thermal velocity and M is the molar mass. The condensation coefficient, σ, is defined as the fraction of incident particles which, after collision with the liquid surface, are absorbed by the liquid. The transfer resistivities are surface properties and must depend on surface variables only. The surface temperature is such a variable. The temperature therefore refers to the surface, and the molar gas density $c_{eq}^{g}(T^{s})$ refers to a density in equilibrium with the liquid at the surface temperature. These choices gave better agreement with results from molecular dynamics simulations of a phase transition of particles with a relatively short-range Lennard-Jones spline potential.[22]

The expressions are most appropriate for hard spheres near the triple point. The number of moles of particles that collide with the liquid surface is proportional to the mean thermal velocity, v_{mp}, times the molar density in the gas, c^{g}. The pre-factor is characteristic for the quantity transported, and follows from a comparison with the flux–force relation. Experimental values of σ between 0.1 and 1 have been reported.[47] The actual value is somewhat controversial. In an ideal gas the most probable thermal velocity is equal to $\sqrt{6/5}$ times the speed of sound. Only the transfer resistivity $r_{\mu\mu}^{s,g}$ depends on the condensation coefficient. Once $r_{\mu\mu}^{s,g}$ is known, there is thus a unique way to find the condensation coefficient by fitting. Polyatomic gases have also been described by kinetic theory,[9] but similarly convenient expressions are not available. The heat of transfer of the surface becomes

$$q^{*s,g} = -\frac{5}{2}\left(\frac{5\pi+16}{25\pi+104}\right)RT^{s}, \tag{8.21}$$

and the thermal conductivity at zero mass flux is

$$\lambda^{s,g} = \frac{100\sqrt{\pi}c_{eq}^{g}(T^{s})Rv_{mp}(T^{s})}{25\pi+104}. \tag{8.22}$$

Neither of these coefficients depends on the condensation coefficient. We see that kinetic theory predicts a negative sign for the heat of transfer, $q^{*s,g}$.

From eqn (8.15) we know that the coefficient has the order of magnitude of an enthalpy. This indicates that the coupling coefficient for heat and mass transfer at interfaces is substantially larger at the surface than in the homogeneous liquid or gas mixtures. This will be further confirmed in the analysis below.

8.4.3 The Sign and Magnitude of the Heats of Transfer at the Surface

We shall see how we can use non-equilibrium thermodynamics to obtain more information on the sign and magnitude of the heat of transfer.[4] This will be accomplished by invoking the more general representation of the entropy production eqn (8.2). In the analysis above we used the stationary nature of the problem to reduce the number of independent flux–force pairs to two. Alternatively one can use all four force flux-pairs that derive from eqn (8.2). The full force–flux matrix has 16 terms. In this matrix, we proceed by neglecting the coupling of forces and fluxes on *different sides of the surface.* One rational for this assumption could be that this interaction is smaller than interactions on the same side. It follows that

$$
\begin{aligned}
\frac{1}{T^{\mathrm{s}}}-\frac{1}{T^{\mathrm{l}}} &= R_{qq}^{\mathrm{s,l}}J'^{\mathrm{l}}_{q}+R_{q\mu}^{\mathrm{s,l}}J,\\
-\frac{\mu^{\mathrm{s}}(T^{\mathrm{s}})-\mu^{\mathrm{l}}(T^{\mathrm{s}})}{T^{\mathrm{s}}} &= R_{\mu q}^{\mathrm{s,l}}J'^{\mathrm{l}}_{q}+R_{\mu\mu}^{\mathrm{s,l}}J,\\
\frac{1}{T^{\mathrm{g}}}-\frac{1}{T^{\mathrm{s}}} &= R_{qq}^{\mathrm{s,g}}J'^{\mathrm{g}}_{q}+R_{q\mu}^{\mathrm{s,g}}J,\\
-\frac{\mu^{\mathrm{g}}(T^{\mathrm{s}})-\mu^{\mathrm{s}}(T^{\mathrm{s}})}{T^{\mathrm{s}}} &= R_{\mu q}^{\mathrm{s,g}}J'^{\mathrm{g}}_{q}+R_{\mu\mu}^{\mathrm{s,g}}J,
\end{aligned}
\tag{8.23}
$$

where we used $J^{\mathrm{l}}=J^{\mathrm{g}}=J$. The coefficients $R_{ij}^{\mathrm{s,l}}$ and $R_{ij}^{\mathrm{s,g}}$ are transfer resistivities. The forces in eqn (8.11) and (8.12) are the differences in the inverse temperature and chemical potential difference across the whole surface, $\Delta_{\mathrm{l,g}}(1/T)$ and $\Delta_{\mathrm{l,g}}\mu(T^{\mathrm{l}})/T^{\mathrm{l}}$ or $\Delta_{\mathrm{l,g}}\mu(T^{\mathrm{g}})/T^{\mathrm{g}}$, respectively. The force–flux relations (8.11) and (8.12) do not contain the temperature and chemical potential of the surface, like eqn (8.23) does. They contain, in view of the Onsager relation, only 3 independent resistivities. Eqn (8.23) contains, in view of the Onsager relations, 6 independent resistivities. The resistivities in eqn (8.11) and (8.12) can be expressed in terms of the resistivities in eqn (8.23).

The heats of transfer, $q^{*\mathrm{l}}$ and $q^{*\mathrm{g}}$, in pure fluids are zero by definition. We assume that $q^{*\mathrm{s,l}}=q^{*\mathrm{l}}$ and $q^{*\mathrm{s,g}}=q^{*\mathrm{g}}$. It follows that:

$$
R_{q\mu}^{\mathrm{s,l}}=R_{\mu q}^{\mathrm{s,l}}=-R_{qq}^{\mathrm{s,l}}q^{*\mathrm{l}}=0 \quad\text{and}\quad R_{q\mu}^{\mathrm{s,g}}=R_{\mu q}^{\mathrm{s,g}}=-R_{qq}^{\mathrm{s,g}}q^{*\mathrm{g}}=0. \tag{8.24}
$$

Eqn (8.23) therefore reduces to

$$\frac{1}{T^{\mathrm{s}}}-\frac{1}{T^{\mathrm{l}}}=R_{qq}^{\mathrm{s,l}}J'^{\mathrm{l}}_{q},\qquad -\frac{\mu^{\mathrm{s}}(T^{\mathrm{s}})-\mu^{\mathrm{l}}(T^{\mathrm{s}})}{T^{\mathrm{s}}}=R_{\mu\mu}^{\mathrm{s,l}}J,$$
$$\frac{1}{T^{\mathrm{g}}}-\frac{1}{T^{\mathrm{s}}}=R_{qq}^{\mathrm{s,g}}J'^{\mathrm{g}}_{q},\qquad -\frac{\mu^{\mathrm{g}}(T^{\mathrm{s}})-\mu^{\mathrm{s}}(T^{\mathrm{s}})}{T^{\mathrm{s}}}=R_{\mu\mu}^{\mathrm{s,g}}J. \tag{8.25}$$

Addition of inverse temperature differences gives

$$\frac{1}{T^{\mathrm{g}}}-\frac{1}{T^{\mathrm{l}}}=R_{qq}^{\mathrm{s,l}}J'^{\mathrm{l}}_{q}+R_{qq}^{\mathrm{s,g}}J'^{\mathrm{g}}_{q}=\left(R_{qq}^{\mathrm{s,l}}+R_{qq}^{\mathrm{s,g}}\right)J'^{\mathrm{g}}_{q}+R_{qq}^{\mathrm{s,l}}\Delta_{\mathrm{vap}}HJ. \tag{8.26}$$

Addition of chemical-potential differences gives

$$\begin{aligned}
&-\frac{\mu^{\mathrm{g}}(T^{\mathrm{s}})-\mu^{\mathrm{l}}(T^{\mathrm{s}})}{T^{\mathrm{s}}}=\left(R_{\mu\mu}^{\mathrm{s,l}}+R_{\mu\mu}^{\mathrm{s,g}}\right)J=\\
&-\frac{\mu^{\mathrm{g}}(T^{\mathrm{l}})-\mu^{\mathrm{l}}(T^{\mathrm{l}})}{T^{\mathrm{l}}}-\left[\frac{\partial}{\partial(1/T)}\left(\frac{\mu^{\mathrm{g}}}{T}-\frac{\mu^{\mathrm{l}}}{T}\right)_{T=T^{\mathrm{l}}}\right]\left(\frac{1}{T^{\mathrm{s}}}-\frac{1}{T^{\mathrm{l}}}\right)=\\
&-\frac{\mu^{\mathrm{g}}(T^{\mathrm{l}})-\mu^{\mathrm{l}}(T^{\mathrm{l}})}{T^{\mathrm{l}}}-R_{qq}^{\mathrm{s,l}}\Delta_{\mathrm{vap}}HJ'^{\mathrm{l}}_{q}=\\
&-\frac{\mu^{\mathrm{g}}(T^{\mathrm{l}})-\mu^{\mathrm{l}}(T^{\mathrm{l}})}{T^{\mathrm{l}}}-R_{qq}^{\mathrm{s,l}}\Delta_{\mathrm{vap}}HJ'^{\mathrm{g}}_{q}-R_{qq}^{\mathrm{s,l}}(\Delta_{\mathrm{vap}}H)^{2}J.
\end{aligned} \tag{8.27}$$

It follows that

$$-\frac{\mu^{\mathrm{g}}(T^{\mathrm{l}})-\mu^{\mathrm{l}}(T^{\mathrm{l}})}{T^{\mathrm{l}}}=R_{qq}^{\mathrm{s,l}}\Delta_{\mathrm{vap}}HJ'^{\mathrm{g}}_{q}+\left(R_{\mu\mu}^{\mathrm{s,l}}+R_{\mu\mu}^{\mathrm{s,g}}+R_{qq}^{\mathrm{s,l}}(\Delta_{\mathrm{vap}}H)^{2}\right)J. \tag{8.28}$$

By comparing eqn (8.26) and (8.28) with eqn (8.11), using (8.9), we conclude that

$$r_{qq}^{\mathrm{s,g}}=R_{qq}^{\mathrm{s,l}}+R_{qq}^{\mathrm{s,g}},\qquad r_{q\mu}^{\mathrm{s,g}}=r_{\mu q}^{\mathrm{s,g}}=R_{qq}^{\mathrm{s,l}}\Delta_{\mathrm{vap}}H,$$
$$r_{\mu\mu}^{\mathrm{s,g}}=R_{\mu\mu}^{\mathrm{s,l}}+R_{\mu\mu}^{\mathrm{s,g}}+R_{qq}^{\mathrm{s,l}}(\Delta_{\mathrm{vap}}H)^{2}. \tag{8.29}$$

Since$R_{qq}^{\mathrm{s,l}}$, $R_{qq}^{\mathrm{s,g}}$ and $\Delta_{\mathrm{vap}}H$ are positive, $q^{*\mathrm{s,g}}$ is negative! This gives an independent confirmation of the result from kinetic theory. The absolute size of the heat of transfer is given by the fraction of the thermal resistivity that is located on the liquid side of the surface times the heat of evaporation, $\Delta_{\mathrm{vap}}H$. It is clear that $q^{*\mathrm{s,g}}$ is negative, large and has no relation to the transported heats in the liquid and the vapour, which both are zero.

It follows from eqn (8.20) that the alternative set of coefficients obey

$$r_{qq}^{\mathrm{s,l}}=r_{qq}^{\mathrm{s,g}}=R_{qq}^{\mathrm{s,l}}+R_{qq}^{\mathrm{s,g}},\qquad r_{q\mu}^{\mathrm{s,l}}=r_{\mu q}^{\mathrm{s,l}}=-R_{qq}^{\mathrm{s,g}}\Delta_{\mathrm{vap}}H,$$
$$r_{\mu\mu}^{\mathrm{s,l}}=R_{\mu\mu}^{\mathrm{s,l}}+R_{\mu\mu}^{\mathrm{s,g}}+R_{qq}^{\mathrm{s,g}}(\Delta_{\mathrm{vap}}H)^{2}. \tag{8.30}$$

From the definitions (8.15) and (8.18) and the relations (8.29) and (8.30), we obtain

$$q^{\star s,g} \equiv \left(\frac{J'^{g}_{q}}{J}\right)_{\Delta_{l,g}T=0} = -\frac{r^{s,g}_{q\mu}}{r^{s,g}_{qq}} = -\frac{R^{s,l}_{qq}}{R^{s,l}_{qq}+R^{s,g}_{qq}}\Delta_{vap}H, \tag{8.31}$$

$$q^{\star s,l} \equiv \left(\frac{J'^{l}_{q}}{J}\right)_{\Delta_{l,g}T=0} = -\frac{r^{s,l}_{q\mu}}{r^{s,l}_{qq}} = \frac{R^{s,g}_{qq}}{R^{s,l}_{qq}+R^{s,g}_{qq}}\Delta_{vap}H. \tag{8.32}$$

The heat of transfer that refers to the heat flux on the liquid side is positive. A general relation which follows from eqn (8.1) and these definitions of heats of transfer is[4]

$$q^{\star s,g} - q^{\star s,l} = -\Delta_{vap}H \tag{8.33}$$

We see that the sign of the enthalpy of the phase transformation decides the sign(s) of the heat of transfer, and that its magnitude is a fraction of the enthalpy of the phase change. The relation (8.33) means that it is not possible to neglect both heats of transfer in the modelling of phase transitions. That will violate the energy balance. It may also violate the Onsager symmetry and thus the second law. Assuming that one of the heats of transfer is neglected, the other must then be associated with all of the enthalpy of evaporation. According to eqn (8.31) or (8.32) this requires that one of the resistivities $R^{s,l}_{qq}$ or $R^{s,g}_{qq}$ is negligible, which is unlikely. Van der Ham *et al.*[48] studied distillation of nitrogen and oxygen in a cryogenic distillation column using the equations above. The neglect of coupling coefficients at the interface led to an error of 11 % in the nitrogen flux and an error of 39 % in the heat flux on the liquid side. Clearly, the habit to neglect these coefficients should change.

Considering eqn (8.20) only, kinetic theory seems to be concerned solely with the transfer resistivities on the vapour side of the surface. This is not correct. The above analysis shows that kinetic theory uses boundary conditions that refer to the whole surface.

8.4.4 Square Gradient Theory, Integral Relations for Interface Transfer Coefficients

The equilibrium square gradient theory gives analytical expressions for the variation in thermodynamic functions through the interfacial region (along the *x*-axis, and in time *t*). It was originally introduced by van der Waals, see ref. 25 and references therein. The expressions are, as the name says, functions of the density and density gradient squared. In addition to the contribution from, *e.g.*, the van der Waals equation of state (A_W), the

Helmholtz energy, A, obtains a contribution from the gradient of the molar density, c, as follows:

$$A(x,t) = A_W(c(x,t), T(x,t)) + \frac{m}{2c(x,t)}\left[\frac{\partial c(x,t)}{\partial x}\right]^2. \tag{8.34}$$

The influence parameter m was independent of density and temperature in van der Waals' first formulation. The set of thermodynamic equations, consistent with this, is solved for the condition of a minimum in the Helmholtz energy for a given total number of particles. This gives the molar density profile for 1- and 2-phase pure fluids. From standard thermodynamic relations it is then possible to derive expressions for thermodynamic variables of interest.[27,30] This gives *equilibrium* profiles of all thermodynamic functions. From the profile of the pressure parallel to the surface we calculate the surface tension. This calculation can be used to calibrate the value of the influence parameter m. It is chosen such that it gives the correct interfacial tension at some typical temperature.

Away from equilibrium, the expressions for the thermodynamic variables remain valid and give the time- and position-dependent profiles of these variables. For an extensive discussion we refer the reader to ref. 27–29 for pure fluids and ref. 30–34 for mixtures. The expressions give, *e.g.*, the enthalpy profile which is needed in eqn (8.35). The excess variables obtained from these functions obey local equilibrium,[28,31] *cf.*, Chapter 4. The square gradient description itself is, as one can see from eqn (8.34), inherently non-local and does consequently not satisfy local equilibrium.

In order to calculate the transfer resistivities for single components and planar interfaces, Johannessen and Bedeaux introduced integral relations.[38] The expressions given below give the inverse temperature and chemical-potential differences across the surface of discontinuity[41] located between 0 and δ. In order to obtain the resistivities across the dividing surface one must subtract the values for the homogeneous phases, as extrapolated to the dividing surface. We refer the reader to ref. 38 for a discussion of this. The integral relations for the transfer resistivities are:

$$\begin{aligned}
r_{qq}^{g} &= \int_0^{\delta} r_{qq}(x)\mathrm{d}x, \\
r_{\mu q}^{g} &= \int_0^{\delta} r_{qq}(x)(H^{g} - H(x))\mathrm{d}x, \\
r_{\mu\mu}^{g} &= \int_0^{\delta} r_{qq}(x)(H^{g} - H(x))^2\mathrm{d}x.
\end{aligned} \tag{8.35}$$

The integrands contain the *local* resistivity to heat transfer, $r_{qq}(x)$, times a function of the enthalpy which varies through the interface. Non-equilibrium molecular dynamics simulations are able to provide the local resistivity to heat transfer. In Figure 8.2 it is given for n-octane according to

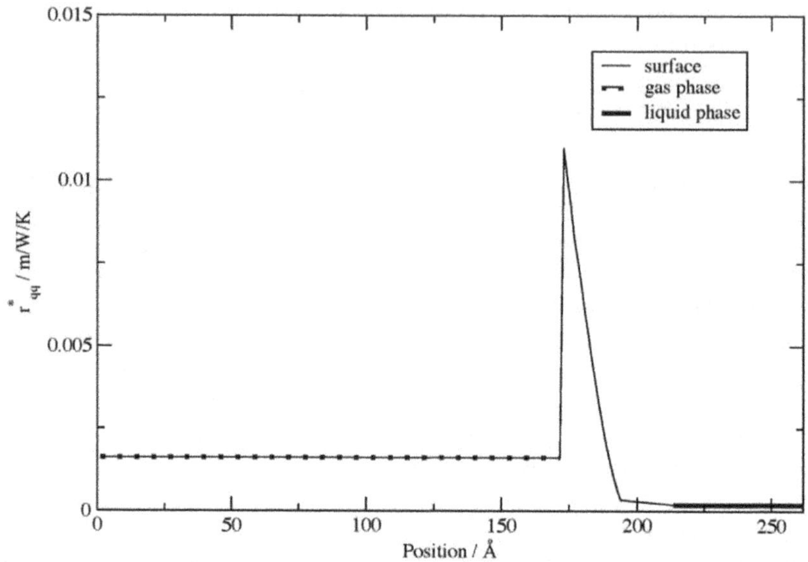

Figure 8.2 The thermal resistivity variation across a vapour (left)–liquid (right) interface computed for *n*-octane according to Simon *et al.*[21] The peak in the resistivity is located on the vapour side of the equimolar surface. Reprinted with permission from *J. Phys. Chem.*[21]

Simon *et al.*[21] The enthalpy profile is obtained from the square gradient model; solved under equilibrium conditions. The local resistivity has also been modelled in the context of the square gradient theory.[27–34] The integral relations were extended to mixtures and to curved surfaces by Glavatskiy and Bedeaux.[30] In the square gradient analysis, one uses a contribution proportional to the square of the density gradient. The location of this peak is important.[21–23,36] Molecular dynamics results for a pure component[21] indicate that the peak shifts away from the equimolar surface towards the vapour side. The curvature dependence of the transfer resistivities depends strongly on the location of this peak.[36]

8.4.5 Coefficients from Simulations, Experiments and Square Gradient Theory

In non-equilibrium molecular dynamics simulations (NEMD) the computer is used to solve Newton's equations for a many particle system, *cf.*, Chapter 11. It was verified by Røsjorde *et al.*[19,20] (Lennard-Jones spline particles) and by Simon *et al.*[21] (a model of *n*-octane) that the surface temperature was the same function of the surface tension, in stationary state simulations and in equilibrium simulations. This implies that the surface is in local equilibrium (*cf.*, Chapter 4).

NEMD has been used to determine the transfer resistivities of a pure fluid of Lennard-Jones particles.[19–22] The Lennard-Jones spline potential was

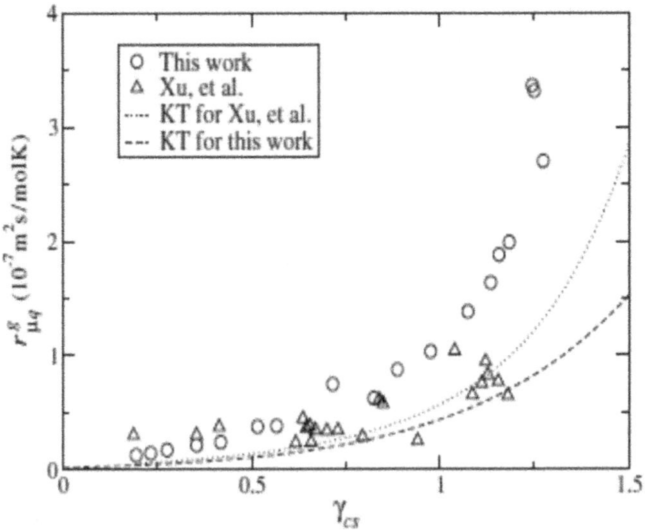

Figure 8.3 The interfacial transfer resistivity $r^{g}_{\mu q}$ for coupling between heat and mass in a one-component system. The resistivity is plotted as a function of the corresponding state (CS) interfacial tension $\gamma_{CS} \equiv \gamma\left(\rho_c^{2/3} k_B T_c\right)^{-1}$, where the subscript c indicates the value at the critical point. Results represented by triangles (from Xu *et al.*[22]) have a cut-off radius of 1.7σ in the potential, where σ is the molecular diameter. Results denoted "This work" (from Ge *et al.*[23]) represent a potential with cut-off radius 2.5σ. The results from kinetic theory (KT) are shown for comparison. Reprinted with permission from *Phys. Rev. E.*[23]

rather short range (the cut-off diameter is 1.7 times the molecular diameter, σ). The resulting transfer resistivities agreed well with the values predicted by kinetic theory, see Figure 8.3. This was to be expected, as kinetic theory applies to hard spheres. For a longer range Lennard-Jones spline potential, with a cut-off diameter 2.5σ, Ge *et al.*[23] found that the main coefficients did not deviate much from kinetic theory. The coupling coefficient or the heat of transfer was, however, about three times larger than the value from kinetic theory. The potential minimum and the particle diameter were equal to the values with the smaller cut-off, meaning that the increase in the coupling coefficient can be associated with the longer-range of the interaction potential.

The values of the transfer resistivities found from NEMD simulations for Lennard-Jones systems and from kinetic theory lead to discontinuities in the temperature at the surface of a few tenths of a degree rather than a few degrees. As shown by Bedeaux and Kjelstrup,[16] it is necessary to have much larger interfacial resistivities in order to explain the temperature differences of several degrees observed by Fang and Ward,[10,11] Phillips and co-workers[12–14] and Badam *et al.*[15]

Additional insight was obtained from the square gradient theory.[25–33] A peak in the thermal resistivity $r_{qq}(x)$ in the interfacial region, *cf.*, Figure 8.2, was chosen in agreement with the NEMD results.[21] With knowledge of this peak, the overall interfacial transfer resistivities were found from eqn (8.35). Wilhelmsen *et al.*[36] used these formulae to find the dependence of the transfer resitivities on surface curvature in pure fluids and mixtures, see Figure 8.4. The surface curvature had a considerable impact on the transfer resistivities for heat and mass transfer in nanometre-sized bubbles and droplets. The heat of transfer behaved similarly, for single and two-component fluids. It depended heavily on the position of the peak.

The few findings for single-fluid evaporation or condensation can be summarized as follows: kinetic theory seems to predict the thermal transfer resistivities well, for particles with short as well as long-range interactions. Kinetic theory considerably underestimates the transfer resistivities for coupling, when the particle interactions are long-range. The heats of transfer of the surface can be estimated as a fraction of the enthalpy of evaporation. The results from square gradient theory find that the heats of transfer are rather insensitive to surface curvature. They do, however, depend on the position of the peak resistance of the thermal resistivity, $r_{qq}(x)$, relative to the

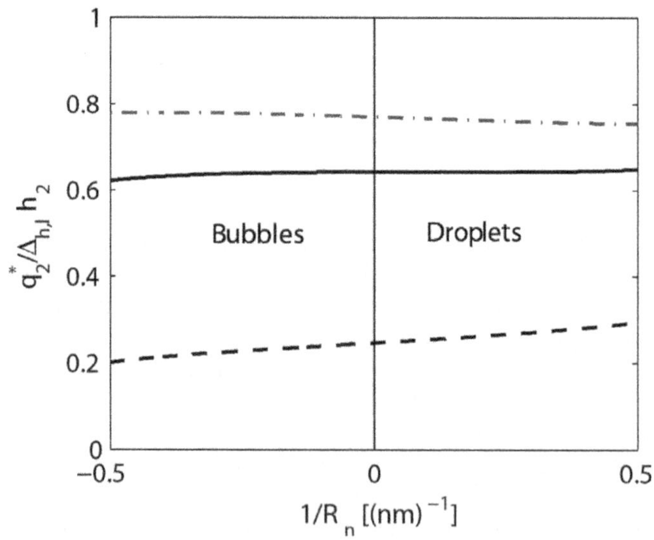

Figure 8.4 The heat of transfer, divided by the enthalpy of hexane, in a hexane–cyclohexane mixture calculated using square gradient theory and the integral relations. Values are shown for droplets and bubbles as a function of interface curvature $1/R_n$. Zero means a flat surface. The thermal resistivity $r_{qq}(x)$ has a peak near the equimolar surface (solid line), on the gas side (dashed line) or on the liquid side (dash-dotted line) of the surface.
Reprinted with permission from *Phys. Chem. Chem. Phys.*[36] Similar results were obtained for a one-component system.

equimolar surface. Experiments indicate that the transfer resistivities can be large compared to predictions from kinetic theory.

8.5 Evaporation or Condensation in Two-component Fluids

In a two-component system, the surface position is conveniently given by the equimolar surface of *one* of the two components. This component is called the solvent from now on. The other component is then the solute. The names are, of course, most appropriate when the overall concentration of the solvent exceeds the concentration of the solute. The equimolar surface of the solvent serves as the frame of reference for the mass fluxes. The solvent flux J_w is then always constant through the surface. The w is short for water, which is a typical solvent. The solute flux is, in general, not constant through the surface due to adsorption of solute at the dividing surface:

$$\frac{d\Gamma_{sol}}{dt} = J_{sol}^{l} - J_{sol}^{g}. \tag{8.36}$$

Adsorption of solute in the surface Γ_{sol} reduces the surface tension. In a stationary state, the concentration of solute and its absorption in the surface are everywhere constant, so that not only J_w, but also J_{sol}, is constant throughout the system. The analysis in this chapter is restricted to this case.

As in the one-component system, the measurable heat flux from the liquid into the surface, J'^{l}_{q}, is not the same as the measurable heat flux out of the surface into the gas, J'^{g}_{q}, because the component enthalpies differ from phase to phase. The total heat flux is the measurable heat flux plus the enthalpy carried by the solvent and the solute fluxes. In the stationary state J_q is constant throughout the system and:

$$J_q = J'^{g}_{q} + H_{w}^{g} J_w + H_{sol}^{g} J_{sol} = J'^{l}_{q} + H_{w}^{l} J_w + H_{sol}^{l} J_{sol}. \tag{8.37}$$

The entropy production for the liquid–vapour interface in the stationary two-component system is equal to:

$$\sigma^s = J'^{l}_{q} \Delta_{l,s}\left(\frac{1}{T}\right) + J'^{g}_{q} \Delta_{s,g}\left(\frac{1}{T}\right) + J_{sol}\left[-\frac{\Delta_{l,g}\mu_{sol,T}(T^s)}{T^s}\right] + J_w\left[-\frac{\Delta_{l,g}\mu_{w,T}(T^s)}{T^s}\right]. \tag{8.38}$$

The chemical potential differences are evaluated at temperature T^s. The heat flux in the liquid can be eliminated using eqn (8.37). Following the procedure outlined in Section 8.3, this results in

$$\sigma^s = J'^{g}_{q} \Delta_{l,g}\left(\frac{1}{T}\right) + J_{sol}\left[-\frac{\Delta_{l,g}\mu_{sol,T}(T^l)}{T^l}\right] + J_w\left[-\frac{\Delta_{l,g}\mu_{w,T}(T^l)}{T^l}\right]. \tag{8.39}$$

where the chemical-potential differences are now evaluated at the temperature of the liquid, T^l. For an ideal vapour we can again use eqn (8.7) to

(8.9), to express the chemical-potential differences in the partial pressures. The force conjugate to J_w is therefore:

$$-\frac{\Delta_{l,g}\mu_{w,T}(T^l)}{T^l} = -R\ln\frac{p_w}{p^*{}_w(T^l)}, \tag{8.40}$$

and the force conjugate to J_{sol} is:

$$-\frac{\Delta_{l,g}\mu_{sol,T}(T^l)}{T^l} = -R\ln\frac{p_{sol}}{p^*{}_{sol}(T^l)}. \tag{8.41}$$

For each component, the partial pressure is divided by the partial pressure of the saturated vapour at the temperature of the liquid, T^l. When the vapour is not ideal, we must use the fugacity instead of the pressure. Expressions for activity coefficients in non-ideal gas and in liquid mixtures can be found in Perry.[49]

The force–flux equations that can be derived from eqn (8.39) are:

$$\begin{aligned}
\Delta_{l,g}\left(\frac{1}{T}\right) &= r_{qq}^{s,g}J'^{g}_{q} + r_{qs}^{s,g}J_{sol} + r_{qw}^{s,g}J_w,\\
-\frac{\Delta_{l,g}\mu_{sol,T}(T^l)}{T^l} &= r_{sq}^{s,g}J'^{g}_{q} + r_{ss}^{s,g}J_{sol} + r_{sw}^{s,g}J_w,\\
-\frac{\Delta_{l,g}\mu_{w,T}(T^l)}{T^l} &= r_{wq}^{s,g}J'^{g}_{q} + r_{ws}^{s,g}J_{sol} + r_{ww}^{s,g}J_w.
\end{aligned} \tag{8.42}$$

In order to simplify the definition of the heats of transfer and of the co-transfer coefficient, it is convenient to also write the fluxes in terms of the forces. This gives

$$\begin{aligned}
J'^{g}_{q} &= l_{qq}^{s,g}\Delta_{l,g}\left(\frac{1}{T}\right) + l_{qs}^{s,g}\left[-\frac{\Delta_{l,g}\mu_{sol,T}(T^l)}{T^l}\right] + l_{qw}^{s,g}\left[-\frac{\Delta_{l,g}\mu_{w,T}(T^l)}{T^l}\right],\\
J_{sol} &= l_{sq}^{s,g}\Delta_{l,g}\left(\frac{1}{T}\right) + l_{ss}^{s,g}\left[-\frac{\Delta_{l,g}\mu_{sol,T}(T^l)}{T^l}\right] + l_{sw}^{s,g}\left[-\frac{\Delta_{l,g}\mu_{w,T}(T^l)}{T^l}\right],\\
J_w &= l_{wq}^{s,g}\Delta_{l,g}\left(\frac{1}{T}\right) + l_{ws}^{s,g}\left[-\frac{\Delta_{l,g}\mu_{sol,T}(T^l)}{T^l}\right] + l_{ww}^{s,g}\left[-\frac{\Delta_{l,g}\mu_{w,T}(T^l)}{T^l}\right].
\end{aligned} \tag{8.43}$$

where the **l**-matrix of transfer conductivities is the inverse of the **r**-matrix. In addition to the heat of transfer of the solvent, we define the heat of transfer for the solute:

$$\begin{aligned}
q^{*s,g}_{w} &\equiv \left(\frac{J'^{g}_{q}}{J_w}\right)_{\Delta T=\Delta\mu_{sol}=0} = \frac{l_{qw}^{s,g}}{l_{ww}^{s,g}},\\
q^{*s,g}_{sol} &\equiv \left(\frac{J'^{g}_{q}}{J_{sol}}\right)_{\Delta T=\Delta\mu_{w}=0} = \frac{l_{qw}^{s,g}}{l_{ss}^{s,g}}.
\end{aligned} \tag{8.44}$$

We also define the *co-transfer coefficient*:

$$\tau_{\rm sol}^{\rm s,g} \equiv \left(\frac{J_{\rm sol}}{J_{\rm w}}\right)_{\Delta T = \Delta\mu_{\rm sol} = 0} = \frac{l_{\rm sw}^{\rm s,g}}{l_{\rm ww}^{\rm s,g}}. \tag{8.45}$$

By introducing the heats of transfer and the co-transfer coefficient, the linear relations (8.43) become:

$$\begin{aligned}
J'^{\rm g}_q &= l_{qq}^{\rm s,g}\Delta_{\rm l,g}\left(\frac{1}{T}\right) - q^{*\rm s,g}_{\rm sol}\, l_{\rm ss}^{\rm s,g}\frac{\Delta_{\rm l,g}\mu_{{\rm sol},T}(T^{\rm l})}{T^{\rm l}} - q^{*\rm s,g}_{\rm w}\, l_{\rm ww}^{\rm s,g}\frac{\Delta_{\rm l,g}\mu_{{\rm w},T}(T^{\rm l})}{T^{\rm l}},\\
J_{\rm sol} &= q^{*\rm s,g}_{\rm sol}\, l_{\rm ss}^{\rm s,g}\Delta_{\rm l,g}\left(\frac{1}{T}\right) - l_{\rm ss}^{\rm s,g}\frac{\Delta_{\rm l,g}\mu_{{\rm sol},T}(T^{\rm l})}{T^{\rm l}} - \tau_{\rm sol}^{\rm s,g}\, l_{\rm ww}^{\rm s,g}\frac{\Delta_{\rm l,g}\mu_{{\rm w},T}(T^{\rm l})}{T^{\rm l}},\\
J_{\rm w} &= q^{*\rm s,g}_{\rm w}\, l_{\rm ww}^{\rm s,g}\Delta_{\rm l,g}\left(\frac{1}{T}\right) - \tau_{\rm sol}^{\rm s,g}\, l_{\rm ww}^{\rm s,g}\frac{\Delta_{\rm l,g}\mu_{{\rm sol},T}(T^{\rm l})}{T^{\rm l}} - l_{\rm ww}^{\rm s,g}\frac{\Delta_{\rm l,g}\mu_{{\rm w},T}(T^{\rm l})}{T^{\rm l}}.
\end{aligned} \tag{8.46}$$

These equations govern evaporation and condensation of a binary mixture.

Kinetic theory also gives expressions for transfer resistivities in two-component systems. We will not give the rather complicated expressions here, but refer the reader to ref. 4. Molecular dynamics simulations were done by Inzoli *et al.*[24] for a two-component fluid mixture with short-ranged Lennard-Jones spline potentials. The particles had the same diameter and mass, but their potential depths differed by a factor 0.8. The resistivities for heat and mass transfer were determined. The coupling coefficients for heat and mass transfer were significant. The results for the heats of transfer, defined with the heat flux on the liquid side, as well as results using the heat flux on the gas side, are shown as functions of the interfacial tension in Figure 8.5 for the two components. The differences in the corresponding values obeyed eqn (8.33).

Phillips and coworkers measured the transfer resistivities for mixtures.[12–14] For a discussion of this work, see ref. 17 and 18.

Also, the square gradient theory was used to calculate transfer resistivities for binary mixtures.[33,36] Three amplitudes were chosen. If they were chosen such that the predictions of kinetic theory for the diagonal transfer resistivities were reproduced, the coupling coefficients differed from kinetic theory not only in size, but also in sign.

Evidence from molecular dynamics simulations[24] gave good agreement with $r_{qq}^{\rm s,g}$ from kinetic theory for short-range potentials. The remaining transfer resistivities did not compare well with the results of kinetic theory, however. Models using the square gradient model reinforced this picture. Further NEMD simulations are therefore needed to bring more clarity. Such simulations should clearly not only be done for short-range potential interactions, but also for longer ranges.

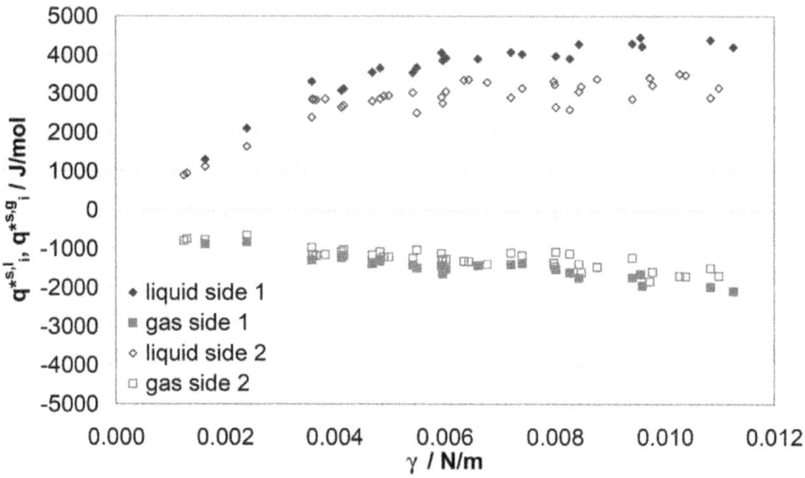

Figure 8.5 The heats of transfer for the liquid side and the gas side heat fluxes, as a function of the surface tension, in a two-component Lennard-Jones mixture.[24] Results are shown for component 1 (filled symbols) and 2 (open symbols).
Reprinted with permission of *Chem. Eng. Sci.*[24]

8.6 Thermodynamic Properties of a Contact Line

In order to give a solid foundation of the description of the motion of the three-phase contact line, a non-equilibrium thermodynamic description of the contact line was developed.[50] It was postulated that during its motion the contact line is a separate thermodynamic system. For this contact line one can define excess masses, momentum and energy densities for which conservation laws are given. Also excess fluxes along the line are defined. These quantities depend on the choice of the dividing surfaces between the phases. These dividing surfaces should be chosen such that they cross each other through the same contact line. There are many such choices. We will not go into the mathematical details. We only observe that the experimental properties of the contact line are independent of the specific choice. It is clear that the physics of the contact line cannot depend on the precise location we choose, as this location is only a mathematical construct. The same is true for the choice of the location of the dividing surfaces.

The Gibbs equation was formulated for the contact line and on the basis of this law the excess entropy production density along the line was constructed. This identifies the conjugate thermodynamic forces and fluxes for the contact line. Linear laws relating these quantities were then given.

An important aspect in the description of the motion of the contact line is that excess densities develop along the surfaces between the phases near the contact line. This leads to apparent slip of the bulk fluids along the dividing surfaces. An advantage is that the velocity field is no longer singular near the contact line as is the consequence when no slip boundary conditions are used. We refer the reader to the work of Shikhmurzaev[51] for a detailed discussion of this. For a treatment of the non-equilibrium thermodynamics of the contact line we refer the reader to ref. 50.

8.7 Concluding Remarks

We have seen above that the coupling resistivity for transfer of heat and mass across an interface can be very substantial. The size of the resulting heat of transfer has the same order of magnitude as the enthalpy of the phase change. Unlike the situation in homogeneous phases, the coupling cannot be neglected. As a consequence, it is not sufficient to use simple transport laws, like Fourier's law, to model transport of heat across surfaces. As we discussed, a neglect of the coupling resistivities violates the first law. The neglect can lead to large errors in particular in the heat flux at the surface.[48] These results will have a bearing on the dynamic modelling of phase transitions.

Molecular dynamics results show that kinetic theory predicts the resistivity to heat transfer correctly. The coupling resistivities for heat and mass transfer are found to differ substantially from kinetic theory predictions for long range interaction potentials, however. They can be substantially larger and even have different signs. Experiments favour larger values of the transfer resistivities. While some coefficients are known for surfaces, almost nothing is known for contact lines.

Given the general nature of the phenomenon and the large discrepancy between experiments and theory, there is a need for more knowledge on these issues. The results above may to some degree also apply to the solid–liquid transition, adsorption and membrane transport studies,[52] see Chapters 10 and 11. In view of the large errors possibly made in present day's modelling of interface heat fluxes, more knowledge is needed. We will welcome more NEMD simulations studies with a variety of long-range potentials. Such studies, in combination with density functional theories and further experiments, should provide more insight.

References

1. P. L. Kapitza, *J. Phys.*, 1941, **4**, 181.
2. R. Byron Bird, W. E. Stewart and E. N. Lightfoot, *Transport Phenomena*, 2nd rev. edn, Wiley, New York, 2007.
3. R. Taylor and R. Krishna, *Multicomponent Mass Transfer*, Wiley, New York, 1993.
4. S. Kjelstrup and D. Bedeaux, *Non-Equilibrium Thermodynamics of Heterogeneous Systems*, Series on Advances in Statistical Mechanics, World Scientific, Singapore, 2008, vol. 16.
5. D. Bedeaux, L. F. J. Hermans and T. Ytrehus, *Physica A*, 1990, **169**, 263.
6. D. Bedeaux, J. A. M. Smit, L. J. F. Hermans and T. Ytrehus, *Physica A*, 1992, **182**, 388.

7. J. W. Cipolla, Jr., H. Lang and S. K. Loyalka, *J. Chem. Phys.*, 1974, **61**, 69.
8. J. W. Cipolla, Jr., H. Lang and S. K. Loyalka, in *Rarified Gas Dynamics*, ed. Beeker and Fiebig, DFVLR-Press, Port-Wahn, Germany, 1974, vol. II, pp. 1–10.
9. C. Cercignani, in *Rarified Gas Dynamics*, ed. S. S. Fisher, AIAA, New York, 1981, part 1, p. 305.
10. G. Fanga and C. A. Ward, *Phys. Rev. E: Stat. Phys., Plasmas, Fluids, Relat. Interdiscip. Top.*, 1999, **59**, 417.
11. C. A. Ward and G. Fang, *Phys. Rev. E: Stat. Phys., Plasmas, Fluids, Relat. Interdiscip. Top.*, 1999, **59**, 429.
12. C. T. Mills and L. F. Phillips, *Chem. Phys. Lett.*, 2002, **366**, 279.
13. C. J. Pursell and L. F. Phillips, *Phys. Chem. Chem. Phys.*, 2006, **8**, 4694.
14. C. T. Mills and L. F. Phillips, *Chem. Phys. Lett.*, 2002, **372**, 615.
15. V. K. Badam, V. Kumar, F. Durst and K. Danov, *Exp. Therm. Fluid Sci.*, 2007, **32**, 276.
16. D. Bedeaux and S. Kjelstrup, *Physica A*, 1999, **270**, 413.
17. H. Struchtrup, S. Kjelstrup and D. Bedeaux, *Phys. Rev. E: Stat., Nonlinear, Soft Matter Phys.*, 2012, **85**, 061201-6.
18. H. Struchtrup, S. Kjelstrup and D. Bedeaux, *Int. J. Heat Mass Transfer*, 2013, **85**, 221.
19. A. Røsjorde, D. W. Fossmo, D. Bedeaux, S. Kjelstrup and B. Hafskjold, *J. Colloid Interface Sci.*, 2000, **232**, 178–185.
20. A. Røsjorde, D. Bedeaux, S. Kjelstrup and B. Hafskjold, *J. Colloid Interface Sci.*, 2001, **240**, 355.
21. J.-M. Simon, S. Kjelstrup, D. Bedeaux and B. Hafskjold, *J. Phys. Chem. B*, 2004, **108**, 7186.
22. J. Xu, S. Kjelstrup, D. Bedeaux, A. Røsjorde and L. Rekvig, *J. Colloid Interface Sci.*, 2006, **299**, 452.
23. J. Ge, S. Kjelstrup, D. Bedeaux, J.-M. Simon and B. Rousseau, *Phys. Rev. E: Stat., Nonlinear, Soft Matter Phys.*, 2007, **75**, 061604–14.
24. I. Inzoli, S. Kjelstrup, D. Bedeaux and J.-M. Simon, *Chem. Eng. Sci.*, 2011, **66**, 4533.
25. J. S. Rowlinson, *J. Stat. Phys.*, 1979, **20**, 197.
26. J. W. Cahn and J. E. Hilliard, *J. Chem. Phys.*, 1958, **28**, 258.
27. D. Bedeaux, E. Johannessen and A. Røsjorde, *Physica A*, 2003, **330**, 329.
28. E. Johannessen and D. Bedeaux, *Physica A*, 2003, **330**, 254.
29. E. Johannessen and D. Bedeaux, *Physica A*, 2004, **336**, 252.
30. K. S. Glavatskiy and D. Bedeaux, *Phys. Rev. E: Stat., Nonlinear, Soft Matter Phys.*, 2008, **77**, 061101.
31. K. S. Glavatskiy and D. Bedeaux, *Phys. Rev. E: Stat., Nonlinear, Soft Matter Phys.*, 2009, **79**, 031608.
32. K. S. Glavatskiy and D. Bedeaux, *J. Chem. Phys.*, 2010, **133**, 234501.
33. K. S. Glavatskiy and D. Bedeaux, *J. Chem. Phys.*, 2010, **140**, 104708.
34. K. S. Glavatskiy and D. Bedeaux, *Phys. Rev. E: Stat., Nonlinear, Soft Matter Phys.*, 2011, **83**, 019902.
35. K. S. Glavatskiy and D. Bedeaux, *J. Chem. Phys.*, 2014, **140**, 104708.

36. Ø. Wilhelmsen, D. Bedeaux and S. Kjelstrup, *Phys. Chem. Chem. Phys.*, 2014, **16**, 10573.
37. E. Magnanelli, Ø. Wilhelmsen, D. Bedeaux and S. Kjelstrup, *Phys. Rev. E*, 2014, **90**, 032402.
38. E. Johannessen and D. Bedeaux, *Physica A*, 2006, **370**, 258.
39. E. Johannessen and D. Bedeaux, *Phys. Fluids*, 2007, **19**, 017104.
40. E. Johannessen, J. Gross and D. Bedeaux, *J. Chem. Phys.*, 2008, **129**, 184703.
41. J. W. Gibbs, *Collected Works*, Dover, London, 1961, vol. 2.
42. A. M. Albano, D. Bedeaux and J. Vlieger, *Physica A*, 1979, **99**, 293.
43. A. M. Albano, D. Bedeaux and J. Vlieger, *Physica A*, 1980, **102**, 105.
44. T. Savin, K. S. Glavatskiy, S. Kjelstrup, H. C. Öttinger and D. Bedeaux, *Eur. Phys. Lett.*, 2012, **97**, 40002.
45. D. Bedeaux, S. Kjelstrup and H. C. Öttinger, *J. Chem. Phys.*, 2014.
46. S. Kjelstrup, T. Tsuruta and D. Bedeaux, *J. Colloid Interface Sci.*, 2002, **256**, 451.
47. T. Ytrehus and S. Østmo, *Int. J. Multiphase Flow*, 1996, **22**, 133.
48. L. V. van der Ham and S. Kjelstrup, *Int. J. Thermodyn.*, 2011, **14**, 179.
49. R. H. Perry and D. W. Green, *Perry's Chemical Engineers Handbook*, McGraw-Hill, New York, 7th edn, 1997.
50. D. Bedeaux, *J. Chem. Phys.*, 2004, **120**, 3744.
51. Y. D. Shikhmurzaev, *J. Fluid Mech.*, 1997, **334**, 211.
52. P. Villaluenga and S. Kjelstrup, *J. Non-Equilib. Thermodyn.*, 2012, **37**, 253.

CHAPTER 9

Non-equilibrium Thermodynamics Applied to Adsorption

JEAN-MARC SIMON

Laboratoire Interdisciplinaire Carnot de Bourgogne, UMR 6303
CNRS-Université de Bourgogne Franche Comté, 9 Av. Savary, FR-21000, Dijon, France
Email: jmsimon@u-bourgogne.fr

9.1 Introduction to Adsorption Phenomena

Many natural or industrial processes involve adsorption steps where mass and heat transfer take place at surfaces. For example, in a simple picture of heterogeneous catalysis, reactants located in a fluid phase are first adsorbed on the solid surface, the adsorbent (see Figure 9.1), and then moved to catalytic sites inside or on the material. Later, the products move to a place on the surface where they desorb. It is common to distinguish between physisorption and chemisorption. The latter implies a chemical reaction between the surface and the adsorbate. In general, large energetic effects are involved in chemisorption; larger than for physisorption. Although there are many similarities, this chapter will be restricted to physisorption.

The adsorption process at the scale of a particle on a porous solid is illustrated in Figure 9.1. It is normally divided in successive steps. First, the adsorbate in the fluid phase moves towards the interface, where it adsorbs. When adsorbed on the external surface of the solid, the particles move or are stuck at particular sites. Depending on the interaction energy with the

Experimental Thermodynamics Volume X: Non-equilibrium Thermodynamics with Applications
Edited by Dick Bedeaux, Signe Kjelstrup and Jan V. Sengers

Published by the Royal Society of Chemistry, www.rsc.org

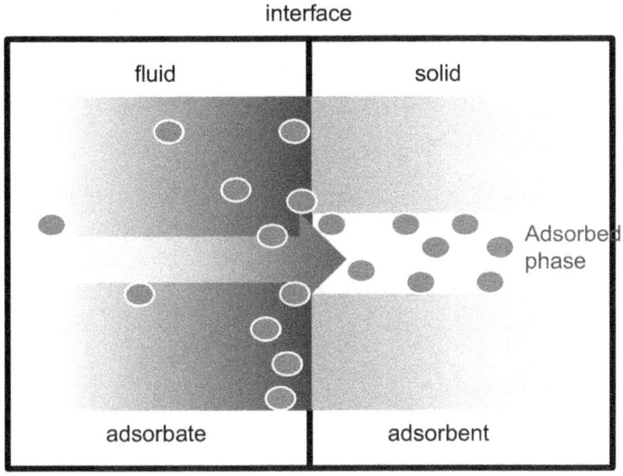

Figure 9.1 Schematic representation of an adsorption process between a fluid phase and a porous solid.

surface, they can desorb back to the fluid phase or be trapped into open channels. Being trapped, they diffuse inside the channel. On the macroscopic scale, we distinguish between three steps, *i.e.*, the transport in the fluid phase, the crossing of the interface, and the transport inside the porous solid. All these processes can be described by non-equilibrium thermodynamics (NET). The aim of this Chapter is to illustrate how NET can be applied to help understand adsorption kinetics for a gas in contact with a microporous material. As an example we use results obtained with molecular dynamics simulations of adsorption of *n*-butane on the zeolite silicalite-1.

We review briefly the kinetics of adsorption and its challenges in this section. The system *n*-butane in contact with silicalite-1 will be described under equilibrium conditions. In Section 9.2, the NET equations will be given. They are used in Section 9.3 to describe the transport of heat and mass in zeolites by using molecular dynamics simulations of processes inside the crystal and across the external surface. In particular it will be shown how systematic use of NET can lead to new insights in adsorption kinetics.

9.1.1 State-of-the-art

Experimentally, adsorption kinetics is treated by considering that the limiting step is diffusion of the adsorbate on the surface of the adsorbent. The mass transfer through the gas phase and in the interface is not considered *per se*. For simple geometries (infinite membranes, spheres, *etc.*) of the adsorbent structure, exact analytical solutions of Fick's law are well known,[1] and the analysis of uptake curves leads directly to Fick's diffusion coefficients of the adsorbed component, *i.e.*, the intra-crystalline diffusion coefficient.

Values of the diffusion coefficients of *n*-alkane chains, adsorbed on zeolite of MFI type are shown in Figure 9.2. They were measured with the transient

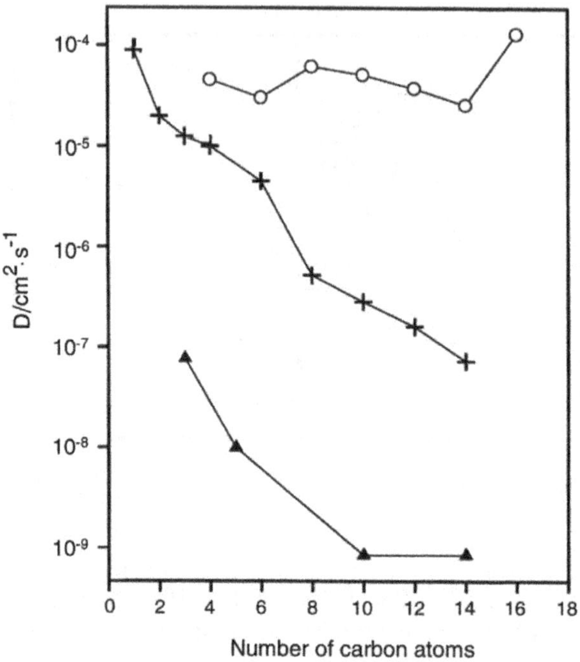

Figure 9.2 Diffusion coefficients of chains of *n*-alkanes adsorbed in a MFI type zeolite at 300 K from different techniques: zero length column (▲) ZLC, quasi-elastic neutron scattering (+) QENS, and equilibrium molecular dynamics (○) MD. Large discrepancies in the values are visible. Reported by H. Jobic.[2] Copyright (2000) by Elsevier.[2] Reproduced with permission.

method; the zero length column (ZLC) method. The results were compared with self-diffusion coefficients obtained under equilibrium conditions from quasi-elastic neutron scattering (QENS) and with results from molecular dynamics simulations. It appears that the values of the diffusion coefficients depend strongly on the method used.[2,3] In general, the values obtained from analysis of the adsorption kinetics (from ZLC) are much smaller than those obtained from equilibrium methods. In Figure 9.2, they are two orders of magnitude smaller, and even smaller, if compared with results from equilibrium molecular dynamics. The discrepancies have been explained by crystal defaults, surface effects, non-isothermal conditions, *etc.*, that are known to have large impacts on the kinetics.[3,4]

Transport phenomena in the gas phase are well known. Intra-crystalline heat and mass transfer is known to some extent. Little attention has been paid until now to heat and mass transfer through the surface. One of the reasons is that analyses of surfaces are hardly accessible by experiments.[4] A possibility to study systematically the surface during adsorption kinetics is offered by molecular simulations. Molecular simulations have been very powerful in the study of equilibrium of adsorbed phases. In particular

Monte-Carlo (MC) simulations in the Gibbs ensemble[5,6] have been used to compute isotherms. To get access to transport coefficients, like the diffusion coefficients given in Figure 9.2, equilibrium molecular dynamics simulations (MD)[6,7] have been used. However, few molecular simulations have focussed on interface transfer coefficients.[8–12] The role and effect of the surface on the adsorption kinetics will be illustrated here by simulation data from a simple system: *n*-butane on silicalite-1.

9.1.2 Adsorption of *n*-Butane on Silicalite-1: The Equilibrium State

Consider the silicalite-1 crystal in contact with a gas phase of *n*-butane. Silicalite-1 is a pure siliceous zeolite of MFI type. It has an orthorhombic *Pnma* crystallographic structure with lattice parameters $a = 2.0022$ nm, $b = 1.9899$ nm and $d = 1.3383$ nm.[13] It exhibits straight channels in the y-(b)-direction and zig-zag channels in the x-(a) and z-(d) directions. Crystals of dimension $L_x = 2a$, $L_y = 3b$, $L_z = 3d$ were simulated with periodic boundary conditions (PBC) in x- and z-directions, at the crystal limits. In the y-direction gas volumes were added such that infinite slabs were simulated. PBC were also applied in the y-direction at the gas limits. In order to mimic adsorption experiments, the number of *n*-butane molecules in the gas was varied. In the molecular dynamics simulations, the system evolved by itself until equilibrium was reached. A thermostat was applied to the gas phase to maintain a constant temperature during the adsorption.[10,11] Atomic models were used for the zeolite, and *n*-butane was modelled as four united atoms of methyl and methylene groups. Intramolecular and intermolecular Lennard-Jones potentials were applied, such that all the atoms were allowed to move. The simulated system is illustrated in Figure 9.3. More simulation details can be found in ref. 8 and 14.

In these simulations it was possible to study equilibrium as well as non-equilibrium conditions. The loading, c, *i.e.*, the number of molecules adsorbed per unit cell, and the gas pressure, p, according to the ideal-gas law, were calculated from knowledge of the number of molecules located in the adsorbed phase and the gas phase. In Figure 9.4 the computed isotherm, $c = f(p)$, has been compared with experimental results and with Monte-Carlo calculations.[8,14,15] The agreement is good, indicating that the phase-equilibrium conditions were well reproduced.

A heat of adsorption, ΔH_{ads}, of -55 kJ · mol^{-1}, has been obtained, also in excellent agreement with experiments.[8] The interaction energy between the crystal and the *n*-butane molecule along the y-direction is shown in Figure 9.5. The interaction potential is periodic, following the periodicity of the crystal. The analysis of the external surface gave a surface thickness around 1.2 nm. Changes in energy and/or loading were well reproduced by the simulation under equilibrium conditions. This is a prerequisite for studying non-equilibrium systems.

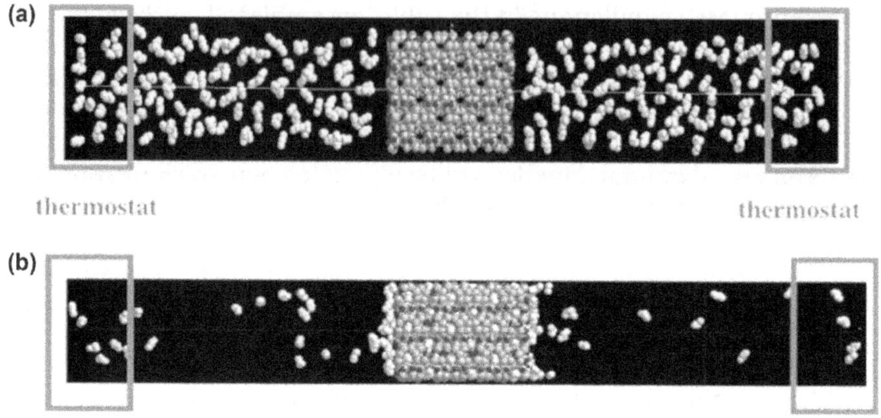

Figure 9.3 Instantaneous configurations of an infinite membrane of silicalite-1 zeolite in contact with a gas of *n*-butane from molecular dynamics simulations at 300 K. The pore openings are oriented towards the straight channels. Silica atoms are shown in layers, with oxygen atoms between. The methyl and methylene groups of the butane are light grey. In (a) the butane molecules are localized in the gas phase and the zeolite is empty. This is the initial configuration of the simulation. After reaching equilibrium conditions (b), butane molecules are adsorbed on the external surface and in the pores of the zeolite. Periodic boundary conditions are applied in all directions. To maintain the temperature at 300 K a thermostat is applied at the boundaries of the gas phases (in the rectangles).

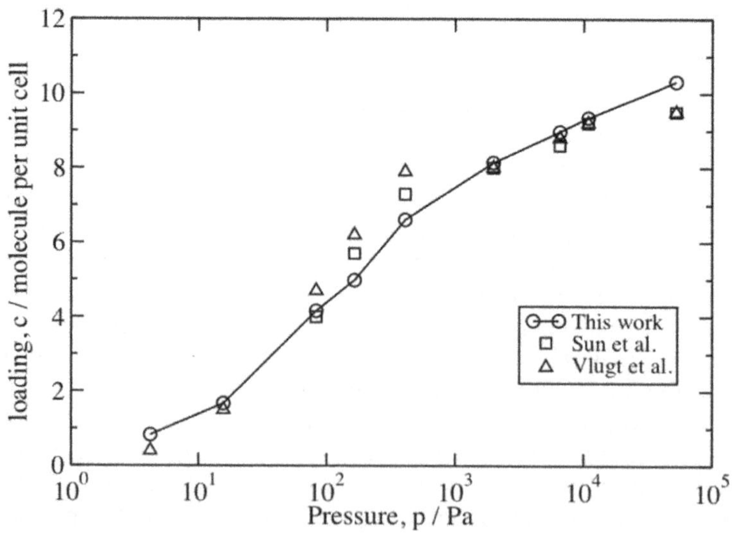

Figure 9.4 Adsorption isotherm from our molecular dynamics with explicit surfaces,[8] from experiments of Sun *et al.*,[15] and Monte-Carlo simulations obtained by Vlugt *et al.*[16] at 300 K.
Copyright (2004) by Taylor and Francis.[8] Reproduced with permission.

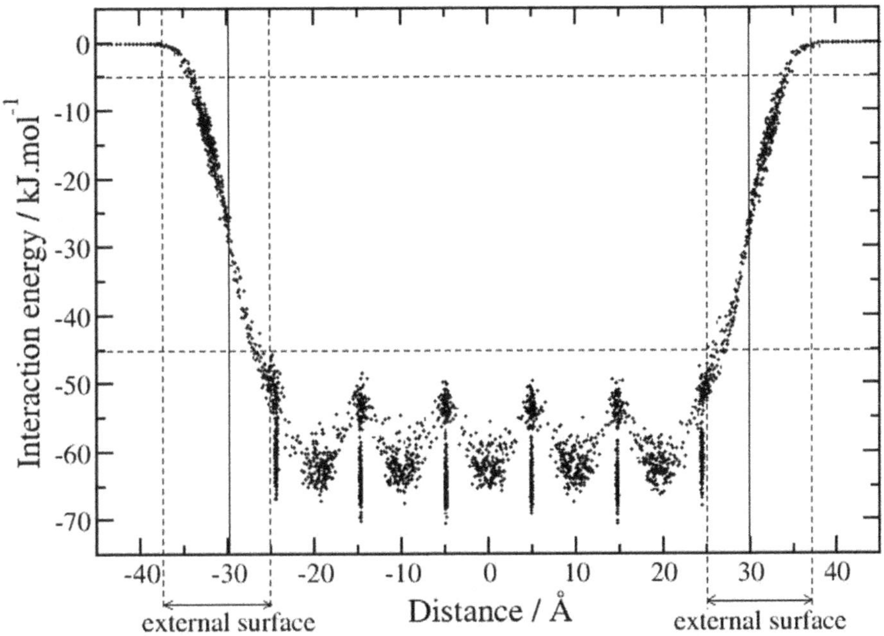

Figure 9.5 Profile of the potential energy between the silicalite and the molecules of *n*-butane at 300 K along the *y*-direction. Inside the crystal the profile of the energy is periodic, corresponding to the crystallographic periodicity of the straight and zig-zag channels and their intersections. In the figure are shown the crystallographic limits of the crystal (bold vertical lines) and the position of the external surfaces (dashed vertical lines), surface thickness is estimated to be around 1.2 nm.[11] For the units along the *x*-axis, 1 Ångstrom = 0.1 nm.
Copyright (2007) by Elsevier.[11] Reproduced with permission.

9.2 Adsorption Process: Non-equilibrium Description

When a system is shifted from equilibrium to non-equilibrium conditions, thermodynamic quantities like temperature, T, or chemical potential, μ, are no longer uniform in the system. As a response to the shift, the system tends to reach another equilibrium state by exchange of mass, heat or momentum, with the surroundings or within itself. There are irreversible processes, and entropy is produced in the system, $dS_{irr} > 0$. The entropy production, dS_{irr}/dt, quantifies the energy dissipated as heat, or the lost work during the time interval dt. In non-equilibrium thermodynamics the total entropy production of a system can be written as an integral over the entire volume where there is local entropy production, σ. Except under stationary-state conditions, the value of σ evolves with time t, like other variables. In the case of the adsorption of *n*-butane on silicalite-1, described in Section 9.1, the total entropy production is the sum of contributions from the gas phase, the adsorbed phase inside the crystal pores and from the interface:

$$\frac{dS_{irr}}{dt} = \int^{V^c} \sigma^c dV + \int^{\Omega^s} \sigma^s d\Omega + \int^{V^g} \sigma^g dV, \tag{9.1}$$

where V and Ω are, respectively, the volume and the surface area. Superscripts c, g and s refer to the crystal phase, the gas phase and the interface, respectively. In the following description, we consider transport along the x-direction perpendicular to the surface, and consider the system to be homogeneous in the other directions, y and z. It has the same area, Ω, whatever the x-position. Under these conditions the entropy production reduces to:

$$\frac{dS_{irr}}{dt} = \Omega \left[\int^{L^c} \sigma^c dx + \sigma^s + \int^{L^g} \sigma^g dx \right], \tag{9.2}$$

where L^c and L^g are thicknesses of the crystal- and gas phase in the x-direction. Non-equilibrium thermodynamics (NET) implies[17] that each local contribution to the entropy production is a sum of products of thermodynamic forces, X_i , and fluxes J_i. The mass flux of the adsorbate is measured relative to the frame of reference given by the crystal lattice and surface, *i.e.*, not the centre-of-mass frame of reference. With transport of one component, *n*-butane, and heat, the expression for σ is:

$$\sigma = J'_q X_q + J X_\mu, \tag{9.3}$$

Here J'_q and J are the x-components of the measurable heat flux and of the mass flux at time t. Following NET, the heat and mass flux can be written as the sum of products of Onsager's transport conductivities, L_{ij}, and thermodynamic forces, X_i:

$$\begin{aligned} J'_q &= L_{qq} X_q + L_{q\mu} X_\mu, \\ J &= L_{\mu q} X_q + L_{\mu\mu} X_\mu. \end{aligned} \tag{9.4}$$

Here L_{qq} and $L_{\mu\mu}$ are direct coefficients, while $L_{q\mu}$ and $L_{\mu q}$ are characteristic of the coupling between heat and mass fluxes. It follows from Onsager's reciprocal relations that the conductivity matrix is symmetric, $L_{q\mu} = L_{\mu q}$. It is practical for stationary states to express the thermodynamic forces as the product of resistivities, r_{ij}, and fluxes:

$$\begin{aligned} X_q &= r_{qq} J'_q + r_{q\mu} J, \\ X_\mu &= r_{\mu q} J'_q + r_{\mu\mu} J. \end{aligned} \tag{9.5}$$

Like for the L_{ij} coefficients, r_{qq} and $r_{\mu\mu}$ are the direct resistivities, while $r_{q\mu}$ and $r_{\mu q}$ are characteristic of the coupling between heat and mass fluxes; by symmetry $r_{q\mu} = r_{\mu q}$.

Both in experiments and in MD simulations, it is common to consider stationary conditions. This has the added advantage that the entropy production for the surface simplifies. Here we restrict ourselves to that case. The entropy production, eqn (9.3) can be written introducing the expression of the thermodynamic forces. For the crystal and gas phases we have:

$$\sigma(x) = J_q'(x)\frac{\mathrm{d}}{\mathrm{d}x}\left[\frac{1}{T(x)}\right] - J(x)\frac{1}{T(x)}\frac{\mathrm{d}}{\mathrm{d}x}[\mu(x)]_T, \tag{9.6}$$

where subscript T means that the chemical potential, μ, of the adsorbate should be differentiated keeping the temperature constant. For the surface, we have

$$\begin{aligned}\sigma^{\mathrm{s}} &= J_q'^{\mathrm{g,c}}\left[\frac{1}{T^{\mathrm{g,c}}} - \frac{1}{T^{\mathrm{c,g}}}\right] - J\left[\frac{\mu^{\mathrm{g,c}}(T^{\mathrm{c,g}}) - \mu^{\mathrm{c,g}}(T^{\mathrm{c,g}})}{T^{\mathrm{c,g}}}\right],\\ &= J_q'^{\mathrm{c,g}}\left[\frac{1}{T^{\mathrm{g,c}}} - \frac{1}{T^{\mathrm{c,g}}}\right] - J\left[\frac{\mu^{\mathrm{g,c}}(T^{\mathrm{g,c}}) - \mu^{\mathrm{c,g}}(T^{\mathrm{g,c}})}{T^{\mathrm{g,c}}}\right].\end{aligned} \tag{9.7}$$

In the first expression, the heat flux in the gas phase is used, while the heat flux in the crystal is used in the second one. The expressions are equivalent, but they lead to different sets of surface resistivity coefficients. The choice of using one instead of the other is mainly dictated by practical considerations, *i.e.*, the possibility to have access more accurate data. Superscript g,c means the gas phase close to the crystal, and c,g means the crystal close to the gas phase.

We see that the entropy production of the interface does not contain properties of the surface.

The equivalence of the expressions can be shown using the fact that the total heat flux, J_q, and the mass flux are the same on both sides of the surface. The measurable heat fluxes are connected at stationary state by:

$$J_q = J_q'^{\mathrm{g,c}} + H^{\mathrm{g,c}}J = J_q'^{\mathrm{c,g}} + H^{\mathrm{c,g}}J, \tag{9.8}$$

where $H^{\mathrm{i,j}}$ is the partial molar enthalpy of the component in phase i near phase j.

9.2.1 The Gas and the Crystal Phase

With the equality $r_{\mu q} = r_{q\mu}$, we can write the fluxes in eqn (9.6) in a stationary state as

$$\begin{aligned}J_q' &= -\lambda\frac{\mathrm{d}T}{\mathrm{d}x} + q^*J,\\ J &= -\frac{1}{r_{\mu\mu}T}\frac{\mathrm{d}}{\mathrm{d}x}\mu_T - \frac{q^*}{r_{\mu\mu}T^2}\frac{\mathrm{d}T}{\mathrm{d}x}.\end{aligned} \tag{9.9}$$

In these expressions, λ is the thermal conductivity and q^*, is the measurable heat of transfer; it gives the amount of heat transferred under constant temperature due to a mass flux:

$$q^* = \left(\frac{J_q'}{J}\right)_{X_q=0} = -\frac{r_{q\mu}}{r_{qq}}, \tag{9.10}$$

and

$$\lambda = -\left[\frac{J_q'}{(\mathrm{d}T/\mathrm{d}x)}\right]_{J=0} = \frac{1}{T^2 r_{qq}}. \tag{9.11}$$

The adsorbate interacts strongly with the crystal. The thermal conductance of a crystal is in general large compared to that of a fluid, and the crystal therefore contributes significantly to the transport of heat through a porous material. As a consequence, during adsorption, the temperature of the material is often considered constant. Transfer of heat will then be due to the mass flux and the heat of transfer, *cf.*, the last term of eqn (9.9). The mass flux can alternatively be expressed in terms of the Fick's diffusion coefficient, D and the thermal diffusion coefficient, D_T:[18]

$$J = -D\frac{\mathrm{d}c}{\mathrm{d}x} - cD_\mathrm{T}\frac{\mathrm{d}T}{\mathrm{d}x}, \tag{9.12}$$

where the loading c of the adsorbate was defined before. At constant temperature the expression reduces to Fick's first law of diffusion. We find:

$$\begin{aligned} D &= \frac{1}{r_{\mu\mu}T}\left(\frac{\mathrm{d}\mu}{\mathrm{d}c}\right)_\mathrm{T}, \\ D_\mathrm{T} &= \frac{q^*}{c r_{\mu\mu} T^2}. \end{aligned} \tag{9.13}$$

The ratio D_T/D is called the Soret coefficient, s_T. For a zero mass flux, the coefficient expresses the change in concentration as a function of a change in temperature:

$$s_\mathrm{T} = \frac{D_\mathrm{T}}{D} = -\frac{1}{c}\left[\frac{\mathrm{d}c}{\mathrm{d}x}\Big/\frac{\mathrm{d}T}{\mathrm{d}x}\right]_{J_k=0} = \frac{q^*}{RT^2\Gamma}, \quad \Gamma = \frac{1}{RT}\left(\frac{\mathrm{d}\mu}{\mathrm{d}\ln c}\right)_\mathrm{T}, \tag{9.14}$$

where Γ is the thermodynamic factor defined by the last equality and R is the gas constant. The Maxwell–Stefan theory[19,20] gives the diffusivity $D^\mathrm{MS} = D/\Gamma$. The expression applies for diffusion of one component. See ref. 20 for a generalisation to n components.

The above description is well defined for the gas phase. For the crystal phase, the density is not uniform when we are looking at sizes smaller than a unit cell.[21] Thermodynamic variables are only defined for length scales larger than the unit cell. In this case and with many other adsorbents, a description using the variation of the chemical potential is highly preferential to a description using the concentration variation. The first choice is therefore the natural choice for describing interface transport.

9.2.2 The Crystal–Gas Interface

From eqn (9.7) we obtain the force–flux relationships. When the measurable heat flux in the gas is used as a variable we have:

$$\begin{aligned}\left[\frac{1}{T^{\mathrm{g,c}}}-\frac{1}{T^{\mathrm{c,g}}}\right] &= r_{qq}^{\mathrm{s}} J_q^{\prime \mathrm{g,c}} + r_{q\mu}^{\mathrm{s,g}} J,\\ \left[\frac{\mu_k^{\mathrm{g,c}}(T^{\mathrm{c,g}})-\mu_k^{\mathrm{c,g}}(T^{\mathrm{c,g}})}{T^{\mathrm{c,g}}}\right] &= r_{\mu q}^{\mathrm{s,g}} J_q^{\prime \mathrm{g,c}} + r_{\mu\mu}^{\mathrm{s,g}} J.\end{aligned} \tag{9.15}$$

Alternatively, when the measurable heat flux of the crystal side is used, we find:

$$\begin{aligned}\left[\frac{1}{T^{\mathrm{g,c}}}-\frac{1}{T^{\mathrm{c,g}}}\right] &= r_{qq}^{\mathrm{s}} J_q^{\prime \mathrm{c,g}} + r_{q\mu}^{\mathrm{s,c}} J,\\ \left[\frac{\mu_k^{\mathrm{g,c}}(T^{\mathrm{g,c}})-\mu_k^{\mathrm{c,g}}(T^{\mathrm{g,c}})}{T^{\mathrm{g,c}}}\right] &= r_{\mu q}^{\mathrm{s,c}} J_q^{\prime \mathrm{c,g}} + r_{\mu\mu}^{\mathrm{s,c}} J.\end{aligned} \tag{9.16}$$

Superscript s,c of the surface resistivities refers to the use of the measurable heat flux in the gas close to the crystal, while s,c refers to the use of the measurable heat flux in the crystal close to the gas phase. The fluxes can be written in the following way by using the gas value of the measurable heat flux:

$$\begin{aligned}J_q^{\prime \mathrm{g,c}} &= -\lambda^{\mathrm{s}}(T^{\mathrm{g,c}}-T^{\mathrm{c,g}}) + q^{*\mathrm{s,g}} J,\\ J &= -\frac{q^{*\mathrm{s,g}}}{R_{\mu\mu}^{\mathrm{s,g}} T^{\mathrm{g,c}} T^{\mathrm{c,g}}}(T^{\mathrm{g,c}}-T^{\mathrm{c,g}}) - \frac{1}{T^{\mathrm{c,g}} R_{\mu\mu}^{\mathrm{s,g}}}[\mu^{\mathrm{g,c}}(T^{\mathrm{c,g}})-\mu^{\mathrm{c,g}}(T^{\mathrm{c,g}})],\end{aligned} \tag{9.17}$$

with $\lambda^{\mathrm{s}} = (r_{qq}^{\mathrm{s}} T^{\mathrm{g,c}} T^{\mathrm{c,g}})^{-1}$, $q^{*\mathrm{s,g}} = -r_{q\mu}^{\mathrm{s,g}}/r_{qq}^{\mathrm{s}}$ and $R_{\mu\mu}^{\mathrm{s,g}} = \left(r_{\mu\mu}^{\mathrm{s,g}} r_{qq}^{\mathrm{s,g}} - r_{q\mu}^{\mathrm{s,g}} r_{\mu q}^{\mathrm{s,g}}\right)/r_{qq}^{\mathrm{s}}$.

With the measurable heat flux on the crystal side we have likewise:

$$\begin{aligned}J_q^{\prime \mathrm{c,g}} &= -\lambda^{\mathrm{s}}(T^{\mathrm{g,c}}-T^{\mathrm{c,g}}) + q^{*\mathrm{s,c}} J,\\ J &= -\frac{q^{*\mathrm{s,c}}}{R_{\mu\mu}^{\mathrm{s,c}} T^{\mathrm{g,c}} T^{\mathrm{c,g}}}(T^{\mathrm{g,c}}-T^{\mathrm{c,g}}) - \frac{1}{T^{\mathrm{g,c}} R_{\mu\mu}^{\mathrm{s,c}}}[\mu^{\mathrm{s,g}}(T^{\mathrm{g,c}})-\mu^{\mathrm{s,c}}(T^{\mathrm{g,c}})],\end{aligned} \tag{9.18}$$

with $q^{*\mathrm{s,c}} = -r_{q\mu}^{\mathrm{s,c}}/r_{qq}^{\mathrm{s}}$ and $R_{\mu\mu}^{\mathrm{s,c}} = \left(r_{\mu\mu}^{\mathrm{s,c}} r_{qq}^{\mathrm{s,c}} - r_{q\mu}^{\mathrm{s,c}} r_{\mu q}^{\mathrm{s,c}}\right)/r_{qq}^{\mathrm{s}}$.

The formulations are equivalent and can be used to derive relations between coefficients.

9.3 Molecular Dynamics Simulation of Adsorption

We proceed to illustrate the force–flux relations for the case of adsorption using molecular dynamics simulations. All results have been found with *n*-butane silicalite-1. The presentation has three parts. The first presents results for the adsorbed phase within the pore structure, using eqn (9.10) to (9.14), the second gives resistivities of the interface from eqn (9.17) and (9.18). In the third part, the dynamic adsorption process will be simulated and discussed in view of the previous results. The results presented in Section 9.3 are mainly taken from ref. 8, 10–12 and 22. More details can be found in these articles and references therein.

9.3.1 Mass and Heat Flux in the Crystal in Stationary State

The system consisted of a crystal of 36 unit cells of silicalite-1, see Figure 9.6. The crystal dimensions were $L_x = 2a$, $L_y = 6b$, $L_z = 3d$, *cf.*, Section 9.1.2. Periodic boundary conditions were applied to mimic transport in an infinite crystal. After an equilibration period, mass and/or heat fluxes were applied to the system in the *b*-direction. The flux was obtained by exchanging *n*-butane molecules and/or kinetic energy between the centre of the simulation box and its boundaries. These zones are visualized in rectangles in

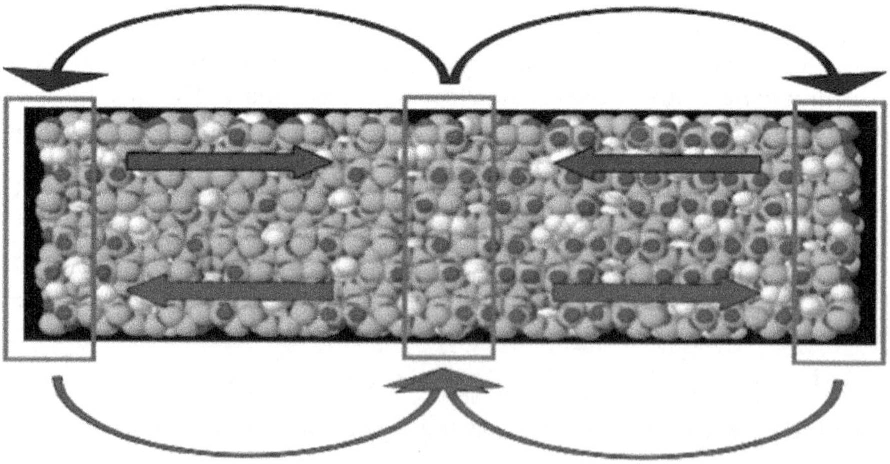

Figure 9.6 Instantaneous configurations from molecular dynamics simulations at 400 K of an infinite crystal of the zeolite silicalite-1 where *n*-butane molecules (grey) are adsorbed (3.89 m.u.c.). The rectangles are buffer zones where *n*-butane molecules are forced to move from the centre to the boundaries of the simulation box, this is illustrated by curved top arrows. This procedure creates an excess of density in the boundaries and a mass flux of butane takes place from the boundaries to the centre, obeying Fick's law of diffusion (large top arrows). Alternatively, by exchange of kinetic energy (bottom curved arrows), a heat flux is created (large bottom arrows) in agreement with the Fourier's law.

Figure 9.6. This is a typical non-equilibrium procedure, where buffer zones are used to create artificially an external flux. As a response, the system creates an internal flux, completing a non-equilibrium loop, which obeys the transport laws of the system. Stationary internal fluxes of heat and mass were obtained by imposing constant external fluxes of heat and mass. Only the total heat flux, J_q, can be held constant, not the measurable heat fluxes, J'_q.

The average temperature of the simulation varied in the range (360 to 400) K. The amount of adsorbed *n*-butane varied from 2 to 6 molecules per unit cell (m.u.c.). The temperature and density profiles obtained with a mass flux (a) and a heat flux (b) are shown in Figure 9.7. The profiles are symmetric with respect to the centre of the box because of the set-up. Data from the two half-boxes were averaged and only half of the box is shown. In the presence of a mass flux, the temperature profile is flat (Figure 9.7a) while the loading varies in a linear way inside the crystal. Results for the Fick diffusion coefficient, D, see eqn (9.12), are shown in Figure 9.8a at a temperature of 360 K. The value of D increases with increasing loading. In general, the mobility of a molecule decreases as the density increases, but this diffusion coefficient reflects both the mobility of the adsorbed phase in the zeolite lattice and the thermodynamic property of the system, as expressed by the thermodynamic factor, Γ, eqn (9.14). The value of Γ was calculated as a function of loading and temperature from the isotherm (*cf.*, Figure 9.4) using the expression $\Gamma = (\mathrm{d}\ln p/\mathrm{d}\ln c)_T$ where p is the gas pressure. The Γ values are shown in Figure 9.8b. The theoretical value at saturation is infinite. The thermodynamic factor show an increase with the loading, which can help explain the variation in D.

When exposed to a constant heat flux, see Figure 9.7b, the system obtained linear profiles of the density and temperature. The slopes were non-zero, and were used to extract the thermal conductivity and the Soret coefficient, s_T, from eqn (9.11) and (9.14). The results are plotted in Figures 9.9 and 9.10. The thermal conductivity was nearly independent of the loading and of the temperature. We obtained the average value 1.46 ± 0.07 $\mathrm{W \cdot m^{-1} \cdot K^{-1}}$ in good agreement with experimental data on silicate material.[23] Silicalite-1 is a good thermal conductor, and the result can be explained by heat transfer almost solely through the crystal.

The s_T-values presented in Figures 9.10a and b decrease as the loading and the temperature increase, in agreement with the inverse trend of D and according to eqn (9.14). The thermal diffusion effect is also commonly quantified by the heat of transfer, eqn (9.10). The heat of transfer, q^*, gives the amount of heat necessary to maintain a uniform temperature, when mass is transported though the system. The dimension is $\mathrm{J \cdot mol^{-1}}$.

The total heat flux, J_q, is used to define the total heat of transfer or the energy of transfer, Q^*. The measurable heat flux is replaced by the total heat flux in eqn (9.10). The coefficients are related *via* the partial molar enthalpy, $Q^* = q^* + H$, see eqn (9.8). The set of coefficients are shown in Figure 9.11 for 360 K, as a function of the loading. The q^*-value increases slightly, while Q^* is roughly constant. Both are near 9 $\mathrm{kJ \cdot mol^{-1}}$

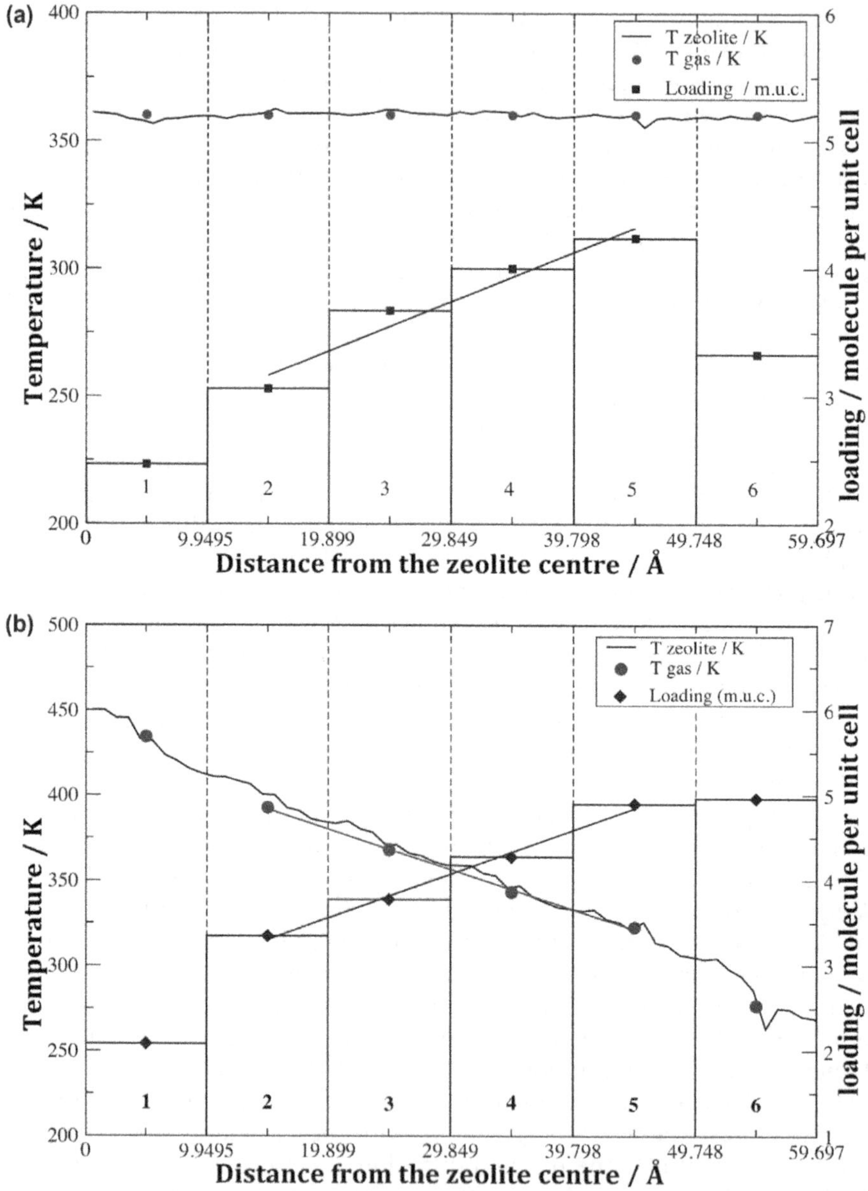

Figure 9.7 Temperature and density profiles, under constant mass flux (a) and heat flux (b). The position on the *x*-axis gives the distance from the centre of the crystal. Straight lines are fitted to the linear parts of the curves. Copyright (2008) by the American Chemical Society.[22] Reproduced with permission.

(H is around $-1.5\ \text{kJ}\cdot\text{mol}^{-1}$). This means that thermal effects associated with the heat of transfer can be significant, although they are often neglected.

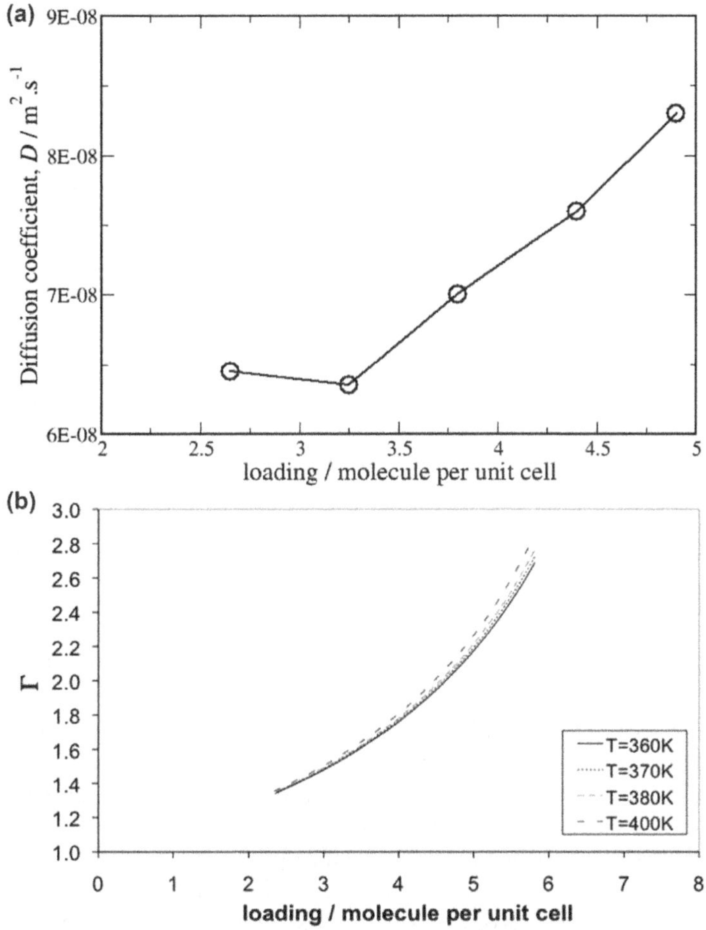

Figure 9.8 Fick's diffusion coefficient at 360 K (a) and thermodynamic factor for different temperatures (b) as function of loading.
Copyright (2008) by the American Chemical Society.[22] Reproduced with permission.

9.3.2 Mass and Heat Flux across the Interface in a Stationary State

When a molecule is crossing the surface from the gas to the pores the partial molar enthalpy changes significantly. The thin surface is a zone where mass and heat fluxes couple strongly. To study this effect, explicit surfaces were simulated and subjected to heat and mass fluxes. Fluxes were generated following the same procedure as for the pure crystal, *cf.*, Section 9.3.1, by exchange of *n*-butane molecules and kinetic energy between buffer zones located at the boundaries of the gas phase and in the centre of the zeolite,[12] see Figure 9.12. The crystal surfaces were mostly flat. Textured surfaces with

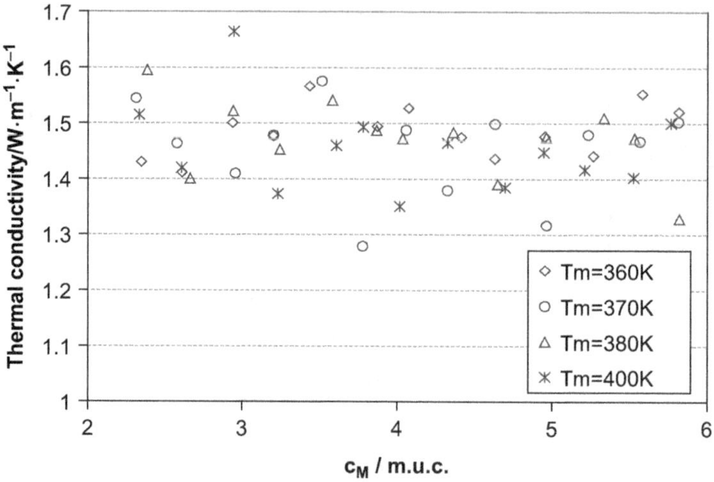

Figure 9.9 Thermal conductivity at mean temperature T_m and loading c_M. Copyright (2008) by the American Chemical Society.[22] Reproduced with permission.

half zig-zag channels were also studied.[12] A comparison was made with a flat surface.

Stationary total heat- and mass-fluxes were simulated, and temperature and loading profiles were determined. Figure 9.13 gives the temperature profile in the presence of a heat flux. The temperatures were constant in the crystal, here about 360 K, and they were increased progressively until 520 K was reached in the gas buffer zone. A large jump of 20 K can be seen at the interface. This reveals a large interface resistivity to heat transfer, large compared to the same property of the gas phase and of the zeolite phase. The interface resistivity to heat transfer is plotted in Figure 9.14 as a function of the gas pressure for different surface temperatures. The value of r_{qq}^s strongly depends on the pressure and weakly on the temperature. For all temperatures, the value decreases as the gas pressure increased. In order to compare the values of r_{qq}^s with the zeolite thermal conductivity given above, ($\lambda \approx 1.5\ \mathrm{W \cdot m^{-1} \cdot K^{-1}}$), we estimated[11] the zeolite resistivity using a surface thickness, L^s, of $1.2 \cdot 10^{-9}$ m and the expression $r_{qq}^s = L^s/T^2\lambda$. We obtained $r_{qq}^s \approx 6 \cdot 10^{-15}\ \mathrm{m^2 \cdot s \cdot J^{-1} \cdot K^{-1}}$ which is two to three orders of magnitude smaller than the surface resistivity plotted in Figure 9.14. This gives an explanation for the large temperature jump at the interface. An analysis of the density of the adsorbed gas at the external surface,[12] showed that the large surface excess resistivity was located to the gas side of the surface. The resistivity on the zeolite side of the surface was one to two orders of magnitude lower than the resistivity on the gas side. This distribution may be largely temperature independent, explaining the trend in Figure 9.14. The surface heat conductivity, however, depended on the temperature, see eqn (9.17).

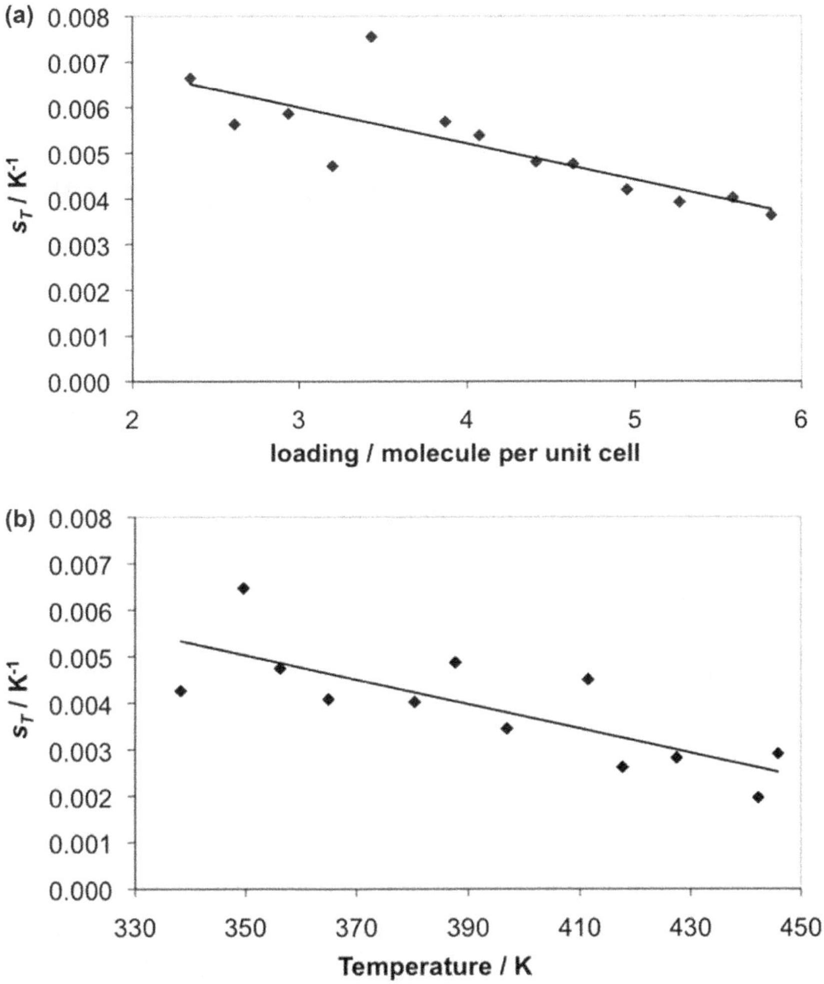

Figure 9.10 Soret coefficients calculated at 360 K for different loadings (a) and at the loading of 3.89 m.u.c. for different average temperatures T_m (b). The lines are a guide to the eye, highlighting the decreasing trend of the coefficient.

It follows that a decrease in temperature in the gas under constant pressure favours transport of heat through the surface.

Values of $R^s_{\mu\mu}$, calculated from eqn (9.17) at constant temperature, are shown in Figure 9.15 as a function of the pressure in the gas phase. Controlled is the measurable heat flux at the gas side (a) or at the zeolite side (b). Similar results are shown for three temperatures. The value of $R^s_{\mu\mu}$ decreases slightly with increasing pressure. No clear temperature dependence was found. The zig-zag textured surface[12] (not shown here) had a smaller mass

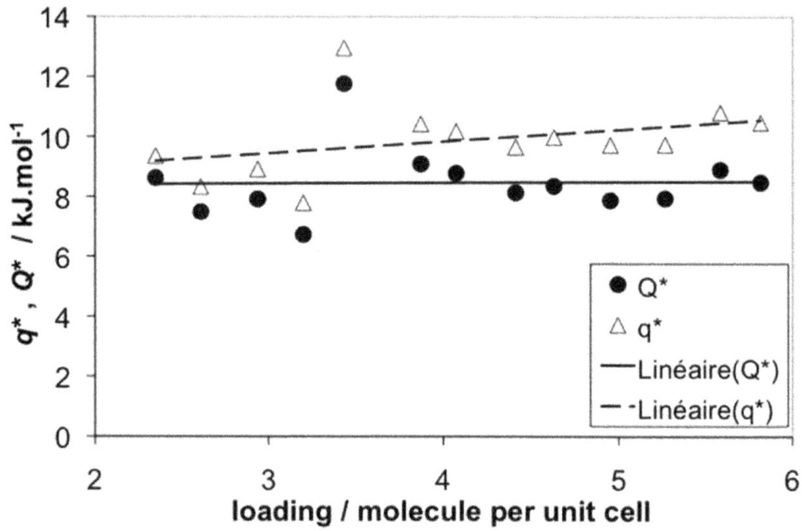

Figure 9.11 Total (Q^*) and measurable heat of transfer (q^*) as a function of the loading at 360 K. Linear fits are shown, despite the large dispersion of data. The slope of q^* is slightly positive while it is nearly zero for Q^*. Copyright (2008) by the American Chemical Society.[22] Reproduced with permission.

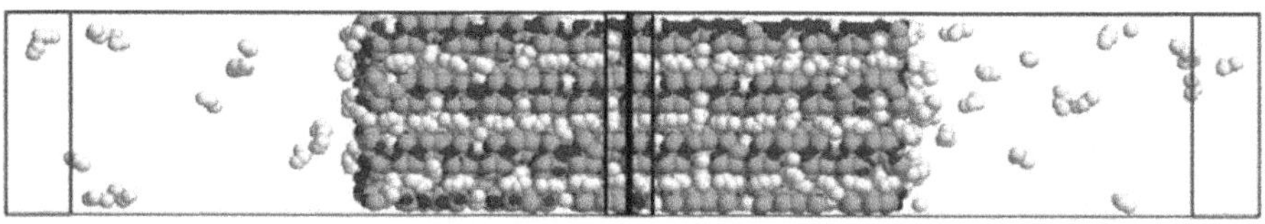

Figure 9.12 Instantaneous configurations from molecular dynamics simulations of an infinite membrane of the zeolite silicalite-1 in contact with a gas of *n*-butane (grey). The butane molecules are located in the gas phase and are adsorbed on the external surface and inside the pores of the zeolite. The rectangles are buffer zones where *n*-butane molecules and kinetic energy are exchanged to create a mass or a total heat flux. Copyright (2009) by Elsevier.[12] Reproduced with permission.

resistivity, meaning that surface roughness may favour mass diffusion into the crystal. No roughness effect was seen in the heat resistivity. Like for the thermal conductivity, we could estimate a mass resistivity using the diffusion coefficients for gas in the zeolite, from $R^{s}_{\mu\mu} = \Gamma RL^{s}/Dc$. The result was $2 \cdot 10^{-4}\,\mathrm{J \cdot s \cdot m^2 \cdot K^{-1} \cdot mol^{-2}}$, one order of magnitude smaller than the surface value. This means that the surface acts not only as a heat resistance, but also as a mass resistance when compared to the zeolite and the gas phases.

The measurable heats of transfer are shown in Figure 9.16 for both heat fluxes used in their determination. In each case, the value decreases slightly as the pressure increases. They depended weakly on the temperature. With

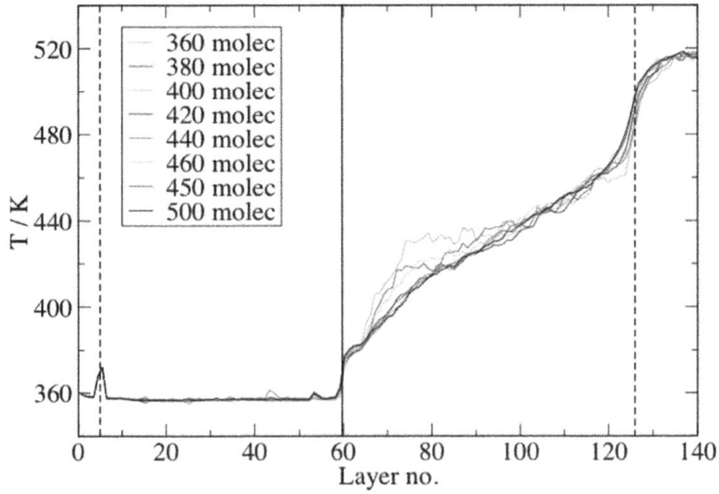

Figure 9.13 Temperature profiles for different quantities of *n*-butane molecules. The layer number is counted from the centre of the crystal. Each layer has a thickness close to 0.1 nm. The position of the crystal boundary is indicated by a vertical line, and the vertical dashed lines give the limits of the buffer zone. The temperature is nearly uniform in the crystal, in the gas phase it increases progressively while a jump can be seen at the surface boundary.
Copyright (2009) by Elsevier.[12] Reproduced with permission.

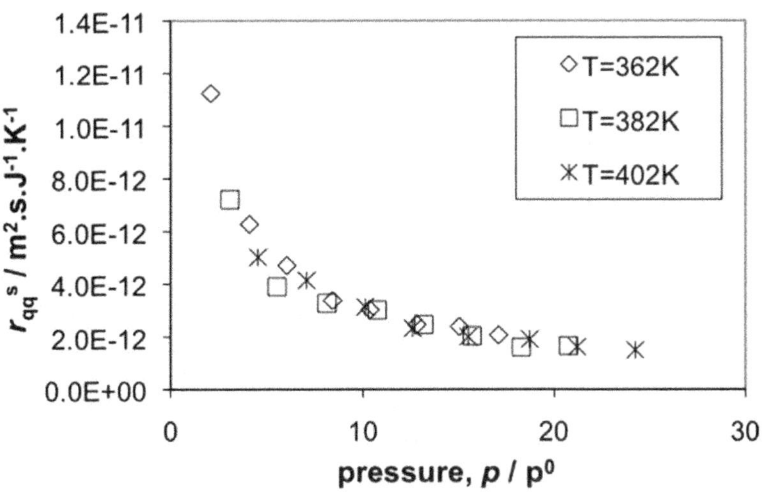

Figure 9.14 Surface resistivity to heat transport as a function of the gas pressure, p, at three different surface temperatures. The standard pressure is $p^0 = 10^5$ Pa.
Copyright (2009) by Elsevier.[12] Reproduced with permission.

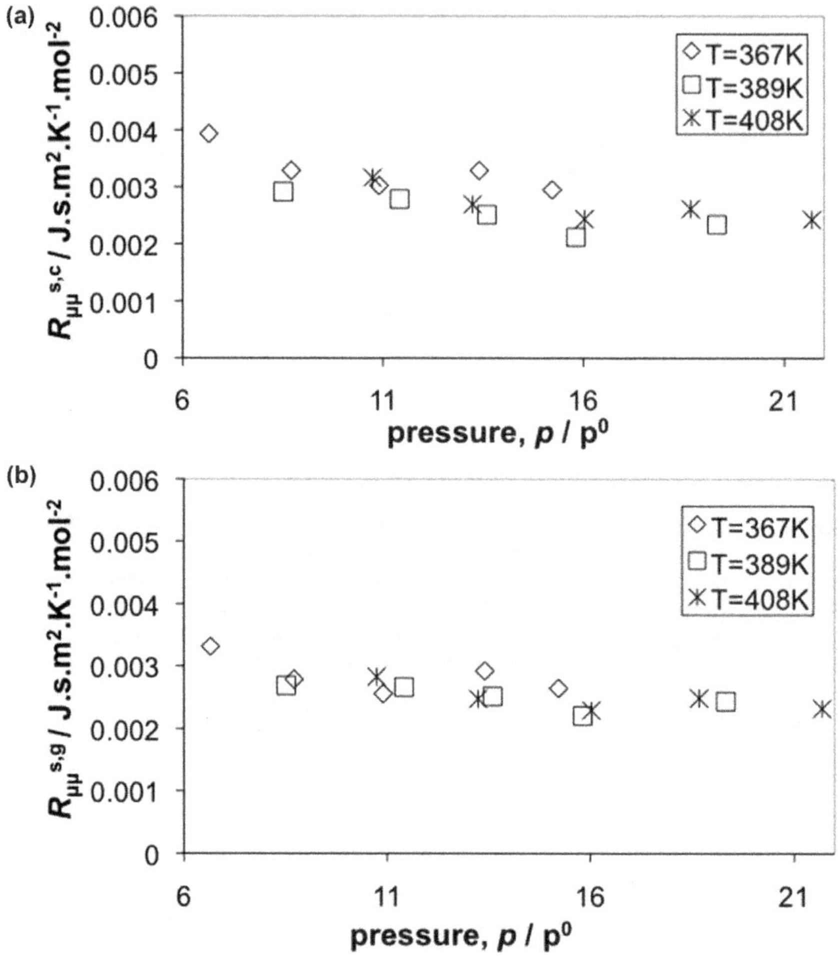

Figure 9.15 Surface resistance to mass transfer as a function of pressure, (a) with the measurable heat flux taken at the gas side and (b) with the heat flux taken at the zeolite side.

the heat flux in the gas, the value is about 5 $\text{kJ} \cdot \text{mol}^{-1}$. The corresponding value with the heat flux in the crystal as a reference is 60 $\text{kJ} \cdot \text{mol}^{-1}$. The difference is the expected enthalpy of adsorption $\Delta_{\text{ads}}H = H^{c,g} - H^{g,c} \approx -55$ $\text{kJ} \cdot \text{mol}^{-1}$. This means that the coupling of heat and mass fluxes at the surface of the zeolite is strong. The effect of this coupling on the adsorption or desorption kinetics is the topic of the following section.

9.3.3 Interface Phenomena during Adsorption Uptake

When molecules are adsorbed, heat is released into the system, *i.e.*, physisorption is an exothermic phenomenon. During adsorption, this heat is

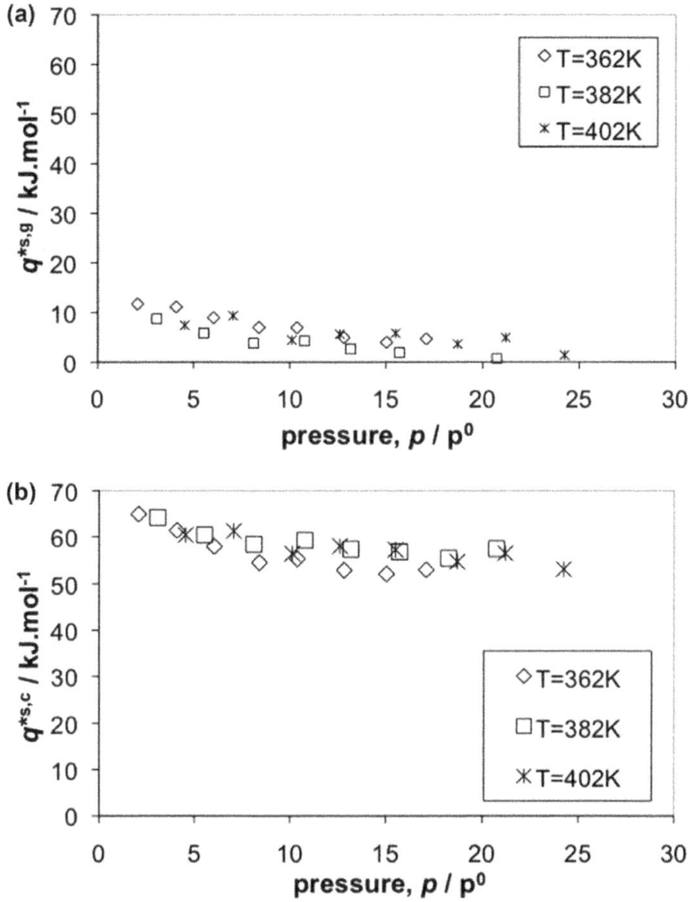

Figure 9.16 Heat transfer as a function of the pressure, (a) with the measurable heat flux taken at the gas side and (b) with the heat flux taken at the zeolite side.
Copyright (2009) by Elsevier.[12] Reproduced with permission.

spreading inside the porous material and its temperature rises. The material cools down to the ambient temperature by radiation or by heat conduction to the surrounding gas. If the characteristic time of the temperature equilibration is smaller than the characteristic time of the adsorption kinetics, the adsorption process can be assumed to be isothermal, otherwise it is a non-isothermal process.[24,25] In the latter case, Ruthven *et al.* have mentioned the role played by the surface as a heat resistance.[26,27] They have also explained that a higher temperature in the zeolite can delay the adsorption kinetics. Only a few authors analysed their experiments taking into account such non-isothermal conditions.[24–28] Although isothermal conditions are justified in many cases, it is important to verify that this is indeed the case, in order to produce reliable results.

To better illustrate non-isothermal effects, molecular dynamics simulations were performed, mimicking the adsorption kinetics of *n*-butane into the silicalite-1, see Figure 9.3.[10,11] The gas was thermostatted at 300 K, such that heat could only be released from the zeolite by transport through the surface and then through the gas phase.[10,11] A typical uptake curve is plotted in Figure 9.17. The figure shows that the process has two stages. It starts with

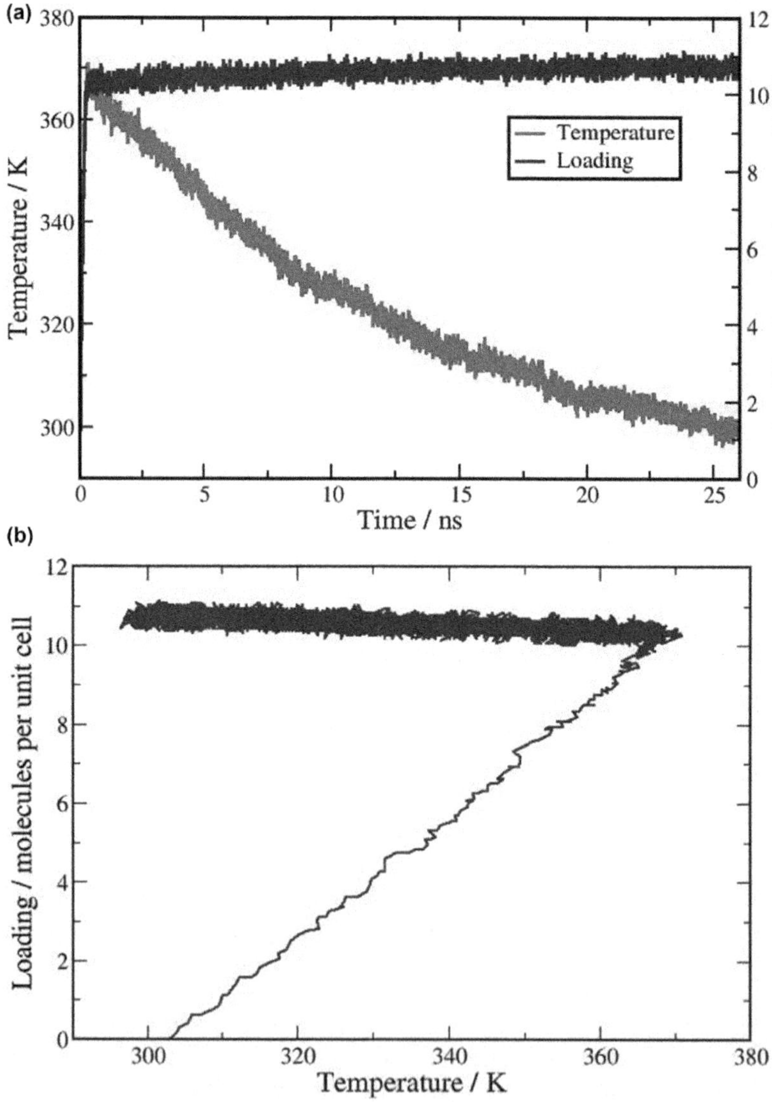

Figure 9.17 (a) Uptake curve for adsorption of *n*-butane molecules into silicalite-1. (b) Evolution of the loading as a function of the zeolite temperature. The evolution of the temperature and the loading are strongly correlated. Copyright (2007) by Taylor and Francis.[10] Reproduced with permission.

a fast regime of 0.2 ns where about 95% of the system's capacity for *n*-butane is filled up, see the very left part of Figure 9.17a. During this process the temperature increased by up to 70 K. The loading was completed in a second slow stage that took about 25 ns (the remaining part of Figure 9.17a). In this stage, the temperature relaxed toward its equilibrium value (300 K). From these two stages it was possible to calculate diffusion coefficients, as is commonly done to analyse experiments, using a simple expression of the integral of the diffusion equation for infinite membrane:[1]

$$1-\left(\frac{c(t)}{c(\infty)}\right)=\exp\left(-\frac{D\pi^2 t}{L^2}\right),\tag{9.19}$$

with t and L the time elapsed and the thickness of the membrane, respectively. We obtained a value of $D=4.6\cdot 10^{-8}\ \mathrm{m^2\cdot s^{-1}}$ for the first stage and $D=4.110\cdot 10^{-10}\cdot \mathrm{m^2\cdot s^{-1}}$ for the later stage. Compared with the diffusion coefficient from the zeolite, Figure 9.8(a), we obtained good agreement with the first stage result, while the result was two orders of magnitude smaller for the second one. The difference was similar to the difference observed experimentally between the ZLC- and QENS methods, see Figure 9.2. These results clearly suggest that the first stage can be governed by the intra-crystalline diffusion. The role of the surface as an explanation of this discrepancy will be discussed below.

As can be seen from Figure 9.17b, the loading and the zeolite temperature are strongly correlated during adsorption. The two mentioned stages are now clearly visible; they can be characterized by a straight line with a positive slope in the first stage, and line with a negative slope in the second stage. The variation take place until equilibrium is reached. These results can be taken as signs of the presence of two different transport regimes.

Temperature profiles obtained during the second stage are shown in Figure 9.18 as a function of time. Results for half the simulation box are shown. The zeolite is on the left-hand side in this figure and the gas is on the right-hand side. Despite large local fluctuations, the temperatures in the gas and in the crystal phases were rather uniform, while at the surface they exhibited large temperature jumps up to 70 K at the beginning of this second stage. These jumps decreased as the system evolved, until equilibrium where they vanished. This behaviour can be explained by an interface heat resistance which is much larger than the heat resistance of the zeolite crystal, as was described in the preceding sections.

The use of non-equilibrium thermodynamics to describe transport in this heterogeneous system sheds new light on non-isothermal adsorption kinetics. The following discussion is based on the expressions of the heat and mass fluxes given in eqn (9.17) and (9.18). In these expressions q^{*s} is positive, so that a mass flux will contribute to the heat flux with a term that has the same direction as the mass flux. In the first stage, a massive mass flux is directed into the crystal, on average 45 000 $\mathrm{mol\cdot m^{-2}\cdot s^{-1}}$. The contribution to the heat flux is proportional to this, and will completely dominate the

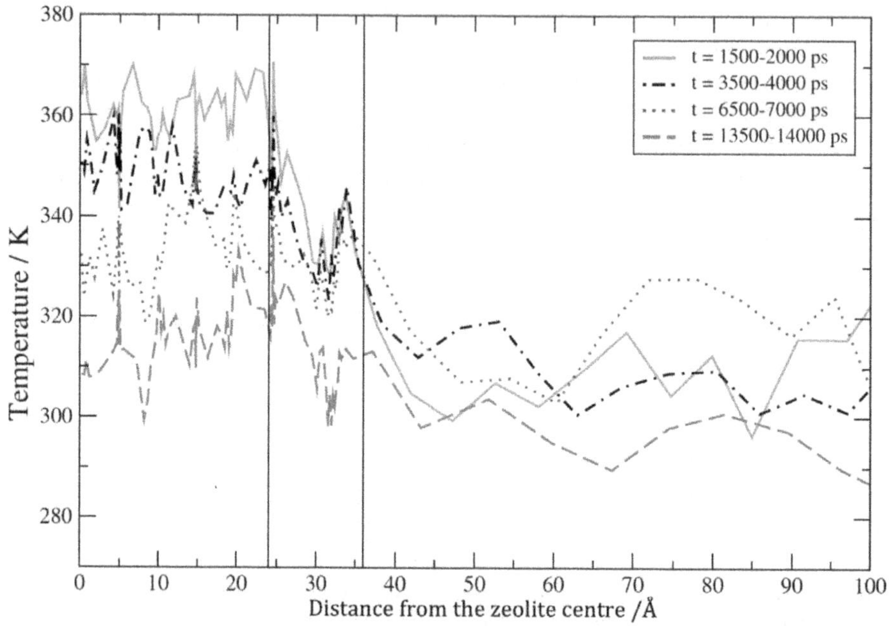

Figure 9.18 Temperature profile along the system for different periods of simulations (they refer to the time axis of Figure 9.17a. The surface zone is delimited by the two vertical lines between 2.4 nm and 36 nm. The crystal and the gas phases are located respectively on the left-hand side and on the right-hand side of the surface. The profile shows a large temperature jump across the surface, up to 70 K.
Copyright (2007) by Elsevier.[11] Reproduced with permission.

Fourier type term. It also agrees with the observed increase in the temperature difference across the surface. At 362 K, $q^{*s,c} \approx 60$ kJ · mol^{-1} and $R^s_{\mu\mu} \approx 0.003$ J · K^{-1} · m^{-2} · s^{-1}. For a temperature difference of 70 K, the contribution of the temperature difference to the mass flux (thermal diffusion) on the crystal side is 13 000 mol · m^{-2} · s^{-1} and about 1000 mol · m^{-2} · s^{-1} to the mass flux at the gas side. These values are much lower than the net mass flux, which indicates that during this first stage the mass flux is rather governed by the difference in the chemical potential across the surface, the second term in eqn (9.17) and (9.18). Intra-crystalline diffusion may still govern the kinetics of the first stage.

During the second stage the mass flux is on average reduced by three orders of magnitude, to around 20 mol · m^{-2} · s^{-1}. The heat flux is directed out of the crystal, and the surface-temperature difference decreases until equilibrium. Compared with the mass flux contribution from the temperature difference calculated above, whichever way it is calculated, we see that the net mass flux is smaller by two to three orders of magnitude. The second stage can therefore be characterized by two contributions which nearly balance each other, *i.e.*, the term with the temperature difference (thermal diffusion) and with the chemical-potential difference (diffusion) across the surface. In addition to the surface heat conductivity, the thermal diffusion properties of the surface perfectly control the influx of molecules into the zeolite. As the temperature difference decreases by conduction (or by radiation), the balance breaks and a small influx can take place. This regime can last until equilibrium is reached. These results illustrate the dominant role of the external surface during adsorption/desorption.

These results have been obtained on systems of nanometre size. In order to predict the behaviour of much larger systems, the transport equations eqn (9.8) must be integrated using the computed resistivities as input data. This procedure was followed by Schnell *et al.*[29] in a study of the transport of *n*-butane across a silicalite-1 membrane, 50 μm thick. Also in this case, temperature jumps were observed according to the non-equilibrium description that we presented in Section 9.2. This means that the surface might well play a role for the total transport properties when adsorption or desorption phenomena take place even on large scales. Data are presently lacking to make a systematic evaluation. Surface transport properties are commonly neglected in chemical engineering applications. This study shows that the assumption should be reconsidered.

9.4 Conclusion

The adsorption of mass and the accompanying heat flux in porous systems can be described with non-equilibrium thermodynamics. The example of *n*-butane adsorption in the zeolite silicalite-1 was used to illustrate the importance of the surface for the adsorption kinetics. Transport coefficients of the adsorbed phase and of the external surface were presented. They revealed a strong coupling coefficient between heat and mass fluxes, characterized by a heat of transfer which is of the order of the adsorption enthalpy. This coupling can explain that adsorption becomes non-isothermal in a non-stationary situation. In a first short stage the kinetics was governed by a variation in the chemical potential across the system. At the same time, the zeolite heated up due to the strong exothermic enthalpy of adsorption. In the second stage the influx was much smaller. There is then nearly a balance between diffusion and thermal diffusion.

The role of thermal diffusion has not been mentioned so far in the literature by others, and we have here seen that it is central to understand non-isothermal adsorption. The results elucidate the exothermic properties of the adsorption and the high interface resistivity to heat transfer mentioned by Ruthven *et al.*[26,27] The system reaches equilibrium by release of heat; the kinetics are not controlled by diffusion, because we are in a non-diffusive regime. The equations presented here for the interface could explain the large variation in experimental diffusion coefficients reported in the literature (*cf.* Figure 9.2). We have shown that the origin of the difference can be related to the heat evolution at the surface. In larger crystals, with lower surface-to-volume ratio, the importance of the slow regime may be reduced, and give more reliable experimental results assuming isothermal conditions.

Acknowledgements

The author acknowledges support for a sabbatical at NTNU in Trondheim from the CNRS and the Norwegian Research Council through the Storforsk Grant No. 167336 to S. Kjelstrup.

References

1. J. Crank, *The Mathematics of Diffusion*, Oxford University Press, London, 1956.
2. H. Jobic, *J. Mol. Catal. A: Chem.*, 2000, **158**, 135.
3. J. Kärger, *Adsorption*, 2003, **9**, 29.
4. C. Chmelik, P. Kortunov, S. Vasenkov and J. Kärger, *Adsorption*, 2005, **11**, 455.
5. D. Frenkel and B. Smit, *Understanding Molecular Simulation: From Algorithms to Applications*, Academic Press, San Diego, 2nd edn, 2002.
6. B. Smit and T. L. M. Maesen, *Chem. Rev.*, 2008, **108**, 4125.
7. R. Krishna, *J. Phys. Chem. C*, 2009, **113**, 19756.
8. J. M. Simon, A. Decrette, J. P. Bellat and J. M. Salazar, *Mol. Simul.*, 2004, **30**, 621.
9. D. A. Newsome and D. S. Sholl, *J. Phys. Chem. B*, 2005, **109**, 7237.
10. J. M. Simon, I. Inzoli, D. Bedeaux and S. Kjelstrup, *Mol. Simul.*, 2007, **33**, 839.
11. I. Inzoli, J.-M. Simon, S. Kjelstrup and D. Bedeaux, *J. Colloid Interface Sci.*, 2007, **313**, 563.
12. I. Inzoli, S. Kjelstrup, D. Bedeaux and J.-M. Simon, *Microporous Mesoporous Mater.*, 2009, **125**, 112.
13. H. Vankoningsveld, H. Vanbekkum and J. C. Jansen, *Acta Crystallogr., Sect. B: Struct. Sci.*, 1987, **43**, 127.
14. I. Inzoli, J.-M. Simon and S. Kjelstrup, *Langmuir*, 2009, **25**, 1518.
15. M. S. Sun, D. B. Shah, H. H. Xu and O. Talu, *J. Phys. Chem. B*, 1998, **102**, 1466.
16. T. J. H. Vlugt, R. Krishna and B. Smit, *J. Phys. Chem. B*, 1999, **103**, 1102.
17. S. Kjelstrup and D. Bedeaux, *Non-Equilibrium Thermodynamics of Heterogeneous Systems*, Series on Advances in Statistical Mechanics, World Scientific, Singapore, 2008, vol. 16.
18. S. R. de Groot and P. Mazur, *Non-equilibrium Thermodynamics*, Dover Publications, New York, 1984.
19. X. Liu, S. Schnell, J.-M. Simon, P. Krüger, D. Bedeaux, S. Kjelstrup, A. Bardow and T. J. H. Vlugt, *Int. J. Thermophys.*, 2013, **34**, 1169.
20. R. Taylor and R. Krishna, *Multicomponent Mass Transfer*, Wiley, New York, 1st edn, 1993.
21. J.-M. Simon and J. M. Rubí, *J. Phys. Chem. B*, 2011, **115**, 1422.

22. I. Inzoli, J.-M. Simon, D. Bedeaux and S. Kjelstrup, *J. Phys. Chem. B*, 2008, **112**, 14937.
23. A. J. H. McGaughey and M. Kaviany, *Int. J. Heat Mass Transfer*, 2004, **47**, 1799.
24. L. Heinke, C. Chmelik, P. Kortunov, D. B. Shah, J. Kärger, S. Brandani, D. M. Ruthven and J. Kärger, *Microporous Mesoporous Mater.*, 2007, **104**, 18.
25. S. Brandani, C. Cavalcante, A. Guimaraes and D. M. Ruthven, *Adsorption*, 1998, **4**, 275.
26. D. M. Ruthven, L. K. Lee and H. Yucel, *AIChE J.*, 1980, **26**, 16.
27. D. M. Ruthven and L. K. Lee, *AIChE J.*, 1981, **27**, 654.
28. J. Giermanska-Kahn, J. Cartigny, E. Cohen de Lara and L. M. Sun, *Zeolites*, 1996, **17**, 365.
29. S. K. Schnell, *Molecular Simulations of Zeolites: Heterogeneous Systems at Equilibrium and Non-Equilibrium*, PhD thesis, TU-Delft, The Netherlands, 2013.

CHAPTER 10

Non-equilibrium Thermodynamics of Aqueous Solution–Crystal Interfaces

F. ELİF GENCELİ GÜNER

İstanbul Teknik Üniversitesi Kimya Mühendisliği Bölümü, 34469, İstanbul, Turkey
Email: gencelie@itu.edu.tr; egenceli@hotmail.com

10.1 Introduction

Crystallisation is the process of formation of solid crystals in a solution, melt or more rarely, deposition directly from a gas. It can be natural or artificial.[1] Crystallisation is also a chemical solid–liquid separation technique, in which mass transfer of a solute from the liquid solution to a pure solid crystalline phase occurs. Industrially, crystallisation occurs in a crystallizer.

Several methods are applied in industry to achieve crystallisation either from melt or solutions. For melt and mostly for solution crystallisation, cooling below the saturation temperature is the applied way to induce crystallisation. In order to provoke crystallisation, the system is brought to a condition away from that prescribed by solid–liquid equilibrium: the system is said to be supersaturated with respect to the crystallizing compound. In cooling crystallisation on a heat-exchanger surface, the super-saturation generally reaches a higher value compared to the bulk super-saturation. Thus the formation of a *scale layer* on the heat exchanger surface becomes inevitable.

Experimental Thermodynamics Volume X: Non-equilibrium Thermodynamics with Applications
Edited by Dick Bedeaux, Signe Kjelstrup and Jan V. Sengers

Published by the Royal Society of Chemistry, www.rsc.org

Scaling is an important limiting factor for the heat transfer in cooling processes. The higher the heat fluxes are that can be maintained before the scale layer is deposited, the higher is the production rate that can be achieved. In order to lower operation costs in industrial crystallizers and to achieve crystallisation process intensification, scale prevention on cooled surfaces is extremely important. Prevention of scaling may be possible by understanding its mechanism better, developing more knowledge about the nucleation and growth conditions on the cooled wall and focusing on the transport processes at the interface between the scale layer that grows upon the cooling surface and the solution.

For a moment consider only the flat parts of the growth layers on a crystal surface and disregard steps and kinks and surface integration phenomena. It can then be expected that coupled heat and mass transfer effects, when calculated with the Onsager equations, in principle will be noticeable and will lead to a temperature jump at the liquid-crystal interface.[2] (The exception is regular thermal conduction at zero mass flux or regular concentration diffusion at uniform temperature.) In thermodynamic descriptions of the dynamic phase transition of pure components or multicomponent mixtures, it is commonly assumed that the temperature and chemical potential are continuous functions through the interface, in spite of the fact that the Kapitza resistance for simple heat transport has been known for a long time.[3] For instance, both the fugacity model[4] and the film model,[5] which are central in the modelling of phase transitions, assume that there is continuity in the temperature and the chemical potential profile across the interface. On the other hand, kinetic theory[5] predicts a jump in the temperature, and kinetic theory can now be supported by molecular dynamics simulations of hard spheres.[6] Badam *et al.*[7] have also measured a temperature jump across the vapour–liquid interface, depending on the rate of evaporation and on the heat transfer rate.

The purpose of this chapter is to help build a knowledgebase for interface transfer coefficients, enabling us to deal with phase transitions in a more precise way. Thus we review recent experimental works on $MgSO_4 \cdot 7H_2O$ crystallisation from $MgSO_4$ aqueous solution[8] and ice crystallisation from pure water[9] on cold surfaces. For growing crystals, measurements of the temperature jump at the interface were investigated. Based on these findings, coupled heat and mass-flux equations from non-equilibrium thermodynamics were defined for crystal growth, to describe the temperature jump at the interface of the growing crystal, and the distribution ratio of heat of crystallisation between the crystal layer and the solution was calculated.

10.2 Crystallisation System

The present crystallisation system is made up of several homogeneous phases separated by interfaces or surfaces. In that sense, *the system* can be said to be heterogeneous. The crystallisation itself is, in the terminology common in the field, of a homogeneous type, however. For the growing crystal on a cooled metal surface the following thermodynamic phases can be distinguished; the homogeneous solution phase, crystal phase, metal phase, and the interfaces between the homogeneous phases: which are the solid–solid interface between the crystal and metal, and the liquid–solid interface between the liquid and crystal.

In a macroscopic description, not only the homogenous phases, but also the interface (between the solution and the crystal) is a separate thermodynamic phase. This applies at equilibrium, but according to Chapter 4 also away from global equilibrium. The thermodynamic properties of the interface is, in both cases, given by the values of the excess densities of Gibbs,[10] see Chapter 1 (Section 1.5) for further explanations. Excess concentrations can be obtained from the integral of the concentration above (or below) the values of the nearby phases. The location of the *equimolar* interface is defined such that the surplus of moles of the component on one side of the interface is equal to the deficiency of moles of the component on the other side of the interface. Calculation and determination of the location of this plane (for excess surface concentration, excess internal energy, excess enthalpy, the surface tension and the excess entropy) is explained elsewhere.[2,11]

A first attempt to use non-equilibrium thermodynamics in the description of crystallisation was made by Ratkje and Flesland[12] in their studies of freeze concentration of ice from an aqueous solution. That study did not take into account the heterogeneous nature of the problem, however, which is essential.[13] General expressions for the excess entropy production rate of an interface were already derived a long time ago[14–16] for curved interfaces that were allowed to move in space and change their curvature. Such an analysis is rather complicated, and simplifications are required. In any case, the expression for the entropy production in the surface predicts a jump in intensive variables, like the temperature, across the surface. Badam *et al.*[7] measured such a temperature jump across the liquid vapour interface and showed the importance of coupling of heat and mass transport, see also Chapters 8 and 9. The aim of the present chapter is to review experimental results that document a temperature jump across the liquid–solid interface during crystallisation.

We shall take advantage of the results of Kjelstrup and Bedeaux,[17] and use their description of coupled transport equations for a planar liquid–vapour interface to derive similar expressions for the liquid–solid transition. It can be documented that this is the only way to describe properly the simultaneous transport of heat and mass, in a way that is consistent with the second law of thermodynamics.[8]

10.3 Experimental

10.3.1 Experimental Set-ups and Procedures

In a crystallisation system, the temperature profile and the transport of heat and mass through an interface was considered. Due to their sizable enthalpies (39.5 kJ · mol^{-1} and 6.01 kJ · mol^{-1}, respectively), $MgSO_4$ aqueous solution–$MgSO_4 \cdot 7H_2O$ crystal and water–ice transformations were chosen and tested in two different experimental set-ups. Experimental conditions for both cases are tabulated in Table 10.1. Further details on these experiments can be found elsewhere.[8,9]

For both cases crystal growths took place on the cold surfaces as schematically shown in Figure 10.1. Differences for each case are noted with an asterisk.

Table 10.1 Experimental conditions.

	$MgSO_4 \cdot 7H_2O$ crystallisation[8]	Ice crystallisation[9]
Liquid composition	29.72 wt % $MgSO_4$	18.2 MΩ ultra-pure H_2O
T_{feed} to crystallizer	37 °C	20 °C
Solution $T_{saturation}$	35.7 °C[18]	0 °C[9]
Nucleation *via*	Seeding	Primary
Temperature measurement		
Cooling side (T^c_{in}, T^c_{out})	ASL F250-PT100 Accuracy: ± 0.01 °C Resolution: ± 0.001 °C Measurement rate: Every 2 s	ASL F250-PT100 Accuracy: ± 0.01 °C Resolution: ± 0.001 °C Measurement rate: Every 1 s
Bulk	ASL F250-PT100 *T* (distance from cryst. surf.): T^1 (1.7 mm) T^2 (6.7 mm) T^3 (11.7 mm)	TLC Brand: Hallcrest SH1LCRR-10C25W Working range: − 7 °C to + 7 °C Measurement rate: Every 1 s
	Accuracy: ± 0.01 °C Resolution: ± 0.001 °C Measurement rate: Every 2 s TLC[a] Brand: Hallcrest R33C1W Working range: 33 to 34 °C Measurement rate: Every 2 s	TLC[b] Brand: Hallcrest SH1LCRR-10C25W Working range: − 7 °C to + 7 °C Measurement rate: Every 2 s
Bulk concentration measurement	ICP-AES/Ion Chromatography Error: ± 2.5 % Density measurement Error ± 0.15 wt % *C* (distance from cryst. surf.): C^1 (1.7 mm) C^2 (6.7 mm) C^3 (11.7 mm)	

[a]Thermochromatic Liquid Crystal under $MgSO_4 \cdot 7H_2O$ crystal.
[b]Thermochromatic Liquid Crystal behind ice crystal.

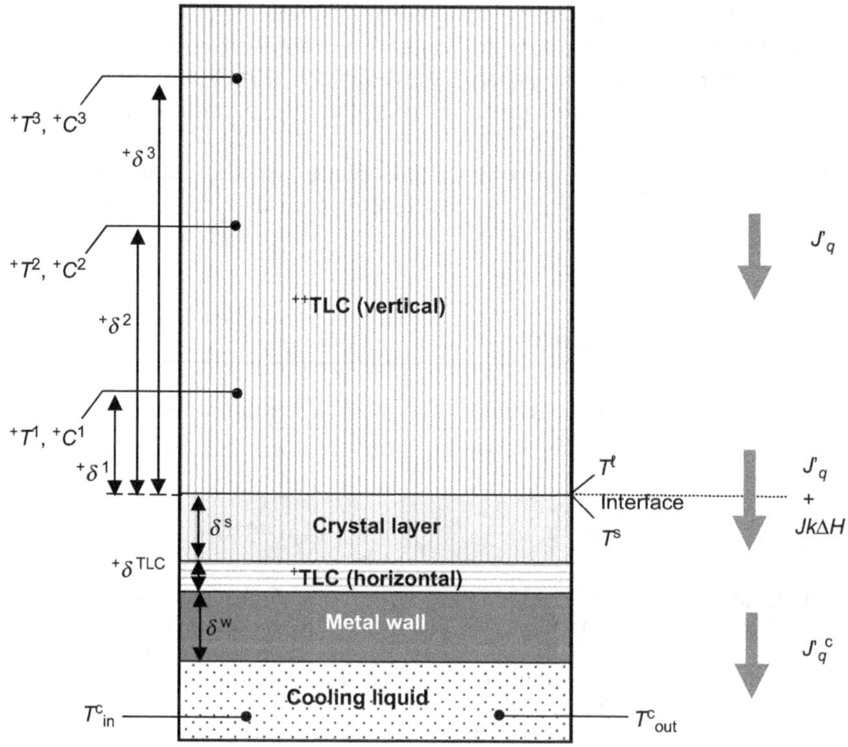

Figure 10.1 Magnified view of the crystallisation surface. (Figure is not drawn proportional to the original.) k: fraction of the enthalpy of crystallisation carried out *via* the cooled surface. T^{ℓ}: Liquid side temperature of the interface. The symbol (Ell) is short for liq. T^{s}: Crystal side temperature of the interface. $^{+}$only for the $MgSO_4 \cdot 7H_2O$ system. $^{++}$only for the ice system.

In two different experimental set-ups, $MgSO_4$ solution (at 37 °C) and pure water (at room temperature) were fed into their own crystallizers and stabilized for 10 minutes. When uniform temperature profiles were achieved, the solutions were cooled down indirectly through the crystallizer's bottom metal wall (for the $MgSO_4$ case this wall was covered with a TLC-Thermochromatic Liquid Crystal sheet). Under steady-state conditions and in the presence of any crystals, the temperature profiles and the heat fluxes from the liquid into the coolant were continuous. When the solutions were cooled below their own saturation temperatures, crystallisation on the cold surface was initiated *via* (i) seeding in the $MgSO_4 \cdot 7H_2O$ case, (ii) primary nucleation for the ice case. As both crystallisation processes are exothermic and generated heat into the system, the temperature profiles and heat fluxes were not continuous anymore. A temperature jump was seen at the locations where crystals nucleate and grow. The crystallisation on the cooled surface was visualized by the naked eye and by digital pictures. The temperature jumps in the interface due to nucleation and growth of $MgSO_4 \cdot 7H_2O$ and ice crystals were confirmed by calculations using the values of TLC and temperature sensor measurements.[8,9]

10.3.2 Data Analysis and Investigations on Interface Temperature Jumps

10.3.2.1 $MgSO_4 \cdot 7H_2O$ Crystallisation Case

10.3.2.1.1 Data Analysis. Measuring the difference between cooling liquid inlet and outlet temperature difference ($T^c{}_{in} - T^c{}_{out}$), and using the metal wall (δ^w), TLC (δ^{TLC}) and crystal layer (δ^s) thicknesses, coolant heat flux and the crystal side temperature of the interface (T^s) were calculated *via* Fourier's law. The liquid side temperature of the interface is notified by T^ℓ and it was determined by linear extrapolation from the temperature sensor readings (T^1, T^2, T^3) located in the bulk (δ^1, δ^2, δ^3) to the interface thickness. TLC temperature values for crystal-growing areas and crystal-free areas were extracted from the pictures. The thickness of the salt crystal (δ^s) growing into the liquid direction and the mass flux (J) of the crystals were estimated using different approaches all of which overlap perfectly well with each other. These approaches were: using image analysis for measuring the crystal size changes in time, weighing the crystal amount at the end of the experiment, and doing concentration measurement in the beginning and at the end of the experiment.

10.3.2.1.2 Temperature Investigations at the Interface. Due to the transparency of $MgSO_4 \cdot 7H_2O$ crystals, it was not possible to determine the crystal borders on the post-processing temperature images. Thus, the temperature profiles of the TLC surface for the crystal-covered areas were detected and compared to the one of crystal-free area. The part of the heat of crystallisation evolved in the crystal side is transferred both under and near the borders of the crystal on the TLC.

Before stationary-state conditions were established, it was possible to have warmer or colder temperature readings under the crystals compared to the surrounding. This is due to the dynamic nature of the crystallisation process in which molecules or ions dissolve and absorb on the crystal surface simultaneously during growth. It should be noted that the heat generated from the crystallisation did not only conduct under the crystal itself, but also evolved in the horizontal direction of the TLC. The TLC temperature reading collected near the crystal borders did thus not represent the temperature reading for the crystal-free area. Based on these facts, the use of single crystal values in the temperature-data collection from TLC readings was avoided. Instead, representative temperature readings were collected for the same moment in time from two different areas, with and without crystals. By using *crystal-group average temperature values*, these little variations due to the optical properties and heat distribution in the horizontal direction of

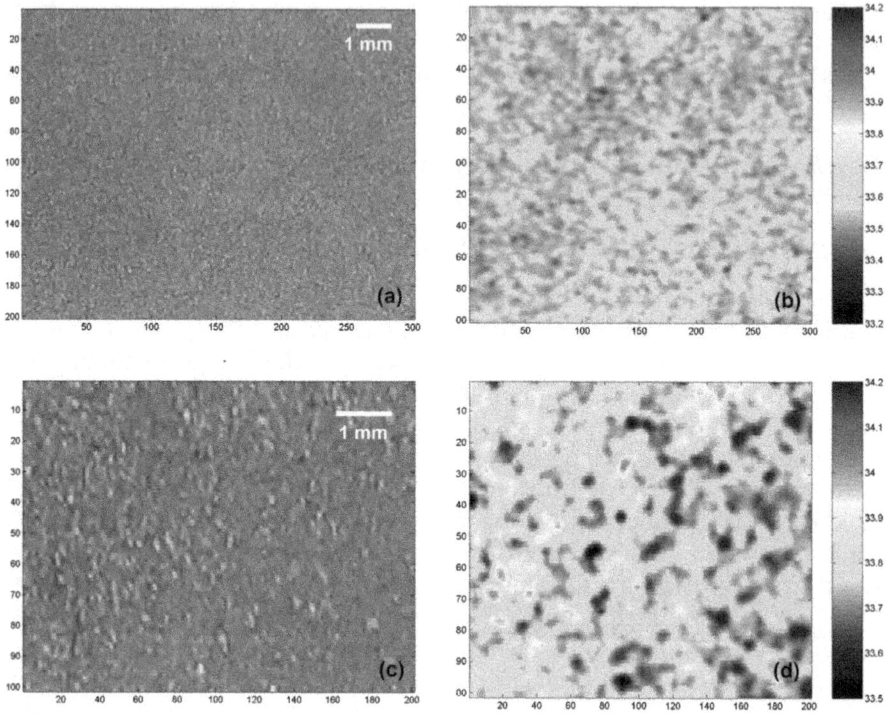

Figure 10.2 Picture of crystal-free (a) and crystal-covered (c) areas on the TLC surface and their post-processing temperature images (b), (d). 1 pixel is 65 μm.
Reprinted with permission from *Cryst. Growth Des.*[8]

the TLC are minimized and included in the accuracy of ± 0.03 °C. In Figure 10.2(a) a photo for the crystal-free area and in (b) its post-processing temperature image is presented. The averaged TLC temperature reading for crystal-free areas for this moment is determined as 33.60 °C. For the same moment, an area covered with crystals and its post-processing temperature image is also presented in Figure 10.2(c) and (d), respectively, with an average temperature reading value of 33.75 °C.[8]

The temperature readings and the results were obtained following the procedure described in the 10.3.2.1.1 Data analysis section. For the entire experiment the temperature jumps across the interface $T^{\ell} - T^{s}$ were all calculated to be 0.2 °C.[8]

During the whole experiment the coolant outlet temperature was higher than the coolant inlet temperature, the measured and calculated wall temperatures were equal to each other, and the crystal-free area on TLC temperature measurements was constant. The crystal shapes, furthermore, had no impact on the TLC post-processing temperature image, which allowed us to use the temperature readings under the crystals. All these facts support the soundness of our experiment.

10.3.2.2 Ice Crystallisation Case

10.3.2.2.1 Data Analysis. The transparency property of water and ice crystals were used to follow the temperature profiles of the TLC temperature readings for both crystal and liquid layers. The ice-surface temperature (T^s) was derived from the linear extrapolation from the TLC sheet readings under the ice layer. The water-side temperature (T^ℓ) was directly read from the TLC sheet readings on the liquid side of ice interface.

The ice thickness (δ^s) growing into the water direction, the ice growth rate and the mass flux (J) of the crystals were estimated from the experimental photos. The heat flux from the liquid side was calculated by using TLC data readings and Fourier's law.

10.3.2.2.2 Temperature Investigations at the Interface. Pure water was cooled below its freezing temperature and primary ice nucleation (0 s) on the cold surface was initiated. When the ice crystallisation process started to generate and release heat into the system, the temperature profile and heat flux became discontinuous. The ice grew with curvature at both ends of the heat exchanger. A curvature arose because in the set-up, the area of the bottom heat exchanger is smaller than the area of the bottom of the crystallizer.[9]

Crystallisation is a dynamic process in which molecules desorb and adsorb on the crystal surface simultaneously and in a sub-pixel size range. For this reason, and due to minor noise in the CCD pixel recordings, recordings of single pixel values from TCL reading were not pursued. Instead, representative temperature readings were collected by averaging the temperature data horizontally in an area of interest (AOI) for each vertical position perpendicular to the heat-exchange surface. This approach minimized small variations due to image noise and micro-plume effects due to convection, giving an accuracy of $\pm$ 0.01 °C.[9] Figure 10.3 (I and II) shows TLC images, magnified AOI and horizontally averaged temperature data for different vertical locations of these AOI for experimental moments 56 and 66 seconds after nucleation occurred.

The ice crystal growth on the cooled surface and the corresponding colour changes were visible to the naked eye and in the digital images. A temperature jump across the interface due to growth of ice crystals was confirmed by calculations of temperatures deduced from the TLC. For two experimental moments, having different ice growth rates, the temperature jumps across the interfaces $T^\ell - T^s$ were calculated to be 1.27 °C and 1.68 °C for the 56 s and 66 s experimental moments, respectively. Ice values are far higher than for the $MgSO_4 \cdot 7H_2O$ crystallisation case with 0.2 °C, but then the heat of crystallisation is also higher for water. During the entire experiment, the coolant outlet temperature had higher values than the coolant inlet temperature. The ice-side interface temperature (T^s) value was also near the ice freezing temperature of 0.00 °C. These observations support the soundness of the measurements and conclusions.

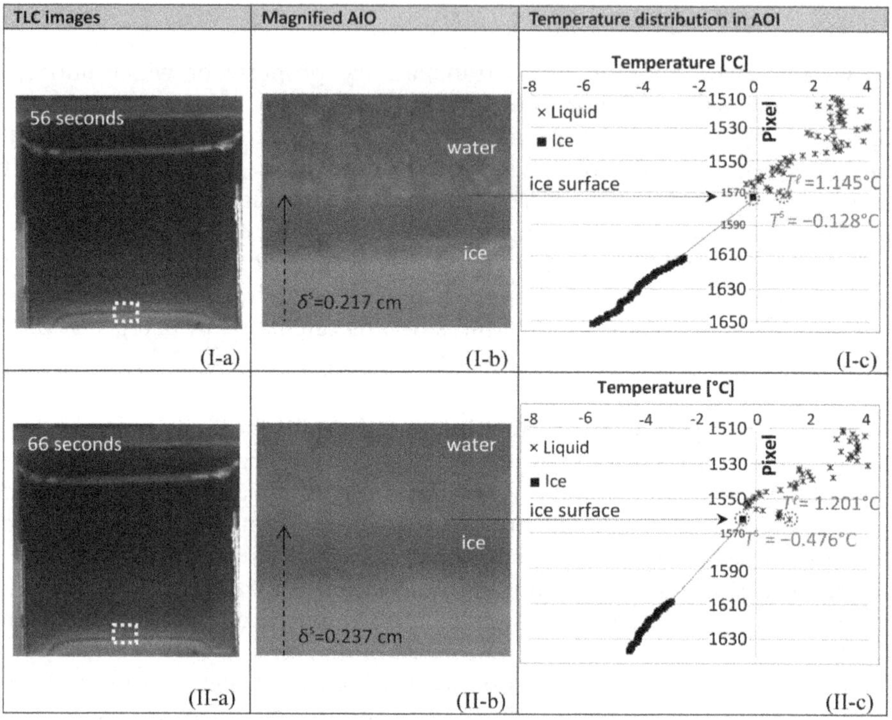

Figure 10.3 TLC images, magnified AIO's, T distribution in AOI's of 56 s and 66 s experimental moments. 1 pixel is 20 μm.
Reprinted with permission from *Chem. Phys. Lett.*[9]

10.4 Heat and Mass Transport, and the Entropy Production Equations for Crystallisation

As shown in the previous section, the experimental[7–9] and new computational[18,19] evidence for a temperature jump across the interface during phase transformation, as well as the possibility of a surface temperature being different from that of the temperature of the surroundings, bring forward a need for a new theoretical description.[20,21] The systematic theory of non-equilibrium thermodynamics is indispensable in this context.[2] The surface is considered as a separate thermodynamic system, with its own excess variables, following Gibbs.[10] The excess entropy production of the surface will then define dynamic boundary conditions, *i.e.*, fluxes and forces at the interface.

In order to analyse crystal growth on the cold surfaces, the heat and mass transfer equations were defined from the entropy production for three different regions (see Figure 10.1), *i.e.*, at points solid (s), liquid (ℓ) and interface (i) to obtain the temperature, concentration and heat and mass flux profiles. The details of these derivations for the crystal and the liquid layer can be found elsewhere.[8] Here we will only elaborate the derivations for the interface.

10.4.1 Interface (i)

The interface is the thin layer between the crystal and the liquid layers. In the crystallisation system, it is the most interesting layer since crystallisation takes place at the interface. The layer is defined as a two-dimensional thermodynamic system by excess densities.

As experimentally proven above, an extrapolation of the measured temperature profiles both from the crystal side and from the liquid side up to their contact point (the interface) shows a temperature discontinuity across the solid–liquid interface. The interface position was chosen as the equimolar surface of the crystal. The heat and mass transport and the entropy production equations for the interface are defined in Table 10.2.

Eqn (10.1) shows that the enthalpy of crystallisation ($\Delta H_{\text{cryst.}}$) is shared between the crystal and the liquid layers. It can partly be transported into the crystal $J'^{\text{s}}_q > 0$ and partly be going back into the mother liquid $J'^{\ell}_q < 0$. The relative fraction of the enthalpy of crystallisation carried by the mass fluxes (J) at uniform temperature is given by the so-called heats of transfer, *cf.*, eqn (10.8b) below.[8]

Table 10.2 Mass and heat transport, and entropy production equations for the interface layer at steady state (i).[a,b]

Heat and Mass transport for 'i'		
Energy conservation for one-dimensional transport:	$J'^{\text{s}}_q + J\cdot H^s = J'^{\ell}_q + J\cdot H^l$ $J'^{\text{s}}_q = J'^{\ell}_q + J\cdot \Delta H_{\text{cryst.}}H,$ $J\cdot(H^{\text{s}} - H^l) = J'^{\ell}_q + J'^{\text{s}}_q < 0.$	(10.1)
Entropy production for 'i'		
At steady state interface entropy production consists: - heat conduction term in to the interface, - heat conduction term out of the interface, - mass transport term across the interface.[2]	$\sigma^{\text{i}} = J'^{\text{s}}_q \Delta_{\text{i,s}}\left(\frac{1}{T}\right) + J'^{\ell}_q \Delta_{\ell,\text{i}}\left(\frac{1}{T}\right) - J\frac{1}{T^{\text{i}}}\Delta_{\ell,\text{s}}\mu_T(T^{\text{i}})$	(10.2)
Eliminating T^{i} from eqn (10.2):	$\sigma^{\text{i}} = J'^{\text{s}}_q \Delta_{\ell,\text{s}}\left(\frac{1}{T}\right) - J\frac{1}{T^{\ell}}\Delta_{\ell,\text{s}}\,\mu_T(T^{\ell}),$	(10.4a)[c]
	$\sigma^{\text{i}} = J'^{\ell}_q \Delta_{\text{s},\ell}\left(\frac{1}{T}\right) - J\frac{1}{T^{\text{s}}}\Delta_{\text{s},\ell}\,\mu_T(T^{\text{s}}).$	(10.4b)[d]

[a] μ_T: Chemical potential evaluated at constant temperature ($\text{J}\cdot\text{K}^{-1}$).
[b] $\Delta H_{\text{cryst.}}$: Enthalpy of crystallization ($\text{J}\cdot\text{mol}^{-1}$).
[c] Used for $MgSO_4$ crystallisation calculations.
[d] Used for ice crystallisation calculations.

The entropy production equation given in eqn (10.2) is valid for crystal growth. Subscript i,s means that the difference is taken between the salt and the interface, subscript ℓ,i means a difference between the interface and the solution; while ℓ,s means across the interface (*i.e.*, between crystal and the solution). As the mass flux of the crystal into and out of the surface is constant, the chemical driving force (μ_T) into and out of the surface were combined. Since $J'^{\mathrm{s}}_q \neq J'^{\ell}_q$, the other terms in eqn (10.2) cannot be combined in a similar way. The expression describes all energy dissipated as heat at the interface.[8]

To have the driving force depend on the interface or surface temperature is not desired as it is not convenient. Thus, the surface temperature and one of the heat fluxes could be eliminated from eqn (10.2) by introducing the energy balance into the entropy production and by using the identity given in eqn (10.3):

$$\frac{1}{T^{\mathrm{i}}}\Delta_{\ell,\mathrm{s}}\mu_T(T^{\mathrm{i}}) = \frac{1}{T^{\mathrm{i}}}\left(\mu^{\mathrm{s}}(T^{\mathrm{i}}) - \mu^{\ell}(T^{\mathrm{i}})\right)$$
$$= \frac{1}{T^{\ell}}\left(\mu^{\mathrm{s}}(T^{\ell}) - \mu^{\ell}(T^{\ell}) + (H^{\mathrm{s}} - H^{\ell})\left(\frac{1}{T^{\mathrm{i}}} - \frac{1}{T^{\mathrm{s}}}\right)\right), \qquad (10.3)$$

where H is the enthalpy ($\mathrm{J \cdot mol^{-1}}$).

The entropy production for the interface could be rewritten as presented in Table 10.2 in eqn (10.4a) and (10.4b), dependent either on the liquid or crystal temperatures at the surface, respectively. These relations will be used in $MgSO_4$ and ice crystallisation cases correspondingly.

The thermodynamic driving forces for transport of heat and mass from the entropy production are given both for $MgSO_4$ and ice cases as seen in Table 10.3.

For $MgSO_4$ crystallisation, the activity of the crystal in the liquid is given at the liquid temperature near the surface, at T^{ℓ}. Since the activity of the crystal is more difficult to obtain, the activity of the liquid that would have been in equilibrium with the crystal at the temperature T^{ℓ} is taken in eqn (10.5a). The activity is calculated from the saturation curve using the Pitzer Model.[22] For the ice crystallisation case, the chemical force in terms of activities or in terms of vapour pressures $\left(a_{\mathrm{w}} = \frac{p_{\mathrm{w}}}{p^{*}_{\mathrm{w}}}\right)$ is presented in eqn (10.6).

On the s-side, there is pure ice. Therefore $p^{\mathrm{s}}_{\mathrm{w}}$ is the vapour pressure of water in equilibrium with ice at the temperature of the ice. On the ℓ-side there is pure water. The vapour pressure of the ℓ-phase is the vapour pressure of the water, also taken at the temperature of the ice T^{S}. Superscript * indicates the pure-phase vapour pressure. The chemical driving force is the degree of super-saturation calculated at T^{S}. A ratio larger than unity may lead to crystallisation, *but not for all conditions*. Whether crystallisation may take place or not, depends also on the value of T^{S} with respect to T^{ℓ}, and the heat of transfer. The vapour pressures of water are known as function of temperature.[23]

Table 10.3 Thermodynamic driving forces for $MgSO_4$ and ice cases.

$MgSO_4$ crystallisation

$$\Delta_{\ell,s}\left(\frac{1}{T}\right)=\frac{1}{T^{\ell}}-\frac{1}{T^{s}}=\frac{\Delta_{s,\ell}T}{T^{\ell}T^{s}}\approx\frac{\Delta_{s,\ell}T}{(T^{\ell})^{2}}=-\frac{\Delta_{\ell,s}T}{(T^{\ell})^{2}},$$

$$-\frac{1}{T^{\ell}}\Delta_{\ell,s}\Delta\mu_{T}(T^{\ell})=-\frac{RT^{\ell}}{T^{\ell}}\ln\frac{a^{s}(T^{\ell})}{a^{\ell}(T^{\ell})}=-R\ln\frac{a^{s}(T^{\ell})}{a^{\ell}(T^{\ell})}. \quad (10.5a)$$

Ice crystallisation

$$\Delta_{s,\ell}\left(\frac{1}{T}\right)=\frac{1}{T^{\ell}}-\frac{1}{T^{s}}=\frac{\Delta_{\ell,s}T}{T^{i}T^{o}}\approx-\frac{\Delta_{s,\ell}T}{(T^{s})^{2}},$$

$$-\frac{1}{T^{s}}\Delta_{s,\ell}\Delta\mu_{T}(T^{s})=-\frac{1}{T^{s}}\left(\mu^{s}(T^{s})-\mu^{\ell}(T^{s})\right), \quad (10.5b)$$

$$-\frac{1}{T^{s}}\Delta_{s,\ell}\Delta\mu_{T}(T^{s})=R\ln\frac{a_{w}^{s}(T^{s})}{a_{w}^{\ell}(T^{s})}=R\ln\frac{\dfrac{p_{w}^{\ell}(T^{s})}{p_{w}^{*}(T^{s})}}{\dfrac{p_{w}^{i}(T^{s})}{p_{w}^{*}(T^{s})}}=R\ln\frac{p_{w}^{\ell}(T^{s})}{p_{w}^{s}(T^{s})}. \quad (10.6)$$

10.4.2 Equations of Transport for Crystal Growth

For the surface there is a linear relation between the thermodynamic forces and the conjugate fluxes. In the steady state, force-flux relations can be written for the $MgSO_4$ crystallisation case between the interface (i) and the crystal side (s); and for ice crystallisation case between the interface (i) and the liquid side (ℓ), as shown in Table 10.4 eqn (10.7a) and (10.7b). These equations express that there are jumps in the intensive variables like the temperature and the chemical potential at the surface and that $T^{i}\neq T^{s},T^{\ell}$. This fact, which is also proven experimentally above, is normally not taken into account in the description of phase transitions.[24,25]

In eqn (10.7a) (and 10.7b), $R_{qq}^{i,s}$ and $R_{\mu\mu}^{i,s}$ ($R_{qq}^{i,\ell}$ and $R_{\mu\mu}^{i,\ell}$) are the two main interface resistivities to heat and mass transfer, respectively; and $R_{q\mu}^{i,s}$ and $R_{q\mu}^{i,s}$ ($R_{q\mu}^{i,\ell}$ and $R_{\mu q}^{i,\ell}$) are the coupling resistivities for the interface. According to Onsager, $R_{q\mu}^{i,s}=R_{\mu q}^{i,s}$ (and $R_{q\mu}^{i,\ell}=R_{\mu q}^{i,\ell}$) are equal. The "heat of transfer" for the crystal side of the surface ($q^{*i,s}$) and liquid side of the surface ($q^{*i,\ell}$) are defined by the coefficient ratios expressed in eqn (10.8a) and (10.8b), respectively.

The stationary state heat flux equations for the crystal and liquid sides are defined in eqn (10.9a) and (10.9b). By substituting eqn (10.1) into eqn (10.7a), using eqn (10.3), the relations between the resistivities[2,26] for the salt side and the liquid side heat flux can be obtained, see the case of evaporation in Chapter 8, eqn (8.13). Eqn (10.10) is derived from eqn (10.8a) and (10.8b) which shows that the assumption of zero cross coefficients violates

Table 10.4 Equations of transport for crystal growth.

MgSO$_4$ crystallisation		Ice crystallisation	
$-\frac{\Delta_{\ell,s}(T)}{(T^{\ell})^2} = R_{qq}^{i,s} J'^{s}_{q} + R_{q\mu}^{i,s} J,$ $-\frac{1}{T^{\ell}} \Delta_{\ell,s}\mu_T(T^{\ell}) = R_{\mu q}^{i,s} J'^{s}_{q} + R_{\mu\mu}^{i,s} J.$	(10.7a)	$-\frac{\Delta_{s,\ell}(T)}{(T^{s})^2} = R_{qq}^{i,\ell} J'^{\ell}_{q} + R_{q\mu}^{i,\ell} J,$ $R\ln\frac{p_w^{\ell}(T^s)}{p_w^{s}(T^s)} = R_{\mu q}^{i,\ell} J'^{\ell}_{q} + R_{\mu\mu}^{i,\ell} J.$	(10.7b)
$q^{*i,s} \equiv \left(\frac{J'^{s}_{q}}{J}\right)_{T^s = T^{\ell}} = -\frac{R_{q\mu}^{i,s}}{R_{qq}^{i,s}}.$	(10.8a)	$q^{*i,\ell} \equiv \left(\frac{J'^{\ell}_{q}}{J}\right)_{T^s = T^{\ell}} = -\frac{R_{q\mu}^{i,\ell}}{R_{qq}^{i,\ell}}.$	(10.8b)
$J'^{s}_{q} = -\frac{1}{R_{qq}^{i,s}}\frac{\Delta_{s,i}(T)}{(T^{\ell})^2} + q^{*,s}J.$	(10.9a)	$J'^{\ell}_{q} = -\frac{1}{R_{qq}^{i,\ell}}\frac{\Delta_{s,\ell}(T)}{(T^{s})^2} + q^{*,\ell}J.$	(10.9b)

thermodynamic laws.[2] The derivation presented in eqn (10.10) is similar to the one for evaporation in Chapter 8 eqn (8.33).

$$q^{*i,s} - q^{*i,\ell} = \Delta H_{\text{cryst.}}. \tag{10.10}$$

10.4.3 Determination of Transfer Resistivities

10.4.3.1 $MgSO_4 \cdot 7H_2O$ Crystallisation Case

The heat of transfer for the salt side divided by the enthalpy of crystallisation ($q^{\star i,s}/\Delta H$) was calculated for the entire experiment. For $MgSO_4 \cdot 7H_2O$ crystallisation, it was reported that typically 70 to 80 % of the heat of crystallisation was transferred into the solid side, whereas 20 to 30 % of this latent heat was transferred back in to the liquid side with a uniform temperature.[8] This is significant information, which shows the error one can make by assuming that all the heat of crystallisation is transferred into the cold side (*i.e.*, to the salt side), as is generally practiced in crystallisation calculations.[24,25] In this case, the error may not seem very big, however it is case dependent, and might well be larger in other systems.

The description allows also for calculation of $R_{\mu\mu}$, the interface mass resistivity, using eqn (10.7a) and the coupling coefficients $R_{q\mu} = R_{\mu q}$. By taking advantage of the Onsager equations, there is a new alternative to resolve the difficulty of finding the temperature jump in the absence of a heat flux. The interface resistivity coefficients for heat transfer (R_{qq}) and mass transfer ($R_{\mu\mu}$) are calculated as $2.1 \cdot 10^{-7}\ \text{m}^2 \cdot \text{K}^{-1} \cdot \text{W}^{-1}$ and $1.26 \cdot 10^3\ \text{J} \cdot \text{m}^2 \cdot \text{s} \cdot \text{K}^{-1}\ \text{mol}^{-2}$, respectively. The coupling resistivity coefficients ($R_{q\mu} = R_{\mu q}$) then become $-3.88 \cdot 10^{-3}\ \text{m}^2 \cdot \text{s} \cdot \text{K}^{-1} \cdot \text{mol}^{-1}$.

The minimum chemical driving force, or degree of super-saturation at the temperature near the phase boundary that is needed to overcome the temperature jump before crystallisation can take place is now possible to calculate. From eqn (10.7a) for no mass flux ($J=0$) the criterion to have $MgSO_4 \cdot 7H_2O$ crystallisation is:

$$\frac{\Delta_{s,\ell}(T)}{T^{\ell}} \leq 0.5 \cdot 10^{-4} \Delta\mu_T(T^{\ell}). \tag{10.11}$$

Such a relation is essential to prevent scale formation on cold surfaces.

As seen in eqn (10.12), a relation between the coolant-side heat flux and mass flux can be derived from eqn (10.7a). By using this relation, just by regulating the coolant heat flux, it is possible to control the salt mass flux (*i.e.* salt scale layer).

$$J = -\frac{1}{T^{\ell}} \Delta\mu_T(T^{\ell}) \frac{1}{R^{i,s}_{\mu\mu}} - \frac{R^{i,s}_{\mu q}}{R^{i,s}_{\mu\mu}} J'^{s}_{q}. \tag{10.12}$$

Eqn (10.12) predicts that a bigger heat flux on the salt side promotes crystallisation. In other words, when the cooling rate is larger, the growth rate of the crystals increases proportionally. Mass transfer onto the crystal surface becomes, however, more difficult if $\Delta_{s,\ell}(T)$ becomes too large. In this case J'^{s}_{q} is reduced at the cost of J'^{ℓ}_{q}.[8]

It is also possible to derive relations between diffusion coefficient (D_{AB}) and mass resistivity ($R_{\mu\mu}$) and the thermal conductivity coefficient (λ) and heat resistivity (R_{qq}), as seen in eqn (10.13) and (10.14). The details of these derivations, in which Δz, R and C refer to distance from the crystal surface, ideal gas constant and concentration, respectively, are elaborated in the work of Genceli *et al.*[8]

$$D_{AB} = \frac{\Delta z R}{R^{i,s}_{\mu\mu} C^{\ell}}. \tag{10.13}$$

$$\lambda = \frac{\Delta z}{R^{i,s}_{qq} (T^{\ell})^2}. \tag{10.14}$$

10.4.3.2 Ice Crystallisation Case

In Figure 10.3 (I-b and II-b), above the ice surfaces, one can see a cloudy bright region. In this layer, the density of water was gradually reduced in the direction of the heat-exchanger. In Figures 10.4 (I-c to II-c) a waving trend of temperature above T^{ℓ} value can be seen for some pixels. We interpreted this as due to the effect of the heat production at the interface; *i.e.*, the effect of $q^{*,\ell}$. Thus, plumes were created by the rapid growth, contributing to the stirring of the liquid layer.

The coupling coefficient ratio of the resistivity to heat transfer at zero ice growth is formulated in eqn (10.7b). The heat of transfer for the liquid (water) side and solid (ice) side divided by the enthalpies of crystallisation ($q^{*\ell}/\Delta H$ and $q^{*s}/\Delta H$, respectively) for different experimental moments are presented in Table 10.5. It was calculated that typically 30 % of the released heat of crystallisation is transferred back into the water side. Under the

Table 10.5 Values of transfer resistivity and enthalpy distribution for two moments of ice experiment.

	Experimental moment		
	56 s	66 s	unit
$T^{\ell} - T^{s}$	1.273	1.676	°C
$R_{qq}^{i,\ell}$	$1.18 \cdot 10^{-7}$	$1.18 \cdot 10^{-7}$	$m^2 \cdot K^{-1} \cdot W^{-1}$
$R_{\mu\mu}^{i,\ell}$	0.23	0.26	$J \cdot m^2 \cdot s \cdot K^{-1} \cdot mol^{-2}$
$R_{q\mu}^{i,\ell} = R_{\mu q}^{i,\ell}$	$-2.01 \cdot 10^{-4}$	$-2.23 \cdot 10^{-4}$	$m^2 \cdot s \cdot K^{-1} \cdot mol^{-1}$
$R_{qq}^{i,s}$	$1.18 \cdot 10^{-7}$	$1.18 \cdot 10^{-7}$	$m^2 \cdot K^{-1} \cdot W^{-1}$
$R_{\mu\mu}^{i,s}$	6.91	7.21	$J \cdot m^2 \cdot s \cdot K^{-1} \cdot mol^{-2}$
$R_{q\mu}^{i,s} = R_{\mu q}^{i,s}$	$-9.10 \cdot 10^{-4}$	$-9.33 \cdot 10^{-4}$	$m^2 \cdot s \cdot K^{-1} \cdot mol^{-1}$
$q^{*\ell}/\Delta H$	−0.28	−0.32	—
$q^{*s}/\Delta H$	0.72	0.68	—

same conditions, the latent heat transferred to the ice side was 70 %. Similar to the situation for $MgSO_4 \cdot 7H_2O$ crystallisation case, the error one can make by assuming that all the heat of crystallisation is transferred to the ice side, has the same order of magnitude. The heat of transfer should not be neglected.

The resistivity to heat transfer ($R_{qq}^{i,\ell}$) was calculated for the last moment before ice nucleation started ($J = 0$). By taking advantage of the Onsager set of equations, it is possible to find the temperature jump in the absence of a heat flux. Using the equality of the coupling coefficients ($R_{q\mu} = R_{\mu q}$) in eqn (10.7b), Values of the interface mass resistivity ($R_{\mu\mu}$) were calculated for four different experimental moments.

By using the heat transfer resistivity and the coupling coefficient in eqn (10.7b), for $J = 0$, we finally obtained the condition for the minimum chemical driving force, (*i.e.*, degree of super-saturation) at the temperature near the phase boundary that is needed to overcome the temperature jump before ice crystallisation can take place.

$$\frac{\Delta_{\ell,s}(T)}{T^{s}} \leq 0.2 \cdot 10^{-5} \Delta\mu_T(T^{s}). \tag{10.15}$$

The relation presented in eqn (10.15) is quite significant for explanation of and possible prevention of scale formation on cold surfaces. It explains that crystallisation may be hindered not only by activation energy, but also by a thermal force acting in an unfavourable way.

10.5 Concluding Remarks

The evidence for a temperature jump between the solid and the liquid side of an interface during crystallisation has been reviewed. So far this has been found for growth of epsomite from $MgSO_4$ aqueous solutions and of ice from pure water on a cold surface. Crystal growth is a common-place phenomenon and the thermodynamic arguments derived for these crystallisation examples will apply to any solid–liquid phase transition. The observations mean also that the common assumption used in the modelling of phase equilibria, that the temperature profile is continuous though the interface, cannot apply in many cases. Crystallisation is mostly an exothermic process, and one would expect that the heat released at the interface during growth is distributed to both the liquid and the solid phases, also at (near) isothermal conditions. The measurements presented in this work confirm this argument. For epsomite and ice crystal growth around 30 % of the heat of crystallisation is transferred back into the liquid side. Such a distribution cannot be modelled with Fourier's law only, similar to the situation for liquid–vapour transitions[2,7,10,14–16,18–21,27] (see Chapter 8). The coupling of fluxes of heat and mass is large at interfaces, meaning that the theory of irreversible thermodynamics is needed.[2] For both cases, values of the interface transfer resistivity were determined and the minimum chemical driving force at the temperature near the phase boundaries that is needed to overcome the temperature jump before crystallisation can take place was calculated. By taking advantage of these relations, it may be possible to control the crystal mass flux (*i.e.*, the scale layer) by just regulating the coolant heat flux.

This knowledge gained can therefore improve on film or fugacity models for the interface, change current modelling of phase transitions and eventually help prevention of crystal growth at unwanted locations in process industry saving considerable capital and operational costs.

Acknowledgements

I would like to acknowledge the support from STW-NWO (VENI Grant 10680). I also would like to thank Signe Kjelstrup, Johan Wåhlin and Mogens Hinge for their scientific input.

References

1. A. Myerson, *Handbook of Industrial Crystallization*, Butterworth-Heinemann, USA, 2nd edn, 2002.
2. S. Kjelstrup and D. Bedeaux, *Nonequilibium Thermodynamics of Heterogeneous Systems,* Series on Advances in Statistical Mechanics, World Scientific, Singapore, 2008, vol. 16.
3. P. L. Kapitza, *Zh. Eksp. Teor. Fiz.*, 1941, **11**, 581.
4. D. Mackay, *Multimedia Environmental Models: The Fugacity Approach*, Lewis Publishers, 2nd edn, 2011.
5. R. B. Bird, W. E. Stewart and E. N. Lightfoot, *Transport Phenomena*, John Wiley & Sons, 2nd edn, 2002.

6. S. Kjelstrup, T. Tsuruta and D. Bedeaux, *J. Colloid Interface Sci.*, 2002, **256**, 451.
7. V. K. Badam, V. Kumar, F. Durst and K. Danov, *Exp. Therm. Fluid Sci.*, 2007, **32**(1), 276.
8. F. E. Genceli, M. Rodriguez Pascual, S. Kjelstrup and G. J. Witkamp, *Cryst. Growth Des.*, 2009, **9**(3), 1318.
9. F. E. Genceli Güner, J. Wåhlin, M. Hinge and S. Kjelstrup, *Chem. Phys. Lett.*, 2015, **622**, 15.
10. J. W. Gibbs, *The Scientific Papers of Journal, Gibbs*, Dover, New York, 1961.
11. D. Bedeaux and S. Kjelstrup, *Int. J. Thermodyn.*, 2005, **8**(1), 25.
12. S. K. Ratkje and O. J. Flesland, *J. Food Eng.*, 1995, **25**, 553.
13. X. C. Chen, P. Chen and K. W. Free, *J. Food Eng.*, 1997, **31**, 395.
14. D. Bedeaux, A. M. Albano and P. Mazur, *Physica A*, 1976, **82**, 438.
15. D. Bedeaux, *Adv. Chem. Phys.*, 1986, **64**, 47.
16. A. M. Albano and D. Bedeaux, *Physica A*, 1987, **147**, 407.
17. S. Kjelstrup Ratkje and D. Bedeaux, *J. Electrochem. Soc.*, 1996, **143**(3), 779.
18. A. Røsjorde, D. W. Fossmo, D. Bedeaux, S. Kjelstrup and B. Hafskjold, *J. Colloid Interface Sci.*, 2000, **232**, 178.
19. J. Ge, S. Kjelstrup, D. Bedeaux, J. M. Simon and B. Rousseau, *Phys. Rev. E*, 2007, **75**, 061604.
20. E. Johannessen and D. Bedeaux, *Physica A*, 2003, **330**, 354.
21. K. Glavatskiy and D. Bedeaux, *Phys. Rev. E*, 2009, **79**, 031608.
22. V. Pillay, R. S. Gaertner, C. Himawan, M. M. Seckler, A. E. Lewis and G. J. Witkamp, *J. Chem. Eng. Data.*, 2005, **50**(2), 551.
23. B. J. Mason, *The Physics of Clouds*, Clarendon Press, Oxford, U.K., 2nd edn, 1971.
24. P. Pronk, *Fluidized Bed Heat Exchangers to Prevent Fouling in Ice Slurry Systems and Industrial Crystallizers*, PhD Dissertation, Delft University of Technology, The Netherlands, 2006.
25. A. Mersmann, *Crystallization Technology Handbook*, Marcel Dekker Inc., New York, 2nd edn, 2001.
26. J. Xu, S. Kjelstrup, D. Bedeaux, A. Røsjorde and L. Rekvig, *J. Colloid Interface Sci.*, 2006, **299**, 455.
27. G. W. Meindersmaa, C. M. Guijtb and A. B. de Haan, *Desalination*, 2006, **187**(1–3), 291.

CHAPTER 11

Membrane Transport

JUAN P. G. VILLALUENGA* AND V. MARÍA BARRAGÁN

Department of Applied Physics I, Complutense University of Madrid, Plaza Ciencias 1, 28040 Madrid, Spain
*Email: jpgarcia@ucm.es

11.1 Introduction

Membrane technology is becoming increasingly important as a separation technology.[1–5] In current mass and heat membrane transport modelling, equilibrium between the membrane and the two adjacent fluid phases is assumed. This assumption implies that the relevant transport variables, such as temperature or chemical potential, are continuous functions at the membrane boundaries. However, it has become increasingly clear that the membrane interface adjacent to the bulk phase of liquid or vapour can pose a separate resistance to transport.[6–12] This fact can be related to the variation in intensive variables across the interface. For instance, a rapid enthalpy drop may introduce an excess resistance to heat and mass transfer, as it has been found for water transport across a membrane.[8,9] One aim of this chapter is to present a more detailed examination of the membrane function if the equilibrium assumption is removed. As Kjelstrup and Bedeaux[13] have shown, the treatment of an interface as a separate system in terms of non-equilibrium thermodynamics offers a possibility to circumvent the assumption of zero driving forces at interfaces. This chapter also aims to investigate the importance of this possibility for a membrane system.

Because of the novelty of the concept of a membrane interface as a thermodynamically separate system, values of the interfacial resistivity coefficients are mostly not available in the literature for transport processes in membranes. In general, these coefficients are expressions of the thermal

Experimental Thermodynamics Volume X: Non-equilibrium Thermodynamics with Applications
Edited by Dick Bedeaux, Signe Kjelstrup and Jan V. Sengers

Published by the Royal Society of Chemistry, www.rsc.org

conductivity, diffusion coefficient and the heat of transfer, among others, for the membrane interface. In some cases the resistivity coefficients for the surfaces are modelled with reference to the membrane values, and the interfacial coefficients are varied to adjust the model predictions with the experimental data. Some attempts to estimate experimentally the interfacial resistivities have also been done. Non-equilibrium molecular dynamics simulations and the kinetic theory of gases have also been used to determine the interface resistivity coefficients. Nevertheless, an effort should be put into a more exact determination of these coefficients to bring the application of non-equilibrium thermodynamics to membrane systems further.

It has been documented in the literature on membrane transport processes that, for instance, a temperature difference across a membrane is able to drive a mass flux. This phenomenon, which is called Soret effect, indicates that thermal and diffusive transport processes can be coupled.[14–16] Coupling means that a mass flux is caused by a temperature difference, and that a heat flux can be originated by a chemical-potential difference alone. However, most of the models used do not consider the possible coupling of the mass flux to the heat flux in a membrane process. Another purpose of this chapter is to describe systematically these reciprocal effects under the framework of non-equilibrium thermodynamics.

11.2 Non-equilibrium Thermodynamics Description of Heat and Mass Transport across a Membrane System

We use the framework of non-equilibrium thermodynamics and the dividing surface as presented in Section 1.4, Chapter 4 and used in Chapter 8 to derive transport equations for mass and heat processes in a membrane system. It is the aim of the present chapter to derive and solve equations for heat and mass transport across a membrane bounded by two fluid phases. In particular, some relevant experimental conditions will be studied to gain insight into two membrane processes: gas permeation and pervaporation. Pervaporation is a membrane separation process in which one or more components of a liquid mixture permeate selectively through a dense membrane while they partially evaporate. The objective is to give a new basis for analysis of experimental results where heat and mass transport occurs in a membrane system.

The main part of a membrane system is the membrane cell, which is basically composed of the membrane separating two fluid phases, *i.e.*, feed and permeate. The region between the feed phase and the membrane defines one of the membrane surfaces, while the region between the membrane and the permeate phase defines the other membrane surface. Thus, two dividing surfaces in a membrane cell can be defined: the feed–membrane interface and the membrane–permeate interface. Both surfaces are indicated by superscript s. The feed and permeate homogeneous phases next to the membrane are indicated by superscript l (left) and r (right), respectively, and the membrane is indicated by superscript m. The width of the interfaces is much smaller than the width of the membrane. To give an idea, a typical membrane thickness can be of the order of several microns, while the interface thickness can be of the order of nanometres. The non-equilibrium thermodynamics approach allows for relatively thin surfaces to be treated as being two dimensional, while thicker layer can be treated as three-dimensional systems.

In the non-equilibrium thermodynamics description of a membrane system, the membrane and two adjacent surfaces are considered as a heterogeneous system, as shown schematically in Figure 11.1. In the simplest description, there are three subsystems: the three-dimensional membrane and its two-dimensional interfaces. In addition, the approach is so versatile that the feed and permeate phases can be easily included in the description, if necessary, by increasing the number of subsystems. We consider the transport of heat and mass through this heterogeneous system. In the interfaces, transport processes are driven by differences in temperature and chemical potential. In the membrane, transport processes are driven by gradients in temperature and chemical potential. For mathematical simplicity, the transport is considered to be one-dimensional, in the direction perpendicular to the membrane. Furthermore, if stationary state operation is assumed, all the properties are dependent only on the coordinate x across the membrane. Moreover, the mass flux and the total heat flux through the system are constant in stationary-state conditions.

One key quantity which appears in the description is the enthalpy difference across the different interfaces. It is convenient to distinguish between the incoming and outgoing sides of each layer as shown in Figure 11.1. The measurable heat flux at the outgoing side of each surface differs from the heat flux on the incoming side of the surface, but they are related by

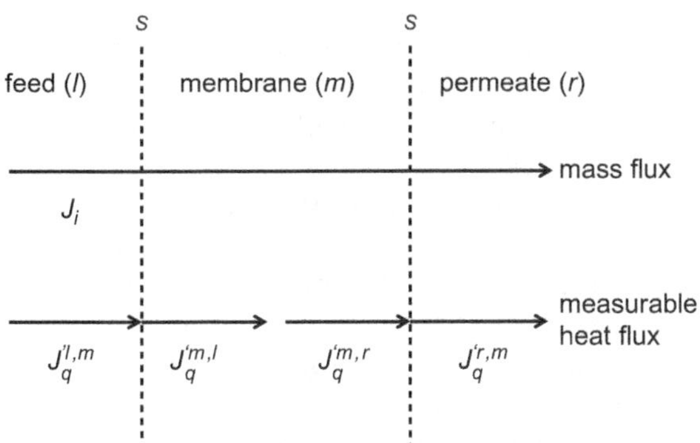

Figure 11.1 A schematic illustration of the system. The mass flux, J_i, is constant in the stationary state, while the measurable heat flux, $J_q^{\prime}$, depends on the position in the system.

the definition of the energy flux:[13] $J_q = J'_q + H_i J_i$. Here, J_q is the total heat flux, J'_q is the measurable heat flux, H_i and J_i are the partial molar enthalpy and the molar flux, respectively, of the permeate component, which is denoted by i. At stationary-state conditions, the constant nature of the total heat flux at the interfaces allows us to relate the incoming and outgoing measurable heat flux in each surface.

As stated above, the notation is based on the three phases (l, m, and r) adjacent to the surfaces (s). A single superscript is used to indicate the relevant subsystem. When a double superscript is used, the first gives the subsystem, and the second gives the nearest subsystem. That is, the first superscript indicates the phase on one side of the surface and the second superscript indicates the adjacent phase. We refer to the original literature for terminology.[13]

At the feed/membrane interface, the incoming heat flux, $J'^{\mathrm{l,m}}_q$, is related to the outgoing heat flux, $J'^{\mathrm{m,l}}_q$, as follows:

$$J'^{\mathrm{l,m}}_q - J'^{\mathrm{m,l}}_q = J_i(H^{\mathrm{m}} - H^{\mathrm{l}}) = J_i \Delta_{\mathrm{l,m}} H, \tag{11.1}$$

where $\Delta_{\mathrm{l,m}}H$ is the enthalpy change at the interface. At the membrane/permeate interface, the incoming heat flux, $J'^{\mathrm{m,r}}_q$, is related to the outgoing heat flux, $J'^{\mathrm{r,m}}_q$, as follows:

$$J'^{\mathrm{m,r}}_q - J'^{\mathrm{r,m}}_q = J_i(H^{\mathrm{r}} - H^{\mathrm{m}}) = J_i \Delta_{\mathrm{m,r}} H, \tag{11.2}$$

where $\Delta_{\mathrm{m,r}}H$ is the enthalpy change at the interface. Constant enthalpy in each phase is assumed in eqn (11.1) and (11.2). To clarify the notation, the measurable heat flux from the incoming side across the surface at the membrane feed side is denoted by $J'^{\mathrm{l,m}}_q$, whereas the measurable heat flux from the outgoing side across the surface at the membrane feed side is denoted by $J'^{\mathrm{m,l}}_q$. Besides, each jump is written as the value to the right minus the value to the left. This choice gives the jumps across the interfaces the same sign as the gradients in the homogeneous phases for increasing or decreasing variables. Then, the subscripts of Δ refer to the two locations between which the difference is taken. As it shown above, $\Delta_{\mathrm{l,m}}H$ is the enthalpy of the permeate component in phase l minus the enthalpy of the permeate component at phase m, whereas $\Delta_{\mathrm{m,r}}H$ is the enthalpy of the permeate component in phase m minus the enthalpy of the permeate component at phase r. Eqn (11.1) and (11.2) indicate that the enthalpy changes at the interfaces create a discontinuity in the measurable heat flux at the membrane boundaries. This discontinuity in the heat flux at the interfaces can lead to jumps in the temperature and the chemical potential across each of the interfaces. When the enthalpy change is large and the thermal conductivity is low at a given interface, it is expected than the temperature and chemical potential jumps are significant. Thus, thermal and chemical equilibrium cannot be assumed between the membrane and the corresponding adjacent fluid phase.

In general, the membrane is a multicomponent homogeneous system, which is composed of the permeate components and the membrane polymer (in most cases, the membranes are polymeric). The polymer can be considered as the solvent, and the permeate components are the solutes. As the polymer is in excess compared to the other components, the former can be used as a frame of reference. The entropy production for the coupled transport of heat and mass can be obtained as the sum of the product of the thermodynamic driving forces and conjugate fluxes. For simplicity, we consider a single-component transport process where there are then two independent driving forces in the membrane phase, one thermal driving force and one diffusional driving force. The generalization to a multicomponent system is straightforward. Accordingly, the entropy production results in

$$\sigma^{\mathrm{m}} = J_q^{\prime \mathrm{m}} \frac{\mathrm{d}}{\mathrm{d}x}\left(\frac{1}{T}\right) - J_i \frac{1}{T}\frac{\mathrm{d}\mu_{i,T}}{\mathrm{d}x}, \tag{11.3}$$

where $J_q^{\prime \mathrm{m}}$ is the measurable heat flux in the membrane, J_i is the molar flux of the permeate component, and T is the temperature. The chemical potential of the permeate component, μ_i, should be differentiated at constant temperature, giving $\mathrm{d}\mu_{i,T}$. We will consider transport along the x-axis only. The driving forces can be written in local-linear form by using the resistivity coefficients, r_{ij}^{m}, as functions of the thermodynamics fluxes as follows:

$$\frac{\mathrm{d}}{\mathrm{d}x}\left(\frac{1}{T}\right) = r_{qq}^{\mathrm{m}} J_q^{\prime \mathrm{m}} + r_{qi}^{\mathrm{m}} J_i, \tag{11.4}$$

$$-\frac{1}{T}\frac{\mathrm{d}\mu_{i,T}}{\mathrm{d}x} = r_{iq}^{\mathrm{m}} J_q^{\prime \mathrm{m}} + r_{ii}^{\mathrm{m}} J_i. \tag{11.5}$$

According to Onsager reciprocal relations, $r_{qi}^{\mathrm{m}} = r_{iq}^{\mathrm{m}}$, see Chapter 1. The resistivity coefficients for coupling of fluxes i and j, r_{ij}^{m}, are related to the measurable quantities in the following way:

$$r_{qq}^{\mathrm{m}} = \frac{1}{\lambda^{\mathrm{m}} T^2}, \tag{11.6}$$

$$r_{qi}^{\mathrm{m}} = -r_{qq}^{\mathrm{m}}\, q_i^{*\mathrm{m}}, \tag{11.7}$$

$$r_{ii}^{\mathrm{m}} = \frac{1}{D_{\mathrm{i}}^{m} T}\left(\frac{\partial \mu_{i,T}}{\partial c_i}\right), \tag{11.8}$$

where λ^{m} is the thermal conductivity of the membrane phase, $q_i^{*\mathrm{m}}$ is the heat of transfer of the permeate component, c_i is the concentration of the permeate component, and D_i^{m} is the diffusion coefficient of the permeate component. The chemical potential of the permeate component, μ_i, should be differentiated at constant temperature. With these variables, the transport equations take the following form:

$$\frac{\mathrm{d}T}{\mathrm{d}x} = -\frac{1}{\lambda^{\mathrm{m}}}\left(J_q'^{\mathrm{m}} - q_i^{*\mathrm{m}} J_i\right), \tag{11.9}$$

$$\frac{\mathrm{d}\mu_{i,T}}{\mathrm{d}x} = \frac{q_i^{*\mathrm{m}}}{\lambda^{\mathrm{m}} T} J_q'^{\mathrm{m}} - \frac{1}{D_i^{\mathrm{m}}}\left(\frac{\partial \mu_{i,T}}{\partial c_i}\right) J_i. \tag{11.10}$$

For a surface with heat and single-component mass transport, two equations are available for the entropy production:[13]

$$\sigma^{\mathrm{s}} = J_q'\Delta\left(\frac{1}{T}\right) - J_i\frac{\Delta\mu_{i,T}}{T} = J_q\Delta\left(\frac{1}{T}\right) - J_i\Delta\left(\frac{\mu_i}{T}\right). \tag{11.11}$$

The difference between these two expressions has been described in detail in the literature.[13,17] Briefly, the expression to the right of the first equality sign gives fluxes and corresponding driving forces which can be experimentally determined. It can be used to analyze experimental and computational results. In contrast, the second equality sign is suitable for integration across the membrane, when the fluxes are constant. It can be solved for calculations of variable profiles at stationary state conditions by using data available in the literature. In each case, the expression of the entropy production defines the relevant thermodynamic fluxes and forces for the surface. When the first expression is applied to the feed/membrane interface, the excess entropy production can be written as:

$$\sigma^{\mathrm{s}} = J_q'^{\mathrm{l,m}}\left(\Delta_{\mathrm{l,s}}\frac{1}{T}\right) + J_q'^{\mathrm{m,l}}\left(\Delta_{\mathrm{s,m}}\frac{1}{T}\right) + J_i\left(-\frac{1}{T^{\mathrm{s}}}\Delta_{\mathrm{l,s}}\mu_{i,T}\right) + J_i\left(-\frac{1}{T^{\mathrm{s}}}\Delta_{\mathrm{s,m}}\mu_{i,T}\right). \tag{11.12}$$

We can choose to eliminate either $J_q'^{\mathrm{l,m}}$ or $J_q'^{\mathrm{m,l}}$ because they are interlinked by the energy balance, see eqn (11.1). Moreover, as the surface temperature T^{s} is normally not known, it is appropriate to use the temperature at the incoming side of the surface, $T^{\mathrm{l,m}}$, or the temperature at the outgoing side of the surface, $T^{\mathrm{m,l}}$. If we assume feed conditions are known from experiments and we want to calculate the temperature and the chemical-potential difference over the interface, the entropy production reduces to:

$$\sigma^{\mathrm{s}} = J_q'^{\mathrm{m,l}}\left(\Delta_{\mathrm{l,m}}\frac{1}{T}\right) + J_i\left(-\frac{1}{T^{\mathrm{l,m}}}\Delta_{\mathrm{l,m}}\mu_{i,T}\right). \tag{11.13}$$

The driving forces can be written in terms of the thermodynamics fluxes by using the resistivity coefficients of the interface, r_{ij}^{s}:

$$\Delta_{\mathrm{l,m}}\left(\frac{1}{T}\right) = r_{qq}^{\mathrm{s}} J_q'^{\mathrm{m,l}} + r_{qi}^{\mathrm{s}} J_i, \tag{11.14}$$

$$-\frac{\Delta_{\mathrm{l,m}}\mu_{i,T}(T^{\mathrm{l,m}})}{T^{\mathrm{l,m}}} = r_{iq}^{\mathrm{s}} J_q'^{\mathrm{m,l}} + r_{ii}^{\mathrm{s}} J_i. \tag{11.15}$$

According to the Onsager reciprocal relations, $r^{s}_{qi} = r^{s}_{iq}$. The resistitivites r^{s}_{ij} can be related to transport parameters for the interface as follows:

$$r^{s}_{qq} = \frac{1}{\lambda^{s}(T^{l,m})^{2}}, \tag{11.16}$$

$$r^{s}_{qi} = -\, r^{s}_{qq} q^{*s}_{i}, \tag{11.17}$$

$$r^{s}_{ii} = \frac{1}{D^{s}_{i} T^{l,m}} \left(\frac{\partial \mu_{i,T}}{\partial c^{s}_{i}} \right), \tag{11.18}$$

where λ^{s} is the interfacial thermal conductivity, q^{*s}_{i} is the interfacial heat of transfer of the permeate component, and D^{s}_{i} is the diffusion coefficient of the permeate component across the interface. With these variables, the transport equations for heat and mass of a single component over the surface have the following form:

$$\Delta_{l,m} T = -\frac{1}{\lambda^{s}} \left(J'^{m,l}_{q} - q^{*s}_{i} J_{i} \right), \tag{11.19}$$

$$\Delta_{l,m} \mu_{i,T}(T^{l,m}) = \frac{q^{*s}_{i}}{\lambda^{s} T^{m,l}} J'^{m,l}_{q} - \frac{1}{D^{s}_{i}} \left(\frac{\partial \mu_{i,T}}{\partial c^{s}_{i}} \right) J_{i}. \tag{11.20}$$

It is important to note that the chemical potential is defined at $T^{l,m}$, which is the temperature on the left-hand side of the interface. The equations will be similar for the membrane/permeate interface.

Although some attempts have been made to estimate experimentally[8] or theoretically[10,18] the interfacial resistivities, expressions or data for the thermal conductivity, diffusion coefficient and the heat of transfer for the interface are mostly not available in the literature. In practice, they can be modelled with reference to the membrane resistivity values. Thus, interfacial conductivities λ^{s} and D^{s}_{i} can be modelled as fractions of the corresponding membrane conductivity as follows:

$$d^{s} \lambda^{s} = \frac{1}{\theta} \lambda^{m}, \tag{11.21}$$

$$d^{s} D^{s}_{i} = \frac{1}{\delta} D^{m}_{i}, \tag{11.22}$$

where d^{s} is the interface thickness, θ and δ are the thermal-conductivity scaling factor and the diffusional scaling factor, respectively. Their meaning is how much the interfacial resistivity is different from the membrane resistivity of the same thickness as the interface. Therefore, when $\theta = 1$ or $\delta = 1$, the interfacial layer has the same resistivity the corresponding part of the membrane and thus no excess resistance. Larger values of θ or δ indicate that the interface has excess resistance. There is some debate about the magnitude of the interfacial resistivity. Monroe *et al.*[8] claim that the interface has a similar resistivity to mass transfer per unit of thickness as the membrane, meaning that the scaling factor for mass transfer is close to unity. In contrast, there are other evidences[10,12,18] that show that surface has an additional resistance, and therefore the values of these coefficients should be greater than one.

One key quantity which appears in this description is the heat of transfer, the ratio between the heat flux and the mass flux when the temperature difference is zero. According to the Onsager symmetry of the coefficients in flux–force relations, this quantity is also responsible for the mass flux due to the temperature gradient. Following Kjelstrup and Bedeaux,[13] the heat of transfer for the whole feed/membrane surface can be estimated as:

$$q_i^{*s} = -k\Delta_{\mathrm{l,m}}H, \tag{11.23}$$

where k is a scaling factor with a value between 0 and 1. The value given by kinetic theory of gases is 0.2 for a liquid–vapour transition. As Glavatskiy *et al.*[12] have pointed out, the enthalpy difference determines the sign of the thermo-diffusive effect. Moreover, the sign of k is decisive for the sign of the mass flux. A similar expression can be used for the heat of transfer in the membrane/permeate interface.

11.3 Applications

11.3.1 Single-component Pervaporation

Pervaporation is a membrane separation process in which one or more components of a liquid mixture permeate selectively through a dense membrane while they partially evaporate. Although the main driving force on a component going through pervaporation is the chemical-potential gradient, obtained by partial-pressure reduction on the permeate side, a temperature gradient may also contribute. In addition, the phase transition from liquid to vapour might well be rate limiting for the overall process, in the case that large enthalpies of evaporation are involved. These facts gave a motivation to apply non-equilibrium thermodynamics to study the pervaporation transport in a membrane.

Recently, Kuhn *et al.*[6] studied the mass and heat transport processes in the pervaporation of pure water in a zeolite-type membrane using the framework of non-equilibrium thermodynamics. The membrane was composed of a NaA-type zeolite deposited on an asymmetric α-alumina support. Kuhn *et al.* divided the transport of water through the zeolite membrane into five steps: (1) diffusion from the bulk to the liquid/zeolite interface, (2) transfer across the liquid/zeolite surface, (3) diffusion in the zeolite layer, (4) transfer across zeolite/support surface, (5) diffusion into the bulk permeate through the support pores. Therefore, they considered a multilayer membrane system composed of five subsystems: the three-dimensional bulk feed, zeolite and support, and two-dimensional interfaces, *i.e.*, the liquid/zeolite surface and the zeolite/support surface.

We continue by giving the mass and heat transport equations for each subsystem. The heat and mass transport in the zeolite can be described by using eqn (11.9) and (11.10). Eqn (11.10) can be rewritten by considering an effective diffusivity D_{im}^{eff} that describes the interaction between the membrane and the permeating component:

$$\frac{1}{RT}\frac{\mathrm{d}\mu_{i,T}}{\mathrm{d}x} = \frac{q_i^{*\mathrm{m}}}{RT^2}\frac{\mathrm{d}T}{\mathrm{d}x} - \frac{J_i}{c_i^{\mathrm{m}} D_{im}^{\text{eff}}}, \tag{11.24}$$

where c_i^{m} is the concentration of component i in the zeolite, and R is the gas constant.

Under the framework of the dusty-gas model, the mass transport in the support can be described by a combination of Knudsen diffusion and viscous flow. As the viscous flow is not molecular diffusion, it has no coupling effects. Following Kuhn *et al.*[6], by using eqn (11.10) together with some assumptions, the drop of the pressure p over the support can be written as

$$\frac{\mathrm{d}p}{\mathrm{d}x} = \frac{J_i - \dfrac{c_{i,\text{pore}} D_{i\text{sup}}^{\text{eff}} q_i^{*\text{sup}}}{RT^2}\dfrac{\mathrm{d}T}{\mathrm{d}x}}{\dfrac{c_{i,\text{pore}} D_{i\,\text{sup}}^{\text{eff}}}{p} + \dfrac{\varepsilon}{\tau}\dfrac{c_{i,\text{pore}} d_p^2}{32\eta}}, \tag{11.25}$$

where $c_{i,\text{pore}}$ is the concentration of component i in the support pores, $D_{i\text{sup}}^{\text{eff}}$ is the effective diffusion coefficient of component i in the support, ε is the support porosity, τ is the support tortuosity, η is the viscosity and d_{p} is the diameter of the support pores. Eqn (11.9) and (11.25) describe the coupled heat and mass transport in the porous support.

Eqn (11.19) can be applied directly to calculate the temperature difference across the liquid/zeolite interface. Similarly, eqn (11.20) could be used to determine the chemical-potential difference across the interface. However, instead of the chemical potential, it is convenient to use the activity a, which is a measurable quantity and is defined as

$$\mu_i(a_i, T) = \mu_i^0(T) + RT \ln a_i(T), \tag{11.26}$$

where μ_i^0 is the chemical potential at the reference state, which is a function of temperature and a_i is the activity of component i, which is defined as the ratio of the fugacity f_i of component i at local conditions and the fugacity f_i^0 at the reference state $a_i = f_i/f_i^0$. For the reference state, the fugacity of liquid water in equilibrium with its vapour at local temperature can be chosen. Hence, the difference in the chemical potential at constant temperature of water over the surface can be written in terms of the activity difference as follows

$$\Delta_{\mathrm{l,m}}\mu_{i,T}(T^{\mathrm{l,m}}) = RT^{\mathrm{l,m}} \ln\left(\frac{a_i^{\mathrm{m,l}}(T^{\mathrm{l,m}})}{a_i^{\mathrm{l,m}}(T^{\mathrm{l,m}})}\right). \tag{11.27}$$

Here, it is should be pointed out that the activity of water at the right-hand side of the surface is considered at $T^{\mathrm{l,m}}$, $a_i^{\mathrm{m,l}}(T^{\mathrm{l,m}})$, instead of at the local temperature $T^{\mathrm{m,l}}$, $a_i^{\mathrm{m,l}}(T^{\mathrm{m},l})$. The subtle difference between the expressions is described in detail in the literature.[13] Using the Gibbs–Helmholtz relation combined with eqn (11.26), the activity of water at the right-hand side of the surface defined at $T^{\mathrm{l,m}}$, $a_i^{\mathrm{m,l}}(T^{\mathrm{l,m}})$, can be corrected to the activity of water at the local temperature, $a_i^{\mathrm{m,l}}(T^{\mathrm{m},l})$, by

$$R\ln\left(\frac{a_i^{\mathrm{m,l}}(T^{\mathrm{m,l}})}{a_i^{\mathrm{m,l}}(T^{\mathrm{l,m}})}\right) = \left(H_i^{\mathrm{m}} - H_i^{0}\right)\frac{\Delta_{\mathrm{l,m}}T}{T^{\mathrm{l,m}}T^{\mathrm{m,l}}}, \tag{11.28}$$

where H_i^{m} and H_i^0 are the partial molar enthalpies of water in the zeolite, and at the reference state, respectively. Following Kuhn *et al.*,[6] by combining eqn (11.20), (11.27) and (11.28), the water activity difference over the liquid/zeolite interface can be estimated as

$$RT^{\mathrm{l,m}}\ln\left(\frac{a_i^{\mathrm{m,l}}(T^{\mathrm{m,l}})}{a_i^{\mathrm{l,m}}(T^{\mathrm{l,m}})}\right) = \frac{q_i^{*\mathrm{s}}}{\lambda^{\mathrm{s}}T^{\mathrm{m,l}}}J_q'^{\mathrm{m,l}} - \frac{1}{D_i^{\mathrm{s}}}\left(\frac{\partial\mu_{i,T}}{\partial c_i^{\mathrm{s}}}\right)J_i + \left(H_i^{\mathrm{m}} - H_i^{0}\right)\frac{\Delta_{\mathrm{l,m}}T}{T^{\mathrm{m,l}}}. \tag{11.29}$$

Therefore, eqn (11.19) and (11.29) can be used to model the mass and heat transport processes across the liquid/zeolite interface. Similar equations can be derived to calculate the differences in temperature and activity across the zeolite/support interface.

In the calculations, Kuhn *et al.*[6] considered a membrane with a zeolite layer thickness of 2 μm deposited on a three-layer asymmetric α-alumina layer, at a feed temperature of 348 K, assuming a water flux of $0.15\ \mathrm{mol\cdot m^{-2}\cdot s^{-1}}$. The solution procedure is described as follows. First, the only contribution to the transport in the liquid boundary layer was assumed to be Fourier-type heat conduction. By using hydrodynamics expressions, the thickness of the liquid boundary layer at the membrane feed side was estimated to be 28.4 μm. This value allowed for calculating of the temperature profile across this layer. Next, eqn (11.19) and (11.29) were used to model the heat and water transport across the liquid/zeolite interface, allowing the calculation of the temperature and activity jumps across this surface. Next, by using eqn (11.9) and (11.24) the temperature and activity gradients can be estimated in the zeolite. From the temperature and activity gradients the temperature and activity profiles were obtained. Next, the temperature and activity jumps across the zeolite/support interface were estimated using eqn (11.19) and (11.29) in a convenient form. Finally, temperature and activity gradients in the support were estimated by using eqn (11.9) and (11.25). The asymmetric α-alumina layer is considered as the support. This procedure yielded the temperature and activity profiles in the complete membrane system, including liquid film and support layers. To elucidate the influence of the coupling and surface effects on mass and heat transport, Kuhn *et al.*[6] also solved the model equations, neglecting the surface and coupling effects.

Figures 11.2 and 11.3 give an enlargement of the temperature and activity profiles, respectively, in the zeolite and at the liquid/zeolite and zeolite/support surfaces. The profiles illustrate the impact of taking the surface effects and the coupling between mass and heat transport into account. Figure 11.2 shows a temperature difference over the membrane system of 1.3 K. Neglecting the coupling effects, results in more than a three-fold lower total temperature difference (0.4 K). The largest temperature drop is observed at the interfaces. Moreover, it is observed that there is a discontinuity in the temperature at the liquid/zeolite and zeolite/support interfaces. The temperature drop at the surfaces dominates the temperature drop over the zeolite layer. At both surfaces a drop in temperature is observed, despite the fact that at the feed side an exothermic process occurs. Approximately 70 % of the temperature drop over the entire system is generated at the surfaces. It was found that the liquid/zeolite surface has a higher contribution than the zeolite/support surface. Figure 11.3 shows that at liquid/zeolite and zeolite/support interfaces, the activity increases with, respectively, 11 % and 6 % of the total difference. The activity difference over the system is similar for the two models. The discontinuities are also observed in the activity profile, albeit much smaller than for the temperature. The positive jumps in activity at the interfaces reveal that there is a molar flux across the surfaces against the activity difference. This means that, at the interfaces, the mass transport is driven by the temperature gradient.

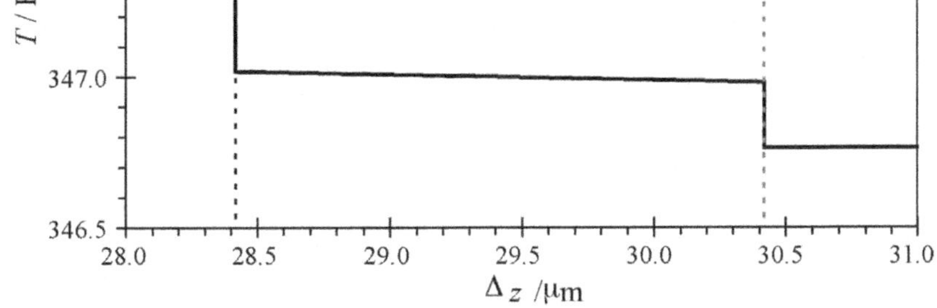

Figure 11.2 Temperature profiles across the zeolite layer and the surfaces calculated by using the non-equilibrium model with coupling and surface effects (solid line), and the model without coupling and surface effects (dashed line). The vertical dashed lines indicate the liquid/zeolite interface and the zeolite/support interface.
Reprinted with permission of *J. Membr. Sci.*[6]

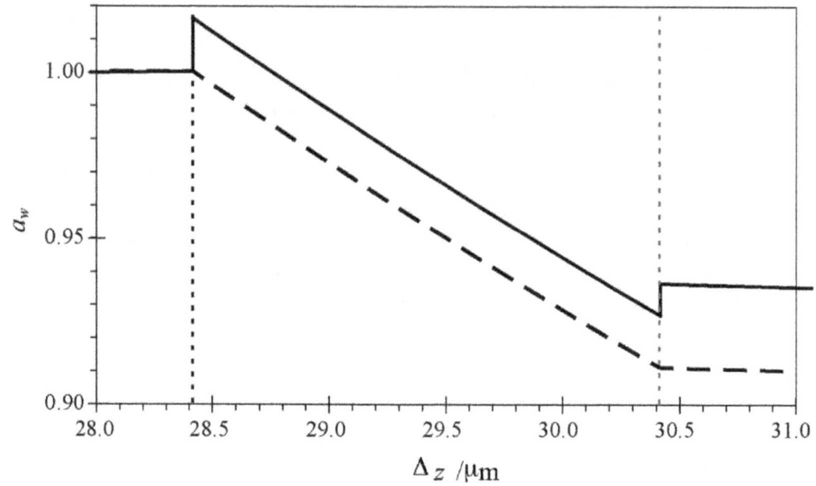

Figure 11.3 Profile of the activity of water across the zeolite layer and the surfaces calculated by using the non-equilibrium model with coupling and surface effects (solid line), and the model without coupling and surface effects (dashed line). The vertical dashed lines indicate the liquid/zeolite interface and the zeolite/support interface.
Reprinted with permission of *J. Membr. Sci.*[6]

11.3.2 Gas Permeation

In a membrane reactor, a chemical reaction and the separation of the resulting products can be performed in one process step. Membrane reactors can be porous and dense. Porous membranes can be zeolite-based membranes,[19] while the palladium membrane is a typical example of a dense membrane.[20] Palladium is especially selective towards transport of hydrogen. In a palladium reactor hydrogen is transported as hydrogen atoms in the palladium, and as hydrogen molecules in the gas phases on each side of the membrane. The heat of adsorption is significant, including the splitting of molecular hydrogen and adsorption at the surface. This dissociative adsorption is likely to influence the mass flux. Although it is known that surface effects can hinder the process, many of the existing models for hydrogen permeation neglect the effect of the membrane surface. Furthermore, while it is known how the temperature affects a chemical reaction, little is known how a temperature gradient affects the reaction rate. With a thin membrane and large heat sinks and sources at the interfaces, such gradients may be large. In addition, the coupling of chemical reactions to fluxes of heat and mass is possible in principle, but has not been described in detail for a membrane reactor system before. These reasons gave Skorpa *et al.*[21] a motivation to apply non-equilibrium thermodynamics to study the transport in a palladium membrane.

Following Skorpa *et al.*,[21] consider a palladium membrane separating two gas phases composed of pure H_2 gas. For simplicity, hydrogen is transported from gas phase to gas phase, and one-dimensional transport in the direction perpendicular to the membrane, are assumed. The membrane is considered to be a homogeneous system, and the region between the gas phases and the palladium defines the two membrane surfaces. Hence, they studied the heat and hydrogen transport processes in a multilayer membrane system composed by three subsystems: the three-dimensional palladium membrane, and two-dimensional interfaces, *i.e.*, the feed gas/palladium surface and the palladium/permeate gas surface. The hydrogen diffusion process in the gas phase layer external to the membrane surfaces is not addressed in their paper.

Prior to membrane transport, hydrogen is adsorbed at the feed membrane surface. The molecular hydrogen is converted into atomic hydrogen in the adsorption process. This dissociative adsorption is written by Skorpa *et al.*[21] as

$$H_2(g) \rightleftarrows 2H(ads). \tag{11.30}$$

The Gibbs-energy difference for the dissociative adsorption at the feed membrane side can be defined as:

$$\Delta_{l,m}G = 2\mu_{H}^{m,l} - \mu_{H_2}^{l,m}. \tag{11.31}$$

In the stationary state, as there is no accumulation of gas in the system, the relation between the flux of atomic hydrogen through the membrane, J_H, and the flux of molecular hydrogen, J_{H_2}, is given by $J_H = 2J_{H_2}$. The total heat flux through the system is also constant, given by eqn (11.1)

$$J_q^{\prime l,m} = J_q^{\prime m,l} + J_{H_2}\Delta_{l,m}H, \tag{11.32}$$

where $\Delta_{l,m}H = 2H_{H}^{m,l} - H_{H_2}^{l,m}$ is the heat of dissociative adsorption in the membrane at the feed side. Constant enthalpy in each layer and symmetric enthalpy effects at both interfaces were assumed. The entropy production for the feed/membrane surface can be obtained by combining eqn (11.13) and (11.31):

$$\sigma^s = J_q^{\prime l,m}\left(\Delta_{l,m}\frac{1}{T}\right) + J_{H_2}\left(-\frac{1}{T^{m,l}}\Delta_{l,m}G\right), \tag{11.33}$$

where $J_q^{\prime l,m}$ is the heat flux evaluated at the left side of the surface. Furthermore, $\Delta_{l,m}G$ is evaluated at the temperature in the membrane close to the surface at the feed side, $T^{m,l}$. The flux–force equations derived from the entropy productions are similar to eqn (11.14) and (11.15):

$$\Delta_{l,m}\left(\frac{1}{T}\right) = r_{qq}^{s}J_q^{\prime l,m} + r_{q\mu}^{s}J_{H_2}, \tag{11.34}$$

$$-\frac{\Delta_{l,m}G}{T^{m,l}} = r_{\mu q}^{s}J_q^{\prime l,m} + r_{\mu\mu}^{s}J_{H_2}. \tag{11.35}$$

The corresponding force–flux relations for the membrane/permeate interface can be derived in a similar manner. The direct coefficients for the resistivity in the surfaces were estimated by multiplying the values of the resistivity coefficient in the rate-determining phase next to the surface with the surface thickness d^s. Hence, as the rate-determining layer for the resistance to gas transfer is the membrane, the mass-resistivity coefficient in the surface $r^s_{\mu\mu}$ is obtained from $r^m_{\mu\mu}$ as $r^s_{\mu\mu} = d^s r^m_{\mu\mu}$. For the resistivity to heat transport, r^s_{qq}, the gas phase next to the surface (feed or permeate) will be rate determining, as the membrane is a well conducting metal. The resistivity in the membrane is neglected. Thus, the heat-resistivity coefficient in the surface r^s_{qq} is obtained from r^g_{qq} as $r^s_{qq} = d^s r^g_{qq}$, where r^g_{qq} is the heat-resistivity coefficient in the gas phase adjacent to the membrane. This latter resistivity coefficient was calculated following a procedure described by Johannessen and Jordal.[22] Finally, the coupling resistivity coefficient in the surface $r^s_{q\mu}$ is estimated by combining eqn (11.19) and (11.23) with the value of $k = 0.2$.

The entropy production for transport of heat and atomic hydrogen in the membrane is given by the general eqn (11.3). Following Skorpa *et al.*,[21] if the measurable heat flux across the membrane is assumed to be constant, the integration of eqn (11.3) gives the following expression:

$$d^m \sigma^m = J'^m_q \left(\Delta_m \frac{1}{T} \right) + J_{H_2} \left(-\frac{2}{T^m} \Delta_m \mu_{H,T} \right), \tag{11.36}$$

where d^m is the membrane thickness and Δ_m means the difference between the right and left side in the membrane phase. The authors contend that this assumption results in a calculation error of less than 2 % for the entropy balance. Accordingly, the force–flux equations for the membrane phase are

$$\Delta_m \left(\frac{1}{T} \right) = r^m_{qq} J'^m_q + r^m_{q\mu} J_{H_2}, \tag{11.37}$$

$$-\frac{2}{T} \Delta_m \mu_{H,T} = r^m_{\mu q} J'^m_q + r^m_{\mu\mu} J_{H_2}. \tag{11.38}$$

The thermal resistivity coefficient r^m_{qq} is related to the thermal conductivity in the membrane phase as given in eqn (11.6). The gas concentration in the membrane is related to the gas pressure in the gas phases by Sieverts' law. Thus, the mass resistivity coefficient $r^m_{\mu\mu}$ can be estimated from the membrane permeability coefficient Π^m as:

$$r^m_{\mu\mu} = \frac{(q^{*m})^2}{\lambda^m T} - \frac{R \ln\left(p^r_{H_2} / p^l_{H_2} \right)}{\Pi^m \left[\left(p^l_{H_2} \right)^{0.5} - \left(p^r_{H_2} \right)^{0.5} \right]}. \tag{11.39}$$

Table 11.1 Estimated resistivity coefficients. Superscript s and m denotes surface and membrane, respectively. Reprinted with permission of *J. Membr. Sci.*[21]

Coefficients	Value	Units
r^{s}_{qq}	$9 \cdot 10^{-13}$	$m^2 \cdot s \cdot J^{-1} \cdot K^{-1}$
$r^{s}_{q\mu}$	$1.6 \cdot 10^{-8}$	$m^2 \cdot s \cdot mol^{-1} \cdot K^{-1}$
$r^{s}_{\mu\mu}$	$4.3 \cdot 10^{-3}$	$J \cdot m^2 \cdot s \cdot mol^{-2} \cdot K^{-1}$
r^{m}_{qq}	$2.7 \cdot 10^{-8}$	$m \cdot s \cdot J^{-1} \cdot K^{-1}$
$r^{m}_{q\mu}$	0	$m \cdot s \cdot mol^{-1} \cdot K^{-1}$
$r^{m}_{\mu\mu}$	$3.79 \cdot 10^{6}$	$J \cdot m \cdot s \cdot mol^{-2} \cdot K^{-1}$

The coupling resistivity coefficient $r^{m}_{q\mu}$ is related to the heat of transfer in the membrane phase q^{*m}_{H} as eqn (11.7) reflects. As the enthalpy of the component to a good approximation is constant in the membrane, it is expected that q^{*m}_{H} is small. This fact permits us to consider that the value of $r^{m}_{q\mu}$ is zero. This means that there is no coupling between the heat and gas fluxes in the membrane. Following Skorpa *et al.*,[21] as the surface has a high resistivity, they increase the value of the set of resistivity coefficients by a factor $\alpha = 10$ or 100. The values of the resistivity coefficients both in the surfaces and in the membrane are listed in Table 11.1.

The effect of coupling and thermal driving forces on the membrane performance was investigated by using a typical set of operating conditions for a 7.2 μm thick membrane.[23] Eqn (11.34) and (11.35) describe the heat and gas transport in the feed/membrane surface, and similar equations describe the transport in the membrane/permeate surface. The transport in the membrane is described by eqn (11.37) and (11.38). Thus, we have heat and mass transport equations for each layer. The solution procedure consists of the integration of the transport equations across each layer, and obtaining $J_{H_2}, J'^{l,m}_{q}$ and temperature and hydrogen activity in the membrane close to the surface at each side. This was done numerically.

Figure 11.4 shows the temperature profile across the membrane calculated for isothermal conditions with a set of surface resistivity coefficients corresponding to a factor $\alpha = 10$. The results show that the temperature jumps at the membrane surfaces are insignificant because the ΔT is less than 0.001 K. Similarly, the jumps in the chemical potential are not important, because the ΔG calculated at the surface are $-8.46\ J \cdot mol^{-1}$ and $-9.13\ J \cdot mol^{-1}$ with and without coupling, respectively. The effect of coupling on the hydrogen fluxes is also negligible.

Despite the small coupling effects, it is interesting to note that the heat flux varies largely between the membrane and its enclosing phases. The measurable heat flux in the membrane, J'^{m}_{q}, is positive and large, $23.24\ kJ \cdot m^{-2} \cdot s^{-1}$. At the feed side, the measurable heat flux, $J'^{l,m}_{q}$,

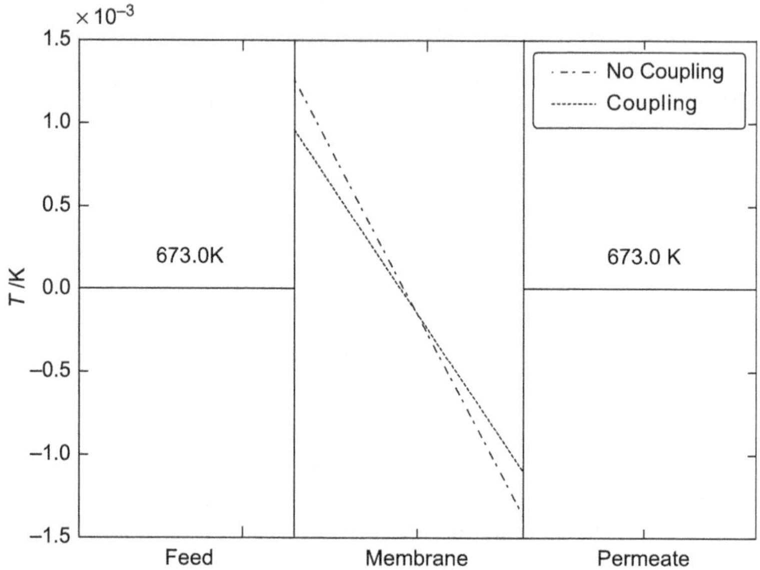

Figure 11.4 Temperature profile across the membrane for identical feed and permeate temperatures.
Reprinted with permission of *J. Membr. Sci.*[21]

is $-6.28\ \mathrm{kJ \cdot m^{-2} \cdot s^{-1}}$, meaning that heat is transported away from the membrane feed side. At the permeate side, the measurable heat flux, $J_q^{\prime \mathrm{r,m}}$, is also $-6.28\ \mathrm{kJ \cdot m^{-2} \cdot s^{-1}}$, meaning that heat is transported towards the membrane permeate side. The discontinuity is due to the enthalpy of dissociative adsorption at the surfaces. These results mean that in order to maintain isothermal conditions, the feed side must be continuously cooled to remove this heat, while on the permeate side we need to supply this amount of heat.

By using the model equations, calculations predict that heating at the permeate side will increase the flux of hydrogen. Thus, the overall temperature differences needed for enhancing the gas flux by 10 % and 25 % were found to be 27.3 K and 72.5 K, respectively, with an initial feed temperature of 673 K. Similarly, an overall temperature difference of −176.5 K will stop the hydrogen flux. This effect is called Soret equilibrium. Thus, cooling the permeate side gives a negative temperature gradient and a decrease in the gas flux.

In order to gain insight into the contribution from coupling of the gas fluxes, the hydrogen fluxes can be expressed as functions of all resistivities and all gradients throughout the system, and divided into direct terms and coupling terms.[21] The model predictions, which are given in Figure 11.5, show that the fraction of the flux caused by the direct terms is constant in spite of temperature changes in the system. Variation with temperature is observed for the total flux, and it is caused by coupling terms. Thus,

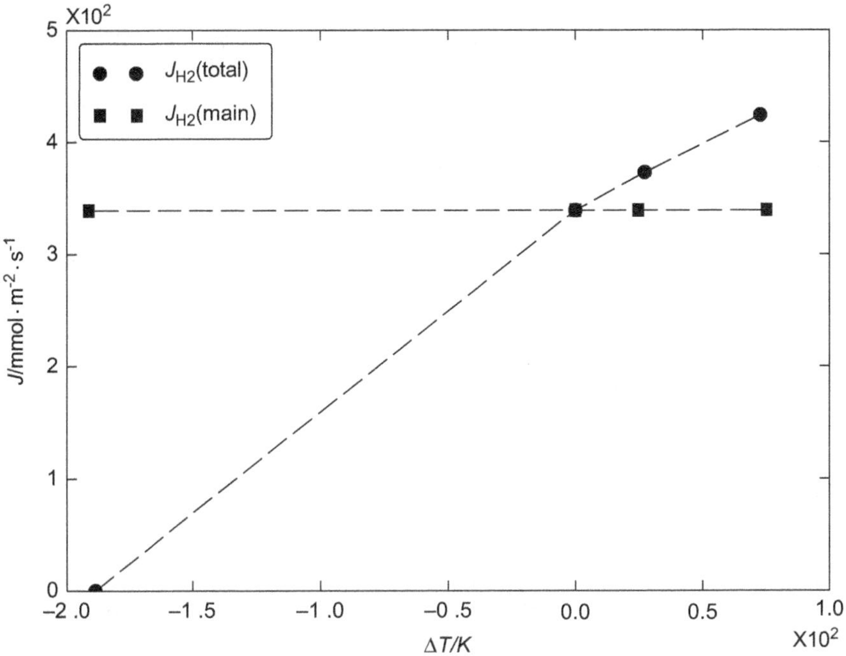

Figure 11.5 Hydrogen flux caused by the direct terms (main) and total hydrogen flux (total) through the membrane as a function of the overall temperature difference.
Reprinted with permission of *J. Membr. Sci.*[21]

application of a temperature gradient gives a significant contribution to the overall mass transfer across the membrane.

Skorpa *et al.*[21] claim that the heat fluxes in the various phases using their non-equilibrium approach differ by orders of magnitude from fluxes that would have been obtained by only using the energy balance in combination with Fick's and Fourier's laws. Therefore, a description without heat and mass coupling can give a serious error in models where the energy balance is fundamental. Conversely, if experimental data are interpreted with a set of equations without coupling effects for determination of transport properties, the determined properties may be incorrect.

11.3.3 Two-component Pervaporation

Coupled heat and mass transport equations to model the pervaporation of a binary mixture in a membrane have been derived based on non-equilibrium thermodynamics by Villaluenga and Kjelstrup.[7] The system is basically composed of the membrane and two adjacent solutions: feed and permeate mixtures. Hence, there are two components (A and B) in the feed and permeate mixtures, whereas in the membrane one really has three components (A, B plus the polymer membrane C), so the natural frame of reference for transport in the membrane system is the membrane itself. Moreover, two dividing surfaces can be defined in the system: the liquid feed/membrane interface and the membrane/vapour permeate interface. In order to analyse the variation in the temperature and the chemical potentials close to and at the interfaces, the heterogeneous system can be divided into three different subsystems: the feed/membrane interface, the membrane, and the membrane/permeate interface. The membrane is considered as a homogeneous system, whereas the two interfaces are considered as surfaces. The equations for the transport of heat and mass in each subsystem can be derived from the corresponding entropy production rate.

For the feed membrane interface, the excess entropy production in the left side or feed-side of the membrane surface can be written as:

$$\sigma^{\mathrm{s}} = J_q^{\prime \mathrm{l}} \Delta_{\mathrm{l,m}}\left(\frac{1}{T}\right) + J_{\mathrm{A}}\left[-\frac{\Delta_{\mathrm{l,m}}\mu_{\mathrm{A},T}(T^{\mathrm{m,l}})}{T^{\mathrm{m,l}}}\right] + J_{\mathrm{B}}\left[-\frac{\Delta_{\mathrm{l,m}}\mu_{\mathrm{B},T}(T^{\mathrm{m,l}})}{T^{\mathrm{m,l}}}\right], \tag{11.40}$$

where $J_q^{\prime \mathrm{l}}$ is the measurable heat flux in the feed, and J_i is the molar flux of component i (i = A, B). We refer to the original literature for terminology.[7,13] The transport equations derived from the expression of the entropy production are:

$$\Delta_{\mathrm{l,m}} T = -\frac{1}{\lambda^{\mathrm{s}}}\left(J_q^{\prime \mathrm{l}} - J_{\mathrm{A}} q_{\mathrm{A}}^{*\mathrm{s}} - J_{\mathrm{B}} q_{\mathrm{B}}^{*\mathrm{s}}\right), \tag{11.41}$$

$$\frac{\Delta_{\mathrm{l,m}}\mu_{\mathrm{A},T}(T^{\mathrm{m,l}})}{T^{\mathrm{m,l}}} = -\frac{q_{\mathrm{A}}^{*\mathrm{s}}}{T^{\mathrm{l,m}} T^{\mathrm{m,l}}}\Delta_{\mathrm{l,m}} T - R_{\mathrm{AA}}^{\mathrm{s}} J_{\mathrm{A}} - R_{\mathrm{AB}}^{\mathrm{s}} J_{\mathrm{B}}, \tag{11.42}$$

$$\frac{\Delta_{\mathrm{l,m}}\mu_{\mathrm{B},T}(T^{\mathrm{m,l}})}{T^{\mathrm{m,l}}} = -\frac{q_{\mathrm{B}}^{*\mathrm{s}}}{T^{\mathrm{l,m}} T^{\mathrm{m,l}}}\Delta_{\mathrm{l,m}} T - R_{\mathrm{BA}}^{\mathrm{s}} J_{\mathrm{A}} - R_{\mathrm{BB}}^{\mathrm{s}} J_{\mathrm{B}}. \tag{11.43}$$

The thermal conductivity λ^{s} in the interface can be estimated with reference to the membrane value by using eqn (11.21). The interfacial heat of transfers $q_{\mathrm{i}}^{*\mathrm{s}}$ can be calculated as a fraction of the enthalpy change $\Delta_{\mathrm{l,m}} H_i$ in the surface according eqn (11.23). As there is a phase transition occurring in the membrane, because the components are in the liquid state in the feed and in the gas state in the permeate, the $\Delta_{\mathrm{l,m}} H_i$ includes the sorption enthalpy and the enthalpy of the phase transition. Following Villaluenga and Kjelstrup,[7] the resistivity coefficients for component fluxes are defined as $R_{ij}^{\mathrm{s}} = r_{ij}^{\mathrm{s}} - r_{iq}^{\mathrm{s}} r_{qj}^{\mathrm{s}} / r_{qq}^{\mathrm{s}}$. The values of the surface resistivity coefficients R_{ij}^{s} are not available in the literature because of the novelty of the concept of the membrane interface as a separate system. Moreover, the estimation of the required surface resistivity parameters is not straightforward. In short, the resistivity coefficients were estimated indirectly from molecular dynamics simulation data and gas kinetic-theory values. We refer the reader to the original paper for details. One task was varying the surface resistivity parameters to adjust the results of the model to the experiment results. For the membrane/permeate surface, the heat and mass transport equations can be derived in a similar manner.

In the membrane phase, there are three independent driving forces: one thermal and two component driving forces. Hence, the entropy production results in

$$\sigma^{\mathrm{m}} = J_q'^{\mathrm{m}} \frac{\mathrm{d}}{\mathrm{d}x}\left(\frac{1}{T}\right) - J_{\mathrm{A}} \frac{1}{T}\frac{\mathrm{d}\mu_{\mathrm{A},T}}{\mathrm{d}x} - J_{\mathrm{B}} \frac{1}{T}\frac{\mathrm{d}\mu_{\mathrm{B},T}}{\mathrm{d}x}, \tag{11.44}$$

where $J_q'^{\mathrm{m}}$ is the measurable heat flux in the membrane. The force–flux equations derived from the entropy production, by choosing as variables some measurable coefficients, can be written as follows:

$$\frac{\mathrm{d}T}{\mathrm{d}x} = -\frac{1}{\lambda^{\mathrm{m}}}\left(J_q'^{\mathrm{m}} - q_{\mathrm{A}}^{*,\mathrm{m}} J_{\mathrm{A}} - q_{\mathrm{B}}^{*,\mathrm{m}} J_{\mathrm{B}}\right), \tag{11.45}$$

$$\frac{c_{\mathrm{A}}}{T}\frac{\mathrm{d}\mu_{\mathrm{A},T}}{\mathrm{d}x} = -\frac{c_{\mathrm{A}} q_{\mathrm{A}}^{*,\mathrm{m}}}{T^2}\frac{\mathrm{d}T}{\mathrm{d}x} - \frac{R_{\mathrm{AA}}^{\mathrm{m}}}{c_{\mathrm{A}}} J_{\mathrm{A}} - \frac{R_{\mathrm{AB}}^{\mathrm{m}}}{c_{\mathrm{B}}} J_{\mathrm{B}}, \tag{11.46}$$

$$\frac{c_{\mathrm{B}}}{T}\frac{\mathrm{d}\mu_{\mathrm{B},T}}{\mathrm{d}x} = -\frac{c_{\mathrm{B}} q_{\mathrm{B}}^{*,\mathrm{m}}}{T^2}\frac{\mathrm{d}T}{\mathrm{d}x} - \frac{R_{\mathrm{BA}}^{\mathrm{m}}}{c_{\mathrm{A}}} J_{\mathrm{A}} - \frac{R_{\mathrm{BB}}^{\mathrm{m}}}{c_{\mathrm{B}}} J_{\mathrm{B}}. \tag{11.47}$$

In the absence of better information, the heat of transfer of water and ethanol in the membrane were estimated from reported data on thermo-osmosis of mixtures of water and methanol in polymeric membranes.[16] Following Villaluenga and Kjelstrup,[7] the resistivity coefficients for component fluxes are defined as $R_{ij}^{\mathrm{m}} = r_{ij}^{\mathrm{m}} - r_{iq}^{\mathrm{m}} r_{qj}^{\mathrm{m}} / r_{qq}^{\mathrm{m}}$. The values of the resistivity coefficients R_{ij}^{m} were estimated by using reported values of the component diffusion coefficients and concentration in the membrane. The values of the resistivity coefficients both in the surfaces and in the membrane are listed in Table 11.2.

The model equations were solved by using thermodynamic data and transport data for a binary mixture of water and ethanol, which is typically

Table 11.2 Estimated resistivity coefficients. Superscript s and m denotes surface and membrane, respectively. Reprinted with permission of *J. Non-Equilib. Thermodyn.*[7]

Coefficients	Value	Units
$R_{\mathrm{AA}}^{\mathrm{s}}$	$1 \cdot 10^{-5}$	$\mathrm{J \cdot m^2 \cdot s \cdot mol^{-2} \cdot K^{-1}}$
$R_{\mathrm{AB}}^{\mathrm{s}}$	$8 \cdot 10^{-4}$	$\mathrm{J \cdot m^2 \cdot s \cdot mol^{-2} \cdot K^{-1}}$
$R_{\mathrm{BB}}^{\mathrm{s}}$	$3 \cdot 10^{-4}$	$\mathrm{J \cdot m^2 \cdot s \cdot mol^{-2} \cdot K^{-1}}$
$R_{\mathrm{AA}}^{\mathrm{m}}$	$2 \cdot 10^{-13}$	$\mathrm{J \cdot s \cdot m^{-5} \cdot K^{-1}}$
$R_{\mathrm{AB}}^{\mathrm{m}}$	$-2.1 \cdot 10^{13}$	$\mathrm{J \cdot s \cdot m^{-5} \cdot K^{-1}}$
$R_{\mathrm{BB}}^{\mathrm{m}}$	$3 \cdot 10^{16}$	$\mathrm{J \cdot s \cdot m^{-5} \cdot K^{-1}}$

encountered in the dehydration of ethanol.[24] For convenience, the activity was used instead of the chemical potential. The solution procedure consists of the integration of the transport equations across each subsystem, and obtaining temperature and component activities in the membrane close to the surface at each side. For the sake of simplicity the system is considered to be in a stationary state. Thus, the water (component A) and ethanol (component B) fluxes are also constant through the system. Further, as the total heat flux, J_q, is constant throughout the system, the following expressions between the measurable heat fluxes in each subsystem can be obtained:

$$J_q'^{\mathrm{l}} = J_q'^{\mathrm{m}} + J_{\mathrm{A}}\Delta_{\mathrm{l,m}}H_{\mathrm{A}} + J_{\mathrm{B}}\Delta_{\mathrm{l,m}}H_{\mathrm{B}}, \tag{11.48}$$

$$J_q'^{\mathrm{m}} = J_q'^{\mathrm{r}} + J_{\mathrm{A}}\Delta_{\mathrm{m,r}}H_{\mathrm{A}} + J_{\mathrm{B}}\Delta_{\mathrm{m,r}}H_{\mathrm{B}}, \tag{11.49}$$

where $J_q'^{\mathrm{l}}$ is the measurable heat flux in the feed side, $J_q'^{\mathrm{m}}$ is the measurable heat flux in the membrane, and $J_q'^{\mathrm{r}}$ is the measurable heat flux in the permeate side. The measurable heat fluxes only can change at the interfaces, where the water and ethanol undergoes an enthalpy change.

Figure 11.6 shows the temperature profile in the complete membrane system, including liquid feed solution and permeate vapour mixture. A linear variation is found across the membrane in the temperature and

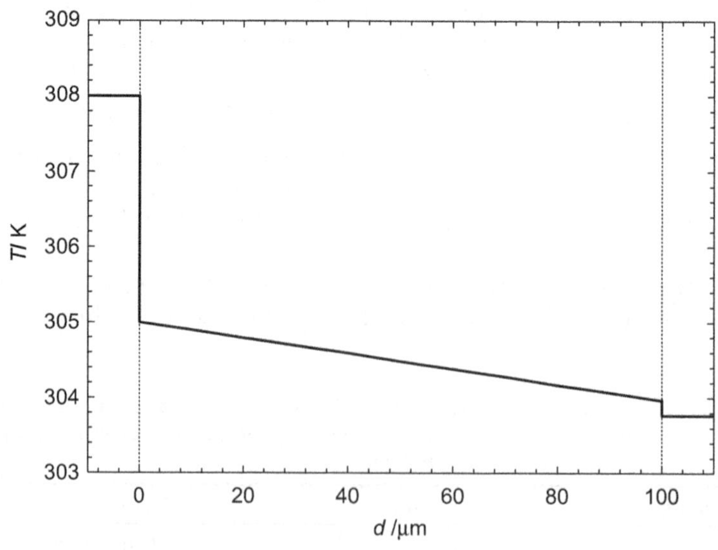

Figure 11.6 Temperature profile across the feed liquid phase, the membrane and the permeate phase. The vertical dashed lines indicate, from left to right, the liquid/membrane interface and the membrane/permeate interface.
Reprinted with permission of *J. Non-Equilib. Thermodyn.*[7]

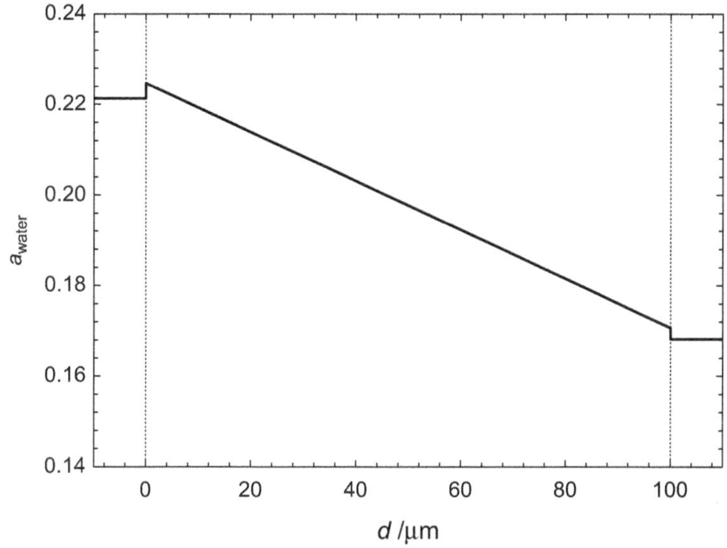

Figure 11.7 Water activity profile across the feed liquid phase, the membrane and the permeate phase. The vertical dashed lines indicate, from left to right, the liquid/membrane interface and the membrane/permeate interface.
Reprinted with permission of *J. Non-Equilib. Thermodyn.*[7]

sizable effects for the temperature at the interfaces. Moreover, model predictions show that the largest temperature drop is observed at the feed/membrane interface, because 71 % of the temperature drop over the system is generated at this surface. The temperature drop at the membrane/permeate interface is only 5 % of the total difference. The temperature difference over the membrane is 24 % of the total difference. These facts indicate that temperature polarization occurring in a pervaporation process may be caused to a larger extent by surface effects.

Figures 11.7 and 11.8 show water and ethanol activity profiles, respectively, across the feed liquid solution, the membrane and the permeate vapour mixture. Unlike the temperature behaviour, the largest activity difference over the entire system is found over the membrane itself. A positive jump of 6 % in the activity of water at the feed/membrane interface is found, the activity shows a drop of 3 % at the membrane/permeate interface, while the activity decreases with 91 % over the membrane. The activity of ethanol decreases 14 % at the feed/membrane interface, there is a decrease of 81 % over the membrane, whereas the activity decreases 5 % at the membrane/permeate interface. Consequently, model predictions show that the activity variations at the feed/membrane and membrane/permeate interfaces are not negligible, indicating that chemical equilibriums cannot be assumed at both sides of the membrane.

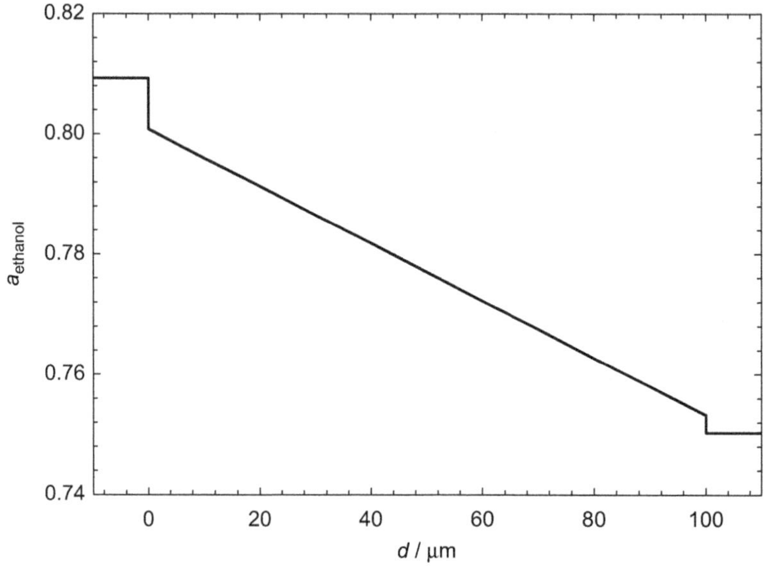

Figure 11.8 Ethanol activity profile across the feed liquid phase, the membrane and the permeate phase. The vertical dashed lines indicate, from left to right, the liquid/membrane interface and the membrane/permeate interface.
Reprinted with permission of *J. Non-Equilib. Thermodyn.*[7]

11.4 Concluding Remarks

Non-equilibrium thermodynamics has been applied to describe transport processes in three different membrane systems considered as heterogeneous systems. The common features of the selected examples of transport phenomena in membranes are basically two-fold. The fact that the membrane interface adjacent to the bulk phase of liquid or vapour pose a separate resistance to mass and heat transport. Fluxes and forces of different nature are coupled both in the bulk phases and at the surfaces of the membrane.

The values of the resistivity coefficients for membrane surfaces are not known for most of the transport processes in membranes, mainly due to the novelty of the concept of the membrane interface as a separate system. As the model predictions depend on the surface resistivities, it is fundamental to address the experimental or theoretical estimation of these coefficients for membrane–interface systems. An effort should be put into a more exact determination of these coefficients for membrane surfaces, to bring the application of non-equilibrium thermodynamics further.

It should be noted that the systematic treatment of membrane systems in terms of non-equilibrium thermodynamics offers a possibility to study other transport processes in membranes, such as membrane distillation, membrane reactors, thermo-osmosis, electro-osmosis, gas permeation, among others.

Acknowledgements

J.P.G.V acknowledges the support from grant UCM/Santander Research (PR6/13-18867) from University Complutense of Madrid.

References

1. B. Van der Bruggen and P. Luis, *Curr. Opin. Chem. Eng.*, 2014, **4**, 54.
2. F. Lufrano, V. Baglio, P. Staiti, V. Antonucci and A. S. Arico, *J. Power Sources*, 2013, **243**, 519.
3. C. Stoquart, P. Servais, P. R. Bérubé and B. Barbeau, *J. Membr. Sci.*, 2012, **1**, 411.
4. P. S. Goh, A. F. Ismail, S. M. Sanip, B. C. Ng and M. Aziz, *Sep. Purif. Technol.*, 2011, **81**, 243.
5. L. Shao, B. T. Low, T. Chung and A. R. Greenberg, *J. Membr. Sci.*, 2009, **327**, 18.
6. J. Kuhn, R. Stemmer, F. Kapteijn, S. Kjelstrup and J. Gross, *J. Membr. Sci.*, 2009, **330**, 388.
7. J. P. G. Villaluenga and S. Kjelstrup, *J. Non-Equilib. Thermodyn.*, 2012, **37**, 353.
8. C. W. Monroe, T. Romero, W. Merida and M. Eikerling, *J. Membr. Sci.*, 2008, **324**, 1.
9. T. Romero and W. Merida, *J. Membr. Sci.*, 2009, **338**, 135.
10. A. Røsjorde, D. Bedeaux, S. Kjelstrup and B. Hafskjold, *J. Colloid Interface Sci.*, 2001, **240**, 355.
11. I. Inzoli, S. Kjelstrup, D. Bedeaux and J. M. Simon, *Chem. Eng. Sci.*, 2010, **66**, 4533.
12. K. Glavatskiy and D. Bedeaux, *J. Chem. Phys.*, 2010, **133**, 234501.
13. S. Kjelstrup and D. Bedeaux, *Non-equilibrium Thermodynamics of Heterogeneous Systems*, Series on Advances in Statistical Mechanics, World Scientific, Singapore, 2008, vol. 16.
14. T. Suzuki, K. Iwano, R. Kiyono and M. Tasaka, *Bull. Chem. Soc. Jpn.*, 1995, **68**, 493.
15. M. Tasaka, T. Mizuta and O. Sekiguchi, *J. Membr. Sci.*, 1990, **54**, 191.
16. J. P. G. Villaluenga, B. Seoane, V. M. Barragán and C. Ruiz-Bauzá, *J. Membr. Sci.*, 2006, **274**, 116.
17. S. Kjelstrup and A. Røsjorde, *J. Phys. Chem. B*, 2005, **109**, 9020.
18. I. Inzoli, J. M. Simon, S. Kjelstrup and D. Bedeaux, *J. Colloid Interface Sci.*, 2007, **313**, 563.
19. J. Coronas and J. Santamaría, *Top. Catal.*, 2004, **29**, 29.
20. S. Yun and S. Oyama, *J. Membr. Sci.*, 2011, **375**, 28.
21. R. Skorpa, M. Voldsund, M. Takla, S. K. Schnell, D. Bedeaux and S. Kjelstrup, *J. Membr. Sci.*, 2012, **394–395**, 131.
22. E. Johannessen and K. Jordal, *Energy Convers. Manage.*, 2005, **46**, 1059.
23. S. K. Gade, P. M. Thoen and D. J. Way, *J. Membr. Sci.*, 2008, **316**, 112.
24. B. Bolto, M. Hoang and Z. Xie, *Chem. Eng. Process.*, 2011, **50**, 227.

CHAPTER 12

Electrochemical Energy Conversion

SIGNE KJELSTRUP* AND DICK BEDEAUX

Department of Chemistry, Norwegian University of Science and Technology, 7491 Trondheim, Norway
*Email: signe.kjelstrup@ntnu.no

12.1 Non-equilibrium Thermodynamics in the Electrochemical Literature

Electrochemical energy conversion is central in any future vision of the world's energy technology, because a modern society needs electric power in so many contexts. Electrochemical energy conversion in, say, batteries, fuel cells or electrolysis cells, has a higher efficiency than many other power producing or consuming energy technologies, and one should understand why in order to take full advantage of it. The conversion takes place on a scale where the reaction Gibbs energy is typically is around $-100\ \text{kJ}\cdot\text{mol}^{-1}$ and gives a voltage of order of magnitude 1 V (*cf.*, eqn 12.1). There is for instance more to gain by considering also Gibbs energies of mixing, like in the mixing of salt water and fresh water,[1,2] or by converting industrial waste heat into electricity.[3] Doing so, we can exploit also the smaller Gibbs energy changes provided by renewable energy sources. This type of electrochemical energy conversions can best be described by NET. In particular formation cells may benefit from a more precise description by NET. This chapter will give examples of equations for both applications.

The broad lines in the development of non-equilibrium thermodynamics (NET) were reviewed in Chapter 1. Classical NET was applied early to

Experimental Thermodynamics Volume X: Non-equilibrium Thermodynamics with Applications
Edited by Dick Bedeaux, Signe Kjelstrup and Jan V. Sengers

Published by the Royal Society of Chemistry, www.rsc.org

electrochemical systems. Works by Katchalsky and coworkers[4,5] and Haase[6] were central. Electrokinetic phenomena in membranes were treated by Laksminarayanaiah,[7] Mickulecki,[8] Staverman,[9] and Førland *et al.*,[10] while Ito *et al.*[11] and Richter *et al.*[12,13] studied molten salt electrolytes in electrochemical cells. Concentration cells with and without membranes were used to test concepts and approximations.[14–16] The Onsager relations were verified,[14] the Nernst–Planck assumption was found to be good for cation-exchange membranes,[15] and the liquid-junction contribution to the cell potential was well predicted from the equations.[16] Membranes exposed to differences in concentration, pressure and temperature were systematically studied by several groups.[7–11,17–21] Continuous and discrete formulations were developed.[5,6,10]

Several books on electrochemistry have presented NET and used Onsager relations, but have not yet taken systematic advantage of the theory, see *e.g.* ref. 22. This is not surprising as a description has been lacking of the coupled phenomena at the electrode surfaces, the most important locations for energy conversion. The heterogeneity of the electrochemical cell and the electrode surface was first addressed in NET in 1996 for the linear regime.[23–26] The theory was later extended to deal with nonlinear regimes, by including mesoscale (internal) variables.[27,28] To introduce mesoscale variables, was suggested already by Prigogine and Mazur,[29] but the idea gained momentum through the works of Rubí and coworkers.[27,30,31] The Butler–Volmer equation was derived in this manner by taking the density of the reacting mixture along the reaction coordinate as internal variable.[27] Linear flux-force relations like eqn (1.3) will apply, but then on the scale of the internal variables. Support for this approach in electrochemistry has been obtained by the method of the general equation for non-equilibrium reversible–irreversible coupling (GENERIC),[32] see Chapter 17 for a description of this method.

The purpose of the chapter is to show how NET can be systematically used to derive formulae for energy conversion, building on some recent works.[25,26] We shall see how the entropy production can be actively used to obtain the electric-potential profile under reversible and irreversible conditions in Sections 12.4 and 12.5. This property can also be used to define the surface overpotential, see Section 12.6. NET provides a common thermodynamic basis for the Nernst and Butler–Volmer equations, including Peltier/Seebeck and Soret/Dufour effects.[33] This is shown in Sections 12.7 and 12.8. With the development of the theory to deal with electrode surfaces,[23–26] the possibility is there to find, not only the electric-potential jump at the surfaces, but also accompanying concentration and temperature variations at the surfaces and in the electrolyte. In order to illustrate the application of the theory, we use a formation cell example presented in Section 12.3, as well as concentration cells and thermoelectric devices generated from this. We start with an overall view of electrochemical energy conversion in Section 12.2.

Hydrodynamic effects can also be important in electrochemistry; in measurements as well as in technical cells in operation. Such effects can also be given a NET framework.[34] This represents another branch of NET, outside the scope here. We start with the common overall description of electrochemical cells to explain what NET can help solve.

12.2 Thermodynamic Properties of a Total Cell

An electrochemical cell in global equilibrium is a dead cell. We are interested in the energy conversion that takes place before this state is reached, a conversion which is driven by the positive entropy change that takes place in the cell and in the surroundings (the second law of thermodynamics). The overall energy available for work is the Gibbs-energy change of the cell, ΔG at constant pressure and temperature (p and T). At reversible conditions, the electromotive force, $E_{j\rightarrow 0}$, is:

$$E_{j\rightarrow 0} = -\frac{1}{nF}\Delta G = -\frac{1}{nF}[\Delta H - T\Delta S], \tag{12.1}$$

where F is Faraday's constant, and n is the number of electrons transferred in the reaction. The symbol j is the electric current density. The Gibbs-energy change has a contribution from the enthalpy change, ΔH, and the entropy change, ΔS. The heat transferred reversibly from the surroundings (abbreviated sur) to the system (abbreviated syst) is the entropy change in the system, $q_{\text{syst,rev}} = T\Delta S = -q_{\text{sur,rev}}$ Under reversible conditions ($j \rightarrow 0$), the entropy change in the system and in the surroundings cancel and there is no entropy production. The enthalpy difference is normally large and negative (order of magnitude $-100\ \text{kJ} \cdot \text{mol}^{-1}$) and contributes much to E. This is illustrated in Figure 12.1.

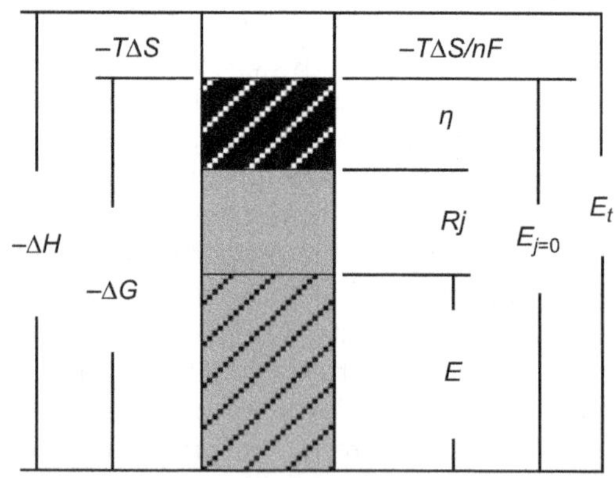

Figure 12.1 Thermodynamic properties (to the left) and corresponding cell potentials and potential losses in a battery or fuel cell (to the right). The thermo-neutral potential (right), E_t, is represented by the negative reaction enthalpy (left), $-\Delta H$. Part of this energy is needed to compensate for the entropy change at a given pressure and temperature, $-T\Delta S$ (white part of column), leaving us with the maximum available work, $-\Delta G$ (left) represented by the reversible potential, $E_{j=0}$. Away from reversible conditions, the cell potential, E (grey part of column with dark stripes), is equal to the reversible potential minus potential drops due to ohmic resistance, Rj (uniformly grey part of column), and overpotential, η (black part of column with white stripes).

But waste heat can be converted to electric energy directly, if the cell reaction has a positive entropy change. In that case $|\Delta G| > |\Delta H|$ because $T\Delta S > 0$. In a salt power plant,[1,2] ΔH is negligible, and the entropy of mixing salt water and fresh water is the only source of $E_{j=0}$.

During operation, the electric potential is reduced to

$$E = E_{j=0} - \eta - Rj, \tag{12.2}$$

where R is the total electric resistance of the whole cell and η is the sum of electrode overpotentials. The cell power is Ej. When $j \neq 0$ there is heat production from these and other irreversible processes. The net entropy change in the system plus surroundings, the entropy production for a one-dimensional system of cross-sectional area Ω, obtains for instance contributions from Joule heat, Rj^2 (R is the total electric resistance) and the (positive) electrode overpotentials, ηj. The major contributions to the total power loss is in fact given by $Rj^2 + \eta j$, so the total entropy change due to irreversible prosesses, $\mathrm{d}S_{\mathrm{irr}}$, per unit of time, $\mathrm{d}t$, to a first approximation, is:

$$\frac{\mathrm{d}S_{\mathrm{irr}}}{\mathrm{d}t} = \frac{\Omega}{T_0}(Rj^2 + \eta j) > 0. \tag{12.3}$$

Here T_0 is the temperature of the surroundings.[25] The reversible heat production plus this heat production gives the *total* heat production, which can be measured in the surroundings.[33] The entropy production in the system and surroundings is always larger than zero for a spontaneous process.[25]

In order to lower the entropy production and increase system efficiency (see Chapter 13), information about local conditions is needed. Where can we locate power losses and what causes them? How can they be made smaller? Non-equilibrium thermodynamics can be used to gain insight into the origin of power losses (entropy production), through a more precise description of eqn (12.3). Through NET we can also answer questions like: will a difference in temperature between the anode and the cathode promote or reduce the potential jump across the cell?

The total entropy production, $\mathrm{d}S_{\mathrm{irr}}/\mathrm{d}t$, for a system of length l in a steady state can also be obtained by integration along the system. At steady state it is equal to the net entropy flow out of the system:

$$\frac{\mathrm{d}S_{\mathrm{irr}}}{\mathrm{d}t} = \Omega \int_0^l \sigma(x)\mathrm{d}x = \Omega\left[J_s^o - J_s^i\right] \geq 0. \tag{12.4}$$

We have here assumed that all transport processes take place in the x-direction (*cf.*, Chapter 1). The contributions to the total entropy production come from $\sigma(x)$. We can use this quantity actively, as we shall see below, to find the electric-potential profile of the cell. We can also use eqn (12.4) to test that a thermodynamic model for any local layer of the cell agrees with the second law of thermodynamics. In a situation where the energy efficiency for power use and power production is in world-wide focus, it may pay to give $\mathrm{d}S_{\mathrm{irr}}/\mathrm{d}t$ more attention (see also Chapter 13).

12.3 A Formation Cell Example

We shall now see, from an example, how NET can be used to derive a simultaneous set of profiles for all intensive variables across a cell, enabling us to calculate the electric work and lost work everywhere. Consider the formation cell with an electrode reversible to Na^+ and a Ag(s)|AgCl(s) electrode reversible to Cl^-. The electrolyte is an aqueous solution of NaCl(aq). With standard notation we write for the formation cell:

$$\mathrm{Na(1)} \mid \mathrm{NaCl(aq)} \mid \mathrm{AgCl(s)} \mid \mathrm{Ag(s)}. \tag{12.5}$$

where l, aq and s are short for liquid, aqueous solution and solid aggregate states. Each homogeneous phase and each interface are now separate thermodynamic subsystems, a three- or two-dimensional one, respectively, see Section 1.4 for definitions. The vertical bar in eqn (12.5) is standardly used to indicate separate phases. The left hand side electrode reaction is:

$$\mathrm{Na(l)} \rightarrow \mathrm{Na^+\ (aq)} + \mathrm{e^-}. \tag{12.6}$$

One mole of Na^+ moves from a uniform sodium electrode (*e.g.*, a stirred amalgam) into the electrolyte per mole of electrons (one Faraday) passing the external circuit from left to right. At the surface of the chloride layer that faces the electrolyte, we have:

$$\mathrm{AgCl(s)} \rightarrow \mathrm{Cl^-(aq)} + \mathrm{Ag^+(s)}. \tag{12.7}$$

This is followed by

$$\mathrm{Ag^+(s)} + \mathrm{e^-} \rightarrow \mathrm{Ag(s)}. \tag{12.8}$$

One mole of Cl^- is produced in the electrolyte, while one mole of silver ions is produced in the chloride, close to this surface. The silver ions are transported to the surface between the chloride layer and the silver. The overall electrode reaction is the sum of eqn (12.6–12.8). The overall *cell* reaction of eqn (12.5) is:

$$\mathrm{Na(l)} + \mathrm{AgCl(s)} \rightarrow \mathrm{Na^+(aq)} + \mathrm{Cl^-(aq)} + \mathrm{Ag(s)}. \tag{12.9}$$

One mole of Ag(s) and of NaCl(aq) are produced while one mole of Na(l) and AgCl(s) are consumed per Faraday passing the cell. The electrolyte is electroneutral, so the concentrations of the sodium and the chloride ions are the same at every position in the electrolyte. When a sodium ion enters the electrolyte from the electrode surface, it is therefore quickly accompanied by a chloride ion which is coming into the electrolyte from the AgCl surface.

The overall performance of such a system is described by eqn (12.1) and (12.2). Details on each surface can be added by applying the Nernst equation, eqn (12.26) below. The electrode surface is the location where most of the energy conversion takes place.

The heat production that can be associated with the energy conversion in the surface is in general not described in the literature. We shall here see how both effects can be captured in the same set of equations with NET.[25,26]

We shall give a local description of the formation cell (12.5), and of the two concentration cells with identical electrodes, either reversible to Na^+, or to Cl^-. In particular, we shall explain and give details of the electric-potential, concentration profiles and temperature profile.

12.4 Entropy Production in Three- and Two-dimensional Subsystems

The electrochemical system is always heterogeneous. The electrolyte and the bulk electrode materials are often relatively uniform and thick. This is why we regard them as homogeneous three-dimensional subsystems. The thinner electrode interfaces can be regarded as two-dimensional systems. When described by excess variables, the surface can be regarded as a two-dimensional autonomous system. We refer the reader to Section 1.4 for a definition of Gibbs excess densities of a surface, and for a discussion of the autonomous nature of the surface, see also Chapter 4.

The purpose of Section 12.4 is to give and discuss expressions for the entropy production (12.9) in the three and two-dimensional subsystems of the example cell. The notation and symbols are the same as used in ref. 25 and 26, see Figure 12.2 for more explanations.

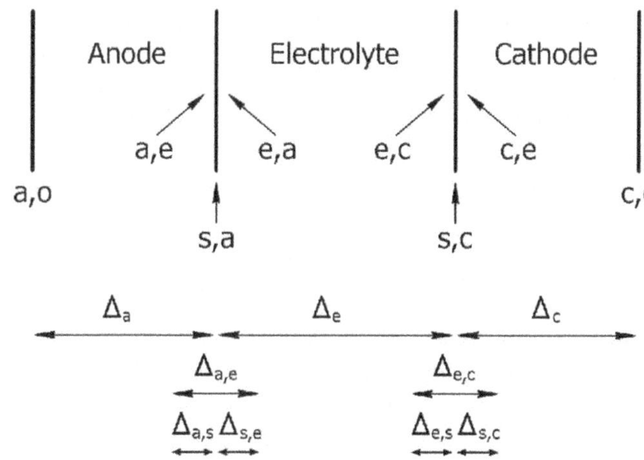

Figure 12.2 Terminology used to describe processes in electrochemical cells according to Kjelstrup and Bedeaux.[25,26] The anode, electrolyte and cathode phases are indicated with superscripts a, e and c, respectively. The electrode surfaces are indicated with superscripts s,a and s,c for the anode and cathode, respectively. A difference Δ with subscript i,j means that the difference is taken between the last (j) and the first (i) phase. A difference symbol with subscript k means that the difference is taken across a homogeneous phase (k). Locations are indicated by superscripts (i,j), where the phase is denoted i, and the adjacent phase is j. The difference symbol Δ will also be used for chemical reactions (see text for explanation).

Kedem and Katchalsky[4,5] proposed to use operationally defined variables in the description of electrochemical cells, motivated by the practical consideration that they are directly linked to measurements. The flux of NaCl, the electric current density and the heat fluxes can all be measured and controlled independently and are such variables. It is nevertheless more common to use as variables the fluxes of charged particles and describe changes in terms of these. The charged particle variables relevant, *e.g.*, for the cell (12.5), are the electrons, the sodium and the chloride ions. We give the entropy production for the various parts of the cell with both sets. We proceed to compare the two descriptions, and show that they lead to different perceptions of the origin of electric power.

12.4.1 Operationally Defined Variables

Consider first the homogeneous parts of the formation cell as illustrated in Figure 12.2. The entropy production in a uniform sodium electrode, phase a, is given by[10,25]

$$\sigma^{\mathrm{a}} = -J_q'^{\mathrm{a}} \frac{1}{T^2}\frac{\partial T}{\partial x} - j\frac{1}{T}\frac{\partial \phi}{\partial x}. \tag{12.10}$$

The variables depend in general on *x*,*y*,*z*,*t*. In the partial derivatives here and below the other three are kept constant. The entropy production has contributions from two flux-force pairs, *cf.* Chapter 1. We have replaced the electric field in eqn (1.16) by the negative gradient in the electric potential ϕ, $E = -\partial\phi/\partial x$. The chosen fluxes are the measurable heat flux $J_q'^{\mathrm{a}}$ and the electric current density across the metal. Similar expressions to eqn (12.10) apply to the pure, solid phases of Ag and AgCl.

The entropy production in the electrolyte has likewise three terms from eqn (1.16):

$$\sigma^{\mathrm{e}} = -J_q'^{\mathrm{e}} \frac{1}{T^2}\frac{\partial T}{\partial x} - J_{\mathrm{NaCl}}^{\mathrm{e}} \frac{1}{T}\frac{\partial}{\partial x}\mu_{\mathrm{NaCl},T}^{\mathrm{e}} - j\frac{1}{T}\frac{\partial \phi}{\partial x}. \tag{12.11}$$

The contributions are from the measurable heat flux across the electrolyte and its conjugate thermal driving force, the flux of NaCl relative to water and the negative gradient in the chemical potential of NaCl over the temperature, and the electric current density and the negative gradient in the electric potential in this material. The measureable heat flux was defined as the total heat flux minus the latent heat (the enthalpy) carried by the components, see ref. 25 for further definition, or Chapter 1, eqn (1.13). The chemical-potential gradient should be evaluated at constant temperature, hence the subscript *T*. The electric potential will be related to the Maxwell potential in Section 12.5 below.

The entropy production in the surface of the Na^+-reversible electrode has contributions from two heat fluxes, the first pointing into the surface on the left-hand side, the other pointing out of the surface on the right-hand side. A term containing the chemical-potential jump of NaCl at constant temperature appears in the absence of equilibrium for adsorption of NaCl at the electrode surface (this term was not included in ref. 25). In a stationary state, the reaction rate, $r^{s,a}$, is uniquely related to the electric current density, in general by $r^{s,a} = j/nF$, where n is the number of electrons involved in the reaction. In our case $n = 1$. This gives[25]

$$\sigma^{s,a} = J_q^{\prime a,e} \Delta_{a,s}\left(\frac{1}{T}\right) + J_q^{\prime e,a} \Delta_{s,e}\left(\frac{1}{T}\right) - J_{NaCl}^{e,a} \frac{1}{T^{s,a}} \Delta_{s,e}\mu_{NaCl,T} - j\frac{1}{T^{s,a}}\left(\Delta_{a,e}\phi + \frac{\Delta_n G^{s,a}}{F}\right), \tag{12.12}$$

where all sub- and super-scripts were defined in Figure 12.2. There is consumption of metal in the electrode and formation of NaCl in the electrolyte next to the metal per unit of charge passing the external circuit. The Gibbs-energy change, $\Delta_n G^{s,a}$, has contributions from these changes in the neutral components (*cf.*, ref. 25 and eqn (12.21)). Subscript n on the difference symbol refer to the neutral components. Their combination gives the electric potential drop, $\Delta_{a,e}\phi$, at reversible conditions, see Section 12.5.

The entropy production in the surface at the AgCl(s)|Ag(s)-electrode is likewise:

$$\sigma^{s,c} = J_q^{\prime e,c} \Delta_{e,s}\left(\frac{1}{T}\right) + J_q^{\prime c,e} \Delta_{s,c}\left(\frac{1}{T}\right) - J_{NaCl}^{e,c} \frac{1}{T^{s,c}} \Delta_{e,s}\mu_{NaCl,T} - j\frac{1}{T^{s,c}}\left(\Delta_{e,c}\phi + \frac{\Delta_n G^{s,c}}{F}\right) \tag{12.13}$$

There is a change in the NaCl content in the electrolyte per unit of charge passing the external circuit due to the electrode reaction and migration of chloride ions from the interface to the electrolyte. The Gibbs-energy change, $\Delta_n G^{s,c}$, in eqn (12.13) reflects the changes in Ag, AgCl and NaCl, *cf.*, eqn (12.21).

It is practical to choose the same frame of reference for all fluxes involved in a cell. A natural frame of reference for processes in an electrochemical cell is the equimolar surface of one of the components in one of the electrodes. The surface of discontinuity of the cathode surface, say, will extend from the electrolyte phase into the Ag phase, see Chapter 17 for a definition of excess variables. The entropy production under isothermal conditions for the various parts of the cell is obtained from the above eqn (12.10) to (12.13) by setting the thermal driving forces equal to zero. The entropy production due to NaCl transport in an isothermal solution, in terms of operationally defined variables is, for instance, given by the second term in eqn (12.11).

We see that the number of fluxes at a surface becomes larger than in a bulk phase, mainly because the discrete description gives fluxes into and out of the surface. Elimination of dependent fluxes, like $r^{s,a}$, is best done in the entropy production. The independent set of fluxes that appears in eqn

(12.12) and (12.13) follows the recommendation of Kedem and Katchalsky for bulk systems.[4,5]

12.4.2 Ionic Fluxes as Variables

Since the properties of charged constituents are more common, we write the entropy production at isothermal conditions also for contributions from the electron flux, the sodium and chloride ion fluxes and the chemical reaction rate. For the electrolyte phase we have:

$$\sigma^{e} = -J^{e}_{Na^+}\frac{1}{T}\frac{\partial}{\partial x}\tilde{\mu}^{e}_{Na^+} - J^{e}_{Cl^-}\frac{1}{T}\frac{\partial}{\partial x}\tilde{\mu}^{e}_{Cl^-}, \tag{12.14}$$

where the driving forces are expressed in terms of electrochemical-potential gradients. The *electrochemical potential* was defined by Guggenheim (see ref. 4) as the combination of the particle chemical potential and the Maxwell potential. For the two ions here, we have

$$\tilde{\mu}^{e}_{Na^+} = \mu^{e}_{Na^+} + F\psi, \quad \tilde{\mu}^{e}_{Cl^-} = \mu^{e}_{Cl^-} - F\psi. \tag{12.15}$$

where ψ is the Maxwell potential. Eqn (12.14) is equivalent to eqn (12.11) for isothermal conditions. A similar expression can be written for the electron. For the isothermal electrode surfaces, we obtain

$$\begin{aligned}\sigma^{s,a} &= -J^{a,e}_{e^-}\frac{\Delta_{a,s}\tilde{\mu}_{e^-}}{T^{s,a}} - J^{e,a}_{Na^+}\frac{\Delta_{s,e}\tilde{\mu}_{Na^+}}{T^{s,a}} - J^{e,a}_{Cl^-}\frac{\Delta_{s,e}\tilde{\mu}_{Cl^-}}{T^{s,a}} - \frac{r^{s,a}}{T^{s,a}}\Delta_r G^{s,a},\\ \sigma^{s,c} &= -J^{e,c}_{Na^+}\frac{\Delta_{e,s}\tilde{\mu}_{Na^+}}{T^{s,c}} - J^{e,c}_{Cl^-}\frac{\Delta_{e,s}\tilde{\mu}_{Cl^-}}{T^{s,c}} - J^{c,e}_{e^-}\frac{\Delta_{s,c}\tilde{\mu}_{e^-}}{T^{s,c}} - \frac{r^{s,c}}{T^{s,c}}\Delta_r G^{s,c}.\end{aligned} \tag{12.16}$$

These equations are equivalent to eqn (12.12) and (12.13). The change in Gibbs energy is the reaction Gibbs energy of eqn (12.6) and (12.7) combined with (12.8), and the difference symbol has subscript *r*. The expressions from eqn (12.6) and (12.7) are:

$$\begin{aligned}\Delta_r G^{s,a} &= -\mu^{a,e}_{Na} + \mu^{e,a}_{Na^+} + \mu^{a,e}_{e^-},\\ \Delta_r G^{s,c} &= -\mu^{s,c}_{AgCl} + \mu^{e,c}_{Cl^-} + \mu^{c,e}_{Ag} - \mu^{c,e}_{e^-}.\end{aligned} \tag{12.17}$$

12.4.3 Relations Between Variables

Compare next the two descriptions of the same phenomena by the different variable sets. The entropy production is the same, so the fluxes of the two descriptions are related.

We have for the anode surface:

$$-J^{a,e}_{e^-} = j/F = r^{s,a} = J^{e,a}_{Na^+} - J^{e,a}_{Cl^-}, \tag{12.18}$$

and

$$J_{Na^+}^{e,a} = J_{NaCl}^{e,a} + t_{Na^+}^{e,a} j/F,$$
$$J_{Cl^-}^{e,a} = J_{NaCl}^{e,a} - t_{Cl^-}^{e,a} j/F, \tag{12.19}$$

where transport number of the ions have been introduced. The transport number of an ion is the fraction of the electric current carried by that ion, giving $t_{Na^+} + t_{Cl^-} = 1$. When eqn (12.18) and (12.19) and the analogous equations for the cathode surface are introduced in (12.16) and the result is compared to eqn (12.12) and (12.13), we find the relations between the intensive variables:

$$F\phi^a = F\psi^a - \mu_{e^-}^a, \quad F\phi^c = F\psi^c - \mu_{e^-}^c,$$
$$F\phi^e = F\psi^e + t_{Na^+}^e \mu_{Na^+}^e - t_{Cl^-}^e \mu_{Cl^-}^e. \tag{12.20}$$

The reaction Gibbs energies for the neutral-component description in Section 12.4.1 become:

$$\Delta_n G^{s,a} = -\mu_{Na}^{a,e} + t_{NaCl}^{e,a} \mu_{NaCl}^{e,a} = -\mu_{Na}^{a,e} + t_{Cl^-}^{e} \mu_{NaCl}^{e,a},$$
$$\Delta_n G^{s,c} = -\mu_{AgCl}^{s,a} + \mu_{Ag}^{c,e} - t_{NaCl}^{e,c} \mu_{NaCl}^{e,c} = -\mu_{AgCl}^{s,a} + \mu_{Ag}^{c,e} + t_{Na^+}^{e} \mu_{NaCl}^{e,c}. \tag{12.21}$$

where $t_{Cl^-}^e = 1 - t_{Na^+}^e$. Eqn (12.21a) accounts for the change in sodium and salt content at the anode, while (12.21b) accounts for the changes at the cathode. The relation between the Gibbs-energy difference $\Delta_r G^{s,a}$ (or $\Delta_r G^{s,c}$) and the corresponding operationally defined reaction Gibbs energy $\Delta_n G^{s,a}$ (or $\Delta_n G^{s,c}$) is similar to the relation between the corresponding potentials, *cf.*, eqn (12.20)

$$\Delta_r G^{s,a} = \Delta_n G^{s,a} + \mu_{e^-}^{a,e} + t_{Na^+}^e \mu_{Na^+}^{e,a} - t_{Cl^-}^e \mu_{Cl^-}^{e,a},$$
$$\Delta_r G^{s,c} = \Delta_n G^{s,c} - t_{Na^+}^e \mu_{Na^+}^{e,c} + t_{Cl^-}^e \mu_{Cl^-}^{e,c} - \mu_{e^-}^{c,e}. \tag{12.22}$$

By introducing eqn (12.21) into eqn (12.22), we obtain eqn (12.17).

12.5 The Surface Potential Jump in the Reversible Limit

Before we examine the cell in operation, we consider the reversible limit. In this limit, $j \rightarrow 0$, the cell is isothermal and we measure the cell emf, $E_{j=0}$, *cf.*, Figure 12.1. Under isothermal conditions, only the last two terms remain in eqn (12.12) and (12.13). The expression for the entropy production and the condition $\sigma^{s,i} = 0$ will give information about the electrode potential jumps. These jumps depend on the variables used, however, see Section 12.4. The cell potential can thus be interpreted in terms of operationally defined potential differences, or equivalently in terms of Maxwell potential differences.

In both cases there are, in principle, contributions from the electrode surfaces and from the electrolyte. The two alternatives give:

$$\begin{aligned}\Delta\phi &= \Delta_{\mathrm{a,e}}\phi + \Delta_{\mathrm{e}}\phi + \Delta_{\mathrm{e,c}}\phi,\\ \Delta\psi &= \Delta_{\mathrm{a,e}}\psi + \Delta_{\mathrm{e}}\psi + \Delta_{\mathrm{e,c}}\psi.\end{aligned} \tag{12.23}$$

The first equation gives the measured cell potential, also called E above. The second gives the difference in the Maxwell potential of the cell. The descriptions are related. We use eqn (12.20) for the conducting leads in the anode and cathode, and obtain:

$$\Delta\phi = \Delta\psi - \Delta\mu_{\mathrm{e}^-}. \tag{12.24}$$

The last term, the chemical-potential difference of electrons in the metallic leads, is considered to be negligible. The measured cell potential can under these conditions be understood as the Maxwell-potential difference. The physical interpretation is, however, particular for the set of variables used.

With operationally defined variables, we write the potential drop of the electrode surfaces in terms of changes in the chemical potentials of the neutral components in the surface. The change in chemical energy gives the electric potential change directly from eqn (12.12) and (12.13) for $\sigma^{\mathrm{s,i}} = 0$:

$$\begin{aligned}\Delta_{\mathrm{a,e}}\phi &= -\frac{1}{F}\Delta_{\mathrm{n}}G^{\mathrm{s,a}} = \frac{1}{F}\left[\mu_{\mathrm{Na}}^{\mathrm{a,e}} - t_{\mathrm{NaCl}}^{\mathrm{e,a}}\mu_{\mathrm{NaCl}}^{\mathrm{e,a}}\right],\\ \Delta_{\mathrm{e,c}}\phi &= -\frac{1}{F}\Delta_{\mathrm{n}}G^{\mathrm{s,c}} = \frac{1}{F}\left[\mu_{\mathrm{AgCl}}^{\mathrm{s,c}} - \mu_{\mathrm{Ag}}^{\mathrm{c,e}} + t_{\mathrm{NaCl}}^{\mathrm{e,c}}\mu_{\mathrm{NaCl}}^{\mathrm{e,c}}\right].\end{aligned} \tag{12.25}$$

The chemical potential of a neutral component is measured with respect to a standard state, which can be the state of pure components at 0.1 MPa. Each term in eqn (12.25) can then be calculated.

In the second scenario, the Maxwell-potential jumps at the electrode are derived from the Nernst equation $\Delta\psi + \Delta_{\mathrm{r}}G/F = 0$.[22,35,36]

$$\begin{aligned}\Delta_{\mathrm{a,e}}\psi &= -\frac{1}{F}\Delta_{\mathrm{r}}G^{\mathrm{s,a}} = \frac{1}{F}\left[\mu_{\mathrm{Na}}^{\mathrm{a,e}} - \mu_{\mathrm{Na^+}}^{\mathrm{e,a}} - \mu_{\mathrm{e^-}}^{\mathrm{a,e}}\right],\\ \Delta_{\mathrm{e,c}}\psi &= -\frac{1}{F}\Delta_{\mathrm{r}}G^{\mathrm{s,c}} = \frac{1}{F}\left[\mu_{\mathrm{AgCl}}^{\mathrm{s,c}} - \mu_{\mathrm{Ag}}^{\mathrm{c,e}} - \mu_{\mathrm{Cl^-}}^{\mathrm{e,c}} + \mu_{\mathrm{e^-}}^{\mathrm{c,e}}\right].\end{aligned} \tag{12.26}$$

In order to calculate the chemical potential of a charged particle, we need yet another standard state; that of an ion in a solution. A difference in single-ion chemical potentials is not measurable.

12.6 The Overpotential at Isothermal Conditions

The bilinear form of the entropy production applies to linear as well as nonlinear processes.[37] Eqn (12.12) and (12.13) apply in the presence of sizable current densities. In this situation, the electrode processes are not reversible, however. The last parentheses can be seen as effective electrochemical driving forces. The effective driving force is the overpotential of the electrode.[22–24,27] With operationally defined variables, we obtain the overpotential as:

$$\eta^{s,a} = \left|\Delta_{a,e}\phi + \frac{1}{F}\Delta_n G^{s,a}\right|. \tag{12.27a}$$

From the alternative description, we obtain

$$\eta^{s,a} = \left|\Delta_{a,e}\psi + \frac{1}{F}\Delta_r G^{s,a}\right|. \tag{12.27b}$$

The electrode overpotential was used already in eqn (12.2) and (12.3). It is defined as a positive quantity, *cf.* eqn (12.3). The overpotential is in both cases (12.27a) and (12.27b) equal to the difference in the electrochemical potential of the products and reactants.[27] In general, $\eta = |\Delta_r \tilde{G}|/nF$, where n is the number of electrons involved in the electrode reaction ($n = 1$ in our case). The expression (12.27b) was given by Newman.[22] He did not make the link to the entropy production, however. The expression (12.3) described the entropy production in terms of overpotential and Ohmic-potential drops. In order to obtain a more precise description of the total entropy production in this equation, the full expression of the local entropy production must be used and integrated out for all parts of the cell.

The overpotential can be determined from impedance measurements in set-ups with two equal electrodes and a reference electrode.[35,38] The potential drop between one electrode and the currentless reference can be measured when the current density between the working and counter electrodes is j. The overpotential is (in the absence of temperature gradients) related to the ohmic resistance of the electrode surface;

$$\eta^{s,a} = jR^{s,a}. \tag{12.28a}$$

Under isothermal conditions, and equilibrium for adsorption at the interface, one obtains the Butler-Volmer equation from mesoscopic non-equilibrium thermodynamics.[27,32]

$$j = j_0[\exp((1 - \alpha)\eta^{s,a}\, F/RT) - \exp(-\alpha\eta^{s,a}/RT)]. \tag{12.28b}$$

Subscript 0 means the equilibrium exchange value of the current density, and α is the transfer factor, used to indicate the position of the activation energy barrier.

Transport phenomena of different tensorial order do not couple (the Curie principle).[28] Therefore, there is no coupling between vectorial phenomena (transports of heat, mass and charge) and scalar phenomena (like chemical reactions) in a homogeneous phase. The surface has its own symmetry properties, however, with other possibilities for coupling. The relevant flux into or through an electrochemical surface is the *scalar* component perpendicular to the surface. This means that the scalar component of heat and mass fluxes can couple, not only to each other, but also to the electrochemical reaction. These coupling effects can be large,[25] see also Chapter 8.

In the general case, we obtain for the anode surface a set of four fluxes linearly related to the forces. The overpotential, which will give the last force, will therefore be coupled to two thermal forces and one chemical driving force. The thermal forces derive from the surface temperature being different from that of the adjacent materials. The chemical driving force relates to a possible lack of equilibrium of NaCl between the electrolyte and the surface. While not all forces are likely to be large, they may be large when the electrode materials conduct less well and the current density is high. These effects have not yet been explored experimentally, but the theoretical results[23–25] are supported by GENERIC.[32]

12.7 Transport Processes in the Electrolyte

We have so far discussed the electric potential profile across the cell under reversible conditions, without giving any flux-force relations.

When applied to the electrode surface, NET gives dynamic boundary conditions, essential for integration of variables through the surface. The boundary conditions contain jumps in the intensive variables of the system; the most important jump here being the electric-potential jump. In principle, also jumps in other intensive variables like the chemical potential or the temperature, may count.

The main purpose of the entropy production is to provide the flux-force relations which determine these jumps. The coupling of the fluxes is central. This section considers transport processes in the electrolyte under isothermal conditions. In the final Section 12.8, we allow for temperature gradients and surface phenomena in the linear regime.

12.7.1 Transport Numbers and Transference Coefficients

Consider an aqueous solution of NaCl bounded by two electrodes. The time-dependence of the salt concentration $c^{\mathrm{e}}_{\mathrm{NaCl}}$ in a formation- or concentration cell satisfies

$$\frac{\partial}{\partial t} c^{\mathrm{e}}_{\mathrm{NaCl}}(x,t) = -\frac{\partial}{\partial x} J^{\mathrm{e}}_{\mathrm{NaCl}}(x,t). \tag{12.29}$$

Here $J^{\mathrm{e}}_{\mathrm{NaCl}}$ is the salt flux in the electrolyte. The positive direction of transport is from left to right. By integrating this equation across the thickness of the electrolyte, from $x=0$ to $x=d_{\mathrm{e}}$, we obtain

$$\int_0^{d_{\mathrm{e}}} \frac{\partial}{\partial t} c^{\mathrm{e}}_{\mathrm{NaCl}}(x,t)\mathrm{d}x = -J^{\mathrm{e}}_{\mathrm{NaCl}}(d_{\mathrm{e}},t) + J^{\mathrm{e}}_{\mathrm{NaCl}}(0,t) = J^{\mathrm{e,a}}_{\mathrm{NaCl}}(t) - J^{\mathrm{e,c}}_{\mathrm{NaCl}}(t). \tag{12.30}$$

The equation describes that the accumulation of salt in the electrolyte per unit of cross-sectional area is the difference of the salt fluxes at the two electrodes. *In a concentration cell, the difference is zero.* In the formation cell, the integrated salt content in the electrolyte increases with j/F moles of NaCl per unit of time and per unit of cross-sectional area. The increase in the salt content in the electrolyte is therefore:

$$\frac{\partial}{\partial t}c^{e}_{NaCl}(x,t) = -\frac{\partial}{\partial x}J^{e}_{NaCl}(x,t) = \frac{j}{d_e F}. \tag{12.31}$$

Eqn (12.29) to (12.31) imply that the salt concentration is *independent* of the position and depends on the time, while the salt flux is *independent* of time and depends on the position.

The relation between the ion fluxes and the salt flux depends, however, on the boundary conditions (the electrodes). When both electrodes are reversible to Cl^-, the component flux of NaCl is everywhere defined by the flux of Na^+. When both are reversible to Na^+, the flux of NaCl is everywhere defined by the (negative) flux of Cl^-:

$$J^{e}_{NaCl} = J^{e}_{Na^+} \quad \text{Concentration cell, } Cl^-\text{-reversible electrodes,} \tag{12.32}$$

$$J^{e}_{NaCl} = J^{e}_{Cl^-} \quad \text{Concentration cell, } Na^+\text{-reversible electrodes.} \tag{12.33}$$

In the formation cell, the fluxes are position dependent, *cf.*, eqn (12.31), and eqn (12.32) and (12.33) do not apply. The relation between the salt and ion fluxes is always as given by eqn (12.19) meaning that possible position dependent terms in the ionic fluxes must cancel.

The transport number of an ion is the fraction of the electric current carried by the ion. The transport numbers of cations and anions can therefore be defined without reference to the boundaries:

$$t^{e}_{Na^+} \equiv F\left(\frac{J^{e}_{Na^+}}{j}\right)_{\partial c^{e}_{NaCl}/\partial x = \partial T/\partial x = 0};\ t^{e}_{Cl^-} \equiv -F\left(\frac{J^{e}_{Cl^-}}{j}\right)_{\partial c^{e}_{NaCl}/\partial x = \partial T/\partial x = 0}. \tag{12.34}$$

The transport numbers do not depend on the position.

The transference coefficient of a neutral component is defined by the ratio of the component flux with the electric current density:

$$t^{e}_{NaCl}(x,t) \equiv F\left(\frac{J^{e}_{NaCl}(x,t)}{j}\right)_{\partial \mu^{e}_{NaCl}(x,t)/\partial x = \partial T/\partial x = 0}. \tag{12.35}$$

Such a definition can also be used *e.g.* for water. Constant chemical potential and temperature mean that the salt concentration is uniform or independent of the position in the electrolyte. Unlike the transport number, the transference coefficient can depend on the boundaries, *e.g.* the electrode that supplies the current. The transference coefficient can be related to the transport numbers. For the concentration cells, we have from eqn (12.32) to (12.33):[4]

$$t^{e}_{NaCl} = t^{e}_{Na^+} \quad \text{for } Cl^-\text{-reversible electrodes,} \tag{12.36}$$

$$t^{e}_{NaCl} = -\,t^{e}_{Cl^-} \quad \text{for } Na^+\text{-reversible electrodes.} \tag{12.37}$$

We see from eqn (12.36) and (12.37) how the transference coefficient changes when the set of electrodes to measure it has changed.

In order to determine the transference coefficient for the formation cell, we start with the gradient of the transference coefficient. We find from eqn (12.31) and (12.35):

$$\frac{\partial}{\partial x} t^{e}_{\mathrm{NaCl}}(x,t) = \frac{F}{j}\frac{\partial}{\partial x} J^{e}_{\mathrm{NaCl}}(x,t) = -\frac{F}{j}\frac{\partial}{\partial t} c^{e}_{\mathrm{NaCl}}(x,t) = -\frac{1}{d_{e}}. \tag{12.38}$$

The electroneutrality condition gives; $c^{e}_{\mathrm{Na^+}}(x,t) = c^{e}_{\mathrm{Cl^-}}(x,t) = c^{e}_{\mathrm{NaCl}}(x,t)$. It follows that

$$\frac{F}{j}\frac{\partial}{\partial x} J^{e}_{\mathrm{Na^+}}(x,t) = -\frac{F}{j}\frac{\partial}{\partial x} J^{e}_{\mathrm{Cl^-}}(x,t) = -\frac{1}{d_{e}}. \tag{12.39}$$

In the formation cell[25,26] $J^{e}_{\mathrm{Na^+}}(x,t) \neq t^{e}_{\mathrm{Na^+}} j/F$ and $J^{e}_{\mathrm{Cl^-}}(x,t) \neq t^{e}_{\mathrm{Cl^-}} j/F$. The salt flux is given by

$$J^{e}_{\mathrm{NaCl}} = t^{e}_{\mathrm{Cl^-}} J^{e}_{\mathrm{Na^+}} + t^{e}_{\mathrm{Na^+}} J^{e}_{\mathrm{Cl^-}}. \tag{12.40}$$

In combination with eqn (12.18), we find as extension of eqn (12.19)

$$J^{e}_{\mathrm{Cl^-}} = J^{e}_{\mathrm{NaCl}} - t^{e}_{\mathrm{Cl^-}} j/F \quad \text{and} \quad J^{e}_{\mathrm{Na^+}} = J^{e}_{\mathrm{NaCl}} + t^{e}_{\mathrm{Na^+}} j/F. \tag{12.41}$$

In the formation cell, the natural boundary conditions are

$$J^{e}_{\mathrm{Cl^-}}(0,t) = 0 = J^{e}_{\mathrm{Na^+}}(d_{e},t). \tag{12.42}$$

This means that

$$J^{e}_{\mathrm{NaCl}}(0,t) = t^{e}_{\mathrm{Cl^-}} j/F \quad \text{and} \quad J^{e}_{\mathrm{NaCl}}(d_{e},t) = -\, t^{e}_{\mathrm{Na^+}} j/F. \tag{12.43}$$

These boundary conditions in combination with eqn (12.31) give the position-dependent salt flux:

$$J^{e}_{\mathrm{NaCl}}(x) = -\left[t^{e}_{\mathrm{Na^+}}\frac{x}{d_{e}} - t^{e}_{\mathrm{Cl^-}}\left(1 - \frac{x}{d_{e}}\right)\right]\frac{j}{F}. \tag{12.44}$$

This expression is new in the description of electrochemical cells.[25,26] The position-dependent transference coefficient becomes:

$$t^{e}_{\mathrm{NaCl}}(x) = -\frac{x}{d_{e}} t^{e}_{\mathrm{Na^+}} + \left(1 - \frac{x}{d_{e}}\right) t^{e}_{\mathrm{Cl^-}}. \tag{12.45}$$

The average transference coefficient of the dissolved salt NaCl formed in the electrolyte of a formation cell is equal to $-0.5\left(t^{e}_{\mathrm{Na^+}} - t^{e}_{\mathrm{Cl^-}}\right)$. The average appears in the Planck potential, *cf.*, Section 12.7.3. At the anode and cathode surface in the formation cell we obtain:

$$t^{e,a}_{\mathrm{NaCl}} = t^{e}_{\mathrm{Cl^-}} \quad \text{and} \quad t^{e,c}_{\mathrm{NaCl}} = -\, t^{e}_{\mathrm{Na^+}}. \tag{12.46}$$

These relations will be used in the expression for the reaction Gibbs energies (12.25), see also eqn (12.64) and (12.65). We have earlier incorrectly interchanged the interpretations (12.46).[25]

12.7.2 The Electrolyte Contribution to the Measured Cell Potential

Each flux in the entropy production is a linear combination of all conjugate forces. The flux equations that follow from eqn (12.11) for isothermal conditions are

$$
\begin{aligned}
J_{\mathrm{NaCl}}^{\mathrm{e}} &= -L_{\mu\mu}^{\mathrm{e}} \frac{1}{T}\frac{\partial}{\partial x}\mu_{\mathrm{NaCl}}^{\mathrm{e}} - L_{\mu\phi}^{\mathrm{e}} \frac{1}{T}\frac{\partial \phi}{\partial x}, \\
j &= -L_{\phi\mu}^{\mathrm{e}} \frac{1}{T}\frac{\partial}{\partial x}\mu_{\mathrm{NaCl}}^{\mathrm{e}} - L_{\phi\phi}^{\mathrm{e}} \frac{1}{T}\frac{\partial \phi}{\partial x}.
\end{aligned}
\tag{12.47}
$$

The Onsager coefficients L_{ij} with superscript e for electrolyte (*cf.*, Figure 12.2) are transport coefficients in a symmetric matrix of coefficients. The electric resistivity of the electrolyte is $r^{\mathrm{e}} = T/L_{\phi\phi}^{\mathrm{e}}$. Simple laws of transport, like Fick's or Ohm's law, are not sufficient, because the coupling coefficients are of the same order of magnitude as the diagonal coefficients. The coupling coefficient enters the definition of the transference coefficient:

$$
t_{\mathrm{NaCl}}^{\mathrm{e}} = F\frac{L_{\mu\phi}^{\mathrm{e}}}{L_{\phi\phi}^{\mathrm{e}}}. \tag{12.48}
$$

In the reversible limit $j \to 0$ we find, using the symmetry of the coefficient matrix (the Onsager relation), the contribution from the electrolyte to the potential gradient

$$
\left(\frac{\partial \phi}{\partial x}\right)_{j\to 0} = -\frac{t_{\mathrm{NaCl}}^{\mathrm{e}}}{F}\frac{\partial}{\partial x}\mu_{\mathrm{NaCl}}^{\mathrm{e}}. \tag{12.49}
$$

The effects of boundary conditions (electrodes) on the transference coefficient can now be compared. When both electrodes are reversible to the Na^+ ion, the ideal contribution from the electrolyte is, with eqn (12.37)

$$
\Delta_{\mathrm{e}}\phi_{j\to 0} = t_{\mathrm{Cl^-}}^{\mathrm{e}} \frac{2RT}{F}\ln\left(\frac{c_{\mathrm{NaCl}}^{\mathrm{e,c}}}{c_{\mathrm{NaCl}}^{\mathrm{e,a}}}\right). \tag{12.50}
$$

When the electrodes are reversible to the Cl^- ion and eqn (12.36) applies, the ideal contribution from the electrolyte is:

$$
\Delta_{\mathrm{e}}\phi_{j\to 0} = -t_{\mathrm{Na^+}}^{\mathrm{e}} \frac{2RT}{F}\ln\left(\frac{c_{\mathrm{NaCl}}^{\mathrm{e,c}}}{c_{\mathrm{NaCl}}^{\mathrm{e,a}}}\right). \tag{12.51}
$$

For the formation cell, we obtain from eqn (12.45) and (12.49):

$$
\Delta_{\mathrm{e}}\phi_{j\to 0} = \frac{1}{F}\int_{\mathrm{e,a}}^{\mathrm{e,c}}\left[\frac{x}{d_{\mathrm{e}}}t_{\mathrm{Na^+}}^{\mathrm{e}} - \left(1-\frac{x}{d_{\mathrm{e}}}\right)t_{\mathrm{Cl^-}}^{\mathrm{e}}\right]\mathrm{d}\mu_{\mathrm{NaCl}} = \frac{1}{2F}\left(t_{\mathrm{Na^+}}^{\mathrm{e}} - t_{\mathrm{Cl^-}}^{\mathrm{e}}\right)\Delta_{\mathrm{e}}\mu_{\mathrm{NaCl}}^{\mathrm{e}} \tag{12.52}
$$

In the integration of eqn (12.52), we assumed that $\mathrm{d}\mu_{\mathrm{NaCl}}^{\mathrm{e}}/\mathrm{d}x = \Delta_{\mathrm{e}}\mu_{\mathrm{NaCl}}^{\mathrm{e}}/d_{\mathrm{e}}$.

12.7.2.1 Numerical Considerations

The expressions for the electrolyte contribution to the cell potential describe the maximum electric energy that can be obtained from charge transfer across an electrolyte under various boundary conditions. The expressions (12.50) and (12.51) describe contributions to the electric potential of a concentration cell. In such cells, the electrodes are identical, and do not add to the cell potential.

The magnitude of the potential contributions from the electrolyte in the examples given here, eqn (12.50) to (12.52), is only some mV. By introducing the transport number of Na^+ equal to 0.396, and the transport number of Cl^- equal to 0.604 in eqn (12.52), we obtain 12 mV for a tenfold increase in concentration across the cell at 300 K. The contributions from eqn (12.50) and (12.51) are somewhat larger.

A membrane electrolyte can significantly increase the potential contribution from the electrolyte. With a transport number of a cation near 1, the concentration cell potential (12.51) becomes −120 mV for the same concentration gradient and temperature. This is the case in the reverse electrodialysis cell,[1] see Figure 12.3. The unit cell contains one cation-conducting and one anion-conducting membrane in series. The gradient in the chemical potential of NaCl across one membrane can contribute 120 mV to the cell potential with electrodes of Ag|AgCl. A membrane potential difference of some 100 mV is small compared to an electric potential in V of a formation cell. But by adding several of the unit cells, pictured in

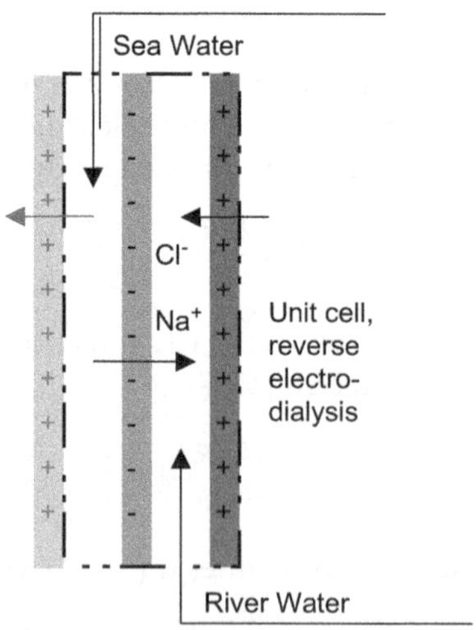

Figure 12.3 The unit cell of a reverse electrodialysis cell used in a saline power plant adapted from ref. 1. By allowing only cations/anions through every second membrane, one can generate charge transport from the difference in chemical potential of NaCl between sea water and river water.

Figure 12.3, into a series of cells, one can reach a sizable potential from this renewable energy source. Membrane costs are still detrimental. The contribution to the cell potential from the electrolyte in a formation cell, eqn (12.52), is smaller when both ions carry charge. In aqueous solutions of KCl, in so-called salt bridges, where both ions contribute equally to the charge transport, the contribution is near 0 mV.[25]

12.7.3 The Planck Potential

The entropy production (12.14) together with (12.15) is the starting point to obtain ionic flux-force relations:

$$J_{Na^+} = -L^{++}\frac{\partial(\mu_{Na^+}+F\psi)}{\partial x} - L^{+-}\frac{\partial(\mu_{Cl^-}-F\psi)}{\partial x},$$
$$J_{Cl^-} = -L^{-+}\frac{\partial(\mu_{Na^+}+F\psi)}{\partial x} - L^{--}\frac{\partial(\mu_{Cl^-}-F\psi)}{\partial x}. \tag{12.53}$$

The Onsager coefficients, L^{ij}, have been related to the ion concentration and mobility:

$$L^{++} = c_{NaCl}u_{Na^+}, \quad L^{-+} = L^{-+} = 0, \quad L^{--} = c_{NaCl}u_{Cl^-}. \tag{12.54}$$

The concentrations of the ions are equal to the salt concentration. These expressions give the Nernst-Planck flux equations for the ions:

$$J_{Na^+} = -c_{NaCl}u_{Na^+}\frac{\partial(\mu_{Na^+}+F\psi)}{\partial x},$$
$$J_{Cl^-} = -c_{NaCl}u_{Cl^-}\frac{\partial(\mu_{Cl^-}-F\psi)}{\partial x}. \tag{12.55}$$

The Maxwell-potential gradient for a concentration cell is found by setting $J^e_{Na^+} = J^e_{Cl^-}$:

$$\frac{\partial}{\partial x}\psi = -\frac{(u_{Na^+}-u_{Cl^-})}{2F(u_{Na^+}-u_{Cl^-})}\left(u_{Na^+}\frac{\partial}{\partial x}\mu_{Na^+} - u_{Cl^-}\frac{\partial}{\partial x}\mu_{Cl^-}\right). \tag{12.56}$$

The chemical potentials of the ions obey

$$\mu^e_{Na^+} = \mu^e_{Cl^-} = \frac{1}{2}\mu^e_{NaCl}. \tag{12.57}$$

By introducing (12.57) into (12.56), we obtain

$$\frac{\partial}{\partial x}\psi = -\frac{(u_{Na^+}-u_{Cl^-})}{2F(u_{Na^+}-u_{Cl^-})}\frac{\partial}{\partial x}\mu^e_{NaCl}. \tag{12.58}$$

Expression (12.58), named after Planck,[39] was obtained already in 1890. The condition $J^e_{Na^+} = J^e_{Cl^-}$, used to derive eqn (12.58), is *not* appropriate for a formation cell, however, as both ionic fluxes are zero when $j=0$, *cf.*, eqn

(12.43). It follows from (12.20) and the relation between the transport numbers, that

$$\Delta_e\phi = \Delta_e\psi + \frac{1}{2F}\left(t^e_{Na^+} - t^e_{Cl^-}\right)\Delta_e\mu_{NaCl}. \tag{12.59}$$

In view of (12.52), we then have for the formation cell

$$\Delta_e\psi = 0. \tag{12.60}$$

The Maxwell potential across the electrolyte of a formation cell is constant! This is remarkable, since the condition applies whether or not a concentration gradient exists in the electrolyte. The result follows from the invariance of the entropy production to the variable set and the assumptions used.

12.8 The Measured Cell Potential

We have shown above how one can make use of the entropy production to find the jumps in electric potential at interfaces and across the electrolyte of a cell. We proceed to combine the contributions of the system single parts to the measurable cell potential, before we study electrochemical cells for non-zero electric currents.

12.8.1 A Formation Cell with a Concentration Gradient

Return to the example of the formation cell, and consider first the description in terms of measurable properties, $\Delta\phi, j, \Delta_e\mu_{NaCl},$ and J^e_{NaCl}. The cell potential is of interest, first with the restriction $j \to 0$. The cell potential is found by adding potential jumps, as we cross the cell:

$$\Delta\phi = \Delta_{a,e}\phi + \Delta_e\phi + \Delta_{e,c}\phi. \tag{12.63}$$

In Figure 12.1 this sum was given the symbol E, which is common in electrochemistry. The reaction Gibbs energies of the neutral components and the transference coefficients, eqn (12.46), give the electrode potential jumps:

$$\Delta_{a,e}\phi = \frac{1}{F}\left(\mu^{a,e}_{Na} - t^e_{Cl^-}\mu^{e,a}_{NaCl}\right), \tag{12.64}$$

$$\Delta_{e,c}\phi = \frac{1}{F}\left(\mu^{s,c}_{AgCl} - \mu^{c,e}_{Ag} - t^e_{Na^+}\mu^{e,c}_{NaCl}\right). \tag{12.65}$$

By adding to these the contribution from the electrolyte, eqn (12.52), we find:

$$\begin{aligned}\Delta\phi &= \frac{1}{F}\left[\mu^{a,e}_{Na} - t^e_{Cl^-}\mu^{e,a}_{NaCl} + \mu^{s,c}_{AgCl} - \mu^{c,e}_{Ag} - t^e_{Na^+}\mu^{e,c}_{NaCl} + \frac{1}{2}\left(t^e_{Na^+} - t^e_{Cl^-}\right)\Delta_e\mu_{NaCl}\right] \\ &= \frac{1}{F}\left[\mu^{a,e}_{Na} + \mu^{s,c}_{AgCl} - \mu^{c,e}_{Ag} - \frac{1}{2}\left(\mu^{e,a}_{NaCl} + \mu^{e,c}_{NaCl}\right)\right].\end{aligned} \tag{12.66}$$

All quantities in this equation can be determined by experiments and a proper choice of standard states. The expression reduces to the Gibbs energy of reaction (12.9), but *only* if the electrolyte is stirred:

$$\Delta\phi = \Delta_{a,e}\phi + \Delta_{e,c}\phi = \frac{1}{F}\left[\mu_{Na}^{a,e} + \mu_{AgCl}^{s,c} - \mu_{Ag}^{c,e} - \mu_{NaCl}^{e}\right], \tag{12.67}$$

We used here the fact that the ionic transport numbers add to unity.

In terms of Maxwell-potential differences, the cell potential E is also equal to:

$$\begin{aligned}\Delta\phi &= \Delta\psi - \Delta\mu_{e^-} = \Delta_{a,e}\psi + \Delta_{e,c}\psi - \Delta\mu_{e^-}\\ &= \frac{1}{F}\left[\mu_{Na}^{a,e} + \mu_{AgCl}^{s,c} - \mu_{Ag}^{c.e} - \frac{1}{2}\left(\mu_{NaCl}^{e,a} + \mu_{NaCl}^{e,c}\right)\right]\end{aligned} \tag{12.68}$$

This right-hand side was derived with the assumptions of

- zero coupling terms in eqn (12.54),
- Nernst-Einstein, that the electric and diffusional mobility of the ions are the same,
- a negligible difference in the chemical potential of electrons.

In this case we obtain the reaction Gibbs energy as given directly by eqn (12.9). The assumptions used can, however, not be tested by experiments.

Consider next the concentration cells, with two sodium-reversible or two chloride-reversible electrodes. The measured cell potential of these cells can be expressed in terms of electrochemical potential differences. We obtain, using eqn (12.20)[5,21,25]

$$\text{For } Na^+ \text{ reversible electrodes: } F\Delta\phi = F\Delta_e\phi = \Delta\tilde{\mu}_{Na^+}, \tag{12.69}$$

$$\text{For } Cl^- \text{ reversible electrodes: } F\Delta\phi = F\Delta_e\phi = -\Delta\tilde{\mu}_{Cl^-}. \tag{12.70}$$

The equations above apply for $j = 0$. In order to solve the dynamic problem, one must solve the flux equations for each layer of the cell. Such equations are given below for the non-isothermal formation cell.

12.8.2 A Non-isothermal Formation Cell

Heat is often well conducted in metal electrodes and in many electrolytes, but an electrochemical cell need not have a uniform temperature. The temperature can become important in two ways; as a variable for the properties involved, and as a gradient or a thermal driving force. We continue to describe the cell potential when the cell is exposed to a temperature gradient. The possible temperature variation in the transport coefficients shall be neglected, however.

12.8.2.1 The Homogeneous Anode Phase

The flux-force relations that result from eqn (12.10) are

$$\begin{aligned} J_q'^{\mathrm{a}} &= -L_{qq}^{\mathrm{a}} \frac{1}{T^2}\frac{\partial T}{\partial x} - L_{q\phi}^{\mathrm{a}} \frac{1}{T}\frac{\partial \phi}{\partial x}, \\ j &= -L_{\phi q}^{\mathrm{a}} \frac{1}{T^2}\frac{\partial T}{\partial x} - L_{\phi\phi}^{\mathrm{a}} \frac{1}{T}\frac{\partial \phi}{\partial x}. \end{aligned} \tag{12.71}$$

The Peltier coefficient is defined by

$$\Pi^{\mathrm{a}}(x) \equiv F\left(\frac{J_q'^{\mathrm{a}}}{j}\right)_{\partial T/\partial x=0} = F\frac{L_{q\phi}^{\mathrm{a}}}{L_{\phi\phi}^{\mathrm{a}}} \equiv -T(x)\left(S_{\mathrm{Na}}^{\mathrm{a,e}} + S_{\mathrm{e}^-}^{*\mathrm{a}}\right). \tag{12.72}$$

The Peltier coefficient is given by the entropy of sodium and the transported entropy of electrons, $S_{\mathrm{e}^-}^{*\mathrm{a}}$. The electric-potential gradient for $j = 0$, is obtained with the Onsager relation:

$$\left(\frac{\partial \phi}{\partial x}\right)_{j\to 0} = -\frac{1}{F}\frac{\Pi^{\mathrm{a}}(x)}{T(x)}\frac{\partial T}{\partial x} = \frac{1}{F}\left(S_{\mathrm{Na}}^{\mathrm{a,e}} + S_{\mathrm{e}^-}^{*\mathrm{a}}\right)\frac{\partial T}{\partial x}. \tag{12.73}$$

Upon integration, we have for $j = 0$:

$$\Delta_{\mathrm{a}}\phi = \frac{1}{F}\left(S_{\mathrm{Na}}^{\mathrm{a,e}} + S_{\mathrm{e}^-}^{*\mathrm{a}}\right)\Delta_{\mathrm{a}}T. \tag{12.74a}$$

The Seebeck coefficient of the anode is defined by

$$\eta_{\mathrm{S}}^{\mathrm{a}} \equiv \frac{\Delta_{\mathrm{a}}\phi}{\Delta_{\mathrm{a}}T} = \frac{1}{F}\left(S_{\mathrm{Na}}^{\mathrm{a,e}} + S_{\mathrm{e}^-}^{*\mathrm{a}}\right). \tag{12.74b}$$

The transported entropies of electrons or holes are small, (1 to 20) $\mathrm{J \cdot K^{-1} \cdot mol^{-1}}$, when compared to thermodynamic entropies of metal or salts, like in eqn (12.73). Semiconductors of the *n* and *p*-type are nevertheless used for thermoelectric-power generation, see Figure 12.4. Research on semiconductors is carried out to increase their ability to convert waste heat into electricity.

12.8.2.2 The Electrolyte

The linear flux-force relations resulting from eqn (12.11) are

$$\begin{aligned} J_q'^{\mathrm{e}} &= -L_{qq}^{\mathrm{e}} \frac{1}{T^2}\frac{\partial T}{\partial x} - L_{q\mu}^{\mathrm{e}} \frac{1}{T}\frac{\partial}{\partial x}\mu_{\mathrm{NaCl},T}^{\mathrm{e}} - L_{q\phi}^{\mathrm{e}} \frac{1}{T}\frac{\partial \phi}{\partial x}, \\ J_{\mathrm{NaCl}}^{\mathrm{e}} &= -L_{\mu q}^{\mathrm{e}} \frac{1}{T^2}\frac{\partial T}{\partial x} - L_{\mu\mu}^{\mathrm{e}} \frac{1}{T}\frac{\partial}{\partial x}\mu_{\mathrm{NaCl},T}^{\mathrm{e}} - L_{\mu\phi}^{\mathrm{e}} \frac{1}{T}\frac{\partial \phi}{\partial x}, \\ j &= -L_{\phi q}^{\mathrm{e}} \frac{1}{T^2}\frac{\partial T}{\partial x} - L_{\phi\mu}^{\mathrm{e}} \frac{1}{T}\frac{\partial}{\partial x}\mu_{\mathrm{NaCl},T}^{\mathrm{e}} - L_{\phi\phi}^{\mathrm{e}} \frac{1}{T}\frac{\partial \phi}{\partial x}. \end{aligned} \tag{12.75}$$

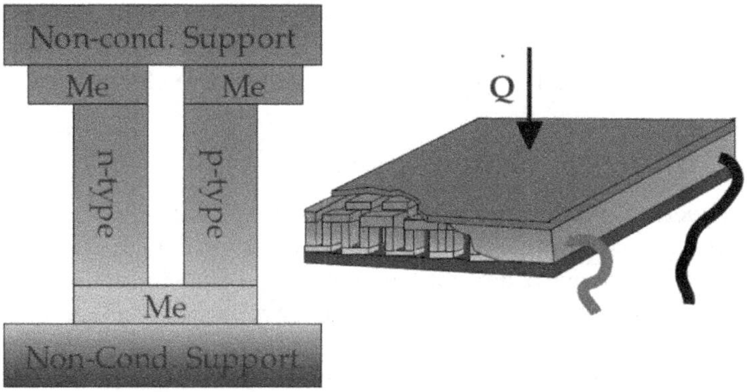

Figure 12.4 Sketch of a thermoelectric element (left) and a module of elements (right), used for conversion of heat Q into electricity. High temperature is indicated in red, low in blue. Charge is conducted in n or p-type semiconductors which have different transported entropies. This generates a potential across a series of n and p-type semiconductors. M.T. Børset is acknowledged for the figure.

The chemical potential is differentiated at constant temperature, see the discussion in Chapter 8. The transference coefficient, $t^{\mathrm{e}}_{\mathrm{NaCl}}(x)$, was given by eqn (12.45) for the formation cell. The definition of the position-dependent Peltier coefficient is, in analogy with (12.45):

$$\begin{aligned}\Pi^{\mathrm{e}}(x) &\equiv F\left(\frac{J'^{\mathrm{e}}_q}{j}\right)_{\partial T/\partial x=\partial\mu^{\mathrm{e}}_{\mathrm{NaCl}}/\partial x=0} = F\frac{L^{\mathrm{e}}_{q\phi}}{L^{\mathrm{e}}_{\phi\phi}} \\ &\equiv T(x)\left[-t^{\mathrm{e}}_{\mathrm{NaCl}}(x)S^{\mathrm{e}}_{\mathrm{NaCl}} + \frac{x}{d_{\mathrm{e}}}t^{\mathrm{e}}_{\mathrm{Na^+}}S^{*\mathrm{e}}_{\mathrm{Na^+}} - \left(1-\frac{x}{d_{\mathrm{e}}}\right)t^{\mathrm{e}}_{\mathrm{Cl^-}}S^{*\mathrm{e}}_{\mathrm{Cl^-}}\right].\end{aligned} \tag{12.76}$$

The Peltier coefficient is given by the entropy of NaCl and the transported entropies of the ions. The transported entropies enter the analysis as integration constants at the surfaces, just as the transport numbers did in the expression for transference coefficient of the salt.

By using these equations, we can define *Soret equilibrium* setting $J^{\mathrm{e}}_{\mathrm{NaCl}}=0$ and $j=0$. Doing this, we can eliminate the electric driving force in eqn (12.75b) and find the gradient in chemical potential as a function of the gradient in temperature. We shall not use this condition here, but rather solve the last equation for a known difference in chemical potential and temperature. The contribution to the cell potential from the electrolyte becomes:

$$\begin{aligned}\Delta_{\mathrm{e}}\phi &= \frac{1}{2F}\left(t^{\mathrm{e}}_{\mathrm{Na^+}} - t^{\mathrm{e}}_{\mathrm{Cl^-}}\right)\Delta_{\mathrm{e}}\mu^{\mathrm{e}}_{\mathrm{NaCl},T} \\ &+ \frac{1}{2F}\left[\left(t^{\mathrm{e}}_{\mathrm{Na^+}} - t^{\mathrm{e}}_{\mathrm{Cl^-}}\right)S^{\mathrm{e}}_{\mathrm{NaCl}} + t^{\mathrm{e}}_{\mathrm{Na^+}}S^{*,\mathrm{e}}_{\mathrm{Na^+}} - t^{\mathrm{e}}_{\mathrm{Cl^-}}S^{*,\mathrm{e}}_{\mathrm{Cl^-}}\right]\Delta_{\mathrm{e}}T.\end{aligned} \tag{12.77}$$

In the last expression, the difference $\Delta_e\mu^e_{NaCl,T}$ is evaluated at $T^{e,c}$.[25] The transported entropies were taken to be independent of the temperature. The pre-factor of $\Delta_e T$ is the Seebeck coefficient of the electrolyte.

12.8.2.3 Numerical Considerations

Large Seebeck coefficients have been observed for electrode reactions which involve gases and/or complex formation of ions.[33,40,41] Selected Seebeck coefficients with aqueous electrolytes, measured in the last century,[42–48] are given in Table 12.1. They all have larger Seebeck coefficients than semiconductor pairs have. The highest value, $-4.17\ \text{mV}\cdot\text{K}^{-1}$, was found for a cell with complexing agents.[48] Ionic liquids with low temperature melting points may provide a possibility to convert low-temperature heat into electricity.[49] Electrochemical cells with ionic liquids can be found to match high- as well as low-temperature sources. Not only can industrial waste heat be used, but geothermal and solar heat sources are also relevant. The second contribution to eqn (12.77) can soon be comparable to the first contribution when large temperature differences are available. Similar to the situation for reverse electrodialysis cells, Figure 12.3, one may consider stacking unit cells.

12.8.2.4 The Anode Surface

Consider the entropy production in the anode surface, with the condition $\Delta_{s,e}\mu_{NaCl,T}=0$. The flux equations become:

$$
\begin{aligned}
J_q^{\prime a,e} &= L^{s,a}_{aa}\Delta_{a,s}\left(\frac{1}{T}\right) + L^{s,a}_{ae}\Delta_{s,e}\left(\frac{1}{T}\right) - L^{s,a}_{a\phi}\frac{1}{T^{s,a}}\left(\Delta_{a,e}\phi + \frac{\Delta_n G^{s,a}}{F}\right),\\
J_q^{\prime e,a} &= L^{s,a}_{ea}\Delta_{a,s}\left(\frac{1}{T}\right) + L^{s,a}_{ee}\Delta_{s,e}\left(\frac{1}{T}\right) - L^{s,a}_{e\phi}\frac{1}{T^{s,a}}\left(\Delta_{a,e}\phi + \frac{\Delta_n G^{s,a}}{F}\right), \qquad (12.78)\\
j &= L^{s,a}_{\phi a}\Delta_{a,s}\left(\frac{1}{T}\right) + L^{s,a}_{\phi e}\Delta_{s,e}\left(\frac{1}{T}\right) - L^{s,a}_{\phi\phi}\frac{1}{T^{s,a}}\left(\Delta_{a,e}\phi + \frac{\Delta_n G^{s,a}}{F}\right),
\end{aligned}
$$

Table 12.1 Selected Seebeck coefficients, Sc, for aqueous electrolytes at 298 K.

Reference	Electrodes	Electrolyte in water	$\text{Sc/mV}\cdot\text{K}^{-1}$
Levin and Bonilla[42]	Ag(s)|AgCl(s)	KCl	0.73
Breck and Agar[43]	3 % Cd amalgam	$CdSO_4$	0.81
Breck *et al.*[44]	Quinhydrone	HCl	− 0.61
Haase and Hoch[45]	Ag(s)|AgCl(s)	KCl	0.83
Kamata *et al.*[46]	Ag(s)	$AgNO_3$	− 0.78
Quickenden and Veron[47]	Pt(s)	$K_4Fe(CN)_6$	1.60
Kuzminsky *et al.*[48]	Cu(s)	Cu(dipy)2Cl, $LiBF_4$	− 4.17

where the Gibbs energy change due to neutral components was given by (12.21). The Peltier coefficients for the bulk phases at the anode surface are transference coefficients for heat:

$$\Pi^{a,e} \equiv F\left(\frac{J_q'^{a,e}}{j}\right)_{\Delta_{a,s}T=\Delta_{s,e}T=\Delta_{s,e}\mu_{NaCl}=0} = F\frac{L_{a\phi}^{s,a}}{L_{\phi\phi}^{s,a}} \equiv -T^{a,e}\left(S_{Na}^{a,e} + S_{e^-}^{*e,a}\right),$$

$$\Pi^{e,a} \equiv F\left(\frac{J_q'^{e,a}}{j}\right)_{\Delta_{a,s}T=\Delta_{s,e}T=\Delta_{s,e}\mu_{NaCl}=0} = F\frac{L_{e\phi}^{s,a}}{L_{\phi\phi}^{s,a}} \equiv -T^{e,a}t_{Cl^-}^{e}\left(S_{NaCl}^{e,a} + S_{Cl^-}^{*a,e}\right). \tag{12.79}$$

We have $\Pi^{a,e} = \Pi^{a}$ (0) and $\Pi^{e,a} = \Pi^{e}$ (0), *cf.*, eqns (12.72) and (12.76). The transported entropy of the electron can be neglected in sums. By using the Onsager relations, the potential difference across the anode surface becomes:

$$\begin{aligned}\Delta_{a,e}\phi &= -\frac{1}{F}\left[\Delta_n G^{s,a} + \Pi^{a,e}\left(\frac{T^{s,a}-T^{a,e}}{T^{a,e}}\right) + \Pi^{e,a}\left(\frac{T^{e,a}-T^{s,a}}{T^{e,a}}\right)\right] \\ &= \frac{1}{F}\left[\mu_{Na}^{a,e}(T^{s,a}) - t_{Cl^-}^{e}\mu_{NaCl}^{e,a}(T^{s,a}) + S_{Na}^{a,e}(T^{s,a}-T^{a,e})\right. \\ &\quad \left. + t_{Cl^-}^{e}\left(S_{NaCl}^{e,a} + S_{Cl^-}^{*e,a}\right)(T^{e,a}-T^{s,a})\right] \\ &= \frac{1}{F}\left[\mu_{Na}^{a,e}(T^{a,e}) - t_{Cl^-}^{e}\mu_{NaCl}^{e,a}(T^{e,a}) + t_{Cl^-}^{e}S_{Cl^-}^{*e,a}(T^{e,a}-T^{s,a})\right].\end{aligned} \tag{12.80}$$

The main contribution to the surface-potential drop is given by the reaction Gibbs energy. The entropy of Na and NaCl refer to different temperatures. In the first equality the expression contains the temperature of the surface, while in the second equality the expression contains the temperature of the anode near the electrolyte (for Na) and of the electrolyte near the anode (for NaCl). In the derivation of the final result, we used the temperature dependence of the chemical potential. For constant temperature, the expression reduces to the one given in eqn (12.68).

12.8.2.5 *The Cathode Surface*

The chemical reaction in the cathode surface produces Cl^- in the electrolyte, while Ag^+ conducts charge in AgCl. The Peltier coefficients are:

$$\Pi^{e,c} \equiv F\left(\frac{J_q'^{e,c}}{j}\right)_{\Delta_{c,s}T=\Delta_{s,c}T=\Delta_{e,s}\mu_{NaCl}=0} = F\frac{L_{e\phi}^{s,c}}{L_{\phi\phi}^{s,c}} \equiv T^{e,c}t_{Na^+}^{e}\left(S_{NaCl}^{e,c} + S_{Na^+}^{*e,c}\right),$$

$$\Pi^{c,e} \equiv F\left(\frac{J_q'^{c,e}}{j}\right)_{\Delta_{c,s}T=\Delta_{s,c}T=\Delta_{e,s}\mu_{NaCl}=0} = F\frac{L_{c\phi}^{s,c}}{L_{\phi\phi}^{s,c}} \equiv T^{c,e}\left(S_{Ag^+}^{*s,c} + S_{AgCl}^{s,c} - S_{Ag}^{c,e}\right). \tag{12.81}$$

The contribution to the cell potential is accordingly:

$$
\begin{aligned}
\Delta_{e,c}\phi &= -\frac{1}{F}\left[\Delta_n G^{s,c} + \frac{\Pi^{e,c}}{T^{e,c}}\Delta_{e,s}T + \frac{\Pi^{c,e}}{T^{c,e}}\Delta_{s,c}T\right] \\
&= -\frac{1}{F}\Big[-\mu^{s,c}_{AgCl}(T^{c,e}) + \mu^{c,e}_{Ag}(T^{c,e}) + t^{e}_{Na^+}\mu^{e,c}_{NaCl}(T^{e,c}) \\
&\quad + t^{e}_{Na^+}S^{*e,c}_{Na^+}\Delta_{e,s}T + S^{*s,c}_{Ag^+}\Delta_{s,c}T)\Big].
\end{aligned} \tag{12.82}
$$

Entropies and transported entropies were again taken constant in the integration. The total cell potential can now be found by adding the parts.

12.9 Conclusion and Perspective

We have seen in this chapter how the cell-potential profile can be determined for reversible conditions, directly from the entropy production. The entropy production defines also the flux equations, to be used to describe a cell in operation. Equations have been given for the bulk phases and the electrode surfaces. The flux equations of the surfaces can be regarded as dynamic boundary conditions. The energy dissipated as heat in the surrounding can be quantified by the cell's entropy production, by integrating across all parts of the cell. This will be the way to obtain a more precise expression for the total entropy production in eqn (12.3).

We have shown how to find flux equations for all parts of an electrochemical cell, to be used for non-zero current density. Local temperature gradients may not be large in a cell made of good conductors. But the use of porous gas electrodes, ceramic electrolytes, semiconductors and large current densities, may change this situation. The sets of equations give a possibility to model heat and mass fluxes that arise from varying and large electric currents and find simultaneous solutions of several variable profiles. For sizable current densities, one then can find temperature and concentration profiles in addition to electric potential profiles. Such profiles have so far been obtained for the polymer fuel cell.[25]

While the sum of the surface-potential drops can be in the order of magnitude of 1 V in a formation cell, they give a zero contribution in concentration cells. But concentration cells can have a contribution from the electrolyte, which also may play a role. Even if this potential difference is an order of magnitude smaller than the potential of a formation cell, it may play a role in renewable energy technology. Concentration cells can, like the name says, exploit concentration differences, but also waste heat. This is done in reverse electrodialysis plants or in thermoelectric generators, respectively, or possibly in combinations of the two. The theory predicts that gas electrodes or complex formers may be beneficial to thermoelectric generators. Such cells may offer alternatives to generators that are now using semiconductors. Concentration cells or thermoelectric cells cannot be precisely described without non-equilibrium thermodynamics. When operationally defined variables are used, one is also able to control assumptions made in the theory by experiment, *cf.*, Section 12.8.1.

Acknowledgements

The authors acknowledge ETH Zürich for guest professorships in 2014.

References

1. O. S. Burheim, J. G. Pharoah, D. A. Vermaas, K. Nijmeijer, B. Hamelers and B. B. Sales, Reverse Electrodialysis, in *Encyclopedia of Membrane Science and Technology*, ed. E. M. V. Hoek and V. V. Tarabara, Wiley, Hoboken, 2013, p. 1482.
2. J. W. Post, J. Veerman, H. V. M. Hamelers, G. J. W. Euverink, S. J. Metz, K. Nijmeijer and C. J. N. Buisman, *J. Membr. Sci.*, 2007, **288**, 218.
3. M. Zerbarjadi, K. Esfarjani, M. S. Dresselhaus, Z. F. Ren and G. Chen, *Energy Environ. Sci.*, 2012, **5**, 5147.
4. A. Katchalsky and P. Curran, *Non-Equilibrium Thermodynamics in Biophysics*, Harvard University Press, Cambridge, MA, 1975.
5. A. Katchalsky and O. Kedem, *Biophys. J.*, 1962, **2**, 53.
6. R. Haase, *Thermodynamics of Irreversible Processes*, Addison-Wesley, Reading, USA, 2nd edn, 1969, Dover, London, 1990.
7. N. Laksminarayanaiah, *Transport Phenomena in Membranes*, Academic Press, New York, 1969.
8. D. C. Mickulecki, W. A. Wiegand and J. S. Shiner, *J. Theor. Biol.*, 1977, **69**, 471.
9. A. J. Staverman, *Trans. Faraday Soc.*, 1952, **48**, 176.
10. K. S. Førland, T. Førland and S. K. Ratkje, *Irreversible Thermodynamics. Theory and Application*, Wiley, Chichester, 3rd edn, 1988, Tapir, Trondheim, 2001.
11. Y. Ito, H. Kaiya, S. Yoshizawa, S. Kjelstrup Ratkje and T. Førland, *J. Electrochem. Soc.*, 1984, **131**, 2504.
12. J. Richter, A. Heller and W. Vreuls, *Ber. Bunsenges. Phys. Chem.*, 1977, **81**, 375.
13. J. Richter and U. Prüser, *Ber. Bunsenges. Phys. Chem.*, 1977, **81**, 508.
14. T. S. Brun and D. Vaula, *Ber. Bunsenges. Physik. Chem.*, 1967, **71**, 824.
15. S. K. Ratkje, T. Holt and M. Skrede, *Ber. Bunsenges. Phys. Chem.*, 1988, **92**, 825.
16. J. Breer, S. K. Ratkje and G.-F. Olsen, *Zeitschr. Phys. Chem.*, 1991, **174**, 179.
17. T. Okada, S. Kjelstrup Ratkje, S. Møller-Holst, L. O. Jerdal, K. Friestad, G. Xie and R. Holmen, *J. Membr. Sci.*, 1996, **111**, 158.
18. T. Okada, S. Møller-Holst, O. Gorseth and S. Kjelstrup, *J. Electroanal. Chem.*, 1998, **442**, 3313.
19. T. Suzuki, Y. Takahashi and M. Tasaka, *J. Membr. Sci.*, 1994, **92**, 85–92.
20. T. Suzuki, Y. Takahashi, R. Kiyono and M. Tasaka, *Colloid Polym. Sci.*, 1994, **272**, 971.

21. S. Kjelstrup Ratkje, M. Ottøy, R. Halseid and M. Strømgård, *J. Membr. Sci.*, 1995, **107**, 219.
22. J. S. Newman, *Electrochemical Systems*, Prentice-Hall, Englewood Cliffs, 2nd edn, 1991.
23. D. Bedeaux and S. Kjelstrup Ratkje, *J. Electrochem. Soc.*, 1996, **143**, 767.
24. S. Kjelstrup Ratkje and D. Bedeaux, *J. Electrochem. Soc.*, 1996, **143**, 779.
25. S. Kjelstrup and D. Bedeaux, *Non-Equilibrium Thermodynamics of Heterogeneous Systems*, Series on Advances in Statistical Mechanics, World Scientific, Singapore, 2008, vol. 16.
26. S. Kjelstrup and D. Bedeaux, in *Springer Handbook of Electrochemistry*, ed. C. Breitkopf and K. S. Lyons, Springer, ch. 4, in press.
27. J. M. Rubí and S. Kjelstrup, *J. Phys. Chem. B*, 2003, **107**, 13471.
28. S. R. de Groot and P. Mazur, *Non-Equilibrium Thermodynamics*. North-Holland, Amsterdam, 1962, and Dover, London, 1984.
29. I. Prigogine and P. Mazur, *Physica*, 1953, **19**, 241.
30. J. M. G. Vilar and J. M. Rubí, *Proc. Natl. Acad. Sci. U. S. A.*, 2001, **98**, 11081.
31. S. Kjelstrup, J. M. Rubí and D. Bedeaux, *Phys. Chem. Chem. Phys.*, 2005, **7**, 4009.
32. D. Bedeaux, S. Kjelstrup and H. C. Öttinger, *J. Chem. Phys.*, 2014, **141**, 124102.
33. O. S. Burheim, S. Kjelstrup, J. G. Pharoah, P. J. S. Vie and S. Møller-Holst, *Electrochim. Acta*, 2011, **56**, 3248.
34. G. D. C. Kuiken, *Thermodynamics for Irreversible Processes*, Wiley, Chichester, 1994.
35. A. J. Bard and L. R. Faulkner, *Electrochemical Methods. Fundamentals and Applications*, Wiley, Hoboken, 2nd edn, 2001.
36. J. O'. M. Bockris, A. K. N. Reddy and M. Gamboa-Aldeco, *Modern Electrochemistry 2A*, Klüver Academic, New York, 2002.
37. J. Ross and P. Mazur, *J. Chem. Phys.*, 1961, **35**, 19.
38. A. K. Meland, D. Bedeaux and S. Kjelstrup, *J. Membr. Sci.*, 2006, **282**, 96.
39. M. Planck, *Ann. Phys. Chem. Neue Folge*, 1890, **39**, 161; M. Planck, *Ann. Phys. Chem. Neue Folge*, 1890, **40**, 561.
40. B. Flem, S. Kjelstrup and Å. Sterten, *Light Met.*, 1996, 203.
41. B. Flem, Q. Xu, S. Kjelstrup and Å. Sterten, *J. Nonequilib. Thermod.*, 2001, **26**, 125.
42. H. Levin and C. F. Bonilla, *J. Electrochem. Soc.*, 1951, **98**, 388.
43. W. G. Breck and J. N. Agar, *Trans. Faraday Soc.*, 1957, **53**, 179.
44. W. G. Breck, G. Cadenhead and M. Hammerli, *Trans. Faraday Soc.*, 1965, **61**, 37.
45. R. Haase and K. Hoch, *Ann. Phys. Chem. Neue Folge*, 1965, **46**, 63.
46. M. Kamata, Y. Ito and J. Oishi, *Electrochim. Acta*, 1986, **31**, 521.
47. T. Quickenden and C. F. Veron, *Sol. Energy*, 1986, **36**, 63.
48. Y. V. Kuzminsky, V. A. Zasukha and G. Y. Kuzminskaya, *J. Power Sources*, 1994, **52**, 231.
49. T. J. Abraham, D. R. MacFarlane and J. M. Pringle, *Energy Environ. Sci.*, 2013, **6**, 2639.

CHAPTER 13

Entropy Production Minimization with Optimal Control Theory

ØIVIND WILHELMSEN,* EIVIND JOHANNESSEN AND SIGNE KJELSTRUP

Department of Chemistry, Norwegian University of Science and Technology, N-7491 Trondheim, Norway
*Email: oivind.wilhelmsen@ntnu.no

13.1 Introduction

Consumption of oil, coal, gas and other energy sources has more than doubled the last fifty years on a world-wide basis. During the same time, atmospheric CO_2 levels have increased with more than 20 %.[1] More energy-efficient processes in the industry are vital to change these trends.

Several measures of process energy efficiency have been proposed in the literature. Most of them compare to the energy input needed in state-of-the-art technologies,[2] so-called optimal energy-management.[3] These studies benchmark the energy input required and assess potentials for reductions of the energy input. However, as emphasized in ref. 2, these methods do not provide information on the amount and localization of the inefficiencies of the process. The only way to obtain such information is to determine the entropy production everywhere. This can give a detailed description of how and where energy is dissipated as heat, information which constitutes the basis for optimization. The systematic theory of non-equilibrium thermodynamics gives therefore the foundation of entropy-production-minimization

Experimental Thermodynamics Volume X: Non-equilibrium Thermodynamics with Applications
Edited by Dick Bedeaux, Signe Kjelstrup and Jan V. Sengers

Published by the Royal Society of Chemistry, www.rsc.org

procedures.[5] Non-equilibrium thermodynamics becomes the core theory, when a local (even molecular) description is sought of an energy converting processes. This is because the dissipated energy (the entropy production) is explicitly formulated in this theory in terms of fluxes and forces (gradients in, *e.g.*, pressure, temperature and composition). In this chapter, we explain how to find the state of minimum entropy production using optimal control theory.[4] The trivial minimum of zero entropy production is only of theoretical interest, as the processes in the industry always occur in a finite amount of time and in process equipment of finite size. Optimal control theory is the ideal tool to identify the most efficient operation of process equipment, where constraints in operating time or equipment size are taken into account.

The issue of process operation with minimum entropy production, was first raised from the mechanical-engineering point of view by Bejan.[5] Among physicists, the school of finite-time thermodynamics has been pioneering in research of these issues, using the method of endoreversible machines.[6–8] This chapter will present insight gained from work on entropy production minimization from a chemical engineering point of view.[9,10] We shall present the hypotheses of Equipartition of Entropy Production (EoEP) and Equipartition of Forces (EoF) as possible approximations to this most energy-efficient state of operation of an industrial process unit (Section 13.5.1).[8,11–15] EoEP has also been found to apply in nature.[16] So-called "highways in state space" for states with minimum entropy production have been observed in chemical reactors[17–20] and in distillation columns.[20] This concept will be further discussed in Section 13.5.2. Entropy production minimization has been applied to distillation columns,[6,21,22] exothermic and endothermic reactors,[1,12,17,23–28] heat exchangers[29,30] and even to processes for paper production.[31] Based on the experience from these works, guidelines to energy-efficient design and operation can be formulated, as we will show in Section 13.5.3. Underlying these results is the fact there exist a ***state of minimum entropy production*** which can serve as an aim for energy-efficient design and operation.

We first explain how the entropy production is related to energy efficiency (Section 13.2) and illustrate this with a simple example (Section 13.3). The systematic methodology to find the state of minimum entropy production with optimal-control theory is then given in Section 13.4. Important results for states with minimum entropy production gained over the last decades will be presented in Section 13.5 before concluding remarks and suggestions for future work will be given in Section 13.6.

13.2 Minimum Entropy Production and Energy Efficiency

The first, η_{I}, and the second, η_{II}, law efficiencies are defined in thermodynamics as:[20]

$$\eta_{\mathrm{I}} = -\frac{w}{q} \quad \text{and} \quad \eta_{\mathrm{II}} = \begin{cases} \dfrac{w}{w_{\mathrm{id}}} & \text{if} \quad w_{\mathrm{id}} < 0 \\ \dfrac{w_{\mathrm{id}}}{w} & \text{if} \quad w_{\mathrm{id}} > 0 \end{cases}. \tag{13.1}$$

Here, q is heat added ***to*** the system, w is the work done ***on*** the system and subscript id refers to the ideal (reversible) limit. The process can either be work-producing (w and $w_{id}<0$) or work-consuming (w and $w_{id}>0$). The first-law efficiency gives a useful estimate of the ability of a process to produce work and can be computed from the energy balance only. In some cases, however, it gives a misleading picture of energy efficiency. Consider, for instance, a perfect Carnot heat engine, which uses the temperature difference between a hot thermal reservoir at temperature, T_h and a cold reservoir at temperature T_c to produce work. By substituting w and q from the Carnot cycle, one obtains, $\eta_I = 1 - T_c/T_h$. In the case where $T_h \to T_c$, the first-law efficiency goes to zero and the formula does hence not make sense. The second-law efficiency is related to the thermodynamic limit of a process, and gives a picture of how close to an ideal performance a given technology is. Since the Carnot engine is a reversible engine, its second-law efficiency is unity for all values of T_h and T_c. For real engines, the efficiency is less than one and measures the distance away from this ideal limit.

By energy efficiency we will in this chapter use the second-law efficiency. The second law of thermodynamics says that all irreversible processes produce entropy. All real processes are irreversible, and the entropy production is an indicator of how much of the work put into or extracted from the process dissipates to heat through irreversibilities. The difference between the real and the ideal work is called the lost work, and is related to the total entropy production for a process, $(\mathrm{d}S/\mathrm{d}t)_{\mathrm{irr}}$, through the Gouy–Stodola theorem:

$$w_{\mathrm{lost}} = w - w_{\mathrm{id}} = T_0\left(\frac{\mathrm{d}S}{\mathrm{d}t}\right)_{\mathrm{irr}}. \tag{13.2}$$

Here, T_0 is the temperature of the environment and w_{lost} is the lost work, which is always positive according to the second law. The second-law efficiency then becomes:

$$\eta_{\mathrm{II}} = \begin{cases} \dfrac{w_{\mathrm{id}} + T_0(\mathrm{d}S/\mathrm{d}t)_{\mathrm{irr}}}{w_{\mathrm{id}}} & \text{if} \quad w_{\mathrm{id}} < 0 \\[2ex] \dfrac{w_{\mathrm{id}}}{w_{\mathrm{id}} + T_0(\mathrm{d}S/\mathrm{d}t)_{\mathrm{irr}}} & \text{if} \quad w_{\mathrm{id}} > 0 \end{cases}. \tag{13.3}$$

We observe from the last equation that when the ideal work is fixed, minimization of the entropy production is equivalent to maximization of the second-law efficiency. For many process units, it is more practical to use other constraints. In a chemical reactor, or in a distillation column, the amount and quality of the produced quantity are relevant constraints. This means that the equivalence between minimization of the entropy production and maximization of the second-law efficiency is lost.[8,20] Eqn (13.2), however, reveals that the total entropy production is the true source of irreversibilities, and is therefore the natural choice as objective function for minimization and measure of energy efficiency.

13.3 The Entropy Production in a Simple Expansion Process

A second-law analysis offers a systematic way to evaluate and compare technologies. At the center of such an analysis is the total entropy production, which can have contributions from heat and mass transfer, flow, chemical reactions and other transport phenomena. To obtain some intuition of what entropy production is, and why it should be minimized, we will first visit the well-known example of isothermal expansion of an ideal gas (Figure 13.1).

A container is filled with n moles of an ideal gas with pressure, $P(t)$, temperature, T_0, and volume, $V(t)$, which depends on the time, t. The container is equipped with a piston and the gas expands isothermally against an external pressure, P_{ext}. Heat, dq, is added to the gas and the work, dw, is done one the gas during a time interval, dt. In standard text-books on thermodynamics, this process is reversible. In that case, the external pressure equals the pressure inside the container at all times, and the process occurs infinitely slowly. The work produced by the gas is then given by minus the area under the solid lines in Figure 13.2. This is the ideal work, *i.e.*, the maximum work this process can produce with the given conditions (the reversible process). Since all real processes occur in a finite duration of time, they produce less work than w_{id}. We assume that the external pressure can be used to *control* the movement of the piston, and that the volume and pressure of the chamber follow the differential equations:

$$\frac{\mathrm{d}V}{\mathrm{d}t} = -\frac{f}{P^2}(P_{\mathrm{ext}} - P) \Leftrightarrow \frac{\mathrm{d}P}{\mathrm{d}t} = \frac{f}{nRT_0}(P_{\mathrm{ext}} - P). \tag{13.4}$$

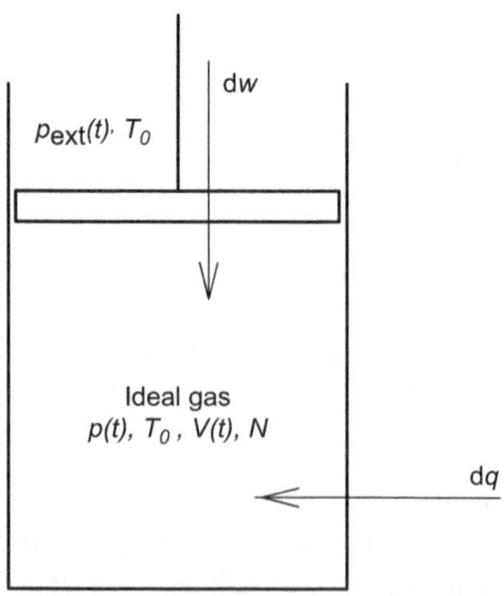

Figure 13.1 Heat is supplied to an ideal gas to perform work through a piston in an isothermal expansion process.

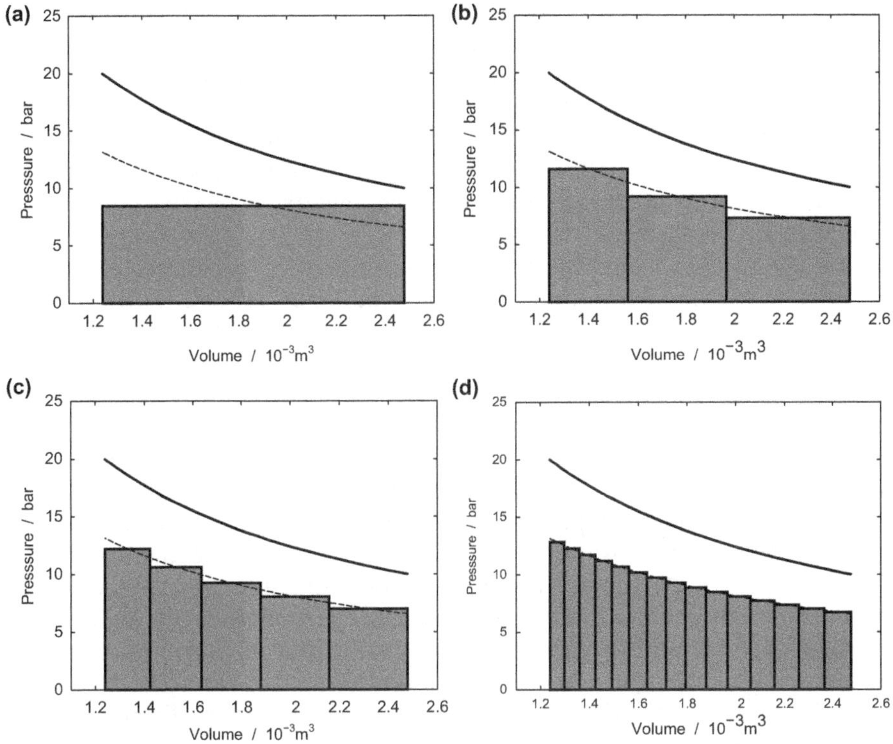

Figure 13.2 Optimal external pressure as function of volume. The grey rectangles represent the step-wise processes which give the smallest entropy production with 1 (a), 3 (b), 5 (c) and 15 (d) steps. The reversible process is given by the solid lines and the pressure which gives minimum entropy production by the dashed lines. Along the *y*-axis, 10 bar = 1 MPa.

Here, f is a constant which describes the friction between the piston and the container walls, n is the number of moles of gas and R is the universal gas constant. Consider first a step-wise process where the gas expands in K steps, with the external pressure being constant in each step, but changing between the steps. The work produced with the *optimal choice of external pressure* and *expansion time* for each step, is given by minus the shaded areas in Figures 13.2 a to d for 1, 3, 5, and 15 steps, respectively. The ideal work is equal to minus the area below the solid lines. The lost work equals the area between the shaded area and the solid line. Since we know the differential equation describing the motion of the piston, we can find the total entropy production.[20]

For the step-wise process, the total entropy production times T_0, is equal to the lost work in the process, confirming the Gouy–Stodola theorem (eqn (13.2)). Details can be found in ref. 20. In this example, the entropy production is the source of the lost work, arising from friction between the piston and the container wall. We are not restricted to step-wise processes in general, and the interesting question is how we can control the external pressure to produce as little entropy as possible during a predefined time interval. This problem can be formulated as an optimal-control problem and solved analytically for the process to give *its state of minimum entropy production*.[20] The resulting pressure *versus* volume curves (dashed lines in Figure 13.2) are always below the reversible limit (solid lines). Moreover, the optimal configuration in this process is characterized by *constant* entropy production.

Reversible processes are hypothetical limiting processes, not achievable in practice. The state of minimum entropy production represents therefore a more *practical limit* for the energy efficiency, since constraints in time and/or space are taken into account. By comparing Figures 13.2 a to d, we see that as the number of steps increases, the optimal configuration becomes more and more similar to the state of minimum entropy production (dashed lines). In the limit of infinitely many steps, the optimal choice of external pressures equals the solution obtained from optimal control theory.

For more complicated processes than this example, concepts such as work, ideal work and lost work are less intuitive. Consider, for instance, a chemical reactor. The chemical reactor either consumes work or produces work, depending on whether heat is added (endothermic reactor) or removed (exothermic reactor). It can be shown that to minimize the work required, or to maximize the work produced in the chemical reactor is equivalent to minimization of the entropy production with fixed inlet and outlet streams.[20] The main purpose of chemical reactors is, of course, to produce chemicals. Nevertheless, when less entropy is produced in the reactor, more work is available for other parts of the process, and the process plant can be operated with higher energy efficiency.

13.4 Identification of the State of Minimum Entropy Production

For most processes, the state of minimum entropy production cannot be found in analytical terms; it must be found numerically. In the chemical process industry, chemical reactors, heat exchangers and distillation columns are central process units.[1,6,17,18,20,27,28,30,32–35] We give an overview of the insight gained from entropy-production-minimization studies of these units. The chemical reactor, with its highly non-linear physical processes will be our main example. Experience from distillation columns and heat exchangers will also be discussed.

The state with minimum entropy production can be determined following the systematic procedure below:

1. Formulate the balance equations for mass, energy and momentum of the process.
2. Derive a consistent local entropy production based on the equations from Step 1 and find proper flux equations from the entropy production.

3. Formulate the optimal control problem and the boundary conditions.
4. Find the state of minimum entropy production numerically by:
   a. Creating a good initial guess, *e.g.*, by a numerical optimization.
   b. Solving the boundary value problem stated in Step 3.

The procedure can be used for any process unit, but we will use the chemical reactor as example, which can be described by a system of ordinary differential equations. The example is then restricted to one dimension which typically is either the spatial axis or the time axis.

13.4.1 Step 1: Conservation Equations of the Process Unit

Chemical reactors often have a tubular design similar to the sketch in Figure 13.3, where a mixture of, for instance, gases flows in the positive z-direction through a bed of catalyst pellets. The gases typically react at the surface of the catalyst and form the desired product. The diameter of the tube is D, the length is L and ξ_j is the degree of reaction j. Heat is transferred from the outside of the chemical reactor at a temperature, $T_a(z)$, where subscript a means ambient. Due to strong turbulence in the catalyst bed, the mixture is well mixed perpendicular to the flow direction. A one-dimensional plug-flow model is then a sufficient description. The mole balances, (eqn (13.5)), are linearly dependent and can be redefined in terms of the degree of reaction (eqn (13.6)). A mathematical model of the process unit should be given in terms of balance equations for moles/mass, energy and momentum, which at steady-state are:

$$\frac{dF_i}{dz} = \Omega\rho_B \sum_{j=1}^{N_r} \nu_{j,i} r_j \quad i = 1, \ldots, N_c, \tag{13.5}$$

$$\frac{d\xi_j}{dz} = \frac{\Omega\rho_B}{F_A^0} r_j \quad j = 1, \ldots, N_r, \tag{13.6}$$

$$\frac{dT}{dz} = \frac{\pi D J_q + \Omega\rho_B \sum_j^{N_r} [r_j(-\Delta_r H_j)]}{\sum_i^{N_c} [F_i C_{p,i}]}, \tag{13.7}$$

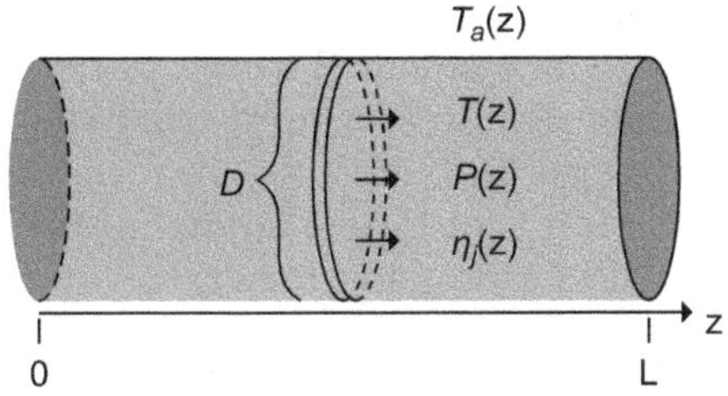

Figure 13.3 Illustration of the chemical reactor.

$$\frac{\mathrm{d}P}{\mathrm{d}z} = -\left(\frac{150\mu}{D_\kappa^2}\frac{(1-\epsilon)^2}{\epsilon^3} + \frac{1.75\rho^0 v^0}{D_\kappa}\frac{1-\epsilon}{\epsilon^3}\right)v, \tag{13.8}$$

where the degree of reaction is defined as:

$$\xi_j = \frac{\text{moles of component A consumed by reaction } j}{\text{moles of component A at the inlet}}. \tag{13.9}$$

Here, F_i is the molar flow rate of component i, Ω is the cross-sectional area of the chemical reactor, ρ_B is the catalyst density, r_j is the rate of reaction j, and $\nu_{j,i}$ is the stoichiometric coefficient of component i in reaction j. N_r is the number of reactions, N_c the number of components, and superscript 0 refers to the inlet. Furthermore, J_q is the heat flux through the reactor wall, $\Delta_r H_j$ is the enthalpy of reaction j and $C_{p,i}$ is the heat capacity of component i. In eqn (13.8), μ is the gas viscosity, ρ^0 is the gas density at the inlet, v^0 is the gas velocity at the inlet, v is the gas velocity, ϵ is the porosity and D_κ the pellet diameter. Component A is a reference component which participates in all the reactions. A more detailed explanation of these equations can be found in ref. 20.

13.4.2 Step 2: Deriving the Local Entropy Production

De Groot and Mazur derived a general expression for the local entropy production.[36] Models are, however, often averaged in space and time, or formulated with a different set of variables, which means that the expression for the local entropy production changes with the model. We show how to derive the local entropy production for the example above using an approach which is also applicable to other cases. The starting point for the derivation is the total entropy balance at steady-state:

$$\begin{aligned}\left(\frac{\mathrm{d}S}{\mathrm{d}t}\right)_{\text{irr}} &= \left(\sum_{i=1}^{N_c} F_i S_i\right)_{\text{out}} - \left(\sum_{i=1}^{N_c} F_i S_i\right)_{\text{in}} - \int_0^L \pi D \frac{J_q}{T_a}\mathrm{d}z \\ &= \int_0^L \left(\frac{\mathrm{d}\sum_{i=1}^{N_c} F_i S_i}{\mathrm{d}z} - \pi D\frac{J_q}{T_a}\right)\mathrm{d}z = \int_0^L \sigma \mathrm{d}z.\end{aligned} \tag{13.10}$$

Here, S_i is the partial molar entropy of component i. The left-hand side is the total entropy production, which equals the entropy flow out (Term 1) minus the entropy flow in (Term 2) minus a term from the entropy transferred across the reactor wall (Term 3). From this, we recognize the local entropy production, σ, inside the integral. Next, the total differential of the entropy flow can be used to find the expression for σ:

$$\sigma = \sum_{i=1}^{N_c}\left[F_i\left(\frac{\partial S_i}{\partial T}\right)_{P,F_i}\frac{\mathrm{d}T}{\mathrm{d}z} + F_i\left(\frac{\partial S_i}{\partial P}\right)_{T,F_i}\frac{\mathrm{d}P}{\mathrm{d}z} + \left(\frac{\partial F_i S_i}{\partial F_i}\right)_{T,P,F_j\neq i}\frac{\mathrm{d}F_i}{\mathrm{d}z}\right] - \pi D\frac{J_q}{T_a}, \tag{13.11}$$

where $S = \sum_{i=1}^{N_c} F_i S_i$ is the total entropy. We assume ideal-gas law:

$$S_i = S_{i,0} - R \ln\left(\frac{P}{P_0} x_i\right). \tag{13.12}$$

Here, P_0 is the reference pressure and $S_{i,0}$ the reference partial molar entropy of component i. This gives the following partial derivatives of the total entropy with respect to temperature, pressure and molar flow rates:

$$\begin{aligned}
\sum_{i=1}^{N_c} F_i \left(\frac{\partial S_i}{\partial T}\right)_{P,F_i} &= \frac{1}{T}\sum_{i=1}^{N_c} [F_i C_{p,i}], \\
\sum_{i=1}^{N_c} F_i \left(\frac{\partial S_i}{\partial P}\right)_{T,F_i} &= -\frac{R}{P}\sum_{i=1}^{N_c} F_i = -\frac{\Omega v}{T}, \\
\left(\frac{\partial \sum_i F_i S_i}{\partial F_i}\right)_{T,P,F_{j\neq i}} &= S_{i,0} - R \ln \frac{P x_i}{P_0}.
\end{aligned} \tag{13.13}$$

By using eqn (13.5), (13.7), (13.8), and (13.13) in eqn (13.11), we obtain after rearrangements and simplifications (more details can be found in ref. 20):

$$\sigma = \Omega \rho_B \sum_{j=1}^{N_r} \left[r_j \left(-\frac{\Delta_r G_j}{T} \right) \right] + \pi D J_q \Delta \frac{1}{T} + \Omega v \left(-\frac{1}{T}\frac{dP}{dz} \right). \tag{13.14}$$

Here, $\Delta_r G_j$ is the Gibbs energy of reaction j and $\Delta\ 1/T = 1/T - 1/T_a$ is the inverse temperature difference across the reactor wall. The first term on the right-hand side is the entropy produced by reactions, the second term is called the thermal entropy production and the third term is entropy produced by pressure gradients or viscous flow.

13.4.3 Step 3: The Optimal Control Theory Formulation

The previous two subsections contain the necessary information to formulate the state of minimum entropy production as an optimal-control problem following Pontryagin *et al.*[4,37] The advantage of optimal control theory compared to a purely numerical optimization,[33,38] is that optimal control theory gives the necessary conditions for a minimum. This means that a minimum can be found robustly and definitely with higher accuracy. We follow standard notation in optimal control theory,[37] and define a set of state variables, $y(z)$. These are governed by differential equations (eqn (13.6) to (13.8)), which restrict the trajectories over which the system can evolve. Furthermore, we have one control variable, namely the temperature outside the chemical reactor at z, $T_a(z)$. We now want to minimize the total entropy produced in the chemical reactor (eqn (13.10)), with the functional argument given by eqn (13.14). Common in optimal control theory is to introduce a Hamiltonian:

$$H = \sigma + \lambda_T \frac{\mathrm{d}T}{\mathrm{d}z} + \lambda_P \frac{\mathrm{d}P}{\mathrm{d}z} + \sum_{j=1}^{N_r} \lambda_{\xi_j} \frac{\mathrm{d}\xi_j}{\mathrm{d}z}. \quad (13.15)$$

Here, $\lambda(z)$, are Lagrange multipliers. The necessary conditions for a minimum are then given by Pontryagins' minimum principle:[4,37]

$$\begin{aligned}
&\frac{\mathrm{d}T}{\mathrm{d}z} = \frac{\mathrm{d}H}{\mathrm{d}\lambda_T} && \frac{\mathrm{d}\lambda_T}{\mathrm{d}z} = -\frac{\mathrm{d}H}{\mathrm{d}T} \\
&\frac{\mathrm{d}\xi_j}{\mathrm{d}z} = \frac{\mathrm{d}H}{\mathrm{d}\lambda_{\xi_j}} && \frac{\mathrm{d}\lambda_{\xi_j}}{\mathrm{d}z} = -\frac{\mathrm{d}H}{\mathrm{d}\xi_j} \qquad j = 1, \ldots, N_r. \quad (13.16) \\
&\frac{\mathrm{d}P}{\mathrm{d}z} = \frac{\mathrm{d}H}{\mathrm{d}\lambda_P} && \frac{\mathrm{d}\lambda_P}{\mathrm{d}z} = -\frac{\mathrm{d}H}{\mathrm{d}P}
\end{aligned}$$

In addition, when there are no constraints on the control variables, the derivative of the Hamiltonian with respect to the control variables should be zero:

$$\frac{\mathrm{d}H}{\mathrm{d}T_a} = 0 \quad \text{for all } z \in [0, L]. \quad (13.17)$$

This algebraic restriction provides an expression for the optimal temperature outside the chemical-reactor wall as function of the state variables and the multiplier functions. If the Hamiltonian does not depend explicitly on z, it is autonomous. In such cases the Hamiltonian is constant along z. For an autonomous system, the Hamiltonian is zero at the optimal reactor length. If state variables are free at the end points of the reactor, and the reactor length is fixed, it means that the multiplier functions at the end of the reactor must be zero. The boundary conditions for the optimal control problem are summarized in Table 13.1.

13.4.4 Step 4: Numerical Solution of the Problem

The optimal control theory formulation for the chemical reactor is difficult to solve numerically, since it forms a coupled system of non-linear differential equations where boundary conditions are given at both ends. It has, however, successfully been solved, first by Johannessen and Kjelstrup.[27]

Table 13.1 Possible boundary conditions for the optimal control problem.

Description	$H(z)$	$x(z)$ at ends	$\lambda(z)$
Fixed reactor length, fixed end-state	Constant	Fixed	(—)
Fixed reactor length, free end-state	Constant	(—)	0
Free reactor length, fixed end-state	0	Fixed	(—)

The algorithm they used has subsequently been applied to many systems. In these works, the boundary-value problem was preceded by a numerical optimization by sequential quadratic programming on a relatively small grid (40 non-uniform grid points) to provide an initial guess for the boundary-value problem solver.[20,33] The boundary-value problem was then solved with a collocation method. The computational time of the cases solved in literature is today in the order of seconds/minutes.[1,17,18,27,28]

13.5 Key Results from the Literature

13.5.1 Equipartition of the Entropy Production and Forces

Johannessen and Kjelstrup formulated the *hypothesis for the state of minimum entropy production* in 2005.[17] The hypothesis was put in a wider context later.[18] It reads:

> *Equipartition of entropy production, but also equipartition of forces, are good approximations to the state of minimum entropy production in the parts of an optimally controlled system that have sufficient freedom to equilibrate internally.*

When all driving forces are controlled, equipartition of entropy production can be proven mathematically.[18,39] The hypothesis has, however, been found to apply also when the number of driving forces is larger than the number of controlled variables. The hypothesis has also been found to apply to processes which are intrinsically non-linear, *e.g.*, chemical reactions and radiative heat transfer.[1] It is only possible to prove equipartition of entropy production (EoEP) in a strict mathematical sense for linear processes,[18,39] as we will see below, or for special cases of non-linear processes.[7] The strict conditions for the mathematical proof are therefore rather limiting for application of the hypothesis, as many transport phenomena are non-linear, with chemical reactions and radiative heat transfer as important examples. A final answer to the surprising fact that EoEP is observed in large parts of chemical reactors being far from equilibrium has thus not been found. For this reason, we recapitulate the proof. To show when EoEP and EoF hold strictly mathematically, we first reformulate the optimal control problem for the chemical reactor with one reaction in terms of vectors and matrices denoted by bold letters/symbols. We have the state variable vector, $\mathbf{y}=[T,P,\xi_1]^T$, the control vector, $\mathbf{u}=[T_a]$, the fluxes, $\mathbf{J}$ and the forces, $\mathbf{x}$:

$$\mathbf{J}(\mathbf{y},\mathbf{u},\mathbf{x}(\mathbf{y},\mathbf{u}))=\left[J_q,\nu,r_1\right]^T \quad \text{and} \quad \mathbf{x}(\mathbf{y},\mathbf{u})=\left[\left(\frac{1}{T}-\frac{1}{T_a}\right),\left(-\frac{1}{T}\frac{\mathrm{d}P}{\mathrm{d}z}\right),\left(\frac{-\Delta_\mathrm{r}G_1}{T}\right)\right]^T. \tag{13.18}$$

The balance equations can now be written in a compact form:

$$\frac{\mathrm{d}\mathbf{y}}{\mathrm{d}z}=\mathbf{A}(\mathbf{y})\mathbf{\Gamma}\mathbf{J}(\mathbf{y},\mathbf{u},\mathbf{x}(\mathbf{y},\mathbf{u})), \tag{13.19}$$

where **A** and **Γ** for the chemical reactor are the following coefficient matrices:

$$\mathbf{A}=\begin{pmatrix} \dfrac{1}{\sum_{i=1}^{N_c}[F_i C_{p,i}]} & 0 & \dfrac{-\Delta_r H_1}{\sum_{i=1}^{N_c}[F_i C_{p,i}]} \\ 0 & -\dfrac{f}{\Omega} & 0 \\ 0 & 0 & \dfrac{1}{F_A^0} \end{pmatrix} \qquad \mathbf{\Gamma}=\begin{pmatrix} \pi D & 0 & 0 \\ 0 & \Omega & 0 \\ 0 & 0 & \Omega\rho_B \end{pmatrix}. \tag{13.20}$$

The Hamiltonian is then:

$$H(\mathbf{y},\,\mathbf{u},\,\lambda)=\mathbf{x}(\mathbf{y},\,\mathbf{u})^{\mathrm{T}}\mathbf{\Gamma}\mathbf{J}(\mathbf{y},\,\mathbf{u},\,\mathbf{x}(\mathbf{y},\,\mathbf{u}))\;+\boldsymbol{\lambda}^{\mathrm{T}}\mathbf{A}(\mathbf{y})\mathbf{\Gamma}\mathbf{J}(\mathbf{y},\,\mathbf{u},\,\mathbf{x}(\mathbf{y},\,\mathbf{u})). \tag{13.21}$$

Here, the first term corresponds to the local entropy production, which is a product of fluxes and forces according to classical non-equilibrium thermodynamics. Many models fit into the general formulation above, for instance heat exchangers, distillation columns and membrane processes. If we assume that we can control all the forces independently, the necessary conditions for a minimum (eqn (13.16)) become:

$$\left(\frac{\partial H}{\partial \mathbf{x}}\right)_{\mathbf{y},\boldsymbol{\lambda}}=(\mathbf{\Gamma}\mathbf{J}(\mathbf{y},\mathbf{x}))^{T}+(\mathbf{x}^{\mathrm{T}}+\boldsymbol{\lambda}^{\mathrm{T}}\mathbf{A}(\mathbf{y}))\left(\frac{\partial(\mathbf{\Gamma}\mathbf{J}(\mathbf{y},\mathbf{x}))}{\partial \mathbf{x}}\right)=0. \tag{13.22}$$

Solving eqn (13.22) for $\boldsymbol{\lambda}^{\mathrm{T}}\mathbf{A}$ and introducing this into the Hamiltonian shows, for an autonomous problem, that the following quantity should be constant through the process:

$$H_s=-(\mathbf{\Gamma}\mathbf{J}(\mathbf{y},\mathbf{x}))^{T}\left(\frac{\partial(\mathbf{\Gamma}\mathbf{J}(\mathbf{y},\mathbf{x}))}{\partial \mathbf{x}}\right)^{-1}(\mathbf{\Gamma}\mathbf{J}(\mathbf{y},\mathbf{x})). \tag{13.23}$$

Where H_s is called the Spirkl–Ries quantity[39] because Spirkl and Ries first proved the above result. It has no simple meaning unless the force–flux relations are linear. Then, $\mathbf{J}(\mathbf{x},\mathbf{y})=\mathbf{L}(\mathbf{y})\mathbf{x}$, and the derivative in the Spirkl–Ries constant is $\mathbf{\Gamma}\mathbf{L}(\mathbf{y})$. We can then show that the Spirkl–Ries quantity reduces to the local entropy production, and EoEP is exactly valid. Next, if the conductivity matrix **L**, and the coefficient matrix, **A**, do not depend on **y**, *i.e.*, are constant, equipartition of entropy production is the same as equipartition of forces. It is evident from the derivations above that the mathematical conditions for EoEP are strict, and even stricter for EoF. But the hypothesis by Johannessen and Kjelstrup applies to far more general cases. Wilhelmsen *et al.*[1] studied a chemical reactor with highly non-linear reactions and a radiative heat flux through the reactor-wall, finding EoEP and EoF as good approximations to the state of minimum entropy production (Figure 13.4).

Figure 13.4 shows how the entropy production varies through the steam reforming reactor for production of hydrogen, in cases where the heat flux has contributions from convection and radiation. The entropy production of the reference case (solid line), where the entropy production is not minimum, is less constant inside the reactor than cases which correspond to

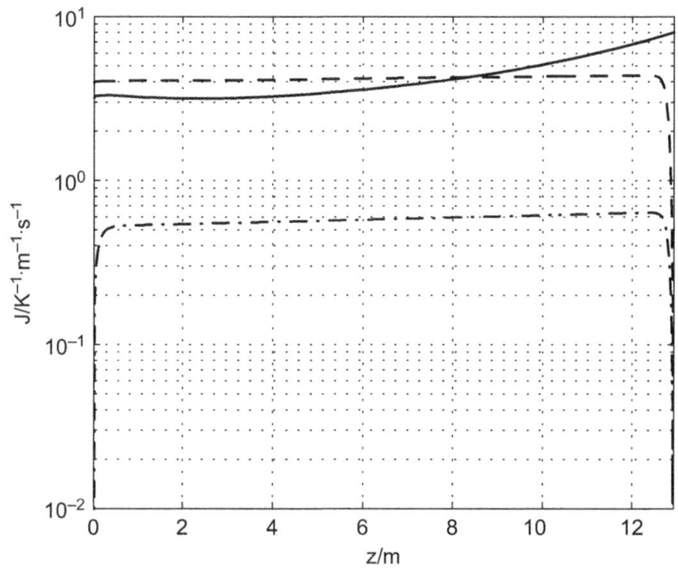

Figure 13.4 The local entropy production for the reference case (solid line), the optimal cases with fixed T_0 (dashed line) and free T_0 (dash-dot line). Along the *x*-axis, the label is z/m and along the *y*-axis, the label is J/K$^{-1}\cdot$m$\cdot$s^{-1}.

minima in the total entropy production, except near the end-points. The dashed and dash-dotted lines represent cases with different boundary conditions, but they all follow EoEP well. Clearly EoEP is not exactly obeyed, but it gives a good approximation to the state of minimum entropy production, confirming the hypothesis.

13.5.2 Highways in State Space

"State space" means the space spanned by state variables such as the temperature, pressure and degrees of conversion. For the chemical reactor, the relation between the degree of conversion, ξ, and the temperature in the process unit at minimum entropy production has been seen to follow a band in state space. As the solution to the optimization problem is obtained with very different boundary conditions, one cannot expect that all results lie on a line. It is thus surprising that the results are located at a band of very limited extension in state space. This band was called a highway in state space,[17,18] resembling a path for efficient driving of cars (efficient process operation). One such highway is illustrated by the thick bottom line in Figure 13.5. In addition, the solutions with minimum entropy production are crowding in on the highway in state space, just as cars crowd in on a real highway. The highway represents the most energy-efficient way to travel a long distance in state space. Numerical evidence for "highways" in state space have been

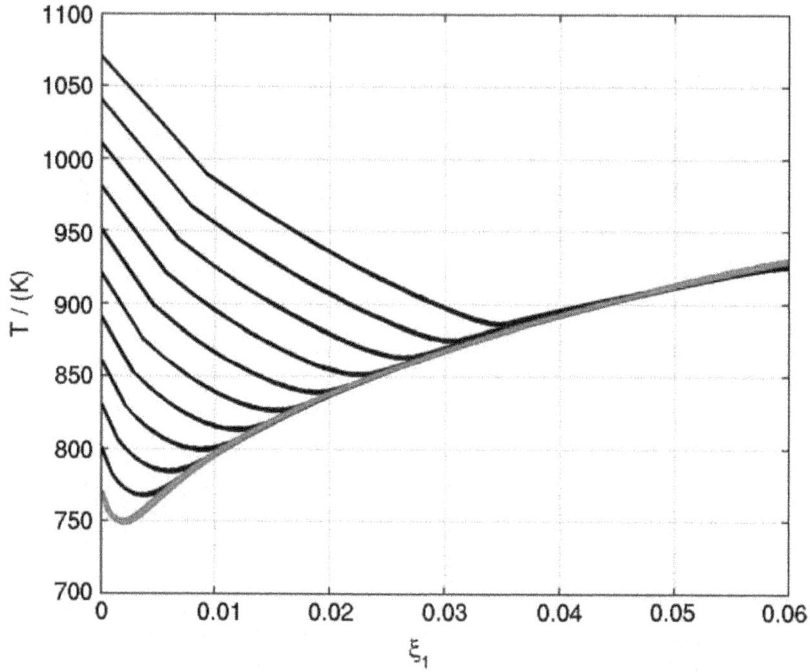

Figure 13.5 The optimal temperature profiles as functions of degree of reaction for several inlet conditions. The highway is exemplified by the thick bottom line.

found for endothermic and exothermic chemical reactors and diabatic distillation columns with minimum entropy production.[20] Knowledge about highways is useful for energy-efficient operation, as the highways show how to run processes energy efficiently, even with changing inlet conditions.

How much of a solution that is covered by the highway, depends on the distance of the starting point from the highway, as illustrated by the lines crowding in on the highway in Figure 13.5. The existence of the highway is non-trivial, and it has some peculiar properties. Along the highway, the entropy production is approximately constant for process intensities *which are not too high*. This means that EoEP is a good approximation to the state of minimum entropy production along the highway.

By investigating the origin of contributions to the entropy production, more insight can be gained. For the solid lines, which crowd into the highway from left to right in Figure 13.5, the local entropy production is dominated by contributions from chemical reactions (see below eqn (13.14) for explanation), and the chemical reactor can be said to operate in a *reaction mode*. Along the highway, however, the thermal contributions dominate the local entropy production, and the chemical reactor can be said to be in a *heat exchange mode*. As discussed in previous work,[19] the highway represents the optimal solution for over 90 % of the reactor length for all the cases shown in Figure 13.5, even if the extent of the highway is rather limited in state space.

Not all systems have highways. The discussion so far has been restricted to one-dimensional plug-flow models. Whether this concept also applies to more dimensions, remains to be seen. For some systems, such as heat exchangers, no highways in state space have been found. In chemical reactors, the highway seems to be associated with a trade-off between thermal and reactive contributions to the local entropy production. This is confirmed by the ratios between the thermodynamic fluxes along the highway as discussed by Wilhelmsen.[1] The systems where highways have been observed have in common that local equilibrium compositions exist. The equilibrium compositions relate to a chemical reaction, such as in the chemical reactors, or to phase equilibria, such as in the distillation column. The main contributions to irreversible phenomena in heat exchangers are thermal and viscous entropy production,[30] where the viscous entropy production is much fixed by the flow rate. The origin of the highway is still unaccounted for. Moreover, is it possible to formulate the highway in mathematical terms? Degenerate classes of optimal control theory solutions, called turnpikes, have similar properties as the highway and may help elucidate the concept.[40]

13.5.3 Rules of Thumb for Energy-efficient Process Design

The discussions in the previous sections, about EoF, EoEP and highways in state space, indicate that solutions with minimum entropy productions have some common properties. These properties can now be exploited to give guidance to design of process equipment which leads to energy-efficient operation. We have seen that the contribution to the entropy production from heat transfer is often the largest source of dissipation, both in chemical reactors, heat exchangers and distillation columns.[1,17,22,30,32,34,41] The first step in a strategy to increase the energy efficiency in these systems should thus be to make the heat transfer as efficient as possible. This can be linked to the statement made by Leites *et al.*[42] in their first commandment: *The driving force of a process must approach zero at all points in a reactor, at all times*. A thermal driving force can be made small by increasing the heat transfer coefficients or the surface area. The interesting question beyond that becomes: what can be done, once the heat transfer has been made as efficient as possible?

In all optimal reactor solutions presented in the literature,[1,17–19,27,28] the optimal solutions enter first a reaction mode at the inlet, before it proceeds into a heat transfer mode of operation in the central part (See Figure 13.5).

It follows for single tubular reactors of length L, that a (close to) adiabatic inlet section, L_1, is an advantage for the total entropy production. Furthermore, the next part, L_2, can best be characterized by equipartition of the entropy production, in some cases also by equipartition of the forces. In other words, finding the optimal solution for a system, translates into a

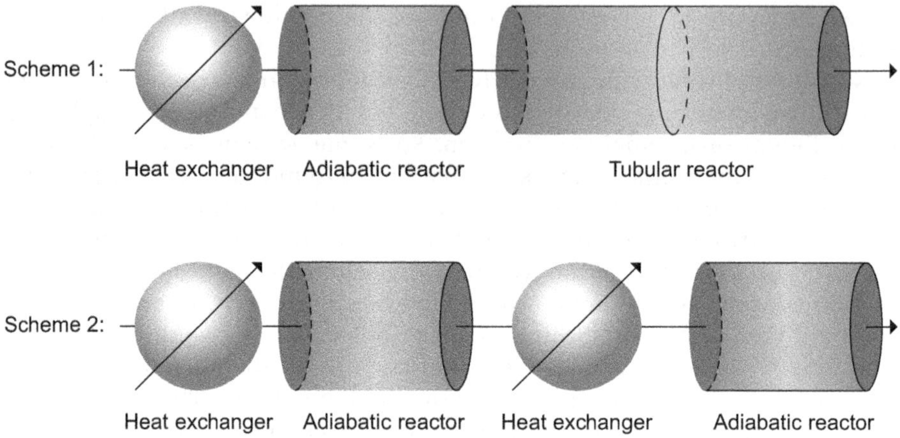

Figure 13.6 Process configurations for energy-efficient reactor design.

procedure where one considers a scheme with separate units, like that illustrated in Figure 13.6. The reactor part of the system consists of two subunits, an adiabatic pre-reactor and a tubular reactor with heat transfer. To complete the system analysis, a heat exchanger is added in front of the adiabatic reactor as in Scheme 1. This system can now be used to account for the trade-off between the contributions to the entropy production, including also the contribution to the entropy production from heat exchange upfront of the reactor system.

The purpose of the heat exchanger (the first item in Scheme 1) is to bring the reacting mixture to the optimal initial temperature. The purpose of the adiabatic reactor, the next unit, is to operate the chemical reactor in reaction mode. Whether it pays, in terms of entropy production, to use Scheme 1 with the reactor in the heat-transfer mode of operation, or to transfer to more discrete units, as illustrated in Scheme 2 in Figure 13.6, depends on the relative values of the heat transfer coefficients. When the heat-transfer coefficients across the reactor tube wall are very low, it is better to use dedicated heat exchangers for heat transfer. It will then be beneficial to split the operation in heat-transfer mode (taken care of by the tubular reactor in Scheme 1) by separate sets of one or more adiabatic reactor stages with interstage heating/cooling, as illustrated in Scheme 2 of Figure 13.6. Two or more heat exchanger–adiabatic reactor pairs may be cost effective and energy efficient as well. This shows that a complex optimal control problem can be reduced, if not avoided, and that the process of finding an energy-efficient reactor design can be simplified significantly. As an example, de Koeijer *et al.* found the second-law optimal path of a four-bed SO_2 converter with intercoolers.[33]

Rules of thumb for energy-efficient design should also be devised for other process equipment to make the accumulated insight from research on entropy production minimization easily accessible to the industry.

13.6 Conclusions and Future Challenges

We discussed the link between the entropy production of a process and the second-law efficiency. The Gouy–Stodola theorem, which contains the entropy production as the fundamental source of energy dissipation, was explained. Next, we presented an approach to systematically formulate and find the state of minimum entropy production with optimal control theory, using a tubular chemical reactor as example.

Important results and insight about states with minimum entropy production from the last decades show how energy-efficient operation and design in the process industry can be obtained.

- Equipartition of Forces (EoF) and Equipartition of Entropy Production (EoEP) cannot be proven in a strict mathematical sense, but are for most cases very good approximations to the state of minimum entropy production, even for highly non-linear processes.
- Some systems have states with minimum entropy production which follow a highway in state space. The entropy production is relatively constant along the highway. The highway has so far been observed only for systems where at least one state-variable has the possibility to equilibrate.
- Entropy production minimization can give rules of thumb for energy-efficient design for chemical reactor systems.

Important future work within the field of *entropy production minimization* will be to further clarify why and when EoF and EoEP are good approximations to this state of operation. The reason why highways emerge should be found. Highly non-linear examples are needed for these investigations. Moreover, the examples should be followed by a mathematical analysis to gain further insight into the theorems. Rules of thumb should be developed for systems other than chemical reactors, such as absorption and membrane processes. Properties of states with minimum entropy production have only been discussed for one-dimensional optimal control problems. Mathematical tools exist that can handle minima constrained also by partial differential equations. More complex systems, beyond the one-dimensional, should be investigated. Such systems may have different properties than we have seen so far for one-dimensional averaged systems.

References

1. Ø. Wilhelmsen, E. Johannessen and S. Kjelstrup, *Int. J. Hydrogen Energy*, 2010, **35**, 13219.
2. M. Margarone, S. Magi, G. Gorla, S. Biffi, P. Siboni, G. Valenti, M. Romano, A. Giuffrida, E. Negri and E. Macchi, *J. Energy Resour. Technol.*, 2011, **133**, 012001.

3. S. M. Svalheim and D. C. King, *Society of Petroleum Engineers*, 2003, *Paper SPE 83993*.
4. L. Pontryagin, V. Boltyanskii, R. Gamkrelidze and E. Mishchenko, *The Mathematical Theory of Optimal Processes*, Pergamon Press, Oxford, 1964.
5. A. Bejan, *Entropy Generation Through Heat and Fluid Flow*, Wiley, Hoboken, 1994.
6. G. de Koeijer, S. Kjelstrup, P. Salamon, G. Siragusa, M. Schaller and K. H. Hoffmann, *Ind. Eng. Chem. Res.*, 2002, **41**, 5826.
7. B. Andresen and J. M. Gordon, *Phys. Rev. E: Stat. Phys., Plasmas, Fluids, Relat. Interdiscip. Top.*, 1994, **50**, 4346.
8. P. Salamon, K. H. Hoffmann, S. Schubert, R. S. Berry and B. Andresen, *J. Non-Equilib. Thermodyn.*, 2001, **26**, 73.
9. S. Kjelstrup, D. Bedeaux, E. Johannessen and J. Gross, *Non-Equilibrium Thermodynamics for Engineers*, World Scientific, Singapore, 2010.
10. S. Kjelstrup, A. Røsjorde and E. Johannessen, in *Chemical Thermodynamics for Industry*, ed. T. Letcher, The Royal Society of Chemistry, Cambridge, 2004, pp. 1–11.
11. E. Sauar, R. Rivero, S. Kjelstrup and K. Lien, *Energy Convers. Manage.*, 1997, **38**, 1777.
12. E. Sauar, S. Kjelstrup and K. Lien, *Comput. Chem. Eng.*, 1997, **21**, 29.
13. E. Sauar, S. Kjelstrup and K. Lien, *Ind. Eng. Chem. Res.*, 1996, **35**, 4147.
14. D. Bedeaux, F. Standaert, K. Hemmes and S. Kjelstrup, *J. Non-Equilib. Thermodyn.*, 1999, **24**, 242.
15. E. Johannessen and S. Kjelstrup, *J. Non-Equilib. Thermodyn.*, 2005, **30**, 129.
16. S. Gheorghiu, S. Kjelstrup, P. Pfeifer and M. O. Coppens, in *Fractals in Biology and Medicine; Mathematics and Biosciences in Interaction*, ed. H. Birkhauser, Springer, Berlin, 2005, pp. 31–42.
17. E. Johannessen and S. Kjelstrup, *Chem. Eng. Sci.*, 2005, **60**, 1491.
18. E. Johannessen and S. Kjelstup, *Chem. Eng. Sci.*, 2005, **60**, 3347.
19. Ø. Wilhelmsen, *The State of Minimum Entropy Production in Reactor Design*, The Norwegian University of Science and Technology, Trondheim, 2010.
20. E. Johannessen, *The state of minimum entropy production in an optimally controlled system*, PhD thesis, The Norwegian University of Science and Technology, Trondheim, 2004.
21. A. Røsjorde and S. Kjelstrup, *Chem. Eng. Sci.*, 2005, **60**, 1199.
22. G. de Koeijer and S. Kjelstrup, *Int. J. Appl. Thermodyn.*, 2000, **3**, 105.
23. S. Kjelstrup, E. Sauar, D. Bedeaux and H. van der Kooi, *Ind. Eng. Chem. Res.*, 1999, **38**, 3046.
24. S. Kjelstrup and T. V. Island, *Ind. Eng. Chem. Res.*, 1999, **38**, 3051.
25. S. Kjelstrup, E. Johannessen, A. Røsjorde, L. Nummedal and D. Bedeaux, *Int. J. Appl. Thermodyn.*, 2000, **3**, 147.
26. L. Nummedal, S. Kjelstrup and M. Costea, *Ind. Eng. Chem. Res.*, 2003, **42**, 1044.
27. E. Johannessen and S. Kjelstrup, *Energy*, 2004, **29**, 2403.

28. L. van der Ham, J. Gross, A. Verkooijen and S. Kjelstrup, *Ind. Eng. Chem. Res.*, 2009, **48**, 8500.
29. L. Nummedal and S. Kjelstrup, *Int. J. Heat Mass Transfer*, 2001, **44**, 2827.
30. E. Johannessen, L. Nummedal and S. Kjelstrup, *Int. J. Heat Mass Transfer*, 2002, **45**, 2649.
31. A. Zvolinschi, E. Johannessen and S. Kjelstrup, *Chem. Eng. Sci.*, 2006, **61**, 3653.
32. G. de Koeijer, A. Røsjorde and S. Kjelstrup, *Energy*, 2004, **29**, 2425.
33. G. de Koeijer, E. Johannessen and S. Kjelstrup, *Energy*, 2004, **29**, 525.
34. L. Nummedal, A. Røsjorde, E. Johannessen and S. Kjelstrup, *Chem. Eng. Proc.*, 2005, **44**, 429.
35. S. K. Ratkje, E. Sauar, E. M. Hansen, K. M. Lien and B. Hafskjold, *Ind. Eng. Chem. Res.*, 1995, **34**, 3001.
36. S. R. de Groot and P. Mazur, *Non-equilibrium Thermodynamics*, Dover, New York, 1984.
37. J. L. Troutman, *Variatonal Calculus and Optimal Control*, Springer, Berlin, 1996.
38. A. E. Bryson and Y. C. Ho, *Applied Optimal Control. Optimization, Estimation and Control*, Hemisphere Publishing Corporation, Washington, 1975.
39. W. Spirkl and H. Ries, *Phys. Rev. E: Stat. Phys., Plasmas, Fluids, Relat. Interdiscip. Top.*, 1995, **52**, 3485.
40. V. I. Gurman, *Automat. Remote Control*, 2003, **64**, 399.
41. L. van der Ham and S. Kjelstrup, *Ind. Eng. Chem. Res.*, 2011, **50**, 9324.
42. I. L. Leites, D. A. Sama and N. Lior, *Energy*, 2003, **28**, 55.

CHAPTER 14

Mesoscopic Non-equilibrium Thermodynamics

IGNACIO PAGONABARRAGA* AND J. MIGUEL RUBÍ

Department of Fundamental Physics, Faculty of Physics, University of Barcelona, C. Martí i Franqués 1, 08028-Barcelona, Spain
*Email: ipagonabarraga@ub.edu

14.1 Introduction

Systems in equilibrium strictly follow the laws of thermodynamics. Despite the disordered motion of large numbers of molecules, the systems can be characterized by a few variables accounting for average properties. Thermodynamics also applies to systems outside equilibrium, in the local equilibrium regime in which the volume elements are considered small thermodynamic systems in equilibrium, as described in Chapter 4. This hypothesis is basic in the classical formulation of non-equilibrium thermodynamics,[1] which is restricted to the linear domain of fluxes and thermodynamic forces.

In the mesoscale domain, the limitation to linear laws is removed and non-linear kinetic laws are obtained. In this chapter we discuss the case of an activated chemical reaction, with and without fluctuations, and the relaxation of dipole orientation in an electric field. The methodology is relevant for Chapters 15 and 16. Chapter 15 discusses the application to nucleation. Chapter 16 discusses the application to biological systems and derives the non-linear kinetic laws characteristic of biochemical processes for enzyme catalysis through the Michaelis–Menten mechanism, energy transduction in proteins and the stretching of DNA molecules. Mesososcopic

Experimental Thermodynamics Volume X: Non-equilibrium Thermodynamics with Applications
Edited by Dick Bedeaux, Signe Kjelstrup and Jan V. Sengers

Published by the Royal Society of Chemistry, www.rsc.org

non-equilibrium thermodynamics provides a general framework for the study of small-scale biological systems far from equilibrium.

An analysis based on the solution of the Boltzmann equation for reactive gases by means of a Chapman–Enskog expansion enables one to describe the transition towards non-equilibrium states.[2] The zeroth-order term in the expansion is the Gaussian equilibrium distribution. The first-order term is a function of the difference between the chemical potentials, the affinity, and temperatures of the initial and final states. It turns out that for large values of the affinity, when the reaction rate is given in terms of fugacity or the law of mass action, the entropy production is given by the product of the reaction rate and the Gibbs energy of the reaction. This is shown in the paper by Ross and Mazur,[2] where they show that the despite these deviations in activated processes the system is in local equilibrium.

How far these forces can move the system away from equilibrium depends not only on the values of the force but also on the nature of the irreversible process. For some transport processes, such as heat conduction or mass diffusion, local equilibrium, see Chapter 4, typically holds even when the systems are subjected to large or even very large gradients. Experiments performed with a nano-motor moving along a carbon nano-tube, under the influence of a temperature difference generated by an electrical current, show that, despite the very large gradients coming into play, of the order of $1\ \mathrm{K \cdot nm^{-1}}$, the relation between force and gradient is still linear.[3] This linearity is also observed in simulations on the orientation of non-polar molecules induced by a temperature gradient.[4] In these cases, linearity does not necessarily imply closeness to equilibrium.

On the other hand, for activated processes such as chemical reactions, nucleation or adsorption processes, in which the rates are given by Arrhenius laws, linearity breaks down at very small values of the affinity which seems to imply that local equilibrium is lost almost immediately. However, Ross and Mazur showed that the dynamics associated to activated processes is compatible with local equilibrium.[2]

For an equilibrium system, the probability distribution of the relevant variables is related to the thermodynamic potential, which in turns coincides with the minimum work needed to establish the state of the system.[5] In far-from-equilibrium situations, however, the probability must be determined from master equations or kinetic equations of the Fokker–Planck type.[6] The latter has the form of a diffusion equation for the probability, valid when the stochastic process is Gaussian. For this type of noise, the stochastic variable varies in a small amount for short time intervals, as in a diffusion process. The connection between mesoscopic dynamics and thermodynamics[7] can be established through the derivation of the Fokker–Planck equation from a diffusion process of the probability compatible with the statistical formulation of the second law.[8] This new approach overcomes the restriction of linearity inherent to non-equilibrium thermodynamics.[9]

Non-equilibrium thermodynamics provides a theoretical framework to explain the collective behaviour of systems out of equilibrium. It is based on the assumption that equilibrium thermodynamic relations hold locally, together with the conservation laws for the corresponding physical quantities. Such an approach, based on the local use of Gibbs equality, allows expressing the local entropy production, introduced in Section 1.2 in Chapter 1, of a system in terms of a product sum of the conjugate thermodynamic fluxes and forces. Fluxes are then considered as a linear combination of all the relevant forces that are compatible by symmetry.

Non-equilibrium thermodynamics is then restricted to the linear response domain in which the response of the system is linear in the perturbation exerted to remove it from equilibrium. Moreover, this theory yields a macroscopic description in terms of average values not accounting for the presence of fluctuations. Whereas the linear approximation holds for transport processes such as heat conduction and mass diffusion, even in the presence of large gradients, it is not appropriate to describe activated processes in which the system enters rapidly in the non-linear domain. Small systems, such as single molecules in a thermal bath, in which fluctuations and non-linearities can be very important, are beyond the scope of this theory. Nonetheless, it is possible to build on the well-established formulation of non-equilibrium thermodynamics to provide a thermodynamic basis to macroscopic non-linear kinetic processes.

In this Chapter we will describe the foundations of mesoscopic non-equilibrium thermodynamics (MNET). In Section 14.2 we analyse the theoretical foundations of MNET and how it can be used to describe the kinetics of mesoscopic systems. Section 14.3 will describe in detail the use of MNET to study chemical kinetics on thermodynamically consistent grounds. This paradigmatic example is illustrative because one can discuss all the conceptual key aspects of MNET and at the same time it is relevant, since chemical kinetics deviates significantly from linear laws assumed in non-equilibrium thermodynamics. We will also emphasize the analogy between the thermodynamic and kinetic perspectives that share a common underlying thermodynamic background on which MNET is based. Section 14.4 discusses a qualitatively different system described by MNET to emphasize the generality of the framework. Section 14.5 considers how to include thermal fluctuations in the framework of MNET, such a contribution is particularly relevant when dealing with mesoscopic systems. The chapter shows that activated processes are not incompatible with a formulation of the fluctuation–dissipation theorem based on local equilibrium, showing how MNET can shed light in fundamental questions associated with non-equilibrium Statistical Physics. The chapter closes with a conclusion Section.

14.2 Statistical Non-equilibrium Thermodynamics of Mesoscopic Systems

Non-equilibrium thermodynamics assumes local equilibrium, *i.e.*, that thermodynamic variables defined locally in space can be interpreted as in global thermodynamic equilibrium, and that the entropy production of any isolated system is always non-negative, see Chapter 4.

In standard non-equilibrium thermodynamics local variables are defined as a function of space and time. Mesoscopic non-equilibrium thermodynamics assumes an extended space that also accounts for additional internal variations of the system. This additional variable space can account, *e.g.*, for conformational states a large molecule can go through, as shown in Figure 14a, or intermediate states when a chemical reaction takes place, or the orientation of an asymmetric molecule. One then includes these additional states into the phase space that the system explores and assumes that the principles on which non-equilibrium thermodynamics is based apply now also to this new extended space. This approach naturally accounts on the same footing for the fluxes of the relevant variables (*e.g.*, mass, momentum or energy) in real space coupled with processes in which the species, at a given position, can change their internal configuration. Mesoscopic non-equilibrium thermodynamics assumes local equilibrium for these additional variables and that they relax in shorter time scales. Such an approach offers a natural means to develop a thermodynamically consistent framework for the dynamics of small systems at a mesoscopic scale.

At the mesoscopic level, the state of the system can be characterized, in general, through the knowledge of the probability density of the relevant coordinate, $P(\gamma,t)$, which depends on γ and the time t, instead of the local spatial densities or concentrations as done in standard non-equilibrium thermodynamics. We use the coordinate γ to describe the non-equilibrated degrees of freedom that characterize the different mesoscopic states or configurations in which the system of interest can be found. Accordingly, the bounds of this coordinate will depend on the nature of the internal degrees of freedom. They can refer, *e.g.*, to the velocity of a particle, the orientation of a spin, the size of a macromolecule; and in general any coordinate whose value defines the state of the system in phase space. For the sake of brevity, the relevant formulae in this section will not specify which non-internal variables are kept constant.

If we want to put forward a statistical description, a natural starting point is the statistical expression of the entropy of the system in terms of its probability, $P(\gamma,t)$, which can be expressed in terms of the Gibbs entropy postulate[1,10]

$$S = S_{\mathrm{eq}} - k_{\mathrm{B}} \int P(\gamma, t) \ln \frac{P(\gamma, t)}{P_{\mathrm{eq}}(\gamma)} \mathrm{d}\gamma, \tag{14.1}$$

where S_{eq} denotes the entropy when the system in γ-space is in equilibrium, and k_{B} stands for Boltzmann constant. The equilibrium probability density, $P_{\mathrm{eq}}(\gamma)$, can be related to the minimum reversible work required to create that state[5] through the expression

$$P_{\mathrm{eq}} = P_{0,\mathrm{eq}} \exp(-\beta \Delta W(\gamma)), \tag{14.2}$$

where $P_{0,\mathrm{eq}}$ stands for the equilibrium probability distribution at a reference configuration with respect to which the minimum reversible work is computed, and $\beta = 1/k_B T$ corresponds to the relevant inverse thermal energy, where T stands for the temperature of the system. Generally, the minimum work can in general be expressed as

$$\Delta W = \sum_i y_i \Delta Y_i \tag{14.3}$$

where y_i is a set of intensive parameters and Y_i their conjugated extensive variables. This general form stands for mechanical, chemical, electrical, magnetic, surface work, *etc.* performed on the system.[11] For example, for the case of a chemical reaction that we will analyze in detail, the relevant intensive parameter is the affinity of the chemical reaction, the difference in chemical potential between products and reactants, and the variation of the extensive parameter corresponds to the variation of number of reactant molecules. The expression of the minimum reversible work (14.3) reduces to the different thermodynamic potentials. For instance, for the case of constant temperature, volume and number of particles, the minimum work coincides with the Helmholtz energy. The statistical mechanics definition of the entropy is therefore crucial to connect thermodynamics with both the mesoscopic description in terms of the probability distribution $P(\gamma,t)$ and the equilibrium behaviour of the system.

The dynamics of the mesoscopic degrees of freedom can be analyzed from the statistical mechanics definition of the entropy. Taking variations in eqn (14.1), one obtains

$$\delta S = -k_B \int \delta P(\gamma, t) \ln \frac{P(\gamma, t)}{P_{\mathrm{eq}}(\gamma)} \mathrm{d}\gamma. \tag{14.4}$$

and we can use conservation of the probability to write the continuity equation

$$\frac{\partial P(\gamma, t)}{\partial t} = -\frac{\partial J(\gamma, t)}{\partial \gamma}, \tag{14.5}$$

where $J(\gamma,t)$ stands for the current in the space of mesoscopic coordinates.

In order to derive an explicit expression for the mesoscopic current, we take the time derivative in eqn (14.4) and use the continuity equation, eqn (14.5), to eliminate the probability time derivative. After a partial integration, one can identify the mesoscopic entropy production, σ, from the evolution equation of the entropy[11]

$$\sigma = -k_B \int J(\gamma, t) \frac{\partial}{\partial \gamma} \left(\ln \frac{P(\gamma, t)}{P_{\mathrm{eq}}(\gamma)} \right) \mathrm{d}\gamma, \tag{14.6}$$

which is expressed in the form of current-force pairs, the latter being the gradients in the space of mesoscopic variables. We will now assume a linear dependence between current and force and establish the linear relationship

$$J(\gamma, t) = -k_B L(\gamma, P(\gamma, t)) \frac{\partial}{\partial \gamma} \left(\ln \frac{P(\gamma, t)}{P_{eq}(\gamma)} \right), \tag{14.7}$$

where $L(\gamma, P(\gamma, t))$ is an Onsager coefficient,[1] which in general depends on the probability interpreted as a state variable in the thermodynamic sense and on the mesoscopic coordinate γ.

The kinetic equation follows by substituting eqn (14.7) into eqn (14.5):

$$\frac{\partial P(\gamma, t)}{\partial t} = \frac{\partial}{\partial \gamma} \left(D(\gamma) P_{eq}(\gamma) \frac{\partial}{\partial \gamma} \frac{P(\gamma, t)}{P_{eq}(\gamma)} \right), \tag{14.8}$$

where the diffusion coefficient, $D(\gamma)$, is defined as

$$D(\gamma) \equiv \frac{k_B L(\gamma, P(\gamma, t))}{P(\gamma, t)}. \tag{14.9}$$

As usually assumed in non-equilibrium thermodynamics, we consider that the Onsager coefficients are in good approximation proportional to $P(\gamma, t)$. Accordingly, the relevant diffusion coefficient no longer depends on $P(\gamma, t)$. Eqn (14.9), which in view of eqn (14.2), can also be written as

$$\frac{\partial P(\gamma, t)}{\partial t} = \frac{\partial}{\partial \gamma} \left(D(\gamma) \frac{\partial P(\gamma, t)}{\partial \gamma} + \frac{D(\gamma)}{k_B T} \frac{\partial \Delta W(\gamma, t)}{\partial \gamma} P(\gamma, t) \right). \tag{14.10}$$

This equation has the form of a Fokker–Planck equation for the evolution of the probability density in γ-space. Hence, MNET provides a systematic and consistent framework to provide thermodynamic understanding to kinetic equations for the probability distribution of mesoscopic systems.

Under the conditions for which the minimum work is given by the Gibbs energy G, $\Delta W \equiv \Delta G = \Delta H - T\Delta S$, where H is the enthalpy, this equation transforms into the Fokker–Planck equation for a system in the presence of a Gibbs energy barrier:

$$\frac{\partial P(\gamma, t)}{\partial t} = \frac{\partial}{\partial \gamma} \left(D(\gamma) \frac{\partial P(\gamma, t)}{\partial \gamma} + \frac{D(\gamma)}{k_B T} \frac{\partial \Delta G(\gamma, t)}{\partial \gamma} P(\gamma, t) \right). \tag{14.11}$$

Therefore, it is consistent to consider kinetic equations for mesoscopic systems in which the barriers are not restricted to being purely energetic. For instance, geometric constrictions lead to entropic barriers,[12] and their effects can also be described in such a kinetic framework.

MNET provides a general formalism able to analyze the dynamics of mesoscopic systems away from equilibrium from the knowledge of the equilibrium probability and once the relevant variables that characterize their phase space are identified. In this way, by knowing the equilibrium thermodynamic potential of a system, one can derive the corresponding kinetic equation.

Since the mesoscopic entropy production can be given a thermodynamic interpretation, it is possible to identify a generalized chemical potential in phase space variables. We may assume that the evolution of these degrees of freedom is described by a diffusion process and formulate the corresponding Gibbs equation

$$\delta S = -\frac{1}{T}\int \mu(\gamma)\delta P(\gamma,t)\mathrm{d}\gamma, \tag{14.12}$$

which resembles the corresponding law proposed in non-equilibrium thermodynamics for a diffusion process in terms of the mass density of particles. Here $\mu(\gamma)$ plays the role of a generalized chemical potential conjugate to the distribution function $P(\gamma, t)$. Comparing eqn (14.12) with the general expression for the change of the entropy, provided by eqn (14.4), yields the expression of the generalized chemical potential

$$\mu(\gamma,t) = k_\mathrm{B}T\ln\frac{P(\gamma,t)}{P_\mathrm{eq}(\gamma)} + \mu_\mathrm{eq}, \tag{14.13}$$

where we have used the fact that, since in equilibrium $\delta S_\mathrm{eq} = 0$, we can refer the generalized chemical potential to its equilibrium value, μ_eq, using the fact that eqn (14.12) in equilibrium vanishes,

$$\delta S_\mathrm{eq} = -\frac{1}{T}\int \mu_\mathrm{eq}\delta P(\gamma,t)\mathrm{d}\gamma = 0. \tag{14.14}$$

Alternatively, exploiting eqn (14.2), we can rewrite the generalized chemical potential in terms of the minimum reversible work,

$$\mu(\gamma,t) = k_\mathrm{B}T\ln\frac{P(\gamma,t)}{P_{0,\mathrm{eq}}} + \Delta W. \tag{14.15}$$

In this reformulation, the "thermodynamic force" driving this general diffusion process is $T^{-1}\,\partial\mu/\partial\gamma$, and the entropy production is given by

$$\sigma(\gamma,t) = -\frac{1}{T}\int J(\gamma,t)\frac{\partial\mu(\gamma,t)}{\partial\gamma}\mathrm{d}\gamma. \tag{14.16}$$

This expression coincides with the entropy production of a diffusion process over a potential landscape in the space of the mesoscopic coordinates. This landscape is conformed by the values of the equilibrium energy associated with each configuration γ. The treatment of a diffusion process in the framework of non-equilibrium thermodynamics can then be extended to the case in which the relevant quantity is a probability density instead of a mass density. This fact shows the close connection between entropy and stochastic dynamics. We will exploit this analogy in the next section to describe how MNET can be exploited to provide a thermodynamically consistent basis for the dynamics of activated processes, and in particular, chemical reactions.

14.3 Mesoscopic Non-equilibrium Thermodynamics for Activated Processes

The general framework put forward in the previous section can be exploited to describe activated processes, such as chemical reactions. This approach will allow us to develop a theory for the thermodynamics that can account for non-linear flux-force relations, and the general underlying structure of such a theoretical framework.

To this end, one can imagine a chemical reaction as a diffusion process in this new, extended, internal space. The diffusion takes place in an energy landscape where the minima correspond to the different reactants and products involved, as shown in Figure 14.1a. Figure 14.1b shows graphically the fact that as the reactants advance gradually along the internal coordinate they convert gradually into the appropriate products at the end of the process.

For concreteness, let us consider the case of a simple chemical reaction

$$2A \rightleftarrows A_2, \tag{14.17}$$

where we introduce an additional coordinate, γ, which identifies the internal states that the reactant has to go through to become product (or *vice-versa*). The local state of the system can be characterized by the concentration, $c(\boldsymbol{r},\gamma,t)$, which generalizes the concept of the local density of reactants and products at position $\boldsymbol{r}$ and time t. γ is a scalar and dimensionless bound coordinate. Without loss of generality, it can be taken to vary between the value $\gamma=0$, which corresponds to the reactant, and $\gamma=1$, which identifies the product. Figure 14.1b shows graphically how the identity of the species along this internal coordinate describes the internal restructuring of the reactants as they become products. This description is complementary to the one developed in Section 14.2 and it can be viewed as a generalization of the general formalism provided by MNET to fields, with a general spatio-temporal dependence.[13] Since we need to account for the fact that chemically reacting species diffuse in space, we consider in this section quantities per unit mass, and hence introduce the mass, m, of the product molecule A_2, as a reference quantity. Due to this choice, the concentration $c(\boldsymbol{r},\gamma,t)$ is dimensionless and integrates out to unity when integrated over γ.

Accordingly, a local thermodynamic description of the reaction process expresses the local specific entropy, $s(\boldsymbol{r},t)$, as

$$s(\boldsymbol{r},t) = -\frac{k_B}{m}\int_0^1 c(\boldsymbol{r},\gamma,t)\ln(c(\boldsymbol{r},\gamma,t))\mathrm{d}\gamma = \int_0^1 c(\boldsymbol{r},\gamma,t)s(\boldsymbol{r},\gamma,t)\mathrm{d}\gamma. \tag{14.18}$$

This expression for the specific entropy identifies the Gibbs entropy as a function of the internal state of the system,

$$s(\boldsymbol{r},\gamma,t) = -\frac{k_B}{m}\ln c(\boldsymbol{r},\gamma,t), \tag{14.19}$$

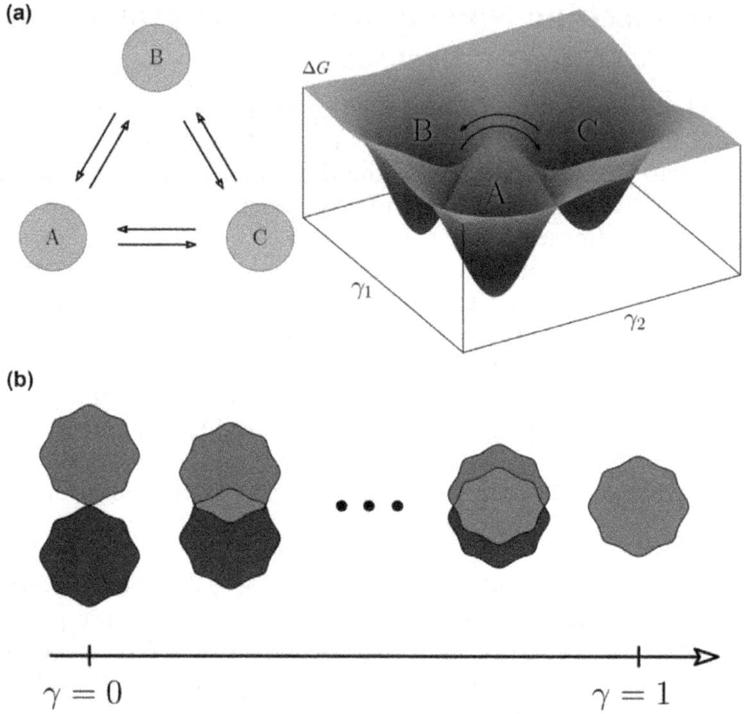

Figure 14.1 A chemical reactions can be treated as a diffusion process through a Gibbs energy barrier that separates the initial and final states of the reaction, which corresponds to the minima of the potential. (a) Transformations of the molecules of a chemical cycle viewed as a diffusion process in an energy landscape. The configurations that identify the transformation from one species to another are described by means of a set of reaction coordinates. The figure depicts an example where three cyclic chemical reactions take place among three different species, A, B and C. The minima of the Gibbs energy correspond to the regions along the internal coordinate that identify these species. The chemical reactions between these species can be described in terms of two internal coordinates, γ_1 and γ_2, due to their cyclic nature. (b) Along a chemical reaction, the molecular structure of a substance transforms progressively until it reaches its final conformation.

from which other thermodynamic quantities for each internal, local configuration, can be derived. For example, the chemical potential μ for each conformation can be expressed

$$\mu(\boldsymbol{r},\gamma,t) = \frac{k_B T(\boldsymbol{r},t)}{m} \ln c(\boldsymbol{r},\gamma,t) + h(\boldsymbol{r},\gamma,t), \tag{14.20}$$

where $h(\boldsymbol{r},\gamma,t)$ refers to the local specific enthalpy in configuration space, without lost of generality. For the example of the chemical reaction, the local specific enthalpy can be understood as the effective energy associated to the

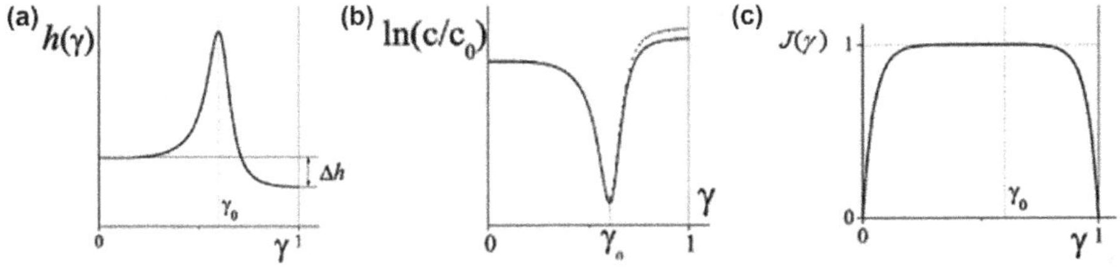

Figure 14.2 Partial enthalpy, concentration of the reacting species and mass flux along the internal coordinate γ. (a) The enthalpy exhibits a maximum at an intermediate configuration γ_0, that separates the region in the internal coordinate space that can be associated to reactants, $\gamma<\gamma_0$, with the one associated to products, $\gamma>\gamma_0$. (b) The structure of the enthalpy correlates with the concentration profiles. (c) The inhomogeneous profile of the species concentration along the internal space is not incompatible with the quasi-steady regime in which the mass flux along the internal coordinate is essentially constant. Only at the boundaries of the internal space does the flux go to zero because species do not leak out of this internal space.

configurational changes the reactants have to go through to become products. Therefore, we can expect it will be characterized by a maximum for an intermediate internal state, γ_0, corresponding to the transition state in chemical kinetics. Figure 14.2 shows a qualitative profile for the enthalpy, characterized by a pronounced maximum associated to the transition state and by a difference between the enthalpy of reactants and products, which acts as a thermodynamic driving force for the chemical reaction. Consistent with non-equilibrium thermodynamics, the local temperature $T(\boldsymbol{r},t)$ is independent of the configuration of the species as they evolve along the internal space. This fact assumes that internal configurations have reached quasi-equilibrium with the local temperature. For simplicity's sake, we will assume the temperature is constant.

The local expression for the Gibbs equation, one of the basis of non-equilibrium thermodynamics, reads now

$$T(\boldsymbol{r},t)\delta s(\boldsymbol{r},t)=\delta h(\boldsymbol{r},t)-\int_0^1 \mu(\boldsymbol{r},\gamma,t)\delta c(\boldsymbol{r},\gamma,t)\mathrm{d}\gamma, \tag{14.21}$$

which can be used to identify the corresponding variation of the entropy in the internal state by using eqn (14.20). Mass conservation implies

$$\frac{\partial c(\boldsymbol{r},\gamma,t)}{\partial t}=-\nabla\cdot\boldsymbol{J}(\boldsymbol{r},\gamma,t)-\frac{\partial r(\boldsymbol{r},\gamma,t)}{\partial\gamma}, \tag{14.22}$$

where $\boldsymbol{J}(\boldsymbol{r},\gamma,t)$ stands for the mass flux to neighbouring spatial positions and $r(\boldsymbol{r},\gamma,t)$ corresponds to the flux of mass along the internal coordinate.

As shown in ref. 14, on which this section is based and which constitutes a more complete and thorough discussion of the MNET description of chemical reactions in spatially varying thermal and concentration gradients, it is possible to derive an expression for the local entropy production by taking the time derivative of the Gibbs relation, eqn (14.21). Disregarding thermal processes for simplicity's sake, assuming that the temperature is constant, and using the entropy balance equation and mass conservation, eqn (14.22), we arrive at

$$\sigma(\boldsymbol{r},\gamma,t) = -\frac{1}{T}\int_0^1 r(\boldsymbol{r},\gamma,t)\frac{\partial\mu(\boldsymbol{r},\gamma,t)}{\partial\gamma}\mathrm{d}\gamma - \frac{1}{T}\int_0^1 \boldsymbol{J}(\boldsymbol{r},\gamma,t)\cdot\nabla\mu(\boldsymbol{r},\gamma,t)\mathrm{d}\gamma. \quad (14.23)$$

This expression constitutes the basis to derive phenomenological relations between thermodynamic fluxes and forces. These relations allow us to close the previous balance equations, and indicate that processes taking place in space and along the internal coordinates have to be treated on equal footing.[10] For the current example, we can derive a close expression for the mass conservation law, eqn (14.22). Assuming the Curie principle, we disregard coupling between variables with different tensorial order. Therefore, we arrive at

$$\boldsymbol{J}(\boldsymbol{r},\gamma,t) = -L(\boldsymbol{r},\gamma,t)\Delta\mu(\boldsymbol{r},\gamma,t),$$

$$r(\boldsymbol{r},\gamma,t) = -L_\gamma(\boldsymbol{r},\gamma,t)\frac{\partial}{\partial\gamma}\frac{\mu(\boldsymbol{r},\gamma,t)}{T}, \quad (14.24)$$

where, as usual, we assume a local relation in time, position and internal conformation. In the previous equations $L(\boldsymbol{r},\gamma,t)$ and $L_\gamma(\boldsymbol{r},\gamma,t)$ stand for the Onsager coefficients, corresponding to the diffusive fluxes in space and along the internal space, respectively. The kinetic coefficients in principle depend on position and/or internal configuration.

Inserting the derived linear relations into the mass conservation relation, eqn (14.22), we obtain the dynamic equation for the concentration of reacting species,

$$\frac{\partial c(\boldsymbol{r},\gamma,t)}{\partial t} = \nabla\cdot L(\boldsymbol{r},\gamma,t)\nabla\mu(\boldsymbol{r},\gamma,t) + \frac{\partial}{\partial\gamma}L_\gamma(\boldsymbol{r},\gamma,t)\frac{\partial}{\partial\gamma}\frac{\mu(\boldsymbol{r},\gamma,t)}{T}, \quad (14.25)$$

which accounts for the change of reactants and products both due to the chemical reaction and to spatial inhomogeneities in their concentrations.

If we integrate the mass balance equation over the internal coordinate, we can identify the mass balance for reactants and products, Since the chemical reaction is characterized by a maximum in the partial enthalpy, $h(\boldsymbol{r},\gamma,t)$, large compared to k_BT as depicted in Figure 14.2, we can identify the reactants with the ensemble of internal configurations, $0<\gamma<\gamma_O$, and products with those configurations where $\gamma_O<\gamma<1$. Moreover, if the shape of $h(\boldsymbol{r},\gamma,t)$ is peaked around γ_O, as shown in Figure 14.2, the reaction rate along the internal coordinate, $r(\boldsymbol{r},\gamma,t)$, is small and one can assume that reactants and products equilibrate fast compared with the transport from reactant to product. Therefore, a quasi-stationary state develops where the reaction flux is independent of the internal configuration. In good approximation one can then write

$$r(\mathbf{r}, \gamma, t) = r(\mathbf{r}, t)\theta(\gamma)\theta(\gamma - 1), \tag{14.26}$$

where θ stands for the Heaviside function. The quasi-steady flux vanishes at the ends of the internal space, consistent with mass conservation, as shown in Figure 14.2. Consistently, the derivative of the reaction rate with respect to γ gives two contributions proportional to a delta function near the end of the reaction coordinate, inside the domain. These contributions correspond to the rate at which reactants transform into products. As a result, integrating the derivative with respect to γ gives the contribution $\pm r(\mathbf{r}, \gamma, t)$. It is also possible to obtain an explicit expression for the flux along the internal coordinate. Using eqn (14.20) and (14.24) and the expression for the quasi-steady flux, eqn (14.26), we can arrive at

$$r(\mathbf{r}, t) = -D(\mathbf{r}, \rho(\mathbf{r}, t))\rho(\mathbf{r}, t)\exp[-\beta m h(\mathbf{r}, \gamma, t)]\frac{\partial}{\partial\gamma}\exp[\beta m\mu(\mathbf{r}, \gamma, t)], \tag{14.27}$$

where we have introduced the density of the reaction complex in state γ, $\rho(\mathbf{r}, \gamma, t)$, and the local mass density, $\rho(\mathbf{r}, t)$. We assume that the total concentration of species along the internal coordinate adds up to unity, so that $\rho(\mathbf{r}, \gamma, t) = \rho(\mathbf{r}, t)c(\mathbf{r}, \gamma, t)$. In eqn (14.27), as usual, we have related the Onsager coefficient, $L(\mathbf{r}, \gamma, \rho(\mathbf{r}, \gamma, t))$, to a diffusion coefficient D along the internal coordinate

$$D(\mathbf{r}, \rho(\mathbf{r}, \gamma, t)) \equiv \frac{k_B L_\gamma(\mathbf{r}, \gamma, \rho(\mathbf{r}, \gamma, t))}{m\rho(\mathbf{r}, \gamma, t)}. \tag{14.28}$$

As noted earlier, we have assumed that the Onsager coefficient $L(\mathbf{r}, \gamma, \rho(\mathbf{r}, \gamma, t))$ is, in good approximation, proportional to $\rho(\mathbf{r}, \gamma, t)$ and therefore the diffusion coefficient no longer depends on $\rho(\mathbf{r}, \gamma, t)$.

Since the flow along the internal coordinate is essentially homogeneous in the quasi-stationary regime, we can integrate the previous expression for the flux, eqn (14.27), to arrive at a simpler expression,

$$r(\mathbf{r}, t) = -D(\mathbf{r}, \rho(\mathbf{r}, t))\rho(\mathbf{r}, t)\frac{\exp[\beta m\mu(\mathbf{r}, 1, t)] - \exp[\beta m\mu(\mathbf{r}, 0, t)]}{\int_0^1 \exp[\beta m h(\mathbf{r}, \gamma, t)]\mathrm{d}\gamma}, \tag{14.29}$$

where reactant and product concentrations enter through the chemical potential. From the expression of the chemical potential in the internal space, eqn (14.20), one arrives at the more explicit relation for the reaction flux

$$r(\mathbf{r}, t) = -D(\mathbf{r}, \rho(\mathbf{r}, t))\rho(\mathbf{r}, t)\frac{c(\mathbf{r}, 1, t)\exp[\beta m h(\mathbf{r}, 1, t)] - c(\mathbf{r}, 0, t)\exp[\beta m\mu(\mathbf{r}, 0, t)]}{\int_0^1 \exp[\beta m h(\mathbf{r}, \gamma, t)]\mathrm{d}\gamma}. \tag{14.30}$$

While eqn (14.30) shows explicitly the dependence of the reaction rate on the concentrations of reactants and products, eqn (14.29) asserts more clearly the thermodynamic meaning of the diffusion process along the internal coordinate associated to the compositional change that reactants suffer during the chemical reaction. As shown in Figure 14.1, the chemical reaction is characterized by the presence of energy barriers as reactants become products. In the context of the simple reaction, eqn (14.17), in which the internal space is one dimensional, the change to products will be controlled by the region around the value of the internal coordinate, γ_0, where the effective energy exhibits its barrier. If this barrier is large compared to the thermal energy, k_BT, reactants and products equilibrate to the local condition independently as the chemical reaction evolves. One can then assume that at each spatial position the chemical potential of species associated to reactants (characterized by $0<\gamma<\gamma_0$) and those associated to the chemical reaction products (characterized by $\gamma_0<\gamma<1$) will have an essentially constant chemical potential, even if the chemical potential for reactants and products will be different (since this difference is the thermodynamic force driving the chemical reaction). One can therefore asume

$$\mu(\boldsymbol{r},\gamma,t)=\mu(\boldsymbol{r},0,t)\theta(\gamma_0-\gamma)+\mu(\boldsymbol{r},1,t)\theta(\gamma-\gamma_0)=\mu_A(\boldsymbol{r},0,t)\theta(\gamma_0-\gamma)\\ +\mu_{A_2}(\boldsymbol{r},1,t)\theta(\gamma-\gamma_0), \tag{14.31}$$

where we have identified the chemical potential at $\gamma=0$ with that of reactants and the chemical potential at the other end of the internal space domain, $\gamma=1$, with that of products. The deviations from local equilibrium in each attraction basin (or interval in the internal space, as defined by the corresponding maxima in the internal space enthalpy) are negligible. However, it is worth noting that the uniformity of the chemical potential in these two attraction basins does not imply that reactant or product concentrations are correspondingly uniform. The concentration along the internal space is highly inhomogeneous, and this variation reflects the cost of changing conformation and identity of the reacting species, as quantified by the enthalpy, as shown in Figure 14.2. Accordingly, we can obtain the concentration of reactants as products integrating the local concentration along the internal space in the corresponding region,

$$c_{2A}(\boldsymbol{r},t)=\int_0^{\gamma_0}c(\boldsymbol{r},\gamma,t)\mathrm{d}\gamma,\quad c_{A_2}(\boldsymbol{r},t)=\int_{\gamma_0}^{1}c(\boldsymbol{r},\gamma,t)\mathrm{d}\gamma. \tag{14.32}$$

Eqn (14.20) provides the local relation between the concentration of reacting species and the chemical potential in the internal space. Therefore, using the local equilibrium assumption, eqn (14.31), which gives us an explicit expression for the chemical potential of reactants and products, we can arrive at an explicit relation between the corresponding concentration of reactants and products and the enthalpy of the reacting species. The local relation in the internal space reads

$$c(\boldsymbol{r},\gamma,t)=\exp[\beta m(\mu_A(\boldsymbol{r},t)-h(\boldsymbol{r},\gamma,t))]\theta(\gamma_0-\gamma)+\exp[\beta m(\mu_{A_2}(\boldsymbol{r},t)\\ -h(\boldsymbol{r},\gamma,t))]\theta(\gamma-\gamma_0), \tag{14.33}$$

which leads to

$$c_{2A}(\boldsymbol{r},t)=c(\boldsymbol{r},0,t)\int_0^{\gamma_0}\exp[\beta m(h(\boldsymbol{r},0,t)-h(\boldsymbol{r},\gamma,t))]\mathrm{d}\gamma,$$

$$c_{A_2}(\boldsymbol{r},t)=c(\boldsymbol{r},1,t)\int_{\gamma_0}^{1}\exp[\beta m(h(\boldsymbol{r},1,t)-h(\boldsymbol{r},\gamma,t))]\mathrm{d}\gamma. \tag{14.34}$$

These expressions are useful because they explicitly relate the reactant and product concentrations to the values of the concentration at the ends of the internal space. This relation allows connecting the concentration of reactants and products along the internal space with the local values required to describe spatial fluxes. It is this connection that provides the means to describe on the same footing spatially varying transport processes and the fluxes along the internal space.

We can then take advantage of the expressions for the concentration of species along the internal space, eqn (14.33), and the corresponding expressions for the concentrations of reactants and products, eqn (14.34), and insert them into the expression for the reaction flux, eqn (14.30). In this way, we can rewrite the reaction flux as

$$r(\boldsymbol{r},t)=\frac{D(\boldsymbol{r},t)\rho(\boldsymbol{r},t)}{\int_0^1\exp[\beta m h(\boldsymbol{r},\gamma,t)]\mathrm{d}\gamma}\left(\frac{c_{2A}(\boldsymbol{r},t)}{\int_0^{\gamma_0}\exp[-\beta m h(\boldsymbol{r},\gamma,t)]\mathrm{d}\gamma}-\frac{c_{A_2}(\boldsymbol{r},t)}{\int_{\gamma_0}^{1}\exp[-\beta m h(\boldsymbol{r},\gamma,t)]\mathrm{d}\gamma}\right), \tag{14.35}$$

which has the form of the usual law of mass action for a chemical reaction

$$r(\boldsymbol{r},t)=k_+(\boldsymbol{r},t)c_A^2(\boldsymbol{r},t)-k_-(\boldsymbol{r},t)c_{A_2}(\boldsymbol{r},t), \tag{14.36}$$

where we have taken into account the stoichiometry of the chemical reaction, eqn (14.17), that implies $c_{2A}=c_A^2$. k_+ and k_- stand for the reaction rates corresponding to the forward and backward reactions of eqn (14.17). Comparing eqn (14.35) and (14.36) allows us to identify the reaction constants in terms of the diffusion process of the reacting species along the internal space

$$k_+(\boldsymbol{r},t) = \frac{D(\boldsymbol{r},t)\rho(\boldsymbol{r},t)}{\int_0^1 \exp[\beta m h(\boldsymbol{r},\gamma,t)]\mathrm{d}\gamma \int_0^{\gamma_0} \exp[-\beta m h(\boldsymbol{r},\gamma',t)]\mathrm{d}\gamma'},$$

$$k_-(\boldsymbol{r},t) = \frac{D(\boldsymbol{r},t)\rho(\boldsymbol{r},t)}{\int_0^1 \exp[\beta m h(\boldsymbol{r},\gamma,t)]\mathrm{d}\gamma \int_{\gamma_0}^{1} \exp[-\beta m h(\boldsymbol{r},\gamma',t)]\mathrm{d}\gamma'}, \tag{14.37}$$

The description of the reaction process in terms of MNET provides a consistent description of chemical kinetics from non-equilibrium thermodynamics. The fact that reacting species evolve in a composition space in which there are typically energy barriers, leads to a non-linear relation between thermodynamic fluxes (the reaction flux) and the thermodynamic force that drives the chemical reaction (the difference in chemical potential between reactants and products, or Gibbs energy of the reaction, $\Delta G(\boldsymbol{r},t) = \mu_{A_2}(\boldsymbol{r},t) - \mu_A(\boldsymbol{r},t)$. This non-linear dependence is spelled out in eqn (14.29) and indicates that there is no contradiction between non-equilibrium thermodynamics and non-linear constitutive relations. This thermodynamic description allows us to rewrite eqn (14.37) as

$$r(\boldsymbol{r},t) = -k_{\mathrm{B}}L_\gamma[1 - \exp(-\beta m \Delta G(\boldsymbol{r},t))], \tag{14.38}$$

which indicates clearly the thermodynamic underlying structure behind chemical kinetics. Eqn (14.38) is a generally valid expression for the reaction rate; only the specific expression for the Gibbs energy will differ depending whether the system is ideal or not. The non-ideality can be expressed, for example, in terms of an activity coefficient, $f(c(\boldsymbol{r},\gamma,t))$, which generalizes the chemical potential to

$$\mu(\boldsymbol{r},\gamma,t) = \frac{k_{\mathrm{B}}T(\boldsymbol{r},t)}{m}\ln f(c(\boldsymbol{r},\gamma,t))c(\boldsymbol{r},\gamma,t) + h(\boldsymbol{r},\gamma,t).$$

The derivation of the law of mass action, eqn (14.36), is not functionally affected because the affinity only modifies the expression of the relevant diffusion coefficient, which reads now

$$D(\boldsymbol{r},t) \equiv \frac{k_{\mathrm{B}}L_\gamma(\boldsymbol{r},\gamma,t)}{m\rho(\boldsymbol{r},\gamma,t)}\left(1 + \frac{\partial \ln f(c(\boldsymbol{r},\gamma,t))}{\partial \ln c(\boldsymbol{r},\gamma,t)}\right).$$

Therefore, the non-ideal nature of the chemically reacting species can be absorbed into the reaction rates, eqn (14.37), without affecting the law of mass action.[9] This fact indicates the robustness and the thermodynamic foundation of this central law of chemical kinetics. In the example we have worked out, this non-linearity emerges from the slow evolution of species in the presence of enthalpic barriers, but a number of different systems can be brought under this same description.

The derivation has also shown that the non-linear relation between fluxes and thermodynamic forces is not inconsistent with the usual description of chemical reactions based on the law of mass action. Rather, the approach from MNET provides a thermodynamic perspective to such kinetic laws. In particular, MNET shows how the reaction coefficients depend on the enthalpy associated to the chemical reaction and that these coefficients have an Arrhenius-like expression. Therefore, the law of mass action is not restricted to ideal systems. MNET shows that the non-ideality remains encoded in the expression of the reaction rates. The extension to coupled equations[9] and cycle reactions[15] is straightforward and the same approach has allowed providing a unified thermodynamic description of the non-equilibrium behaviour of molecular motors.[16]

14.4 Other Internal Variables

While in the previous sections we have focused on the description of chemical reactions, the general idea of including additional variables that enlarge the thermodynamic space is very flexible and fruitful. This general idea has been applied to understand a wide variety of systems, ranging from Langmuir adsorption kinetics[17] to the inertia contribution in the dynamics of gas molecules or to biophysical processes. MNET has proved very fruitful to describe energy transduction in biological processes.[18] The energy dissipation in biological pumps and the impact of their lack of efficiency are characterized through slipping.[19] MNET naturally accounts for the nature of mesoscopic and small scale systems. This fact has been exploited, in a biophysical context, to propose a unified thermodynamic and kinetic description of single molecule processes[20] or the description of molecular motors.[16] The ability to derive kinetic equations from a thermodynamically consistent approach has made it possible to derive well-known kinetic expressions, such as the Butler–Volmer and Nernst equations,[21] or analyze basic principles, such as the origin of the fluctuation-dissipation theorem in the presence of energy barriers.[22] MNET has also provided a consistent framework to derive kinetic equations for transport processes where barriers have an entropic, rather than an energetic nature,[23] a situation of interest when dealing with the kinetics and transport of systems under strong confinement.

As an enlightening simple example, we consider the situation in which the molecules of a given system are characterized by a dipole of magnitude p per unit of mass.[1] In the presence of an orienting external field, say an electric field of magnitude E, the dipoles, of mass m, tend to orient along the electric field and the equilibrium density distribution of dipoles has the form

$$\rho^{\mathrm{eq}}(\varphi) = C\exp[\beta mpE\cos\varphi], \tag{14.39}$$

where ϕ represents the angle that the dipole makes with the applied field and C stands for a constant associated to the normalization of ρ^{eq}. For this system, the minimum work, introduced generically in eqn (14.3), reads $\Delta W = mpE\Delta(\sin\phi)$, which identifies $\sin\phi$ as the relevant intensive variable and the applied electric field E as its conjugated extensive variable. We can take advantage of the fact that the chemical potential in equilibrium is constant, and by combining eqn (14.20) and (14.39) identify the relevant enthalpy

$$h(\varphi) = -pE\cos\varphi - \frac{k_B T}{m}\ln\sin\varphi + h_0, \tag{14.40}$$

where h_0 is a reference value of the enthalpy. From eqn (14.40) we can identify the relevant chemical potential that controls the dynamics of dipoles in the presence of an external electric field

$$\mu(\varphi,t) = \frac{k_B T}{m}\ln\frac{\rho(\varphi,t)}{\sin\varphi} - pE\cos\varphi + h_0. \tag{14.41}$$

If we consider the dipole orientation, φ, as the internal space that characterizes the different conformations in which dipoles can evolve, we can write down a constitutive relation between the flux of dipoles in this internal space and the inhomogeneities in chemical potential. This can be expressed as gradients of eqn (14.41) whenever the dipole density distribution deviates from its equilibrium shape, eqn (14.39). The expression for the chemical potential is analogous to eqn (14.20) for the chemical potential including the dynamics along the internal variable, in the absence of spatial dependence. Therefore, the entropy production for this system will be formally equivalent to eqn (14.23). Accordingly, the constitutive relation for the corresponding diffusive flux, eqn (14.28), now reads

$$J(\varphi,t) = -L(\varphi,\rho(\varphi,t))\frac{\partial\mu(\varphi,t)}{\partial\varphi}. \tag{14.42}$$

Inserting the chemical potential in orientation space, eqn (14.41), in the expression for the corresponding flux, and using the continuity equation in the internal space

$$\frac{\partial\rho(\varphi,t)}{\partial t} = -\frac{\partial J(\varphi,t)}{\partial\varphi}, \tag{14.43}$$

leads to

$$\frac{\partial\rho(\varphi,t)}{\partial t} = \frac{\partial}{\partial\varphi}\left[\sin\varphi\left(D\frac{\partial}{\partial\varphi}\frac{\rho(\varphi,t)}{\sin\varphi} + \nu\rho(\varphi,t)pE\right)\right], \tag{14.44}$$

where we have introduced the mobility, $\nu(\phi,\rho(\phi,t))$, and corresponding diffusion coefficient D in orientation space

$$\nu(\varphi,\rho(\varphi,t)) = \frac{L(\varphi,\rho(\varphi,t))}{\rho(\varphi,t)};\ D(\varphi,\rho(\varphi,t)) = \frac{k_B T}{m}\nu(\varphi,\rho(\varphi,t)). \tag{14.45}$$

In the previous expressions we have used again the fact that usually the mobility and diffusion coefficients have a weaker dependence on density than the kinetic coefficients directly associated to the phenomenological laws.

Eqn (14.44) describes the relaxation and diffusion of dipoles in the presence of an electric field, and was derived first by Debye.[24,25] This evolution equation is of the Fokker–Planck type. In general, MNET provides Fokker–Planck equations in the relevant internal space. This general structure has been exploited to analyze the study of non-equilibrium processes with a thermodynamically sound basis in a variety of systems; it has been used, *e.g.*, to analyze colloidal rheology.[26] As we will describe in the next section, there is a deeper connection that relates dynamics of this kind to the Fokker–Planck equation for probability densities and indicates that MNET is a consistent theory to describe the dynamics of mesoscopic systems.

14.5 Fluctuating Mesoscopic Non-equilibrium Thermodynamics

In non-equilibrium thermodynamics one can include additive random fluxes that satisfy the fluctuation-dissipation theorem. These additional fluxes, based on local equilibrium, allow deriving, for example, the fluctuating hydrodynamic equations of a liquid proposed by Landau and Lifshitz,[5] as described in Chapter 2.

MNET shares its general structure with that of standard non-equilibrium thermodynamics. Therefore, one can naturally account for the effect of hydrodynamic fluctuations in the dynamics beyond linear constitutive relations in non-equilibrium thermodynamics. Following the method proposed in non-equilibrium thermodynamics to include hydrodynamic fluctuations, as described in Chapter 2, we can assume that there is local equilibrium in the linear laws in the parameter space that includes the internal variables and derive from them the effective, corresponding, random fluxes. From these local, random fluxes along the internal configurational space, we can deduce the effective random fluxes associated to the relevant spatial variables, *e.g.*, reactant and product concentrations in a chemical reaction, when the internal variables are integrated.

We will now derive the corresponding fluctuating fluxes in the case of a chemical reaction, building on the description put forward in Section 14.3. Again, the generalization to other cases of internal variables is straightforward. In the model chemical reaction analyzed in Section 14.3, species diffuse in space and undergo a chemical reaction. Associated with these processes, in MNET one has fluxes of species in space and fluxes along the internal coordinate that quantify the chemical process. According to the concepts underlying fluctuating hydrodynamics, one considers additional contributions to the corresponding fluxes that reflect the random nature of molecular motion, and are described mathematically as stochastic variables.[12] This fact implies that the species densities are also stochastic variables. The phenomenological relations, eqn (14.24), are then generalized to

$$\boldsymbol{J}(\boldsymbol{r},\gamma,t) = -L(\boldsymbol{r},\gamma,t)\nabla\frac{\mu(\boldsymbol{r},\gamma,t)}{T} + \delta\boldsymbol{J}^r(\boldsymbol{r},\gamma,t),$$

$$r(\boldsymbol{r},\gamma,t) = -L_\gamma(\boldsymbol{r},\gamma,t)\frac{\partial}{\partial\gamma}\frac{\mu(\boldsymbol{r},\gamma,t)}{T} + \delta r^r(\boldsymbol{r},\gamma,t), \tag{14.46}$$

which identify the random fluxes corresponding to spatial diffusion, $\delta\boldsymbol{J}^r(\boldsymbol{r},\gamma,t)$, and to the diffusion along the internal space, $\delta r^r(\boldsymbol{r},\gamma,t)$, that accounts for the chemical reaction. In order to specify appropriately the evolution under the action of these random fluxes, we have to specify completely the stochastic properties of these random fluxes, and not only their first moment. In equilibrium, these random fluxes are generated by a set of Gaussian stochastic processes. For such processes, only their first and second moments are required to specify them completely, as described in Chapter 2. Due to spatial isotropy, the first moments of these random fluxes are zero, while their second moments have an amplitude that ensure that the Fluctuation-Dissipation theorems (FDT), as discussed in Chapter 2, are fulfilled.[5]

$$\langle \delta\boldsymbol{J}^r \rangle = \langle \delta r^r \rangle = 0$$

$$\langle \delta J^r_\alpha(\boldsymbol{r},\gamma,t)\delta J^r_\alpha(\boldsymbol{r}',\gamma',t')\rangle = 2k_B L(\boldsymbol{r},\gamma,t)\delta_{\alpha\beta}\delta(\boldsymbol{r}-\boldsymbol{r}')\delta(\gamma-\gamma')\delta(t-t'),$$

$$\langle \delta r^r(\boldsymbol{r},\gamma,t)\delta r^r(\boldsymbol{r}',\gamma',t')\rangle = 2k_B L_\gamma(\boldsymbol{r},\gamma,t)\delta(\boldsymbol{r}-\boldsymbol{r}')\delta(\gamma-\gamma')\delta(t-t') \quad (14.47)$$

In these equations the delta functions express the fact that all these random fluxes correlate locally and do not have any memory. Due to the different vectorial nature of the diffusive fluxes in real space and the diffusion random fluxes along the internal space, there are no correlations between these two fluxes

$$\langle \delta r^r(\boldsymbol{r},\gamma,t)\delta\boldsymbol{J}^r(\boldsymbol{r},\gamma,t)\rangle = 0 \quad (14.48)$$

In general, one needs to identify all random fluxes associated with the relevant variables and take into account that random fluxes of the same tensorial character will in general be coupled. Therefore, MNET provides a general framework that accounts for all potential coupling between the dynamics along internal space and spatial processes. The amplitudes in eqn (14.47) and (14.48) are proportional to the Onsager coefficients. These coefficients determine in particular the relaxation to equilibrium when the system is perturbed from its global equilibrium state. Since the Onsager or kinetic coefficients depend, in principle, on the state of the system, in eqn (14.47) and (14.48) these coefficients must be evaluated in the reference equilibrium state.

Although the initial formulation of fluctuating hydrodynamics was proposed for systems in global equilibrium, the same conceptual approach has been successfully generalized to out of equilibrium systems, with the only assumption that the same relations and structure of the random fluxes can be now applied assuming local equilibrium.[12] Chapter 3 describes the potential of this description to understand the origin of long-range spatial correlations in non-equilibrium steady states. We can follow the same approach and assume that local equilibrium holds in the internal space. Once the properties of the random fluxes have been specified, one can introduce eqn (14.46) into eqn (14.22). These equations are then conservation laws for the fluctuating densities that determine completely their dynamics. In particular, since the random fluxes enter additively, the average behaviour of the densities is not affected by the random fluxes; these will only affect the correlations of the fluctuating densities. In Section 14.3 we have seen how the law of mass action emerges naturally from the diffusion of species along the internal state. Analogously, we can follow the same procedure to identify what is the effective stochastic flux for reactants and products if we integrate out the internal variables. To this end, using the fact that fluctuating fluxes enter additively, we can express the instantaneous fluctuating flux along the internal coordinate, using eqn (14.27) as

$$r(\boldsymbol{r},t) = -D(\boldsymbol{r},t)\rho(\boldsymbol{r},t)\exp[-\beta m h(\boldsymbol{r},\gamma,t)]\frac{\partial}{\partial\gamma}\exp[\beta m\mu(\boldsymbol{r},\gamma,t)] + \delta r^{r}(\boldsymbol{r},\gamma,t). \tag{14.49}$$

In the quasi-steady regime, where the flux is essentially uniform along the internal space, we can integrate eqn (14.49) to arrive at

$$r(\boldsymbol{r},t) = -D(\boldsymbol{r},t)\rho(\boldsymbol{r},t)\frac{\exp[\beta m\mu(\boldsymbol{r},1,t)] - \exp[\beta m\mu(\boldsymbol{r},0,t)]}{\int_0^1 \exp[\beta m h(\boldsymbol{r},\gamma,t)]\mathrm{d}\gamma} + \delta\tilde{r}^{r}(\boldsymbol{r},t), \tag{14.50}$$

where $\delta\tilde{r}^{\mathrm{r}}(\boldsymbol{r},t)$ stands for the net fluctuating reacting flux associated with the chemical reaction. Its stochastic properties can be derived from those of the underlying, local random fluxes as given in eqn (14.47). The net fluctuating reacting flux depends linearly on $\delta r^{r}(\boldsymbol{r},\gamma,t)$, namely,

$$\delta\tilde{r}^{r}(\boldsymbol{r},t) \equiv \frac{\int_0^1 \exp[\beta m h(\boldsymbol{r},\gamma,t)]\delta r^{r}(\boldsymbol{r},\gamma,t)\mathrm{d}\gamma}{\int_0^1 \exp[\beta m h(\boldsymbol{r},\gamma,t)]\mathrm{d}\gamma}, \tag{14.51}$$

and it is therefore also a Gaussian stochastic process with zero mean. Its second moment, can be expressed as

$$\langle\delta\tilde{r}^{r}(\boldsymbol{r},t)\delta\tilde{r}^{r}(\boldsymbol{r}',t')\rangle = \frac{\int_0^1\int_0^1 \exp[\beta m(h(\boldsymbol{r},\gamma,t) + h(\boldsymbol{r}',\gamma',t'))]\langle\delta r^{r}(\boldsymbol{r},\gamma,t)\delta r^{r}(\boldsymbol{r}',\gamma',t')\rangle\mathrm{d}\gamma\mathrm{d}\gamma'}{\left(\int_0^1 \exp[\beta m h(\boldsymbol{r},\gamma,t)]\mathrm{d}\gamma\right)^2}. \tag{14.52}$$

Using eqn (14.47), we can obtain

$$\langle\delta\tilde{r}^{r}(\boldsymbol{r},t)\delta\tilde{r}^{r}(\boldsymbol{r}',t')\rangle = \frac{2D(\boldsymbol{r},t)\rho(\boldsymbol{r},t)}{\left(\int_0^1 \exp[\beta m h(\boldsymbol{r},\gamma,t)]\mathrm{d}\gamma\right)^2}\int_0^1 \exp[2\beta m h(\boldsymbol{r},\gamma,t)]c(\boldsymbol{r},\gamma,t)\mathrm{d}\gamma \tag{14.53}$$

in terms of the concentrations of the reacting species along the internal space. Eqn (14.33) and (14.34) allow us to express these concentrations in terms of those of the reactants and products,

$$c(\boldsymbol{r},\gamma,t)=c_{2A}\frac{\exp[-\beta m h(\boldsymbol{r},\gamma,t)]}{\int_0^1 \exp[-\beta m h(\boldsymbol{r},\gamma,t)]\mathrm{d}\gamma}\theta(\gamma_0-\gamma) + c_{A_2}\frac{\exp[-\beta m h(\boldsymbol{r},\gamma,t)]}{\int_0^1 \exp[-\beta m h(\boldsymbol{r},\gamma,t)]\mathrm{d}\gamma}\theta(\gamma-\gamma_0). \tag{14.54}$$

This expression allows us to rewrite the correlations of the fluctuating fluxes as

$$\langle\delta\tilde{r}^r(\boldsymbol{r},t)\delta\tilde{r}^r(\boldsymbol{r}'t')\rangle=2\left[k_+(\boldsymbol{r},t)c_{2A}\frac{\int_0^{\gamma_0}\exp[\beta m h(\boldsymbol{r},\gamma,t)]\mathrm{d}\gamma}{\int_0^1\exp[\beta m h(\boldsymbol{r},\gamma,t)]\mathrm{d}\gamma} + k_-(\boldsymbol{r},t)c_{A_2}\frac{\int_0^{\gamma_0}\exp[\beta m h(\boldsymbol{r},\gamma,t)]\mathrm{d}\gamma}{\int_0^1\exp[\beta m h(\boldsymbol{r},\gamma,t)]\mathrm{d}\gamma}\right]\delta(\boldsymbol{r}-\boldsymbol{r}')\delta(t-t') \tag{14.55}$$

in terms of the forward and backward reaction rates obtained in Section 14.3, eqn (14.37). If the diffusion along the internal coordinate is characterized by the presence of an energy barrier, as we have assumed by identifying γ_0, then the fractions appearing in eqn (14.54) contribute mostly from the region of conformational changes around that value. Accordingly, we can obtain

$$\langle\delta\tilde{r}^r(\boldsymbol{r},t)\delta\tilde{r}^r(\boldsymbol{r}'t')\rangle=2[k_+(\boldsymbol{r},t)c_{2A}+k_-(\boldsymbol{r},t)c_{A_2}]\delta(\boldsymbol{r}-\boldsymbol{r}')\delta(t-t'). \tag{14.56}$$

In Section 14.3 we showed that chemical kinetics can be expressed also in terms of the Gibbs energy difference between reactants and products as the (non-linear) driving force. Accordingly, we can rewrite eqn (14.55) as

$$\langle\delta\tilde{r}^r(\boldsymbol{r},t)\delta\tilde{r}^r(\boldsymbol{r}',t')\rangle=k_B L_\gamma[1-\exp(-\beta m\Delta G(\boldsymbol{r},t))]\delta(\boldsymbol{r}-\boldsymbol{r}')\delta(t-t'). \tag{14.57}$$

The correlation of the random chemical fluxes, eqn (14.55), coincides with the expression derived by Keizer[27] from a kinetic approach of fluctuating chemical kinetics. The derivation from MNET emphasizes the relevance of the underlying thermodynamic structure of the system and allows for expressing the correlations in eqn (14.56), which stresses again the fact that the amplitude of the correlations is a function of the difference in chemical potential between reactants and products.

This description of fluctuations in MNET can be generalized to other types of chemical reactions and to study the coupling of chemical reactions to spatial gradients in non-equilibrium steady states. In the first analysis that described how to describe fluctuations in MNET, the approach was already extended to a set of coupled chemical reactions.[9] Fluctuating MNET puts the fluctuating chemical-reaction fluxes on the same footing as any other fluctuating flux (here we have mentioned explicitly the case of the diffusive fluxes). This approach has been used to study the nature of (long-range) correlations of a reacting binary mixture subject to a temperature gradient.[28]

It is insightful to compare the initial expression for the correlation of reacting fluxes along the internal space, eqn (14.46), and the expression in terms of the overall reaction fluxes, eqn (14.55). The first expression satisfies detailed balance as the species change their conformation as they advance along the reaction. The correlation of the net random reaction fluxes, as given in eqn (14.55), on the other hand does not have the expected form satisfying detailed balance. This change shows that coarse graining can lead to a violation of the fluctuation-dissipation theorem. MNET shows that by identifying the appropriate phase space that accounts for all the relevant dynamic variables, fluctuation-dissipation theorems based on local equilibrium do hold.

The theorem which holds in γ-space fails due to the increased coarsening of the description. When the dynamics are analyzed in terms of diffusion in the reaction coordinate, the system progressively passes from one state to the other, which supports the assumption of local equilibrium in γ-space and consequently makes the formulation of mesoscopic non-equilibrium thermodynamics possible. However, when we describe the system over time intervals longer that the characteristic time over which the particles can explore the energy variations along the internal space, we are only capturing the activated process, which is not near equilibrium and accordingly, the fluctuation-dissipation theorem, does not hold.

This is a fundamental feature of non-equilibrium statistical dynamics and it is found in many systems where activated dynamics of one form or another are present. Glass dynamics constitutes a paradigmatic example. Even the standard reference system analyzed by Kramers, a particle diffusing in a bistable potential, defined in γ-space, will experience this same difference in the description of the fluctuating fluxes depending whether one follows the diffusion of a particle along the degree of freedom or one directly focuses on the jump kinetics between the two potential minima.[29]

14.6 Conclusions

Thermodynamics furnishes a multi-scale description of many-particle systems: a global description adequate for equilibrium systems, a local description for systems in local equilibrium[1] and a mesoscopic description for systems in local equilibrium at the mesoscale.[30]

In this chapter, we have shown how to extend the use of thermodynamic concepts into the mesoscopic domain where fluctuations and non-linearities play an important role. The probabilistic interpretation of thermodynamics together with probability conservation laws can be used to obtain kinetic equations for the mesoscopic degrees of freedom. The approach we have presented starts from the mesoscopic equilibrium behaviour and adds all the dynamic details compatible with the second law of thermodynamics and with the conservation laws and symmetries of the system. From the knowledge of the equilibrium properties of a system, it is straightforward to obtain Fokker–Planck equations for its dynamics. The coefficients entering the dynamic equations can be obtained from experiments or microscopic theories.

We have shown that the existence of local equilibrium depends on the set of variables used to characterize the system. When an activated process is described not just in terms of the initial and final states but through its reaction coordinate, local equilibrium holds. Increasing the dimensionality of the space of thermodynamic variables, by including as many dimensions as non-equilibrated degrees of freedom, leads to local equilibrium in the enlarged space and allows the use of non-equilibrium thermodynamics at shorter time scales in which fluctuations are still present. We can thus conclude that many kinetic processes, such as nucleation,[31] chemical reactions[10] or active transport,[32] which have been assumed to be far away from equilibrium because of their intrinsic non-linear nature, take place at local equilibrium when a finer description is adopted. MNET opens the way to study their dynamics in terms of kinetic equations of the Fokker–Planck type.

Mesoscopic non-equilibrium thermodynamics provides a systematic method to establish such a link between thermodynamics and randomness, obtaining the stochastic dynamics of a system directly from its equilibrium properties. A wide variety of situations can be studied under the framework of this theory, including non-linear transport in the presence of potential barriers, activated processes, slow relaxation phenomena, and basic processes in biomolecules, such as translocation and stretching. For example, Chapter 15 shows the potential of MNET to understand nucleation processes. The many possibilities of MNET to describe basic mechanisms in small biological systems are discussed in Chapter 16.

Acknowledgements

The authors acknowledge financial support from MINECO (Spain) and DURSI under projects FIS2011-22603 and 2014SGR-922, respectively, and to the *Icrea Acadèmia* program.

References

1. S. R. de Groot and P. Mazur, *Non-Equilibrium Thermodynamics*, Dover, London, 1984.
2. J. Ross and P. Mazur, *J. Chem. Phys.*, 1961, **35**, 19.
3. A. Barreiro, R. Rurali, E. R. Hernández, J. Moser, T. Pichler, L. Forró and A. Bachtold, *Science*, 2008, **320**, 775.

4. F. Römer, F. Bresme, J. Muscatello, D. Bedeaux and J. M. Rubí, *Phys. Rev. Lett.*, 2012, **108**, 105901.
5. L. D. Landau and E. M. Lifshitz, *Statistical Physics*, Pergamon Press, New York, 1989.
6. R. Zwanzig, *Non-equilibrium Statistical Mechanics*, Oxford University Press, Oxford, 2001.
7. D. Reguera, J. M. Rubí and J. M. G. Vilar, *J. Phys. Chem. B*, 2005, **109**, 21502.
8. A. Pérez-Madrid, J. M. Rubí and P. Mazur, *Physica A*, 1994, **212**, 231.
9. A. Pérez Madrid, I. Pagonabarraga and J. M. Rubí, *Physica A*, 1997, **237**, 205.
10. N. G. van Kampen *Stochastic Processes in Physics and Chemistry*, North-Holland, Amsterdam, 1987.
11. J. M. G. Vilar and J. M. Rubí, *Phys. Rev. Lett.*, 2008, **100**, 020601.
12. J. M. Ortiz de Zárate and J. V. Sengers, *Hydrodynamic Fluctuations in Fluids and Fluid Mixtures*, Elsevier, Amsterdam, 2006.
13. A. Pérez-Madrid and J. M. Rubí, *Physica A*, 2001, **298**, 177.
14. D. Bedeaux, I. Pagonabarraga, J. M. Ortiz de Zárate, J. V. Sengers and S. Kjelstrup, *Phys. Chem. Chem. Phys.*, 2010, **12**, 12780.
15. J. M. Rubi, D. Bedeaux, S. Kjelstrup and I. Pagonabarraga, *Int. J. Thermophys.*, 2013, **34**, 1214.
16. S. Kjelstrup, J. M. Rubí, I. Pagonabarraga and D. Bedeaux, *Phys. Chem. Chem. Phys.*, 2013, **15**, 19405.
17. I. Pagonabarraga and J. M. Rubí, *Physica A*, 1992, **188**, 553.
18. J. M. Rubí, M. Naspreda, S. Kjelstrup and D. Bedeaux, *J. Non-Equilib. Thermodyn.*, 2007, **32**, 351.
19. S. Kjelstrup, J. M. Rubí and D. Bedeaux, *Phys. Chem. Chem. Phys.*, 2005, **23**, 4009.
20. J. M. Rubí, D. Bedeaux and S. Kjelstrup, *J. Phys. Chem. B*, 2007, **111**, 9598; J. M. Rubí, D. Bedeaux and S. Kjelstrup, *J. Phys. Chem. B*, 2006, **110**, 12733.
21. J. M. Rubí and S. Kjelstrup, *J. Phys. Chem. B*, 2003, **107**, 13471.
22. A. Pérez-Madrid, D. Reguera and J. M. Rubí, *Physica A*, 2003, **239**, 357.
23. D. Reguera and J. M. Rubí, *Physica A*, 2001, **64**, 061106.
24. P. Debye, *Polar Molecules*, Dover Pub. Inc., New York, 1929.
25. B. J. Berne and R. Pecora, *Dynamic Light Scattering: With Applications to Chemistry, Biology and Physics*, Wiley, New York, 1976.
26. D. Bedeaux and J. M. Rubí, *Physica A*, 2002, **305**, 360.
27. J. Keizer, *Statistical Thermodynamics of Non-equilibrium Processes*, Springer, Berlin, Germany, 1987.
28. D. Bedeaux, J. M. Ortiz de Zárate, I. Pagonabarraga, J. V. Sengers and S. Kjelstrup, *J. Chem. Phys.*, 2011, **135**, 124516.
29. M. Naspreda, D. Reguera, A. Pérez-Madrid and J. M. Rubí, *Physica A*, 2005, **351**, 14.
30. J. M. G. Vilar and J. M. Rubí, *Proc. Natl. Acad. Sci. U. S. A.*, 2001, **98**, 11081.
31. D. Reguera and J. M. Rubí, *J. Chem. Phys.*, 2003, **119**, 9877.
32. S. Kjelstrup, J. M. Rubí and D. Bedeaux, *J. Theor. Biol.*, 2005, **234**, 7.

CHAPTER 15

Non-equilibrium Thermodynamics for Nucleation Kinetics

DAVID REGUERA* AND J. MIGUEL RUBÍ

Departament de Física Fonamental, Facultat de Física,
Universitat de Barcelona, Martí i Franquès 1, 08028 Barcelona, Spain
*Email: dreguera@ub.edu

15.1 Introduction

Matter appears in Nature in different physical states called phases. The transformation between these phases constitutes one of the most amazing phenomena that surround us and that we are all familiar with. The condensation of steam into clouds or liquid drops, the formation of ice or the boiling of water, are just common examples of phase transitions that are important in a wide variety of scientific and technological fields. The mechanism that controls the initiation and the fate of most phase transitions is called nucleation, and tries to explain the formation of the first embryos or *nucleus* of the new phase in the initial metastable phase.[1–6]

Nucleation triggers most first order phase transitions such as condensation (*i.e.* the formation of a liquid from a supersaturated vapour), crystallization, cavitation and boiling (the formation of vapour bubbles in an overstretched or superheated liquid). Given the ubiquitous occurrence of phase transitions, nucleation is crucial in a wide scope of scientific and technological fields that range from nuclear events to the formation of planets and galaxies. Fascinating examples include the development of new

Experimental Thermodynamics Volume X: Non-equilibrium Thermodynamics with Applications
Edited by Dick Bedeaux, Signe Kjelstrup and Jan V. Sengers

Published by the Royal Society of Chemistry, www.rsc.org

materials, atmospheric phenomena like precipitation, haze and aerosol formation, the stability of pharmacological compounds, many protein aggregation-induced diseases like cataracts, Alzheimer or sickle cell anemia, the self-assembly of viral capsids, the cryopreservation of food, the damage by cavitation in pumps and propellers, or the explosive vaporization of liquefied gases. This widespread occurrence has stirred an intense and longstanding interest in the study of nucleation since the first investigations of Fahrenheit on metastability in the eighteen century.[7] However, nucleation remains one of the few fundamental classical problems that it is not completely understood. The reason resides in the intrinsic non-equilibrium nature of the problem.

To better understand the peculiarities of this phenomenon, let us focus on one of the most familiar manifestations of nucleation: the formation of ice. Ice is the thermodynamically most stable phase of water at normal pressure below 0 °C. However, it is possible to keep liquid water undercooled at temperatures below zero degrees for long times in a metastable state. The transformation of liquid water into ice starts by the formation by thermal fluctuations of small aggregates of liquid molecules forming tiny crystal-like clusters. The generation of these clusters is favoured by the energy gain associated to the fact that ice is the most stable phase (*i.e.* it has a lower chemical potential). But it also involves an energetic cost associated to the formation of an interface between the incipient solid phase and the metastable liquid. The competition between these two terms originates an energetic barrier that has to be surmounted for the new phase to appear (see Figure 15.1), and that constitutes the ultimate reason of the long term prevalence of metastable phases. For very small crystals, the surface penalty dominates, and they tend to dissolve back into the liquid phase. In very large crystals, the bulk energy gain overcomes the surface penalty and favours their spontaneous growth. Therefore, there exist a special size of these incipient crystallites, called the *critical size* or the critical cluster that signals the frontier between growth and decay of the new phase. Clusters of the new phase have to reach at least this critical size to trigger the formation of the new phase, and this critically-sized cluster constitutes the embryo and most important entity in the process of phase transformation. The energy required in its formation is known as the *nucleation barrier*, and the rate at which critical-sized clusters are formed is called *nucleation rate*, and its prediction constitutes one of the major goals of nucleation theories.

For more than a century, our understanding of nucleation has been dominated by the "Classical Nucleation Theory" (CNT), developed by the pioneering works of Volmer and Weber,[8] Farkas,[9] Becker and Döring,[10] Frenkel,[3] and Zeldovich,[11] among others. This classical picture provides a reasonable and simplified picture of nucleation that can be used to make straightforward predictions of rates for any substance. For many years, CNT was thought to be enough to describe nucleation. But the development in the last decades of accurate experimental techniques to measure nucleation

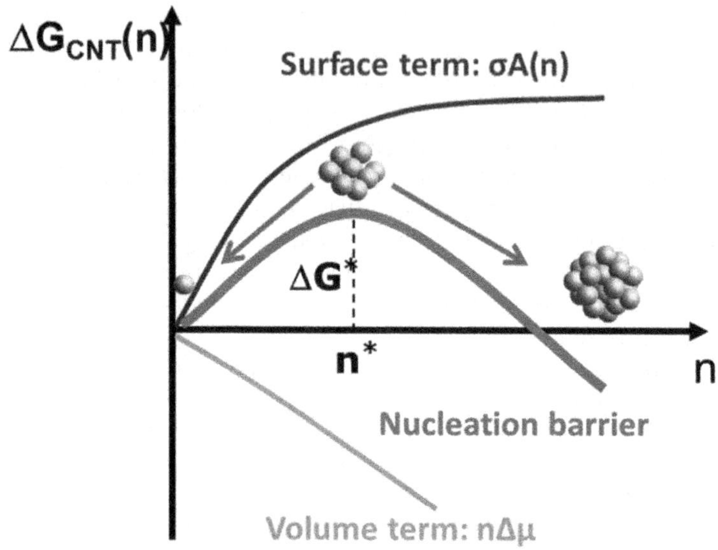

Figure 15.1 Plot of the Gibbs energy landscape for cluster formation according to Classical Nucleation Theory, $\Delta G_{CNT}(n)$, resulting from the addition of bulk and surface contributions. The reaction coordinate is in this case the number of molecules in a cluster n. The maximum height of the barrier or *nucleation barrier*, ΔG^*, is located at a special value of the size known as *critical size* n^*.

rates revealed the severe limitations of this classical picture and stirred a renewed interest in the field. Non-classical theories, including phenomenological, kinetic or density functional techniques have been developed with the aim of overcoming the limitations of CNT.[1,6,12] Despite significant advances in our comprehension of the qualitative mechanisms of nucleation, we are still very far from being able to predict accurately and quantitatively the occurrence of this phenomenon, not even for the simplest case of condensation of a noble gas as Argon, where the discrepancies between theory and experiments can reach more than 20 orders of magnitude![13]

One of the main difficulties in the study of nucleation stems from its intrinsic non-linear, non-equilibrium and nano/mesoscopic nature. That is why non-equilibrium thermodynamics constitutes the appropriate framework to deal with this problem. In particular, in this chapter we will see how mesoscopic non-equilibrium thermodynamics (MNET),[14,15] described in Chapter 14, can provide a proper description of the kinetics of nucleation phenomena and shed light on many of the controversies and limitations surrounding CNT.

For simplicity, we will focus most of our discussion on the particular case of condensation of single component vapours, since crystallization or other instances of nucleation offer additional complications more often related to the thermodynamic rather than to the non-equilibrium aspects of the problem.

The chapter is organized as follows. First, in Section 15.2, we briefly review the classical description of the kinetics and thermodynamics of nucleation. Section 15.3 is then devoted to the derivation of the kinetic equation describing nucleation in the framework of MNET. This equation will be later used in Section 15.4 as the basis to develop a novel and efficient technique to analyze experiments and simulations of nucleation and extract all relevant parameters. Next, we will see how the same ideas can be extended and used to describe the influence of non-isothermal (Section 15.5) and pressure effects (Section 15.6) or the presence of thermal or velocity gradients (Section 15.7). Finally, a brief discussion of the main conclusions and perspectives will close this chapter.

15.2 Kinetics and Thermodynamics of Nucleation: Classical Nucleation Theory

As mentioned in the introduction, nucleation theories aim at describing the formation of the first clusters of the new phase that, when exceeding a certain critical size, trigger the phase transformation. In the classical nucleation theory of condensation, the formation of these clusters is considered as a sort of chemical reaction where clusters containing n molecules can grow or shrink by the addition or loss of individual molecules. Accordingly, the population or number density $f(n,t)$ of clusters of a given size n at time t is described by the following master equation:

$$\frac{\partial f(n,t)}{\partial t} = k^{+}(n-1)f(n-1,t) + k^{-}(n+1)f(n+1,t) - k^{+}(n)f(n,t) - k^{-}(n)f(n,t), \tag{15.1}$$

where $k^{+}(n)$ and $k^{-}(n)$ are, respectively, the rate of attachment and detachment of individual molecules to a cluster of size n. The rate of attachment can be properly described by kinetic theory of gases as $k^{+}(n) = \frac{p}{\sqrt{2\pi k_B T m}} A(n)$, *i.e.* the rate of collisions of gas molecules of mass m at pressure p to a cluster of surface area $A(n)$. But it is more difficult to estimate accurately the evaporation rates. That is the reason why it is common to resort to detailed balance considerations to re-express $k^{-}(n)$ in terms of the known rate of attachment $k^{+}(n)$ and the equilibrium distribution $f_{eq}(n)$ as

$$k^{-}(n+1) = \frac{f_{eq}(n)}{f_{eq}(n+1)} k^{+}(n) = k^{+}(n) e^{-\frac{\Delta G(n) - \Delta G(n+1)}{k_B T}}, \tag{15.2}$$

where $\Delta G(n)$ is the reversible work of formation of a cluster of n molecules at constant pressure p and temperature T, which are the usual conditions at which nucleation takes place. In this way we have converted a complicated kinetic problem into a thermodynamic problem of evaluating the proper Gibbs energy of formation of a cluster. This is still not an easy question, since the clusters of interest for nucleation are unstable and contain only a few molecules. Nevertheless, in the context of CNT these clusters are modelled as tiny spherical objects with a sharp interface and having the same properties as the homogeneous bulk phase. With these simplifying assumptions, that are the core of the so-called *capillarity approximation*, it is then straightforward to express the Gibbs energy of formation of a cluster of n molecules as the sum of a volume and interfacial terms

$$\Delta G(n) = n\Delta\mu + \sigma A(n), \tag{15.3}$$

where $\Delta\mu$ is the difference in chemical potentials between the new and the metastable phase (*i.e.* between the liquid and the vapour phase in the case of condensation), σ is the surface tension of a flat interface of the liquid, and $A(n)$ is the surface area of the spherical droplet containing n molecules. The competition between these two terms gives rise to a barrier, with a maximum located at the critical size n^* and a height that defines the nucleation barrier $\Delta G^* \equiv \Delta G(n^*)$ (see Figure 15.1). Further assuming that the vapour is ideal and the liquid incompressible with a volume per molecule v_l, one gets very simple explicit expressions for these two quantities

$$\Delta G^*_{\mathrm{CNT}} = \frac{16\pi}{3}\frac{v_l^2\sigma^3}{\Delta\mu^2}, \tag{15.4}$$

$$n^* = \frac{32\pi}{3}\frac{v_l^2\sigma^3}{\Delta\mu^3}. \tag{15.5}$$

It is important to emphasize that, given the small nature of the nucleating clusters, the proper thermodynamic work of formation depends on the external control variables and can lead to interesting surprises and finite size effects in the case of closed systems.[4,16,17,52,53]

The previous discussion solves the equilibrium part of the problem. However, nucleation is intrinsically a kinetic problem. In the context of CNT, the kinetics of nucleation and the nucleation rate is commonly obtained by either summing up the set of master equations eqn (15.1) or by approximating it into a continuous Fokker–Planck-like equation. In the following section, we will show how this equation can be rigorously derived in the framework of MNET.

15.3 Nucleation Kinetics using MNET

The main distinctive characteristic of nucleation is that it is an activated process: a free energy barrier has to be surmounted to form a large enough cluster that can then grow spontaneously. Many non-equilibrium processes in Nature bear this activated character, whose defining trait is a highly nonlinear response to the driving forces. This nonlinear aspect makes conventional non-equilibrium thermodynamics, described in Chapter 1, inapplicable to describe accurately activated processes.

In this section, we will see how MNET, introduced in Chapter 14, constitutes an ideal framework to derive the proper kinetic equations describing nucleation phenomena. As indicated in Chapter 14, MNET is a powerful, systematic and simple theory to describe the kinetics of non-equilibrium processes occurring at a mesoscopic scale in terms of arbitrary coordinates or degrees of freedom. It combines the systematic rules of non-equilibrium thermodynamics (NET) with the flexibility of Statistical Mechanics in describing the state of any system in terms of its probability distribution.

Let us assume that the state of the system can be fully specified by a single or, in general, a set of variables or coordinates that will be denoted by γ. These coordinates can represent the orientation of a spin, the length of a chain, the velocity of Brownian particle, or in the case of nucleation the radius or number of molecules of a cluster, as shown in Figure 15.1. In MNET, the out of equilibrium evolution of the system is then conceived as a kind of diffusion over the equilibrium landscape $\Delta W(\gamma)$, that represents the minimum reversible work required to create that state of the system (see Figure 15.2a). This thermodynamic potential landscape dictates the equilibrium distribution, according to the standard Boltzmann's expression

$$f_{\mathrm{eq}}(\gamma) \sim \mathrm{e}^{-\frac{\Delta W(\gamma)}{k_{\mathrm{B}}T}}. \tag{15.6}$$

The key point of MNET is the connection between Statistical Mechanics and NET, established through the statistical mechanics definition of the entropy given by Gibbs' entropy postulate[15,18]

$$\delta S = -k_{\mathrm{B}} \int \delta f(\gamma, t) \ln \frac{f(\gamma, t)}{f_{\mathrm{eq}}(\gamma, t)} \mathrm{d}\gamma, \tag{15.7}$$

where $f(\gamma,t)$ represents the probability that the system is at state γ at time t. In general, the evolution in time of this probability is governed by the continuity equation

$$\frac{\partial f(\gamma, t)}{\partial t} = -\frac{\partial}{\partial \gamma} J(\gamma, t), \tag{15.8}$$

where $J(\gamma,t)$ is a generalized current or density flux in γ-space that has to be specified. By taking the time derivative of eqn (15.7), inserting eqn (15.8), and making an integration by parts, one obtains the entropy production

$$\sigma_S = \frac{\delta S}{\delta t} = -k_{\mathrm{B}} \int J(\gamma, t) \frac{\partial}{\partial \gamma} \left(\ln \frac{f(\gamma, t)}{f_{\mathrm{eq}}(\gamma, t)} \right) \mathrm{d}\gamma, \tag{15.9}$$

which has the usual form of a product of a current, $J(\gamma,t)$, and a generalized thermodynamic force $-k_{\mathrm{B}} \frac{\partial}{\partial \gamma} \left(\ln \frac{f(\gamma, t)}{f_{\mathrm{eq}}(\gamma, t)} \right)$, expressed in terms of the variations of the probability with respect to the equilibrium distribution.

Following the standard procedure of NET, we assume a linear phenomenological relation between the flux and the force

$$J(\gamma,t) = -k_{\mathrm{B}}L(\gamma, f(\gamma))\frac{\partial}{\partial\gamma}\left(\ln\frac{f(\gamma,t)}{f_{\mathrm{eq}}(\gamma,t)}\right), \tag{15.10}$$

by introducing the phenomenological coefficient $L(\gamma, f(\gamma))$, which may in general depend on the state of the system. By substituting this expression for the current into the continuity equation, eqn (15.8), we obtain the Fokker–Planck equation governing the evolution of the probability density

$$\frac{\partial f(\gamma,t)}{\partial t} = \frac{\partial}{\partial\gamma}\left(D(\gamma)f_{\mathrm{eq}}(\gamma,t)\frac{\partial}{\partial\gamma}\left(\frac{f(\gamma,t)}{f_{\mathrm{eq}}(\gamma,t)}\right)\right), \tag{15.11}$$

where

$$D(\gamma,t) \equiv \frac{k_{\mathrm{B}}L(\gamma, f(\gamma))}{f(\gamma,t)}, \tag{15.12}$$

is a generalized diffusion coefficient. The Fokker–Planck equation can be alternatively written as

$$\frac{\partial f(\gamma,t)}{\partial t} = \frac{\partial}{\partial\gamma}\left(D(\gamma)\mathrm{e}^{-\beta\Delta W(\gamma)}\frac{\partial}{\partial\gamma}\left(f(\gamma,t)\mathrm{e}^{\beta\Delta W(\gamma)}\right)\right), \tag{15.13}$$

where $\beta = (k_{\mathrm{B}}T)^{-1}$, or as

$$\frac{\partial f(\gamma,t)}{\partial t} = \frac{\partial}{\partial\gamma}\left(D(\gamma)\frac{\partial f(\gamma,t)}{\partial\gamma} + \frac{D(\gamma)}{k_{\mathrm{B}}T}\frac{\partial\Delta W(\gamma)}{\partial\gamma}f(\gamma,t)\right), \tag{15.14}$$

that clearly evidences the drift-diffusion nature of this equation. The first term represents the diffusion, and the second term the drift due to a thermodynamic force $-\dfrac{\partial\Delta W(\gamma)}{\partial\gamma}$.

The general form of this equation makes it applicable to a wide variety of non-equilibrium problems that include for instance chemical reactions, entropic transport or the dynamics of single biomolecules.[15] Let us particularize it for the case of nucleation, using the number of molecules in a cluster n, as our reaction coordinate. As described in the introduction, nucleation can be viewed as a diffusion process over a free energy landscape. In the case of nucleation at constant pressure, temperature and number of molecules, the proper thermodynamic potential or minimum reversible work $\Delta W(n)$ is the Gibbs energy of formation of a cluster of a given size $\Delta G(n)$ (see Figure 15.1). With this prescription in eqn (15.14), the corresponding kinetic equation for the evolution of the population of clusters becomes

$$\frac{\partial f(n,t)}{\partial t} = \frac{\partial}{\partial n}\left(k^{+}(n)\frac{\partial f(n,t)}{\partial n} + \frac{k^{+}(n)}{k_{\mathrm{B}}T}\frac{\partial\Delta G(n)}{\partial n}f(n,t)\right). \tag{15.15}$$

The previous expression constitutes the classical Frenkel–Zeldovich equation that is the basis of the study of nucleation kinetics,[3] where the diffusivity coefficient in size space $D(n) \equiv k^+(n)$ is just the rate of attachment of molecules to a cluster of size n. It is remarkable that this equation was also derived by Lothe using a similar procedure inspired by NET.[19]

It is worth emphasizing that, using MNET, it is possible to derive descriptions of the kinetics of nucleation using alternative variables or reaction coordinates. For instance, one could use the radius of the cluster instead of the number of molecules, or a global order parameter describing the global degree of crystallization, such as the one used by Frenkel and coworkers in their simulations.[20,21] More refined and elaborated descriptions in terms of a density functional, resembling Dynamical Density Functional Theory, or using a hydrodynamic description in terms of density and velocity fields, can also be successfully implemented, as reviewed in ref. 54.

With these Fokker–Plank-like equations, one obtains a complete time dependent description of nucleation that accounts for transient effects and the potential influence of an initial distribution of pre-existing clusters. Careful analysis of transient and non-stationary effects in nucleation can be found in ref. 1, 5 and 22. However, for most practical purposes and in experiments, the quantity of central interest is the steady-state nucleation rate, that in the case of a significantly high nucleation barrier, is given by the simple expression

$$J_{\mathrm{CNT}} = \frac{N_1}{V} k^+(n^*) Z e^{-\beta \Delta G^*}, \tag{15.16}$$

where N_1 is the total number of monomers, V is the volume, and

$$Z = \sqrt{\frac{|\Delta G''(n^*)|}{2\pi k_{\mathrm{B}} T}}, \tag{15.17}$$

is the Zeldovich factor, a correction associated with the local curvature at the top of the barrier that accounts for the possibility that nearly critical-sized clusters dissociate back to the solution.

The steady state nucleation rate adopts the standard expression of an activated process, depending exponentially on the height of the nucleation barrier, *i.e.* the work of formation of critical-sized clusters. These critical clusters are unstable objects made typically by a few molecules, and its formation is a rare event of stochastic nature. That is the reason why the accurate evaluation of nucleation rates, critical cluster sizes and nucleation barriers, is challenging not only in experiments but also in molecular simulations. Nevertheless, the activated nature of the process and its conception as a generalized diffusion process over a thermodynamic potential landscape in the framework of MNET has opened the door to novel analysis and simulation techniques, described in the following section.

15.4 Novel Simulation Techniques to Study Nucleation

One of the main advantages of having recast the kinetics of nucleation into the general framework of stochastic and activated processes is the possibility of using the vast knowledge acquired in these fields to develop novel techniques to analyze nucleation.

In particular, one of the important concepts in the theory of stochastic processes is the mean first-passage time (MFPT) that for activated processes is ultimately related to the steady-state rate of barrier crossing.[23] In a general one-dimensional case, the MFPT $\tau(\gamma_0;a,b)$ is defined as the average time required for a system initially at γ_0 to leave the region $[a,b]$ for the first time (see Figure 15.2). Its value depends on the nature of the boundary

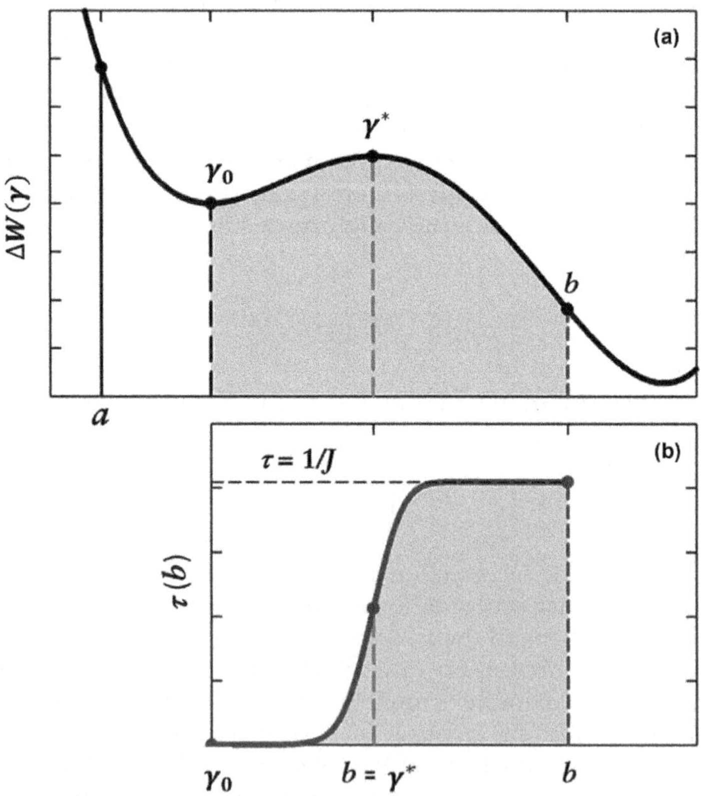

Figure 15.2 Schematic representation of (a) the minimum reversible work landscape $\Delta W(\gamma)$ of a general activated process and (b) its corresponding mean first passage time $\tau(b)$ as a function of the location of the absorbing boundary b. The inflection point of $\tau(b)$ signals the top of the barrier (*i.e.* the critical size), and the inverse of its plateau provides the steady-state nucleation rate.
Adapted from ref. 24.

conditions. For nucleation, the variable γ will represent, for instance, the number of molecules n of the largest cluster in the system and the proper boundary conditions are reflecting at size n_1 (typically, $n_1 = 1$) and absorbing at size n_b. In that case, the explicit expression of the MFPT becomes

$$\tau(n_b) \equiv \tau(n_0;\, n_1, n_b) = \int_{n_0}^{n_b} \frac{1}{D(y)} \mathrm{d}y\, \mathrm{e}^{\beta\Delta G(y)} \int_{n_1}^{n_b} \mathrm{d}z\, \mathrm{e}^{-\beta\Delta G(z)}, \tag{15.18}$$

where by using the notation $\tau(n_b)$ we want to emphasize that we will focus on the behavior of the MFPT in terms of n_b for a fixed starting size n_0. This behavior exhibits two important characteristics for sufficiently high activation barriers $\beta\Delta G^* \gg 1$. First, $\tau(n^*)$, *i.e.* the average time required to reach the critical size for the first time, is related to the steady state nucleation rate as $J = 1/(2\tau(n^*))$, where the factor 1/2 reflects the fact that clusters at the top of the barrier have a 50 % chance of falling to either side. Second, the critical cluster size is very approximately located at the inflection point of the curve $\tau(n_b)$, as shown by Figure 15.2.

More importantly, for sufficiently high barriers, the MFPT in eqn (15.18) can be evaluated analytically using the steepest descent approximation, yielding[24]

$$\tau(n_b) = \frac{\tau_J}{2}(1 + \mathrm{erf}\,[(n - n^*)c]), \tag{15.19}$$

where n^* is the critical cluster size, $\mathrm{erf}(x) = 2\int_0^x \mathrm{e}^{-x^2} \mathrm{d}x/\sqrt{\pi}$ is the error function, $c = Z\sqrt{\pi}$ is the local curvature around the top of the barrier, and $\tau_J = 1/(JV)$ is the inverse of the steady-state nucleation rate. This simple expression offers an accurate and simple way to obtain all relevant kinetic information of a nucleation phenomenon, namely the nucleation rate J, critical cluster size n^*, and Zeldovich factor Z, by simple evaluating the MFPT as a function of cluster size and then fitting it to eqn (15.19). This procedure can be implemented in different types of molecular simulations (*e.g.* Molecular Dynamics (MD), Brownian Dynamics, Kinetic Monte Carlo...) as well as in experiments. It has been in fact extensively used in the recent literature to study different aspects of condensation,[25] homogeneous[26] and heterogeneous crystallization[27] or cavitation[28,29] in a wide variety of systems.

In a simulation, the MFPT can be evaluated in practice by monitoring the size of the largest cluster present in the system and the time $t_i(n)$ at which this largest cluster reaches each particular size n for the first time. The mean first passage time $\tau(n)$ for any size n is then obtained by averaging $t_i(n)$ over R realizations of the simulations with different initial configurations, namely: $\tau(n) = \sum_{i=1}^{R} t_i(n)/R$. Figure 15.3 shows one example obtained in a MD simulation of the condensation of Lennard-Jones argon at $T = 70$ K, using $R = 200$ independent realizations. The resulting curve can be very accurately fitted by eqn (15.19), providing accurate values of J, n^* and Z.

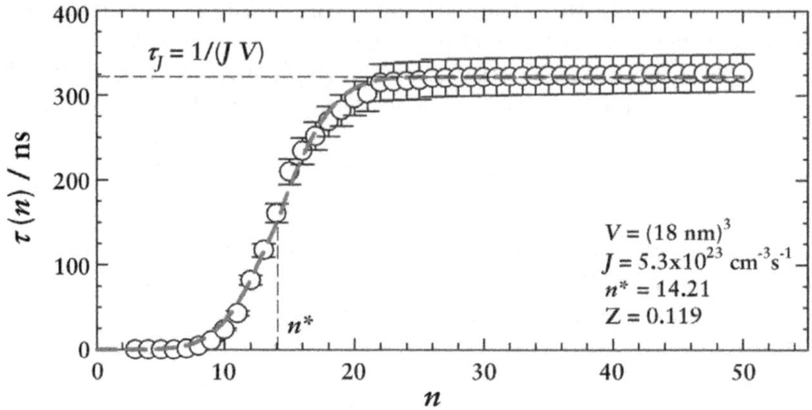

Figure 15.3 Mean first passage time as a function of the cluster size n obtained in MD simulations of condensation of Lennard-Jones argon at temperature $T = 50$ K. $N = 343$ atoms were simulated inside a cubic container of volume $V = (18\,\mathrm{nm})^3$. The results were obtained with 200 repetitions and have been fitted to eqn (15.19) (red dashed line), to obtain accurate values of the nucleation rate J, critical cluster size n^* and Zeldovich factor Z indicated in the inset.
Figure adapted from ref. 24.

However, one can even go one step further and use the MFPT to reconstruct the full Gibbs energy landscape of cluster formation $\Delta G(n)$ directly from a dynamic simulation or experiment. The ingredients required to accomplish that are just two: the MFPT, $\tau(n)$, and the steady state probability distribution for the largest cluster, $P_{\mathrm{st}}(n)$ that can be simply obtained by making a histogram of the size of the largest cluster accumulated in all simulation runs at a given set of fixed conditions (see Figure 15.4). Combining these two ingredients, it is possible to reconstruct the Gibbs energy landscape of cluster formation for any interval of sizes $n_1 \leq n \leq b$, by calculating first[30]

$$B(n) = -\frac{1}{P_{\mathrm{st}}(n)}\left[\int_n^b P_{\mathrm{st}}(n')\mathrm{d}n' - \frac{\tau(b) - \tau(n)}{\tau(b)}\right], \tag{15.20}$$

and then using the expression

$$\beta\Delta G(n) = \beta\Delta G(n_1) + \ln\left(\frac{B(n)}{B(n_1)}\right) - \int_{n_1}^{n} \frac{\mathrm{d}n'}{B(n')}. \tag{15.21}$$

In the previous equations n_1 is a reference size, typically $n_1 = 1$, and b represents an absorbing boundary up to which we sample both $P_{\mathrm{st}}(n)$ and $\tau(n)$. The details of the derivation of eqn (15.20) and (15.21) starting from the Fokker–Planck eqn (15.15), can be found in ref. 30. As an example, Figure 15.4 shows the MFPT, $\tau(n)$, the steady state cluster size distribution $P_{\mathrm{st}}(n)$ and free energy of cluster formation $\Delta G(n)$ reconstructed using

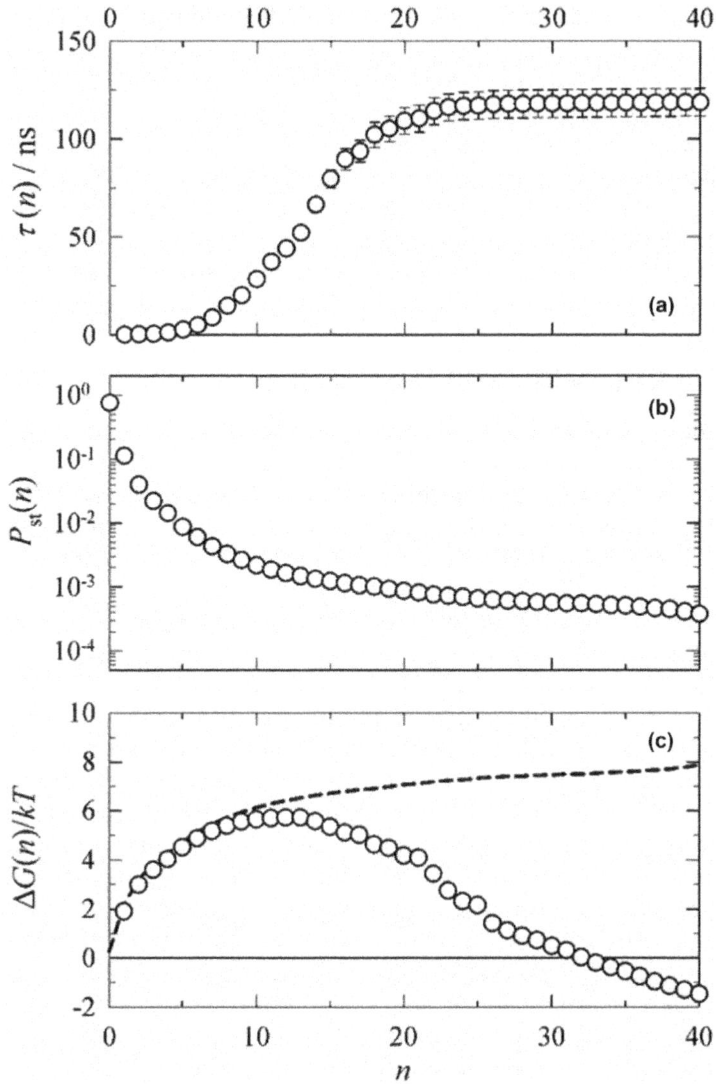

Figure 15.4 (a) Mean first-passage times $\tau(n)$ and (b) steady-state probability distribution $P_{st}(n)$ as function of the largest cluster size n obtained from 300 realizations of MD simulations with $N=343$ Lennard-Jones argon atoms in a volume $V=(11\,\text{nm})^3$ at $T=70$ K. (c) Kinetic reconstruction of the free energy of formation of the largest cluster obtained from the MD simulations using eqn (15.20) and (15.21) (symbols), compared to $-\ln P_{st}(n)$ (dashed line), which would be the standard way to get the free-energy landscape from a given equilibrium probability distribution $P_{eq}(n)$.

Reprinted with permission from ref. 30. Copyright 2008 American Chemical Society.

eqn (15.20) and (15.21) from a simulation of LJ condensation at $T = 70$ K. It is worth emphasizing that the energy landscape reconstructed using this technique has been shown to agree with the results obtained with more sophisticated methods such as umbrella sampling.[31]

With these novel techniques, it has been possible to obtain complete thermodynamic and kinetic information of different nucleation phenomena, as well as to unveil several controversial aspects of nucleation such as the proper definition of liquid clusters in the context of condensation,[32] the validity of CNT at extreme conditions,[33] or the importance of non-isothermal effects and pressure of carrier gas described in the following two sections.

15.5 Non-isothermal Nucleation

The advantages of treating nucleation in the context of non-equilibrium thermodynamics become more evident in the case of dealing with additional influences or couplings in the process. NET offers a systematic framework where these influences can be properly and rigorously incorporated in the dynamic description of the problem. Perhaps the best example is the accounting of non-isothermal effects in nucleation.

The formation of a liquid drop in a metastable vapour or a crystal in a supercooled liquid involves a significant release of energy, associated to the latent heat. For very small clusters or at rapid conditions of formation, this may lead to a significant change in the temperature of the nucleating clusters. The problem is that nucleation is extremely sensitive to the temperature. More precisely, both the equilibrium vapour pressure and the evapouration rate depend exponentially on the value of temperature, leading to huge variations of the nucleation rate with tiny changes of temperature. For instance, nucleation rates in the condensation of argon change by more than 25 orders of magnitude upon varying the temperature just by 5 K. Given the extreme sensitivity of nucleation rates to temperature, non-isothermal conditions and the unavoidable thermal fluctuations during nucleation were suggested as a possible explanation of the huge discrepancies between the predictions of CNT and experiments.[34,35]

A proper NET framework to account for non-isothermal effects and energy fluctuations for nucleating clusters was developed in 1966 in a remarkable work by Feder, Russell, Lothe, and Pound[36] that we will now adapt to the terminology of MNET. The key idea was to incorporate the energy E of the clusters as a second important variable required for a proper description of the state of a nucleating cluster. In terms of these two variables, a cluster of size n can evolve in size-space by the addition or loss of an individual molecule, and this change of size is accompanied by an increase or decrease of its energy by an amount related to the latent heat. In addition, a cluster of a fixed size n can change its energy by collisions with vapour molecules (that do not end up in the cluster) or other carrier gas molecules present in the system, as depicted schematically in Figure 15.5. Thus, the evolution in time of the distribution of clusters characterized in terms of these two variables,

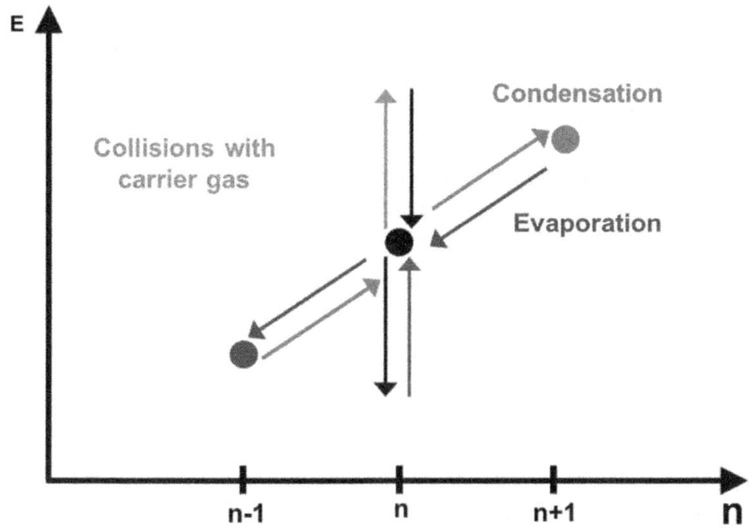

Figure 15.5 Schematic representation of the evolution of a cluster of size n in size and energy space. Collisions with carrier gas molecules can alter the energy of the cluster without changing its size. The addition and loss of a single molecule not only change the cluster size but also its energy by an amount related to the release or absorption of latent heat.

the number of molecules n and their energy E, will now be dictated by the 2D continuity equation

$$\frac{\partial f(n,E,t)}{\partial t} = -\nabla \cdot \boldsymbol{J}, \tag{15.22}$$

where $\boldsymbol{J}=(J_n, J_E)$ is a two dimensional flux composed by the current in size J_n and in energy space J_E. Following the ideas of NET, a linear phenomenological relation can be proposed between the generalized flux $\boldsymbol{J}$ and the thermodynamic force $\nabla \ln \frac{f(n,E,t)}{f_{eq}(n,E)}$, yielding

$$\boldsymbol{J} = -\mathfrak{D} f_{eq} \cdot \nabla\left(\frac{f}{f_{eq}}\right), \tag{15.23}$$

where

$$\mathfrak{D} = zA(n)\begin{pmatrix} 1 & q \\ q & q^2+b^2 \end{pmatrix}, \tag{15.24}$$

is a two dimensional generalized diffusion coefficient. In the previous expression, $z=p/\sqrt{2\pi m k_B T_0}$ is the rate of collisions to a cluster of size n, $A(n)$ is the surface area of the cluster, T_0 is the temperature of the heat bath,

$$q = h - \frac{k_B T_0}{2} - \sigma\frac{\partial A(n)}{\partial n}, \tag{15.25}$$

is the energy increase upon the addition of a molecule, given by the latent heat h corrected by the energy spent in increasing the area of the cluster, and

$$b^2 = \left(c_V + \frac{1}{2}k_B\right)k_B T_0^2 + \frac{z_c}{z}\left(c_{V,c} + \frac{1}{2}k_B\right)k_B T_0^2, \qquad (15.26)$$

is the typical amplitude of energy fluctuations due to collisions with their own vapour molecules (first term) or carrier gas molecules (second term), with $z_c = p_c/\sqrt{2\pi m k_B T_0}$ the collision frequency of carrier gas molecules, p_c the pressure of the carrier gas, and c_V and $c_{V,c}$ the specific heats of the condensable and carrier gas vapours, respectively. For ideal gases, the last expression can be simplified to

$$b^2 = 2k_B^2 T_0^2\left(1 + \frac{N_c}{N}\sqrt{\frac{m}{m_c}}\right), \qquad (15.27)$$

where N_c and N are the number of carrier gas and condensable molecules and m, m_c their respective masses. Note that, somehow counter intuitively, lighter carrier gases are more effective as heat baths, due to their higher collision frequency.

It is reasonable to assume that equilibration in the energy space will proceed faster that in size space. By assuming $J_E = 0$, and solving for J_n one obtains a simple expression for the non-isothermal nucleation rate

$$J_{\text{noniso}} = \frac{b^2}{b^2 + q^2} J_{\text{iso}}, \qquad (15.28)$$

in terms of the isothermal nucleation rate J_{iso}. Thus, non-isothermal effects essentially depend on the ratio between the energy provided by the latent heat, q, and the energy that is removed by collisions b. When $q/b \ll 1$, nucleation proceeds under nearly isothermal conditions, whereas for $q/b \gg 1$, significant deviations are expected, as shown in Figure 15.6.

By assuming that the equilibrium distribution of clusters in size and energy space was Gaussian, and by locating the saddle point in the effective 2D barrier towards nucleation, Feder *et al.* also derived a simple expression for the change in energy and temperature of the nucleating clusters

$$\Delta T = T_0 \frac{q}{b^2 + q^2}\left(-\frac{\partial \Delta G(n)}{\partial n}\right). \qquad (15.29)$$

The previous expression predicts that clusters larger than the critical size, for which $\frac{\partial \Delta G(n)}{\partial n} < 0$, tend to be hotter than the bath temperature, whereas clusters smaller than the critical size tend to be colder. This striking prediction is still the subject of much controversy, having its roots in the still unsolved issue of the proper definition of temperature and its fluctuations for small systems.[37–44]

The importance of non-isothermal effects, the influence of temperature fluctuations, and the efficiency of different thermostats was tested in ref. 45

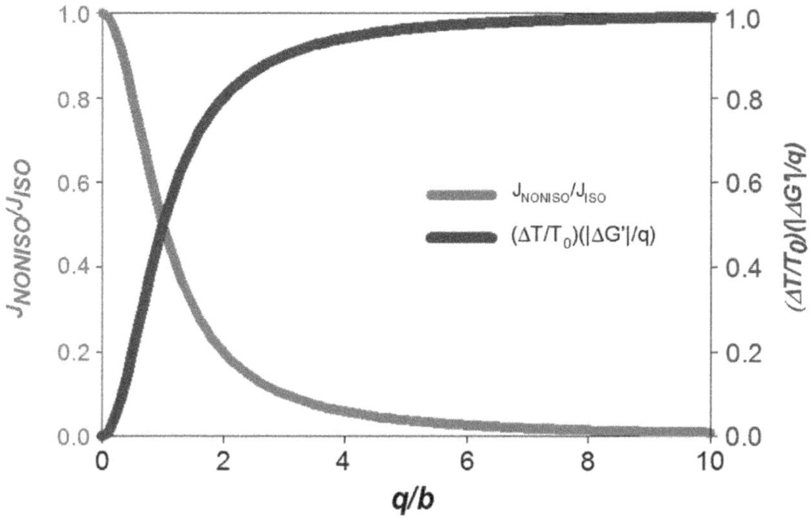

Figure 15.6 Dependence of the non-isothermal nucleation rate and the normalized temperature shift according to classical non-isothermal nucleation theory as a function of the ratio q/b.
Figure adapted from ref. 45.

using extensive MD simulations and the accurate analysis techniques described in the previous section. The results of these simulations for non-isothermal nucleation rates and cluster temperatures could be almost quantitatively explained by the predictions of Feder *et al.* Somewhat surprisingly, the use of different thermostats did not have a significant influence on the nucleation rates. More importantly, non-isothermal effects lead to a significant heating up of the average temperature of the cluster that can reach tens of degrees for large post-critical clusters. Nevertheless, in accordance to the predictions of eqn (15.28), they only lead to a decrease of nucleation rates of at most 2 to 3 orders of magnitude for systems with large values of the latent heat and in the absence of thermalizing carrier gas molecules. This decrease of nucleation rates can be considered modest, compared to the extreme sensitivity of nucleation rates to the bath temperature indicated at the beginning of this section.

Concerning cluster temperatures, the average temperature of all cluster sizes is always higher than the bath temperature, but its distribution for a given size is non-Gaussian, with a peak or most probable temperature which seems to be lower than the bath temperature for subcritical clusters and hotter for post-critical clusters, in accordance with the predictions of Feder *et al.* (see Figure 15.7). It is worth remarking that recent works have emphasized the fact that the previous results can also be fitted by an alternative definition of cluster temperature that makes it coincide with the average temperature.[43,44]

Alternative NET studies of non-isothermal effects, focusing on the global mass and energy balance, rather than on the local energy and size

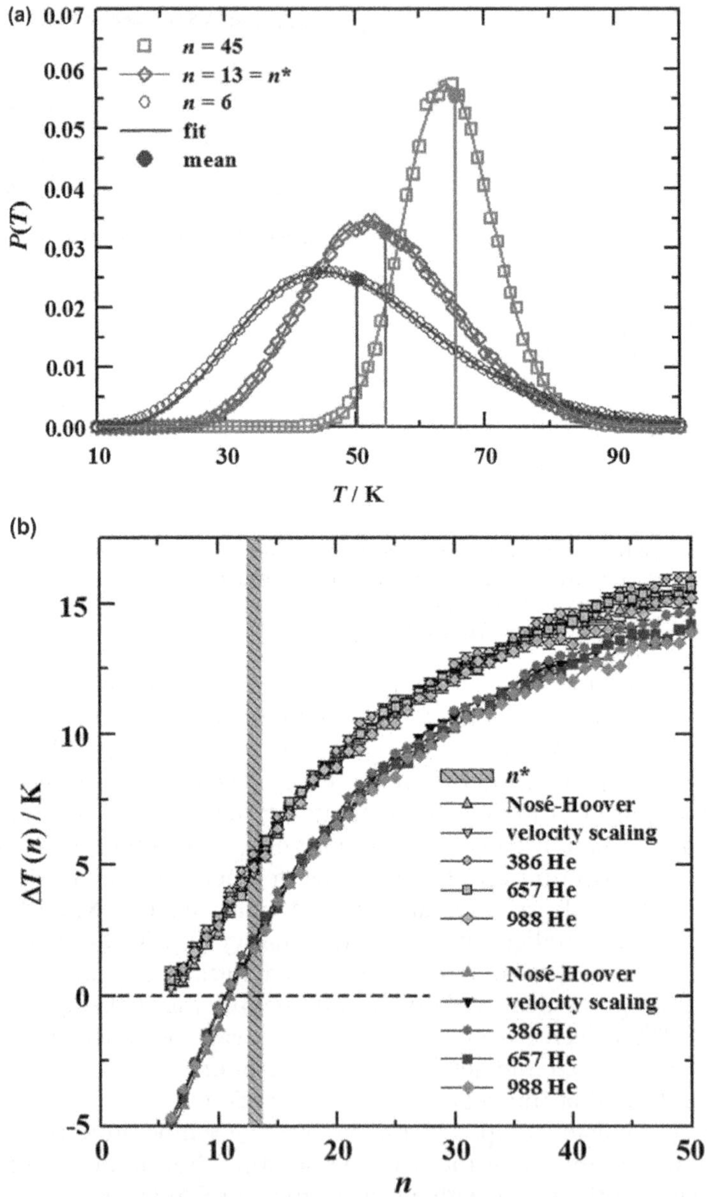

Figure 15.7 (a) Cluster temperature distribution for sub-critical (blue circles), critical (green diamonds), and post-critical (red squares) sized clusters obtained from MD simulations of LJ argon condensation at $S = 869$ and $T_0 = 50$ K. For small cluster sizes, the distribution is non-Gaussian with a most probable temperature that differs from the mean temperature. (b) Deviation from the bath temperature of the average (black symbols) and most probable (coloured symbols) cluster temperatures as a function of the cluster size n obtained from MD simulations using different thermostats and number of He carrier gas molecules.
Reproduced with permission from ref. 45. Copyright 2007, American Institute of Physics.

distribution of the individual clusters have also been provided, successfully explaining non-isothermal variations of the nucleation rate in terms of measurable heat fluxes.[46]

15.6 The Influence of Carrier Gas Pressure on Nucleation

One of the most puzzling and controversial issues in the experiments of nucleation was the role of the carrier gas. This carrier gas was a physically and chemically inert gas (typically Ar, Xe or He) that was added to the condensable species with the main role of getting rid efficiently of the latent heat released during the nucleation and thus guaranteeing proper isothermal conditions. As such, the pressure p_c (measuring the amount of added gas) of the carrier gas was expected to show no influence on the measured nucleation rates. However, experimental measurements using different substances and techniques reported contradictory and puzzling results: either no effects (typically in nucleation pulse chambers), an increase (in diffusion and sometimes in laminar flow chambers) or a decrease (mostly in laminar flow chambers) of nucleation rates with the pressure of the carrier gas.

This puzzling influence of the carrier gas was explained in ref. 47 in terms of accounting properly for the simplest influence of an inert carrier gas. First, the formation of a liquid drop in the presence of a carrier gas requires an extra work associated to the pressure–volume (pV) work against the ambient pressure. This leads to a different height of the nucleation barrier

$$\Delta G^*_{\mathrm{pV}} = \frac{16\pi}{3}\frac{v_l^2\sigma^3}{\Delta\mu_{\mathrm{eff}}^2}, \tag{15.30}$$

due to the pV work that, for an ideal vapour and incompressible liquid, is accounted for by introducing an effective chemical potential $\Delta\mu_{\mathrm{eff}} = k_{\mathrm{B}}T\ln\frac{p}{p_{\mathrm{eq}}} + v_l(p + p_{\mathrm{c}} - p_{\mathrm{eq}})$. This yields to a modified nucleation rate $J_{\mathrm{pV}} = K\exp(-\beta\Delta G^*_{\mathrm{pV}})$.

The second main influence of the carrier gas is in the thermalization of the nucleating clusters, whose rate of formation under general non-isothermal conditions was given by eqn (15.28). By combining these two effects, a simple expression for the influence of the pressure of the carrier gas on the nucleation rate was derived

$$\frac{J_{\mathrm{PE}}}{J_{\mathrm{CNT}}} = \frac{b^2}{b^2 + q^2}\frac{J_{\mathrm{pV}}}{J_{\mathrm{CNT}}}. \tag{15.31}$$

These two effects have an opposite influence, leading to a nontrivial dependence of nucleation rate on the pressure of the carrier gas which can be positive, negative or null, depending on the conditions, as evidenced in Figure 15.8.

This prediction was tested against MD simulations of condensation of Lennard-Jones vapour in the presence of different amounts of carrier gas at different temperatures, showing a perfect agreement, as illustrated in Figure 15.8b. Thus the proper accounting of the equilibrium (*i.e.* pV work) and non-equilibrium (*i.e.* non-isothermal) effects of the carrier gas was the key to solve this controversial issue.

15.7 Nucleation in the Presence of Gradients

The analysis developed in the previous sections has been focused on the simplest case of homogeneous, isothermal and isotropic nucleation. But the real process very often occurs in a media which in general has spatial, thermal or velocity non-homogeneities, which in turn may exert a relevant influence in the process. One example is the case of polymer crystallization, which may occur in the presence of strong thermal gradients and mechanical stresses, demanding a study of the process under non-isothermal and inhomogeneous conditions.

One of the advantages of using MNET to study nucleation is that it is relatively straightforward to deal with the presence of non-homogeneities, thus yielding a more realistic model of nucleation and crystallization. In this section, we will focus on the simplest cases in which the medium may affect the kinetics, namely the presence of temperature gradients,[48] and the impact of flow and stresses in the nucleation process.[49]

In the case of homogeneous isotropic nucleation, we can leave spatial dependencies aside as the process occurs identically at any point of the system. However, when the system is inhomogeneous the conditions controlling condensation and crystallization vary from point to point of the material, and therefore a local description of the process must be considered. It is important to remark that nucleation involves two clearly differentiated length scales. Nucleation occurs on a mesoscopic scale, while thermodynamic quantities as pressure, temperature, density, *etc.* vary on a longer scale and can be considered as locally uniform for the nucleation events.

Taking these considerations into account, a local description of nucleation using MNET, was developed in ref. 48 and applied to describe nucleation in a temperature gradient. The resulting Fokker–Planck equation governing the evolution of the cluster distribution function in spatially inhomogeneous systems, in the *diffusion regime*, is

$$\begin{aligned}\frac{\partial f_c}{\partial t} &= \nabla \cdot (D_0 \nabla f_c) + \nabla \cdot \left(D_{th} \frac{\nabla T}{T} f_c \right) \\ &\quad + \frac{\partial}{\partial n} \left(k^+(n) \left(\frac{\partial f_c}{\partial n} + \beta \frac{\partial \widetilde{\Delta G}(n)}{\partial n} f_c \right) \right),\end{aligned} \tag{15.32}$$

where D_0 is the spatial diffusion coefficient, D_{th} is the thermal diffusion coefficient and $f_c(n,\boldsymbol{x},t)$ represents the distribution of clusters. The previous

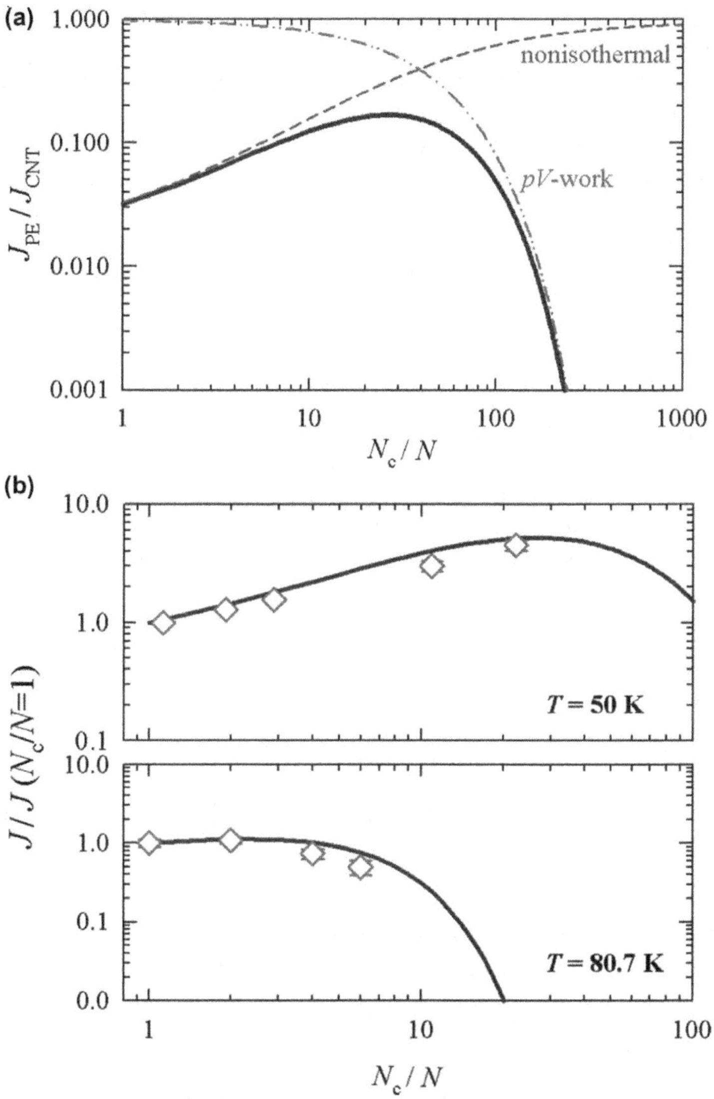

Figure 15.8 (a) Deviation of the nucleation rate from the CNT prediction due to the full pressure effect, eqn (15.31), as a function of the ratio of carrier gas over vapour molecules (solid line) arising from the two contributions of non-isothermal effects (dashed line, eqn (15.28)), and pV work (dash-dotted line) for argon at 50 K and $S = 869$. (b) Comparison of MD simulation results with theoretical prediction of the pressure effect, eqn (15.31).
Figures adapted from ref. 47.

equation accounts for the effects of diffusion, thermal diffusion and cluster formation, represented by the three terms on the right hand side, respectively.

Using this equation, it is possible to perform a detailed analysis of the influence of non-homogeneities and thermal gradients in real experiments for condensation and polymer crystallization. In the case of condensation in thermal[50] and laminar flow[51] diffusion cloud chambers, nucleation turns out to be not significantly affected by diffusion and thermal diffusion effects, when experiments are performed at normal conditions. However, in rarefied media, as in the upper atmosphere or for substances with low equilibrium vapour pressures, non-isothermal conditions can become extremely relevant. In the case of polymer crystallization, their low thermal conductivities, which set up very large gradients, and high values of the Soret coefficient increase the relevance of thermal diffusion leading to important alterations in real crystallization.

The presence of flows or, in general, stresses in the system constitutes another situation that can influence the nucleation process. This factor is particularly relevant in polymer crystallization, which often involves mechanical processing of the melt, such as extrusion, shearing or injection.

The description of the process of nucleation in the presence of flows was carried out in the framework of MNET in ref. 49. The resulting equation, in the diffusion regime, is

$$\frac{\partial f_c}{\partial t} = -\nabla \cdot (f_c \boldsymbol{v}_0) + \nabla \cdot (\mathfrak{D} \cdot \nabla f_c) + \frac{\partial}{\partial n}\left(k^+(n)\left(\frac{\partial f_c}{\partial n} + \beta \frac{\partial \widetilde{\Delta G}(n)}{\partial n} f_c\right)\right), \quad (15.33)$$

where $\boldsymbol{v}_0$ is the stationary velocity profile, and

$$\mathfrak{D} = D_0 \left[\mathbb{I} - \left(\frac{\overrightarrow{\overrightarrow{\eta_B}}}{p} \cdot \nabla \boldsymbol{v}_0 \right)^0 \right], \quad (15.34)$$

is the effective diffusion coefficient which becomes modified by the presence of the flow. In the previous expression, D_0 is the diffusion coefficient of the melt, $\overrightarrow{\overrightarrow{\eta_B}} = D_0 f_c \left(\mathbb{I} + \frac{m \mathscr{L}_{ux}}{f_c T} \right)$ is the Brownian viscosity, and $\mathscr{L}_{ux}$ is the friction tensor that couples the spatial and velocity currents.[49]

The main effects that the presence of a shear flow exerts on the nucleation process can be summarized as follows. On one hand, the flow alters the transport and consequently the evolution of the growing clusters distribution function, which has implications in the effective nucleation and growth rate. On the other hand, the presence of a shear flow changes the spatial diffusion coefficient of the clusters, as shown by eqn (15.34). Since the rate of addition of molecules to a cluster is roughly proportional to the diffusivity of the molecules, variations of the diffusion coefficient directly affect the value of the nucleation rate. Moreover, the presence of the flow destroys the isotropy of the system and leads to a distinction between growth rates (and diffusion) in different directions. Typical values of the parameters controlling this correction imply that this effect is not very important for condensation. However, the high viscosity and the peculiarities of polymer crystallization suggest that the presence of a shear flow may promote drastic changes in the process, as has been observed experimentally. At a more microscopic level, shear can also enhance the destruction of clusters.[55]

15.8 Conclusions

In this chapter we have shown how the application of NET can be very useful in the description of nucleation phenomena. In particular NET at the mesoscopic level provides a quite convenient framework to analyze and describe the kinetics of nucleation in terms of a set of relevant variables or reaction coordinates, such as the radius or number of molecules of a cluster, or the global degree of crystallization of a sample. This description, leading to a generalized Fokker–Planck like equation, has been the basis of novel techniques to accurately characterize nucleation phenomena in simulations and experiments. Additionally, one of the main advantages of NET over other non-equilibrium approaches is the rigorous and simple procedure to incorporate coupled effects and external influences, such as non-isothermal effects, temperature or velocity gradients. The resulting equations have been invaluable to understand better nucleation phenomena and to shed light on some of the controversies that still surround it. Important problems, such as the role of curvature on heat transfer, non-accommodation effects, or microscopic influences of shear on cluster formation, lie beyond the scope of the present work. A proper accounting of other equilibrium and non-equilibrium aspects will be the key to achieve the golden goal of being capable of having quantitatively accurate predictions of nucleation rates, thus solving an important problem that has puzzled scientists for centuries.

Acknowledgements

The work was partially supported by the Spanish MINECO through Grant No. FIS2011-22603.

References

1. P. G. Debenedetti, *Metastable Liquids: Concepts and Principles*, Princeton University Press, Princeton, 1996.
2. D. Kashchiev, *Nucleation: Basic Theory with Applications*, Butterworth-Heinemann, Oxford, 2000.
3. J. Frenkel, *Kinetic Theory of Liquids*, Dover, New York, 1955.
4. F. F. Abraham, *Homogenous Nucleation Theory: The Pretransition Theory of Vapour Condensation*, Academic Press, New York, 1974.
5. K. F. Kelton and A. L. Greer, *Nucleation in Condensed Matter*, Elsevier, Amsterdam, 2010.
6. V. I. Kalikmanov, *Nucleation Theory*, Springer, Heidelberg, 2013.
7. D. G. Fahrenheit, *Philos. Trans. R. Soc. London*, 1724, **33**, 78–84.

8. M. Volmer and A. Weber, *Z. Phys. Chem.*, 1926, **119**, 277–301.
9. I. Farkas, *Z. Phys. Chem.*, 1927, **125**, 236–242.
10. R. Becker and W. Döring, *Ann. Phys.*, 1935, **24**, 719–752.
11. J. B. Zeldovich, *Acta Physicochim. URSS*, 1943, **18**, 1–22.
12. A. Laaksonen, V. Talanquer and D. W. Oxtoby, *Annu. Rev. Phys. Chem.*, 1995, **46**, 489–524.
13. K. Iland, J. Wölk, R. Strey and D. Kashchiev, *J. Chem. Phys.*, 2007, **127**, 154506.
14. D. Reguera, *J. Non-Equilib. Thermodyn.*, 2004, **29**, 327–344.
15. D. Reguera, J. M. Rubi and J. M. G. Vilar, *J. Phys. Chem. B*, 2005, **109**, 21502–21515.
16. D. Reguera, R. K. Bowles, Y. Djikaev and H. Reiss, *J. Chem. Phys.*, 2003, **118**, 340–353.
17. J. Wedekind, D. Reguera and R. Strey, *J. Chem. Phys.*, 2006, **125**, 214505.
18. S. de Groot and P. Mazur, *Non-equilibrium Thermodynamics*, Dover, New York, 1984.
19. J. Lothe, *J. Chem. Phys.*, 1966, **45**, 2678–2680.
20. P. Rein ten Wolde, M. J. Ruiz-Montero and D. Frenkel, *J. Chem. Phys.*, 1996, **104**, 9932–9947.
21. D. Reguera, J. M. Rubi and A. Pérez-Madrid, *J. Chem. Phys.*, 1998, **109**, 5987–5993.
22. V. A. Shneidman, *Sov. Phys. Tech. Phys.*, 1987, **32**, 76–80.
23. P. Hänggi, P. Talkner and M. Borkovec, *Rev. Mod. Phys.*, 1990, **62**, 251–341.
24. J. Wedekind, R. Strey and D. Reguera, *J. Chem. Phys.*, 2007, **126**, 134103.
25. J. Wedekind, J. Wölk, D. Reguera and R. Strey, *J. Chem. Phys.*, 2007, **127**, 154515.
26. E. B. Moore and V. Molinero, *Nature*, 2011, **479**, 506–508.
27. B. Scheifele, I. Saika-Voivod, R. K. Bowles and P. H. Poole, *Phys. Rev. E: Stat., Nonlinear, Soft Matter Phys.*, 2013, **87**, 042407.
28. J. L. F. Abascal, M. A. Gonzalez, J. L. Aragones and C. Valeriani, *J. Chem. Phys.*, 2013, **138**, 084508.
29. V. G. Baidakov and K. S. Bobrov, *J. Chem. Phys.*, 2014, **140**, 184506.
30. J. Wedekind and D. Reguera, *J. Phys. Chem. B*, 2008, **112**, 11060–11063.
31. S. E. M. Lundrigan and I. Saika-Voivod, *J. Chem. Phys.*, 2009, **131**, 104503.
32. J. Wedekind and D. Reguera, *J. Chem. Phys.*, 2007, **127**, 154516.
33. J. Wedekind, G. Chkonia, J. Wölk, R. Strey and D. Reguera, *J. Chem. Phys.*, 2009, **131**, 114506.
34. I. Ford and C. Clement, *J. Phys. A Math. Gen.*, 1989, **22**, 4007–4018.
35. J. Barrett, C. Clement and I. Ford, *J. Phys. A. Math. Gen.*, 1993, **26**, 529–548.
36. J. Feder, K. Russell, J. Lothe and G. Pound, *Adv. Phys.*, 1966, **15**, 111–178.
37. R. Mcfee, *Am. J. Phys.*, 1973, **41**, 230–235.
38. R. Mcfee, *Phys. Today*, 1973, **41**, 1212.
39. C. Kittel, *Am. J. Phys.*, 1973, **41**, 1211–1212.
40. C. Kittel, *Phys. Today*, 1988, **41**, 93.

41. B. B. Mandelbrot, *Phys. Today*, 1989, **42**, 71.
42. R. McGraw and R. A. LaViolette, *J. Chem. Phys.*, 1995, **102**, 8983–8994.
43. J. C. Barrett, *J. Chem. Phys.*, 2011, **135**, 096101.
44. G. S. Boltachev and J. W. P. Schmelzer, *J. Chem. Phys.*, 2010, **133**, 134509.
45. J. Wedekind, D. Reguera and R. Strey, *J. Chem. Phys.*, 2007, **127**, 064501.
46. J. H. ter Horst, D. Bedeaux and S. Kjelstrup, *J. Chem. Phys.*, 2011, **134**, 054703.
47. J. Wedekind, A.-P. Hyvärinen, D. Brus and D. Reguera, *Phys. Rev. Lett.*, 2008, **101**, 125703.
48. D. Reguera and J. M. Rubi, *J. Chem. Phys.*, 2003, **119**, 9877–9887.
49. D. Reguera and J. M. Rubı, *J. Chem. Phys.*, 2003, **119**, 9888–9893.
50. M. M. Rudek, J. L. Katz, I. V. Vidensky, V. Ždímal and J. Smolík, *J. Chem. Phys.*, 1999, **111**, 3623.
51. K. Hämeri and M. Kulmala, *J. Chem. Phys.*, 1996, **105**, 7696–7704.
52. Ø. Wilhelmsen, D. Bedeaux, S. Kjelstrup and D. Reguera, *J. Chem. Phys.*, 2014, **141**, 071103.
53. Ø. Wilhelmsen and D. Reguera, *J. Chem. Phys.*, 2015, **142**, 064703.
54. D. Reguera, J. M. Rubí and A. Pérez-Madrid, *Physica A*, 1998, **259**, 10–23.
55. R. Blaak, S. Auer, D. Frenkel and H. Löwen, *Phys. Rev. Lett.*, 2004, **93**, 068303.

CHAPTER 16

Mesoscopic Non-equilibrium Thermodynamics in Biology

ANDERS LERVIK*[a] AND J. MIGUEL RUBÍ[b]

[a] Department of Chemistry, Norwegian University of Science and Technology, 7491 Trondheim, Norway; [b] Departament de Física Fonamental Facultat de Física, Universitat de Barcelona, Barcelona, Spain
*Email: anders.lervik@ntnu.no

16.1 Introduction

In this chapter, we discuss the application of mesoscopic non-equilibrium thermodynamics (see Section 4.2.4 and Chapter 14) to biological systems. The biological entities within the biological system may span a wide range of length scales, from proteins to organisms and entire populations, and we will in this chapter limit the discussion to the molecular length scale. At this scale, fluctuations are important and we will discuss how the mesoscopic approach to non-equilibrium thermodynamics is able to capture the stochastic behaviour. Further, chemical reactions are important in biological systems and we will show how the mesoscopic approach accounts for the non-linear flux–force relations of chemical reactions. This is an example of where the classical non-equilibrium thermodynamic methods fail, except close to equilibrium. The classical non-equilibrium thermodynamic methods for biological systems, see the pioneering work of Kedem, Katchalsky and Curran[1–4] and the later work of Caplan and Essig,[5] assumes linear relations between the thermodynamic fluxes and forces. This assumption may be acceptable for many transport phenomena under many conditions, but not for chemical reactions.[6] In general, a reaction rate is a non-linear function of the driving force for the reaction, as inherent in the law of mass

Experimental Thermodynamics Volume X: Non-equilibrium Thermodynamics with Applications
Edited by Dick Bedeaux, Signe Kjelstrup and Jan V. Sengers

Published by the Royal Society of Chemistry, www.rsc.org

action. This limits the use of classical non-equilibrium thermodynamic methods to conditions close to equilibrium.

We will in this chapter show how this limitation can be lifted and discuss some applications of mesoscopic non-equilibrium thermodynamics to biochemical reactions, to energy conversion in biochemical systems and to single-molecule stretching experiments, before we conclude the chapter with a short discussion. The examples will show how mesoscopic non-equilibrium thermodynamics can be used to study biochemical reactions, to study coupling effects in biochemical systems and finally how it can be applied to stochastic systems and single molecules.

Before we discuss the application to biochemical reactions, we exemplify and introduce the mesoscopic approach by analysing a single chemical reaction. This example highlights the difference between the mesoscopic and classical non-equilibrium approaches, and we will build on the results in the following sections.

16.1.1 The Mesoscopic Approach to Chemical Reactions

Biochemical processes typically consist of several consecutive chemical reactions, often in a cycle, and may involve many products and reactants. In order to introduce the mesoscopic framework, we first consider a simpler case, where only one reaction is occurring. This allows us to focus on the main differences between the classical and mesoscopic approaches. The example we consider here is based on the introductory example in Chapter 14. However we simplify the discussion here and we do not consider explicit position and/or time-dependence in the following. We refer the reader to Chapter 14 for a more general discussion.

We consider a single first-order reaction where a species, A, is transformed into another species, B,

$$\mathrm{A} \rightleftharpoons \mathrm{B}. \tag{16.1}$$

The entropy production, σ, is given by,[6]

$$\sigma = -J\frac{A}{T}, \tag{16.2}$$

where J is the reaction rate, A is the affinity and T the temperature. The affinity is here given by the chemical potential μ_i of the two species: $A = \mu_\mathrm{B} - \mu_\mathrm{A}$. The corresponding linear law, according to the classical non-equilibrium thermodynamic theory, is,[6]

$$J = -l\frac{A}{T}, \tag{16.3}$$

where l is the Onsager coefficient. Experimentally, it was early found[7] that such linear laws are only valid for chemical reactions close to equilibrium, and that an exponential relation is more general.[7] The entropy production in eqn (16.2) is expressed as a bilinear form and can be expressed as a quadratic

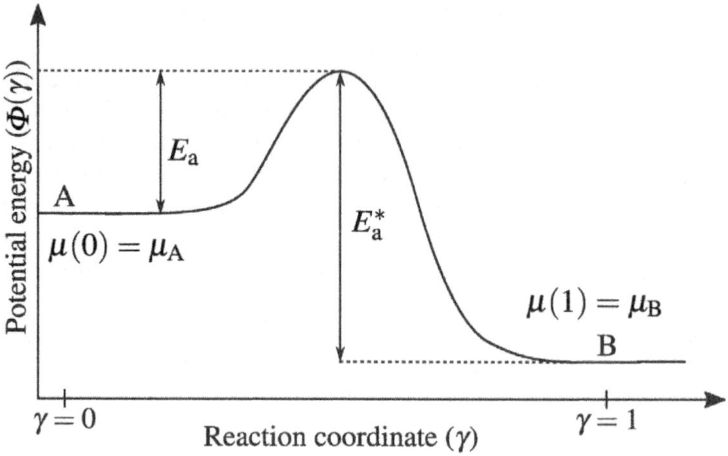

Figure 16.1 Illustration of the potential energy (Φ) landscape of an elementary reaction as a function of the reaction coordinate (γ). Arbitrary units are used for the potential energy and reaction coordinate. The initial configuration can be found at the plateau to left while the final configuration to the right. The activated complex is located at the maximum of the energy. The activation energies for the forward reaction (E_a) and for the backward reaction (E_a^*) are indicated in the figure. The boundary values for the chemical potentials are also given.

form by inserting eqn (16.3) into eqn (16.2). This form is valid up to first order in the Chapman–Enskog expansion of the probability distribution function. Deviations of this expression are observed when higher order terms in the expansion are included, *i.e.*, for higher values of the affinity: when the system is very far from equilibrium.[8]

In the mesoscopic approach, the reaction is studied on a finer scale by introducing a mesoscopic coordinate – the reaction coordinate, $0 \leq \gamma \leq 1$, which is assumed to completely describe the state of the reaction. In this picture, the reaction is described as a flux or diffusion in a potential-energy landscape, $\Phi(\gamma)$, and the state is described with the probability, $P(\gamma)$, of being in the state given by γ. This is similar to the treatment in chemical reaction kinetics, where the transition from an initial to a final configuration can be modelled as a continuous transformation across an activation energy barrier.[9] The situation is illustrated in Figure 16.1.

The chemical potential can be found as a function of the reaction coordinate as,[10]

$$\mu(\gamma) = RT \ln P(\gamma) + \Phi(\gamma), \tag{16.4}$$

where R is the gas constant and the end-point values are, $\mu(0) = \mu_A$ and $\mu(1) = \mu_B$. Time-dependence can be introduced by allowing the different quantities to explicitly depend on time. For the purpose of this example, we suppress this explicit time-dependence. On the mesoscopic scale, the entropy production,[10]

$$\sigma(\gamma) = -\frac{1}{T}J(\gamma)\left(\frac{\partial\mu(\gamma)}{\partial\gamma}\right)_T, \tag{16.5}$$

identifies the mesoscopic flux as $J(\gamma)$ and the corresponding mesoscopic force as,

$$-\frac{1}{T}\left(\frac{\partial\mu(\gamma)}{\partial\gamma}\right)_T = -\frac{R}{f}\left(\frac{\partial f(\gamma)}{\partial\gamma}\right)_T, \tag{16.6}$$

where we have introduced the fugacity, $f(\gamma) = \exp(\mu(\gamma)/RT)$. Here, and in the following text, the subscript on the parenthesis enclosing the partial derivatives identifies the variable(s) held constant. The *linear* flux–force relation that can be inferred from the entropy production on the *mesoscopic scale* is then,

$$J(\gamma) = -\frac{L(\gamma)R}{f(\gamma)}\left(\frac{\partial f(\gamma)}{\partial\gamma}\right)_T = -D\left(\frac{\partial f(\gamma)}{\partial\gamma}\right)_T, \tag{16.7}$$

where $L(\gamma)$ is the mesoscopic Onsager coefficient and $D = LR/f$ is interpreted as a diffusion constant. For the quasi-stationary case where the flux is independent of γ, which is fulfilled for large activation barriers, integration over the mesoscopic space gives (for constant D),

$$J = -D\int_0^1\left(\frac{\partial f(\gamma)}{\partial\gamma}\right)_T \mathrm{d}\gamma = -D(f(1) - f(0)), \tag{16.8}$$

or in terms of the affinity,

$$J = D\exp(\mu_A/RT)(1 - \exp(A/RT)). \tag{16.9}$$

We now see that the linear relation assumed on the mesoscopic scale, leads to a *non-linear* flux–force relation on the macroscopic scale. Further, for small values of the driving force, $A/RT \ll 1$, the linear relation from classical non-equilibrium thermodynamics, eqn (16.3), is recovered,

$$\begin{aligned} J &= D\exp(\mu_A/RT)(1 - \exp(A/RT)) \\ &\approx D\exp(\mu_A/RT)\left(-\frac{A}{RT}\right) = -l\frac{A}{T}, \end{aligned} \tag{16.10}$$

i.e., the classical result is obtained as a limiting behaviour of the mesoscopic result.

The mesoscopic expression is also in agreement with a purely kinetic consideration: In this picture, the reaction rate is given by the concentrations of the two species, c_i, and the rate constants k_+ and k_- in the forward and backward directions,

$$J = k_+ c_A - k_- c_B. \tag{16.11}$$

This rate can be rewritten by introducing the equilibrium concentrations, c_i^{eq}, (corresponding to $J=0$) and the chemical potentials given by, $\mu_i=\mu_i^0+RT\ln c_i$,

$$J=k_{+}c_{\mathrm{A}}\,(1-\exp(A/RT)). \tag{16.12}$$

Here, we have assumed that the mixture is ideal. In principle we could use activities rather than concentrations. In either case, the reaction rate is non-linear and coincides with the mesoscopic expression. By comparing the reaction rates close to equilibrium we can also connect the different kinetic coefficients,

$$l\cdot R=k_{+}c_{\mathrm{A}}=D\exp(\mu_{\mathrm{A}}/RT), \tag{16.13}$$

which shows the relation between the Onsager coefficient, the kinetic coefficient and the mesoscopic coefficient.

This example shows how the mesoscopic approach is able to lift the linear relations in the classical non-equilibrium thermodynamic approach to the non-linear domain. In the derivation we have assumed a quasi-stationary state, which is fulfilled for a large activation barrier.

The example also shows that the results from the kinetic approach based on the law of mass action can be reproduced. In the following section we will consider applications of the mesoscopic approach to a more realistic and complex biochemical reaction.

16.2 Biochemical Reactions

Biochemical reactions and enzyme catalysis are often described using a kinetic approach based on the work of Hill.[11,12] In this framework, enzyme catalysis is described as a cycle reaction and the law of mass action is applied for each step in the cycle. The resulting reaction rates are then given as non-linear functions of the kinetic rate constants and the driving forces. Close to equilibrium, these results reproduce the linear results from classical non-equilibrium thermodynamics. The entropy production can also be obtained as a sum of bilinear fluxes and forces.[12]

As we will show in this section, mesoscopic non-equilibrium thermodynamics can be used to analyse biochemical cycles enzyme catalysis. The cycle diagrams introduced by Hill simplify the description of the biochemical cycle reactions and are helpful for the mesoscopic description. We therefore include this description alongside the mesoscopic description. The advantage of going to the mesoscopic scale compared to the kinetic cycle approach, is the ability to describe coupling with other driving forces, such as a thermal driving force,[13] in a consistent way.[14] This will be further exemplified in the following section, in the context of energy conversion in biological systems. In the present section we will analyse the well-known Michaelis–Menten kinetics[15,16] as an example of a biochemical reaction. This kinetic scheme describes a situation with rapid binding of substrate to a catalytic site followed by a slower conversion to a product. Because of the central role played by these equations in biology, their 100 year anniversary have been recently celebrated.[17]

16.2.1 Michaelis–Menten Kinetics

The Michaelis–Menten scheme describes the conversion of a substrate (S) into a product (P), catalysed by an enzyme (E). This can be pictured as involving two main steps: the binding of the substrate to an enzyme–substrate complex (ES) and the subsequent conversion and release of the product. A more detailed scheme may involve several steps for the binding, conversion and release; however, the simplified two-step mechanism is able to capture the general features of many enzyme-catalysed reactions. The mechanistic scheme describing this two-step operation is,[18]

$$E + S \underset{k_{-1}}{\overset{k_1}{\rightleftharpoons}} ES \underset{k_{-2}}{\overset{k_2}{\rightleftharpoons}} E + P, \tag{16.14}$$

where k_i and k_{-i} are the rate constants in the forward and backward direction for reaction i. In the original work of Michaelis and Menten,[15,17] the first reaction was assumed to be rapid, and treated as in equilibrium. The equilibrium constant for dissociation, $K_d = k_{-1}/k_1 = c_E c_S/c_{ES}$, can then be used together with a mass balance for the amount of the enzyme to relate the concentrations of the different species involved in the first step. For the second reaction, the rate in the forward direction was assumed to be much greater than the rate of the backward reaction which is fulfilled for small k_{-2} and/or low concentration of the product. The reaction rate (in units of $\mathrm{mol \cdot L^{-1} \cdot s^{-1}}$), v, is then given by,

$$v = \frac{dc_P}{dt} = \frac{v_{max} c_S}{K_d + c_S}, \tag{16.15}$$

where $v_{max} = k_2(c_{ES} + c_E)$ is the limiting rate obtained as $c_S \to \infty$. For low concentrations, $v \sim v_{max} c_S / K_d$, the reaction is second order (due to the concentration dependence in v_{max}). For higher concentrations, $v \sim v_{max}$, the reaction is first order and exhibits saturation as shown in Figure 16.2 (left).

As mentioned above, the diagram method introduced by Hill can be used to simplify the description of biochemical reaction cycles and the cycle diagrams are also useful for the mesoscopic description. In Figure 16.2 (right) we show the equivalent cycle diagram description of the Michaelis–Menten scheme. From this diagram, the reaction rate can be directly obtained[19] following the methodology of Hill[12] and making the same assumptions as given above. In the mesoscopic approach, we will use this diagram to identify boundary conditions for the chemical potentials for the different reaction coordinates.

In the mesoscopic description, we introduce two reaction coordinates, one for each of the steps in scheme (16.14) (or equivalently in Figure 16.2

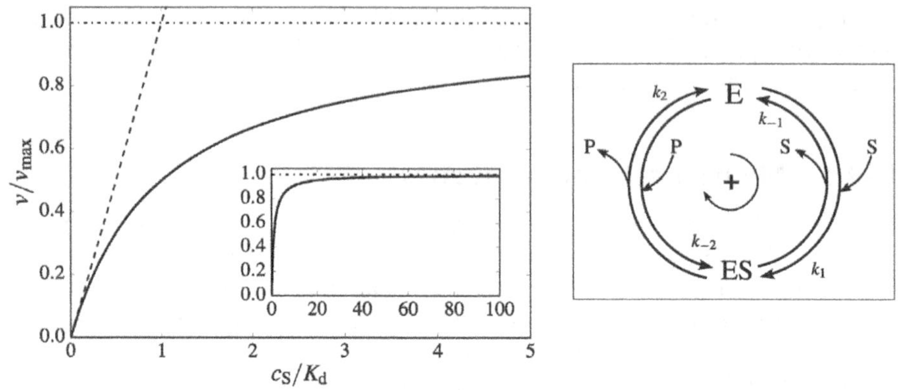

Figure 16.2 (Left) Illustration of Michaelis–Menten kinetics. The scaled reaction rate, v/v_{max}, is shown (solid line) as a function of the substrate concentration scaled with the dissociation constant, c_S/K_d. The linear dependence for low concentrations is shown with the dashed line, and the saturation (limiting behaviour) for higher concentrations is shown with the dash-dotted line. The saturation for large concentration is shown in the inset. (Right) The cycle description of Michaelis–Menten kinetics. The enzyme may exist in two states, E and ES, and the rate constants for the different steps are indicated in the figure. The substrate (S) is bound in the forward E→ES step (rate constant k_1) and the product (P) is released in the forward ES→E step (rate constant k_2).

(right)): γ_1 for the first step and γ_2 for the second. The state is then described by the probability density, $P(\gamma_1,\gamma_2,t)$, where we allow for a time (t) dependence. The two reactions are uncoupled so that,

$$P(\gamma_1,\gamma_2,t)=P_1(\gamma_1,t)P_2(\gamma_2,t), \tag{16.16}$$

and the corresponding chemical potential is,

$$\mu(\gamma_1,\gamma_2,t)=\mu_1(\gamma_1,t)+\mu_2(\gamma_2,t), \tag{16.17}$$

with,

$$\mu_i(\gamma_i,t)=RT\ \ln\ P_i(\gamma_i,t)+\Phi_i(\gamma_i), \quad \text{for } i=1,2. \tag{16.18}$$

The corresponding entropy production is,

$$\sigma(\gamma_1,\gamma_2,t)=\sigma_1(\gamma_1,t)+\sigma_2(\gamma_2,t), \tag{16.19}$$

where,

$$\sigma_i(\gamma_i,t)=-\frac{1}{T}J_i(\gamma_i,t)\left(\frac{\partial\mu(\gamma_i,t)}{\partial\gamma_i}\right)_t, \tag{16.20}$$

and we infer the linear laws as in the example in the introduction,

$$J_i(\gamma_i,t)=-\frac{L_i(\gamma_i)R}{f_i(\gamma_i)}\left(\frac{\partial f_i(\gamma_i,t)}{\partial\gamma_i}\right)_t=-D_i\left(\frac{\partial f_i(\gamma_i,t)}{\partial\gamma_i}\right)_t. \tag{16.21}$$

Following the approach in the introductory example, integration over γ_i for a quasi-stationary case with $J_i(\gamma_i,t) = J_i(t)$ gives,

$$J_i(t) = -\frac{D_i}{\int d\gamma_i \exp(\Phi_i(\gamma_i)/RT)}\left[\exp\left(\frac{\mu_i(1,t)}{RT}\right) - \exp\left(\frac{\mu_i(0,t)}{RT}\right)\right]. \tag{16.22}$$

Since we have assumed high activation-energy barriers, $E_{a,i} \gg RT$, we also approximate the integral over the potential energy as,

$$\exp\left(\frac{E_{a,i}}{RT}\right) = \int d\gamma_i \exp(\Phi_i(\gamma_i)/RT). \tag{16.23}$$

The boundary conditions can be read directly from the cycle diagram in Figure 16.2 (right),

$$\begin{aligned} \mu_1(0,t) &= \mu_E + \mu_S, \qquad \mu_1(1,t) = \mu_{ES}, \\ \mu_2(0,t) &= \mu_{ES}, \qquad \mu_2(1,t) = \mu_E + \mu_P. \end{aligned} \tag{16.24}$$

This gives the fluxes along the two reaction coordinates as,

$$\begin{aligned} J_1(t) &= -D_1 \exp\left(-\frac{E_{a,1}}{RT}\right)\left[\exp\left(\frac{\mu_{ES}(t)}{RT}\right) - \exp\left(\frac{\mu_E(t)+\mu_S(t)}{RT}\right)\right], \\ J_2(t) &= -D_2 \exp\left(-\frac{E_{a,2}}{RT}\right)\left[\exp\left(\frac{\mu_E(t)+\mu_P(t)}{RT}\right) - \exp\left(\frac{\mu_{ES}(t)}{RT}\right)\right], \end{aligned} \tag{16.25}$$

which describe the kinetics in scheme (16.14). In order to obtain the Michaelis–Menten kinetics, we consider first the change in concentrations of substrate, product and the two enzyme states,

$$\begin{aligned} \frac{dc_P}{dt} &= J_2(t), \qquad \frac{dc_S}{dt} = -J_1(t), \\ \frac{dc_{ES}}{dt} &= J_1(t) - J_2(t), \qquad \frac{dc_E}{dt} = -J_1(t) + J_2(t). \end{aligned} \tag{16.26}$$

The reaction rate is here equal to the rate of formation of the product, J_2, and we see that $dc_{ES}/dt + dc_E/dt = 0$, such that $c_{ES} + c_E$ is conserved and equal to a constant, say c_0. Next we consider the stationary state where $J_1 = J_2$ and together with the relation between c_{ES} and c_E we can solve for these two concentrations and obtain

$$J_1 = \frac{c_0 I_2 f^0_{ES}}{\dfrac{I_1 + I_2}{I_1}\dfrac{f^0_{ES}}{f^0_E} + f_S + \dfrac{I_2}{I_1} f_P}(f_S - f_P), \tag{16.27}$$

where we have introduced the short-hand notation $I_i = D_i / \int d\gamma_i \exp(\Phi_i(\gamma_i)/RT)$ and introduced the fugacities of the substrate and product (f_P and f_S) and the standard-state fugacities of the enzyme (f^0_{ES} and f^0_E). This expression describes the stationary state reaction rate, without making Michaelis–Menten approximations.

The first approximation in the Michaelis–Menten scheme corresponds to $I_2/I_1 \ll 1$. This can be seen by comparing the activation energies for the two reactions. The second approximation corresponds to neglecting f_P compared to f_S. The final expression then follows from eqn (16.27),

$$\nu = J_1 = \frac{c_0 I_2 f^0_{ES}}{\dfrac{f^0_{ES}}{f^0_E f^0_S} + c_S} c_S = \frac{\nu_{max}}{K_d + c_S} c_S, \tag{16.28}$$

where we have introduced the concentrations and we identify $\nu_{max} = c_0 I_2 f^0_{ES}$ and the (equilibrium) dissociation constant $K_d = f^0_{ES}/(f^0_E f^0_S)$. The final rate in eqn (16.28) coincides with the kinetic expression for the Michaelis–Menten rate, see eqn (16.15).

We can rewrite eqn (16.27) by introducing two parameters, $\beta = c_P/c_S$ and $\alpha = I_2/I_1 \approx (D_2/D_1) \cdot e^{-(E_{a,2}-E_{a,1})}$,

$$\frac{\nu}{\nu_{max}} = \frac{(c_S/K_d)(1-\beta)}{(1+\alpha) + (c_S/K_d)(1-\beta) + (1+\alpha)\beta(c_S/K_d)}. \tag{16.29}$$

Michaelis–Menten kinetics are recovered when $\beta = 0$ and $\alpha \ll 1$ and for the limiting behaviour we now obtain, $\nu \sim \nu_{max}\left(\dfrac{1-\beta}{1+\alpha\beta}\right)$ for large c_S and $\nu \approx \nu_{max}(c_S/K_d)\left(\dfrac{1-\beta}{1+\alpha}\right)$ for small c_S. This shows that in the presence of substantial amounts of the product P (when β is not negligible) or when the approximation $\alpha \ll 1$ is not satisfactory, the reaction rate will be reduced, compared to the reaction rate obtained if eqn (16.28) is satisfied.

The example in this section shows how the mesoscopic approach can be used to describe the kinetics of biochemical reactions. This approach has been used to describe more complex cycle reactions[14,20] and the main advantage of going to the mesoscopic framework compared to the kinetic framework is that coupling to other driving forces can be included in a consistent way.[13,14,16] We elucidate this in the next section in the context of energy conversion in biochemical systems.

16.3 Energy Conversion in Biochemical Systems

Biological motors and pumps are in need of non-equilibrium conditions in order to transform chemical energy into mechanical energy to be able to transport ions through protein channels. How the mechanism of energy conversion takes place under those conditions constitutes a basic problem in physical biology. We will show in this section that a large concentration gradient of one type of ions may induce motion of another type of ions in a direction opposed to its gradient. This motion is referred to as active transport.[12,21,22] We will also consider another example of active transport: the case of a protein pumping ions across a membrane by utilizing the ATP hydrolysis.

16.3.1 Energy Conversion and Coupling of Ion Concentration Gradients

The mechanism whereby ions are transported by exploiting concentration gradients uses the energy released by the structural change of the enzyme E (Figure 16.3).[12] Ions of type B bind to the enzyme forming the complex EB. The subsequent binding of an ion A to this complex makes the state of the protein unstable and transforms into the state E′. In this state, the binding sites change their position, as indicated in Figure 16.3b) and both ions are released to the outer side of the membrane. Suppose that the concentrations of the species are such that: $[A]_{out} > [A]_{in}$ and $[B]_{in} \gg [B]_{out}$ which implies that

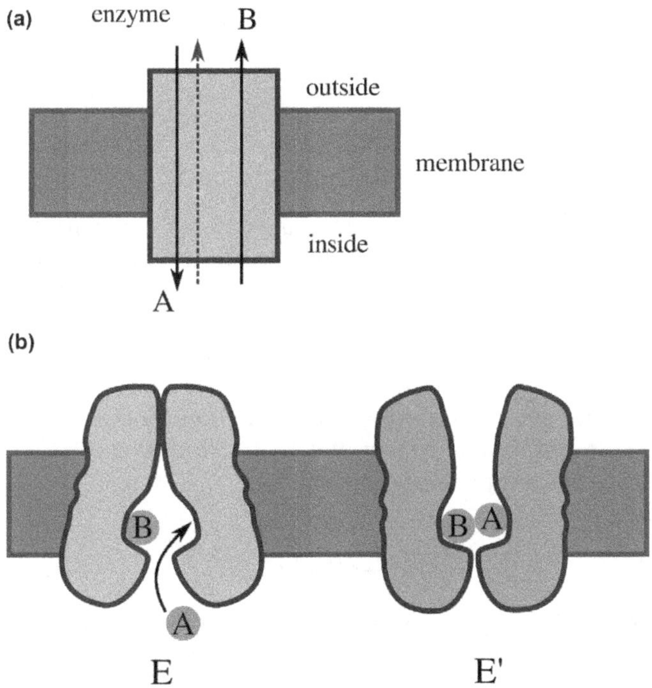

Figure 16.3 Translocation of ions A and B across a protein channel. (a) Black (solid) arrows indicate the direction of transport of the ions in the spontaneous process in which $[A]_{out} > [A]_{in}$ and $[B]_{in} \gg [B]_{out}$. When the translocation follows the protocol given in eqn (16.30), the flip of the enzyme between the states E and E′ changes the transport direction of ions A, as indicated by the red dashed arrow. (b) Illustration of the mechanism. Binding sites for ions (indicated with the circled letters A and B) changes when the structure of the enzyme transforms from E (on the left) to E′ (on the right). Initially, in state E, the binding sites are exposed to the "inside" of the membrane. B binds to the enzyme and subsequent binding of A makes the enzyme state unstable. The conformation changes to enzyme state E′ where the binding sites are exposed to the opposite side ("outside") and the ions are released.

if the motion is spontaneous, B ions go from *in* to *out* whereas A ions move from *out* to *in*. When the transport of both types of ions is coupled, A ions go from *in* to *out*, against its concentration gradient. It is then possible to use a large concentration gradient of B to induce active transport of A molecules.

The different steps of the process can be expressed by means of the following kinetic transformations:

$$\begin{aligned} &B_{in} + E \rightarrow EB, \\ &A_{in} + EB \rightarrow AEB, \\ &AEB \rightarrow AE'B, \\ &AE'B \rightarrow A_{out} + E'B, \\ &E'B \rightarrow B_{out} + E', \end{aligned} \tag{16.30}$$

which can be summarized as:

$$A_{in} + B_{in} + E \rightarrow A_{out} + B_{out} + E'. \tag{16.31}$$

This equation shows that A ions move against its concentration gradient using the energy released by the enzyme due to the binding of B ions. The overall effect is that the presence of a concentration gradient of B induces active transport of A.

The ion-translocation process can be explained on thermodynamic grounds by means of mesoscopic non-equilibrium thermodynamics.[10] Each reaction of (16.30) can be described as a diffusion process along a reaction coordinate, as exemplified in Sections 16.1.1 and 16.2. The entropy production on the mesoscopic scale is given by,

$$T\sigma = -\sum_i J_i \left(\frac{\partial f_i}{\partial \gamma_i}\right)_T \geq 0, \tag{16.32}$$

where the inequality results from the second law expressed locally in the γ_i-space. Since the reactions are independent, each reaction current is given by,

$$J_i = -D_i \Delta f_i, \tag{16.33}$$

where D_i is the diffusion coefficient of the i-th reaction which is positive due to the positive nature of the entropy production. Each reaction coordinate is normalized to one. This relation shows that the reaction rate and the fugacity differences have different signs. For forward reactions, as the ones involved in the conversion process, fugacity differences are thus negative. From the schemes given in eqn (16.30) or (16.31), and (16.32), one then infers the inequality,

$$f_{A_{in}} - f_{A_{out}} > f_{B_{out}} - f_{B_{in}}. \tag{16.34}$$

If the right hand side term is positive thus the left hand side is positive too. This inequality shows that the presence of a concentration difference of B may change the transport direction of A.

16.3.2 Energy Conversion in Biomolecular Pumps

Energy conversion can also take place, for instance, when a concentration gradient is established and maintained by a pumping protein utilizing the energy of a chemical reaction. One specific example of this is the calcium pumping protein, Ca^{2+}-ATPase, which transports calcium ions across vesicle membranes by coupling the mass transport to a chemical reaction – the adenosine triphosphate (ATP)-reaction. The Ca^{2+}-ATPase belongs to a large family of membrane transport proteins, the P-type ATPases which all make use of the hydrolysis of ATP in order to drive vectorial transport. Since these proteins are embedded in membranes they are effectively embedded in a surface; and at a surface, the normal component of a vectorial flux (*e.g.*, the mass flux), will have scalar symmetry and may then couple to the scalar reaction rate.[23,24]

The Ca^{2+}-ATPase is particularly interesting as experiments indicate that coupling to a temperature gradient may also be possible.[25–27] These effects needs to be captured by the theoretical model; however, a heat flux is not readily included in a purely kinetic description of the operation of this pump. These effects can, on the other hand, be introduced in a mesoscopic model: A simple mesoscopic model of this protein[20,28] has two mesoscopic coordinates, γ_r and γ_d, which describe the state of the ATP-reaction and the state of the calcium-ion transfer, respectively. In this view, the operation of the pump can be pictured as a diffusion process on a two-dimensional potential energy landscape. Equivalently, one can consider a two-state Hill diagram,[20] where the two states are connected with three different paths, as illustrated in Figure 16.4.

By following the framework of mesoscopic non-equilibrium thermodynamics, one can obtain equations that describe the operation of this protein. In particular the rate of the ATP-reaction, ν_{ATP}, is obtained as,

$$\nu_{ATP} = -D_{rr}\left[1-\exp\left(-\frac{\Delta_r G(T^{in})}{RT^{in}}\right)\right] + D_{rd}\left[1-\exp\left(\frac{\Delta\mu_{Ca/2H}(T^{in})}{RT^{in}}\right)\right] - \frac{D_{rq}}{R}\left(\frac{1}{T^{out}}-\frac{1}{T^{in}}\right), \tag{16.35}$$

where D_{rr}, D_{rd} and D_{rq} are the generalized diffusion coefficients describing the process, $\Delta_r G$ is the change in Gibbs energy for the reaction and $\Delta\mu_{Ca/2H}$ is the change in chemical potential energy on exchanging ions (calcium and hydrogen) across the membrane. Here, the superscript in/out indicates that the temperature is evaluated at the inside/outside of the membrane. Similar

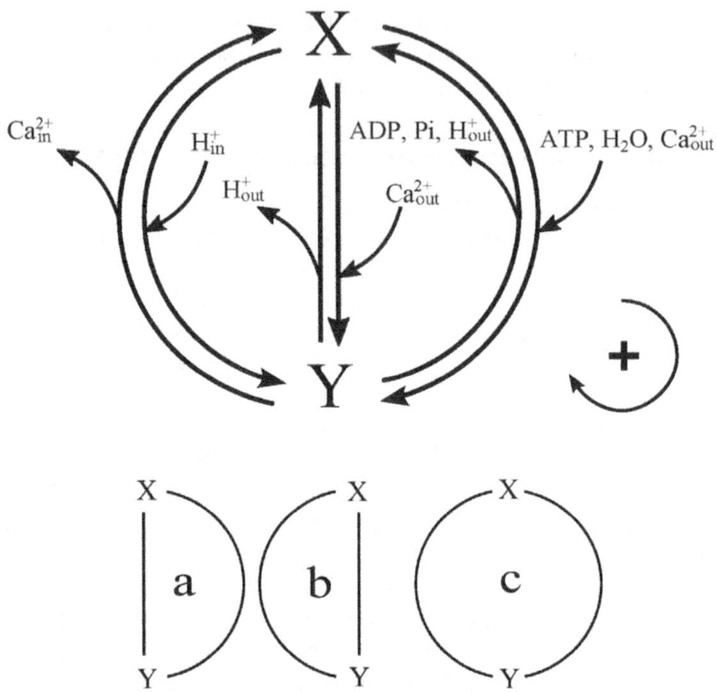

Figure 16.4 Cycle description of the Ca^{2+}-ATPase pump. The protein can be pictured as existing in two states, labelled X and Y, while the operation is described using three different cycles, a, b and c. The top picture shows the binding and release of the species involved in the process, while the lower picture shows the corresponding cycles: cycle a corresponds to the ATP-reaction without the movement of ions; cycle b corresponds to movement of ions between the inside (in) and outside (out) of the membrane (this can typically correspond to leakage through the membrane); finally, cycle c corresponds to the normal operation of the pump – the ATP-reaction proceeds and calcium ions are transported from the outside to the inside of the membrane.

expressions were obtained for the flux of calcium ions, J_{Ca}, (with generalized diffusion coefficients D_{dr}, D_{dd} and D_{dq}) and the measurable heat flux, J'_q(with generalized diffusion coefficients D_{qr}, D_{qd} and D_{qq}).[28] These 9 coefficients describe the operation of the pump.[29] For instance, D_{dr} and D_{rd} describe the coupling of the reaction and the mass flow, *i.e.*, the active transport, while D_{qr} and D_{qd} describe reversible heat flow, connected to the reaction and mass flow. The coefficients D_{rq} and D_{dq} describe how a temperature difference across the membrane may drive the reaction and transport of ions, and the remaining coefficients describe the direct coupling between the fluxes and the corresponding driving forces (*e.g.*, mass transport and the difference in chemical potential).

For this particular protein, the mesoscopic equations describing the operation have been used to investigate the efficiency of the operation under laboratory conditions. It was then found[30] that the efficiency was relatively low (<13 %) for the conditions studied. However, this case may be artificially low due to the experimental conditions and the pump may have a higher efficiency *in vivo*.[30] Both this study[30] and the study of Kjelstrup *et al.*[29] show the usefulness of the mesoscopic framework for interpreting experimental data. Further, this framework can be used as a foundation for design of new experiments, *e.g.*, by clarifying the experimental conditions[29] for obtaining the coefficients describing the transport. Such coefficients could be tabulated and used to characterize the different pumps and molecular machines.

16.4 Single-molecule Stretching

Mesoscopic non-equilibrium thermodynamics can also be used to study the response of a single molecule to an external force. A typical situation is the stretching of DNA molecules by means of an optical tweezer. Two facts are observed in DNA stretching experiments:

(i) The macromolecule is in mechanical equilibrium, consequently inertial effects are negligible. Typical forces exerted on the particles are of the order of 10 pN.[31] When the size of the macromolecule is of the order of $r \approx 10^{-9}$ m, the host liquid density is $\rho \approx 10^3\,\mathrm{kg \cdot m^{-3}}$, and the viscosity is $\eta \approx 10^{-3}\,\mathrm{kg \cdot m^{-1} \cdot s^{-1}}$, one has,

$$\frac{dF}{dt} \approx 100\frac{\mathrm{nN}}{\mathrm{s}}, \tag{16.36}$$

and,

$$\frac{F_{\mathrm{char}}}{\tau_{\mathrm{char}}} \sim \frac{\eta^2/\rho}{m/\varepsilon} \sim \frac{10^{-9}\mathrm{N}}{10^{-12}\mathrm{s}} = 10^3\,\frac{\mathrm{N}}{\mathrm{s}}, \tag{16.37}$$

where F_{char} is a characteristic force, τ_{char} a characteristic time-scale, m is the mass of the molecule and ε the friction coefficient. Therefore, variations in time of the force are very small and inertial effects can be neglected.[32]

(ii) Fluctuations are important. This conclusion follows from the estimate of the work done by the applied force: $\Delta W = F\Delta x$. For typical values of the force $F \approx 10$ pN, and of the displacement $\Delta x \approx 400$ nm,[31] the work is $\Delta W \approx 10k_BT$, where k_B is Boltzmann's constant and T the temperature. This shows that thermal fluctuations whose associated energy is of the order of k_BT play a role in the kinetics. One then needs a probabilistic description in terms of the probability $P(x,t)$, where x is the elongation of the macromolecule. The molecule is immersed in a liquid at rest which acts as a thermal bath. This implies that the noise is Gaussian and consequently that the evolution of the probability density is governed by a Smoluchowski equation which is a diffusion equation for the probability density.

To obtain this equation, we will use the framework of mesoscopic non-equilibrium thermodynamics. The starting point is the Gibbs entropy,

$$S - S_0 = -k_B \int P \ln \frac{P}{P_0} \mathrm{d}x, \tag{16.38}$$

where S_0 and P_0 are the entropy and the probability distribution of a reference state. Taking variations of this equation, one obtains,

$$\delta S = \int \delta s \mathrm{d}x = -k_B \int \delta P \ln \frac{P}{P_0} \mathrm{d}x = \int \left(-\frac{\mu}{T} \delta P\right) \mathrm{d}x. \tag{16.39}$$

Here s is the local entropy and μ the chemical potential defined in eqn (16.4), where $\Phi = V_0 - Fx$ consists of the energy barrier of the molecule V_0 and the work done by the external force. If the force is exerted by an optical trap, as the one in Figure 16.5, it is given by $F = -K(x - X_t)$ where K is a constant and X_t the time-dependent position of the center of the trap.

Comparing the integrands of the second and fourth terms of eqn (16.39) one obtains the local thermodynamic relation,

$$\mathrm{d}s = -\frac{\mu}{T} \mathrm{d}P. \tag{16.40}$$

From this equation one can easily derive the entropy production rate of the stretching of the molecule,

$$\sigma = -\frac{1}{T} J \left(\frac{\partial \mu}{\partial x}\right)_{t,T}, \tag{16.41}$$

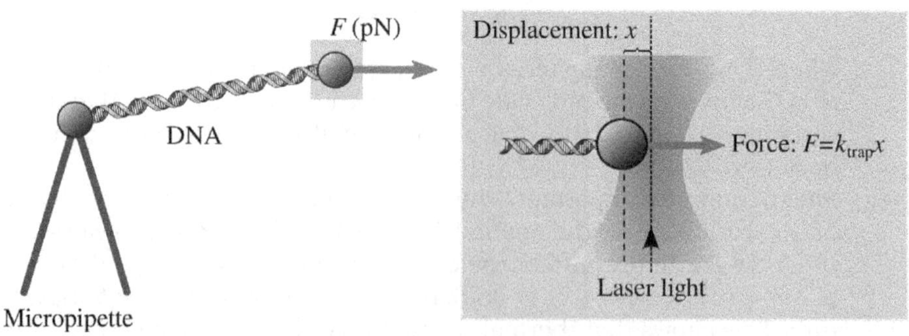

Figure 16.5 DNA stretching experiment. One of the ends of the DNA molecule is attached to a micropipette, as shown in the illustration to the left. The force is exerted by an optical tweezer, as illustrated to the right. Two types of experiments can be performed: isometric and isotensional.[33] In the former the length of the molecule is kept fixed and the force may fluctuate. Just the opposite occurs in the latter. Here k_{trap} is the force constant associated with the trap, x is the displacement and F is the resulting force.

and subsequently the expression for the current,

$$J = -\frac{L}{T}\left(\frac{\partial \mu}{\partial x}\right)_{t,T}. \tag{16.42}$$

Using this expression together with eqn (16.4) in the probability conservation law one obtains the Smoluchowski equation for isothermal conditions,

$$\left(\frac{\partial P}{\partial t}\right)_x = \frac{\partial}{\partial x} D\left[\left(\frac{\partial P}{\partial x}\right)_t + \frac{P}{k_B T}\left(\frac{\partial \Phi}{\partial x}\right)_t\right]. \tag{16.43}$$

The role played by the fluctuations can now be analyzed from the solution of eqn (16.43),

$$P(x,t) = \frac{1}{(2\pi Dt)^{1/2}} \exp\left[-\frac{(x + F_{\text{tot}}/\varepsilon)^2}{2Dt}\right], \tag{16.44}$$

where $F_{\text{tot}} = -(\partial \Phi/\partial x)_{t,T}$. Comparing the forward and backward probabilities one obtains,[34]

$$\frac{P(x,t)}{P(-x,t)} = \exp\left(\frac{F_{\text{tot}} x}{k_B T}\right). \tag{16.45}$$

For $F_{\text{tot}} x \gg k_B T$, the bead follows the force: $P(-x,t) \ll 1$. When this condition is not fulfilled, the backward probability can be of the same order as the forward probability: this is the regime of large fluctuations.

When fluctuations are important, the work is a fluctuating quantity. In the case in which the intrinsic potential is harmonic: $V_0 = (1/2)kx^2$, with k an elastic constant, the average work is given by,[35]

$$\left\langle e^{-\beta W}\right\rangle = \frac{(k+K)}{[K(2k+K)]^{1/2}} e^{-\frac{k(k+K)}{2(2k+K)} x_{\text{eq}}^2}, \tag{16.46}$$

where the equilibrium position of the bead is given by,

$$x_{\text{eq}} = \frac{K}{k+K} X_t. \tag{16.47}$$

In the case in which $K \gg k$, fluctuations are very small $x_{\text{eq}} \approx X_t$ and,

$$\left\langle e^{-\beta W}\right\rangle = e^{-\beta \frac{k}{2} x_{\text{eq}}^2} = e^{-\Delta G}. \tag{16.48}$$

Gibbs-energy differences then follow from measurements of the work. In the case of large fluctuations, when the above inequality is not satisfied, eqn (16.48) is no longer valid.[35]

Stretching of a single RNA molecule has also been analysed under the perspective of mesoscopic non-equilibrium thermodynamics.[36] Moreover, this theory has been applied to analyse the kinetics of molecular motors.[37] When and how thermodynamics can be used to describe small-scale biological systems is discussed by Rubí *et al.*[33]

16.5 Discussion

In this article, we have shown how mesoscopic non-equilibrium thermodynamics can be used to study biological systems operating under far--from-equilibrium conditions. This new approach, that extends the scope of non-equilibrium thermodynamics to the mesoscopic domain, is able to completely characterize kinetic processes occurring in small-scale biological systems.

Classical non-equilibrium thermodynamics carries out an analysis of these processes in terms of only two states: the initial and the final states. This coarse graining description only provides linear relationships for the rates in terms of the chemical-potential differences, giving laws that are only approximations to the actual biochemical kinetics. In the more detailed scenario we propose, one assumes that the transformation, instead of being viewed as a sudden switch, occurs *via* many small intermediate jumps leading the system through a virtual continuum of states. When a non-equilibrium thermodynamics scheme is applied, not to the overall transformation, but to these small steps, the resulting linear contributions to the rate integrate to provide the observed non-linear behaviour. We have also shown how the mesoscopic approach can be used to investigate coupling between different fluxes such as a reaction rate, a mass and a heat flow and the corresponding forces.

The proposed thermodynamic description of the kinetic process is valid when the intermediate states persist in the time scale considered, and therefore can be considered as thermodynamic states. The system equilibrates locally along the reaction coordinate. The transformation can thus be viewed as a diffusion process over the activation barrier. This situation is encountered in many biological processes, such as is the ones analyzed in this article.

Acknowledgements

AL thanks Signe Kjelstrup for discussions regarding the Michaelis–Menten scheme and Stein Almo for essential computer assistance.

References

1. A. Katchalsky and O. Kedem, *Biophys. J.*, 1962, **2**, 55.
2. O. Kedem and A. Katchalsky, *Biochim. Biophys. Acta*, 1958, **27**, 229.
3. O. Kedem and A. Katchalsky, *J. Gen. Physiol.*, 1961, **45**, 143.
4. A. Katchalsky and P. F. Curran, *Nonequilibrium Thermodynamics in Biophysics*, Harvard University Press, Cambridge, MA, 1965.
5. S. R. Caplan and A. Essig, *Bioenergetics and Linear Nonequilibrium Thermodynamics. The Steady State*, Replica Books, 2001.
6. S. R. de Groot and P. Mazur, *Non-equilibrium Thermodynamics*, Dover, New York, 1984.
7. I. Prigogine, P. Outer and C. Herbo, *J. Phys. Chem.*, 1948, **52**, 321.
8. J. Ross and P. Mazur, *J. Chem. Phys.*, 1961, **35**, 19.

9. S. Glasstone, K. J. Laidler and H. Eyring, *The Theory of Rate Processes: The Kinetics of Chemical Reactions, Viscosity, Diffusion and Electrochemical Phenomena*, McGraw-Hill Book Company, New York, 1941.
10. D. Reguera, J. M. Rubí and J. M. G. Vilar, *J. Phys. Chem. B*, 2005, **109**, 21502.
11. T. L. Hill, *Free Energy Transduction in Biology: The Steady-state Kinetic and Thermodynamic Formalism*, Academic Press, New York, 1977.
12. T. L. Hill, *Free Energy Transduction and Biochemical Cycle Kinetics*, Springer-Verlag, New York, 1989.
13. A. Lervik, D. Bedeaux and S. Kjelstrup, *Eur. Biophys. J.*, 2013, **42**, 321.
14. J. M. Rubí, D. Bedeaux, S. Kjelstrup and I. Pagonabarraga, *Int. J. Thermophys.*, 2013, **34**, 1214.
15. L. Michaelis and M. L. Menten, *Biochem. Z.*, 1913, **49**, 333.
16. A. Lervik, S. Kjelstrup and H. Qian, *Phys. Chem. Chem. Phys.*, 2015, **17**, 1317–1324.
17. K. A. Johnson and R. S. Goody, *Biochemistry*, 2011, **50**, 8264.
18. M. B. Jackson, *Molecular and Cellular Biophysics*, Cambridge University Press, Cambridge, 2006.
19. D. A. Beard and H. Qian, *PLoS One*, 2007, **2**, e144.
20. A. Lervik, D. Bedeaux and S. Kjelstrup, *Eur. Biophys. J.*, 2012, **41**, 437.
21. S. Kjelstrup, J. M. Rubí and D. Bedeaux, *Phys. Chem. Chem. Phys.*, 2005, **7**, 4009.
22. J. M. Rubí, M. Naspreda, S. Kjelstrup and D. Bedeaux, *J. Non-Equilib. Thermodyn.*, 2007, **32**, 351.
23. A. Lervik and S. Kjelstrup, *Eur. Phys. J. Spec. Top.*, 2013, **222**, 143.
24. D. Bedeaux, *Adv. Chem. Phys.*, 1986, **64**, C109.
25. L. de Meis, M. Bianconi and V. Suzano, *FEBS Lett.*, 1997, **406**, 201.
26. L. de Meis, *J. Biol. Chem.*, 2001, **276**, 25078.
27. L. de Meis, *J. Membr. Biol.*, 2002, **188**, 1.
28. D. Bedeaux and S. Kjelstrup, *Phys. Chem. Chem. Phys.*, 2008, **10**, 7304.
29. S. Kjelstrup, D. Barragán and D. Bedeaux, *Biophys. J.*, 2009, **96**, 4376.
30. A. Lervik, F. Bresme, S. Kjelstrup and J. M. Rubí, *Biophys. J.*, 2012, **103**, 1218.
31. C. Bustamante, S. B. Smith, J. Liphardt and D. Smith, *Curr. Opin. Struct. Biol.*, 2000, **10**, 279.
32. R. D. Astumian, *Am. J. Phys.*, 2006, **74**, 683.
33. J. M. Rubí, D. Bedeaux and S. Kjelstrup, *J. Phys. Chem. B*, 2006, **110**, 12733.
34. M. H. Vainstein and J. M. Rubí, *Phys. Rev. E: Stat., Nonlinear, Soft Matter Phys.*, 2007, **75**, 031106.
35. J. M. G. Vilar and J. M. Rubí, *Phys. Rev. Lett.*, 2007, **100**, 020601.
36. J. M. Rubí, D. Bedeaux and S. Kjelstrup, *J. Phys. Chem. B*, 2007, **111**, 9598.
37. S. Kjelstrup, J. M. Rubí, I. Pagonabarraga and D. Bedeaux, *Phys. Chem. Chem. Phys.*, 2013, **15**, 19405.

CHAPTER 17

Dynamics of Complex Fluid–Fluid Interfaces

LEONARD M. C. SAGIS[a,b]

[a] Physics and Physical Chemistry of Food, Department AFSG, Wageningen University, Bornse Weilanden 9, 6708 WG Wageningen, The Netherlands;
[b] ETH Zurich, Department of Materials, Polymer Physics, Vladimir-Prelog-Weg 1-5/10, 8093 Zurich, Switzerland
Email: leonard.sagis@wur.nl

17.1 Complex Fluid–Fluid Interfaces

Fluid–fluid interfaces are ubiquitous in nature and our everyday existence. We find them in foam and emulsions in food, pharmaceutical, and personal care products, or in biological systems such as cells, lung alveoli, and the tear film on our eyes. Fluid–fluid interfaces are considered complex when the components forming the interface self-organize into complex microstructures, after adsorption to the interface. Some examples of typical microstructures observable in complex fluid–fluid interfaces are two-dimensional (2D) gels, 2D glasses, 2D liquid crystalline phases, and 2D emulsions and dispersions.[1] Gels and glasses are formed mainly in protein, polymer, or colloidal particle stabilized interfaces. Anisotropic colloidal particles and protein fibrils[2] are examples of surface active materials which can form 2D liquid crystalline phases. Two-dimensional emulsions and dispersions can be formed when mixtures of incompatible surface active component are present at the interface.

A common characteristic of multiphase systems with complex fluid–fluid interfaces is that their dynamic behaviour is often dominated by the thermodynamic and mechanical properties of their interfaces. For this reason

Experimental Thermodynamics Volume X: Non-equilibrium Thermodynamics with Applications
Edited by Dick Bedeaux, Signe Kjelstrup and Jan V. Sengers

Published by the Royal Society of Chemistry, www.rsc.org

such systems can be considered to be soft Interface Dominated Materials (IDMs). Simple interfaces are (typically) stabilized by low-molecular-weight surfactants that do not self-organize at the interface. In isothermal multiphase systems which contain such interfaces, the surface tension tends to be the only relevant interfacial parameter affecting flow and deformation. In soft IDMs with complex interfaces additional surface properties may affect dynamic behaviour, and these effects may even be much larger than the effects of surface tension. For example, in dispersions of vesicles[3–5] or in phase-separated biopolymer systems[6–12] the dynamics is significantly affected by the bending rigidity of the interface. Stability and flow behaviour of emulsions and foam in food products,[13–16] or dynamic behaviour of microbubbles in ultrasound diagnostics,[17,18] are often significantly affected by surface shear and dilatational properties. The latter is the resistance of an interface against all-sided compression.[19] The dependence on these mechanical surface parameters tends to cause significant nonlinearities in the response of soft IDMs to applied deformations or temperature and concentration gradients. When an IDM is perturbed from its equilibrium state, this may affect the microstructure of the interface, and as a result surface properties will change, leading to effects such as strain softening or strain hardening. Such effects have been observed in polymer stabilized interfaces,[20,21] protein fibril-stabilized interfaces,[22–24] air–water interfaces stabilized by oligosaccharide fatty esters,[25,26] and particle-stabilized interfaces.[27–30]

Currently, very few models are available which can adequately describe the effects of deformations on surface structure and surface properties, and the effects these changes have on the overall dynamics of a multiphase system.[1] Non-equilibrium Thermodynamics (NET) can be an important tool to fill this knowledge gap. Several NET frameworks capable of describing multiphase systems are currently available. One of the first frameworks to be extended to multiphase systems was the Classical Irreversible Thermodynamics (CIT) framework,[31–33] by Bedeaux and co-workers, in the late 70s and early 80s of the previous century.[34–37,64] In its original form, the multiphase CIT framework can deal only with simple viscous interfaces. But when combined with the theory of internal variables,[38] the CIT framework can produce also constitutive models for complex interfaces displaying nonlinear viscoelastic behaviour.[39]

Other frameworks, such as the Extended Irreversible Thermodynamics (EIT) framework,[40,41] or the GENERIC framework[42,43] (General Equation for Non-equilibrium Reversible–Irreversible Coupling), have also been extended to describe multiphase systems with complex interfaces.[44–50] In the former Maxwell-type equations can be constructed for the surface stress tensor, and Cattaneo-type equations[51] can be constructed for the surface heat and mass flux vectors. The GENERIC framework is particularly useful for constructing structural models for the surface-stress tensor, which link nonlinear stress-deformation behaviour directly to the time evolution of the microstructure of the interface.[48,49]

In this chapter we present an overview of how NET can be used to model the behaviour of complex fluid–fluid interfaces. We start with a brief discussion on the definition and choice of surface excess variables, and proceed with the conservation principles for multiphase systems with surface excesses associated with their dividing surfaces. We show that the introduction of surface excess variables in the conservation principles leads to a set of differential balance equations (often termed jump balances[19]), for the surface mass density, surface momentum, surface energy, and surface entropy. We then proceed by discussing how constitutive models can be constructed for the fluxes appearing in these jump balances; respectively, the CIT, EIT, and GENERIC frameworks. We conclude with a summary of the most important results, and an outlook on future directions for this field.

17.2 Surface Excess Variables

Interfaces in multiphase systems are in fact three-dimensional regions of finite thickness in which the properties of the system change rapidly but continuously from their value in one bulk phase, to their value in the adjoining bulk phase. The two main frameworks for describing interfaces in multiphase systems are the Gibbs dividing surface model,[52] and the phase field (or diffuse interface) model.[53,54] In the former, a two-dimensional surface, often referred to as a dividing surface, is placed sensibly within the interfacial region, and all bulk fields are extrapolated up to this dividing surface. The difference between the actual and extrapolated fields is accounted for by assigning surface excess fields to the interface. In the phase field model the interface is modelled as a three-dimensional thin layer in which densities and material properties vary continuously. The choice between these two approaches is largely (but not exclusively) determined by the scale of the multiphase system we seek to describe. Problems in which the characteristic length scales of the system, L, are of the same order of magnitude as the thickness of the interfacial layers, h, such as bubble coalescence phenomena, thin film rupture, multiphase flow in nano- or microdevices, or the early stages of phase separation in immiscible polymer blends, are more conveniently described in terms of the phase field model. Systems in which the length scales are much larger than the interfacial thickness, so $h/L \ll 1$, are typically modelled using the Gibbs dividing surface approach. The latter will be the focus of this chapter.

17.2.1 Surface Variables for Simple Interfaces

As mentioned above, in the Gibbs dividing surface model the time evolution of the dividing surface is described by a set of differential equations, or jump balances, for the system excess variables associated with this surface. The first step in deriving these balances is to choose the appropriate system variables to describe the interface. Let us first consider a simple multicomponent system with N components, in which neither the bulk phases nor the dividing surfaces have a complex microstructure. For such a system the state of the bulk phases can be described with the set of fields

$$\{\rho, \boldsymbol{m}, \bar{u}, \rho_{(1)}, \ldots, \rho_{(N-1)}\}. \tag{17.1}$$

Here ρ is the bulk overall mass density, $\boldsymbol{m} = \rho\boldsymbol{v}$ is the bulk momentum density, $\boldsymbol{v}$ is the bulk velocity, $\bar{u}$ is the internal energy per unit volume, and $\rho_{(J)}$ ($J = 1, \ldots, N-1$) are the mass densities of the individual components in the mixture. Each of the bulk variables in eqn (17.1) may have an excess associated with it. Let us denote this set of surface excess variables as

$$\{\rho^{\mathrm{s}}, \boldsymbol{m}^{\mathrm{s}}, \bar{u}^{\mathrm{s}}, \rho^{\mathrm{s}}_{(1)}, \ldots, \rho^{\mathrm{s}}_{(N-1)}\}. \tag{17.2}$$

Here ρ^{s} is the overall surface mass density, $\boldsymbol{m}^{\mathrm{s}} = \rho^{\mathrm{s}}\boldsymbol{v}^{\mathrm{s}}$ is the surface momentum density, $\boldsymbol{v}^{\mathrm{s}}$ is the surface velocity field, $\bar{u}^{\mathrm{s}}$ is the surface internal energy per unit area, and $\rho^{\mathrm{s}}_{(J)}$ ($J = 1, \ldots, N-1$) are the surface mass densities of component J. With this choice of variables, the number of surface variables is equal to the number of bulk variables, which for a system at equilibrium would be a violation of Gibbs' phase rule. To resolve this issue we will take a closer look at the definition of the surface excess fields. They are related to the actual bulk fields by ($\psi = \rho, \boldsymbol{m}, \bar{u}, \rho_{(J)}$):

$$\psi^{\mathrm{s}} = \int_{-d}^{0} (\psi - \psi^{\mathrm{I}})\mathrm{d}z + \int_{0}^{d} (\psi - \psi^{\mathrm{II}})\mathrm{d}z. \tag{17.3}$$

Here z is the coordinate perpendicular to the interface, and $2d$ denotes the thickness of the interface. This integral is equal to the shaded regions in Figure 17.1.

When none of the components adsorb preferentially at the interface, the profiles for overall mass density, momentum density, and the individual component densities, will be similar to those depicted on the left-hand side of Figure 17.1. In this situation the value of the surface excesses associated

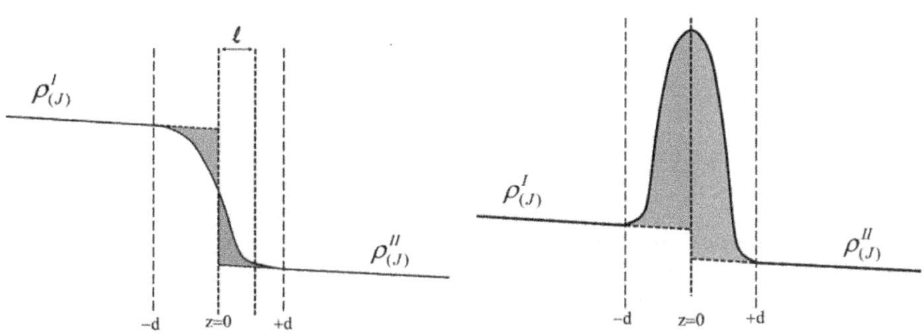

Figure 17.1 Density profiles for component J across the interface. The left side displays a typical profile when this component does not preferentially accumulate at the interface, and the right side depicts a profile for a surface-active component.

with these variables is very sensitive to the choice for the location of the dividing surface. We may even choose a location where one or more of the excess variables are identical to zero. For example, we could choose the location of the dividing surface such that $\rho^{\mathrm{s}}=0$. Note that with this particular choice of the location of the dividing surface, we reduce the number of surface variables by one, and Gibbs' phase rule is satisfied.

In view of their sensitivity with respect to the choice of the location of the dividing surface, densities like the surface mass, momentum, internal energy, or entropy density have been referred to as "ambiguous" surface variables.[45] When we displace the location of the dividing surface by a distance ℓ, we find a new value for these excess variables, related to the old value by

$$\psi^{\mathrm{s}'}=\psi^{\mathrm{s}}+\ell(\psi^{\mathrm{I}}-\psi^{\mathrm{II}}). \tag{17.4}$$

Öttinger *et al.*[45] have suggested that the choice of the location of the dividing surface can be viewed as a gauge degree of freedom, and that eqn (17.4) be viewed as a gauge transformation. We can use the latter equation to construct a set of gauge-invariant variables, in the following way: when in the gauge $\rho^{\mathrm{s}}=0$ we displace the location of the dividing surface by a distance ℓ, we obtain a new value for the surface mass density, equal to

$$\rho^{\mathrm{s}'}=\ell(\rho^{\mathrm{I}}-\rho^{\mathrm{II}}). \tag{17.5}$$

Similarly, we find a new value for the momentum density (using the fact that in the gauge $\rho^{\mathrm{s}}=0$ we also have $\boldsymbol{m}^{\mathrm{s}}=0$)

$$\boldsymbol{m}^{\mathrm{s}'}=\rho^{\mathrm{s}'}\boldsymbol{v}^{\mathrm{s}'}=\ell(\rho^{\mathrm{I}}\boldsymbol{v}^{\mathrm{I}}-\rho^{\mathrm{II}}\boldsymbol{v}^{\mathrm{II}}). \tag{17.6}$$

Eliminating ℓ from eqn (17.6) using eqn (17.5) gives us

$$\boldsymbol{v}^{\mathrm{s}}=\frac{\rho^{\mathrm{I}}\boldsymbol{v}^{\mathrm{I}}-\rho^{\mathrm{II}}\boldsymbol{v}^{\mathrm{II}}}{\rho^{\mathrm{I}}-\rho^{\mathrm{II}}}. \tag{17.7}$$

We see that the surface velocity is independent of the particular choice of the location of the dividing surface, and hence is a gauge-invariant variable. Similarly, we can introduce $u^{\mathrm{s}}=\bar{u}^{\mathrm{s}}/\rho^{\mathrm{s}}$ (the surface internal energy per unit mass), and $\omega^{\mathrm{s}}_{(J)}=\rho^{\mathrm{s}}_{(J)}\big/\rho^{\mathrm{s}}$ (the surface mass fraction of component J), which satisfy

$$u^{\mathrm{s}}=\frac{\rho^{\mathrm{I}}u^{\mathrm{I}}-\rho^{\mathrm{II}}u^{\mathrm{II}}}{\rho^{\mathrm{I}}-\rho^{\mathrm{II}}},\quad \omega^{\mathrm{s}}_{(J)}=\frac{\rho^{\mathrm{I}}\omega^{\mathrm{I}}_{(J)}-\rho^{\mathrm{II}}\omega^{\mathrm{II}}_{(J)}}{\rho^{\mathrm{I}}-\rho^{\mathrm{II}}}. \tag{17.8}$$

These values are clearly independent of the particular choice for the dividing surface. Hence, as an alternative to eqn (17.2) we could introduce a set of surface variables given by

$$\{\rho^{\mathrm{s}},\boldsymbol{v}^{\mathrm{s}},u^{\mathrm{s}},\omega^{\mathrm{s}}_{(1)},\ldots,\omega^{\mathrm{s}}_{(N-1)}\}. \tag{17.9}$$

When one or more components in the system adsorb preferentially at the interface, their density profile across the interface may look similar to the image on the right-hand side of Figure 17.1. In this situation their excess is fairly insensitive to the exact choice of the location of the dividing surface. For such systems we typically do not fix the location of the dividing surface by setting $\rho^{s} = 0$, since this would require some of the surface densities (and hence mass fractions) to be negative. Since we generally prefer to work with positive-valued concentrations, the location of the dividing surface is typically fixed by setting $\rho^{s} = \rho^{s}_{\infty}$, where the latter is a constant, chosen such that all surface densities are positive. Alternatively, the location can be fixed by setting the surface density of one of the non-adsorbing components to zero. In both conventions the number of surface variables is reduced by one, and Gibbs' phase rule is valid at equilibrium. Note that away from equilibrium, in an evolving system, we can fix the location of the dividing surface in this way, only at some reference time $t = t_{r}$. For all times $t > t_{r}$ we must allow for all surface variables (including the one we set to zero to select a particular gauge) to change, by both in-plane transport processes, and exchange with the bulk phases. Hence for evolving systems we must always work with the full set of surface variables given in either eqn (17.2) or in eqn (17.9).

17.2.2 Surface Variables for Complex Interfaces

For complex multiphase systems the set of bulk and surface variables introduced in the previous section may not be adequate to describe the dynamics of the system. We may need to include scalar, vectorial, and tensorial structural variables in these sets, which describe any changes in the microstructure of the system, induced by an applied deformation or temperature gradient. Our set of surface variables would then be given by

$$\{\rho^{s}, \boldsymbol{m}^{s}, \bar{u}^{s}, \rho^{s}_{(1)}, \ldots, \rho^{s}_{(N-1)}, \Gamma^{s}_{1}, \ldots, \Gamma^{s}_{n}, \boldsymbol{c}^{s}_{1}, \ldots, \boldsymbol{c}^{s}_{m}, \boldsymbol{C}^{s}_{1}, \ldots, \boldsymbol{C}^{s}_{k}\}. \tag{17.10}$$

Here Γ^{s}_{n} denote scalar structural variables, $\boldsymbol{c}^{s}_{m}$ denote vectorial variables, and $\boldsymbol{C}^{s}_{k}$ denote variables of a tensorial nature. Note that the lower-case subscripts n, m, and k denote the various structural variables needed to describe the interface, and are not to be confused with variables associated with a specific component in the mixture (which are denoted with upper-case subscripts). The vectorial and tensorial variables are all defined on the surface, but are not restricted to the class of tangential surface vectors and tensors. An example of a system for which structural surface variables could be included, are systems with interfaces stabilized by rigid anisotropic particles. For such systems it would be sufficient to include an additional scalar variable for the surface concentration of particles (either a surface density, a surface mass fraction, or an area fraction), and a single tensorial variable, representing the average orientation of the particles.[39,48–50] Similarly, for polymer stabilized interfaces we could introduce a scalar variable for the local surface segment density, and a tensor field describing the orientation and stretching of the segments. For interfaces stabilized by components forming a single-domain 2D liquid crystalline phase, a vectorial variable representing the director field could be introduced.[55] In systems where the surface-active species form a 2D crystalline phase, the (deformed) lattice vectors of the crystal structure could be included as surface variables. Alternatively, we could include scalar variables in eqn (17.10), representing al relevant scalar products of the lattice vectors,[56,57] or several tensorial fields equal to the dyadic products of these vectors. As we will see in Section 17.4, the inclusion of this type of structural variables in the set of surface variables will lead to additional time-evolution equations for these variables.[39,48–50,55–57]

17.3 Conservation Principles for Surface Mass, Momentum, and Energy

In this section we will discuss the conservation principles for mass, momentum, and energy for a multicomponent multiphase system with excess variables associated with its dividing surfaces. These quantities were introduced in Section 1.4 and discussed in Chapter 4. We start with the principle of conservation of mass. This principle requires that the total mass of a multiphase system is constant in time. For a multiphase system with mass associated with the dividing surfaces, we can express this requirement as

$$\frac{\mathrm{d}}{\mathrm{d}t}\left[\int_R \rho \mathrm{d}V + \int_\Sigma \rho^{\mathrm{s}} \mathrm{d}\Omega\right] = 0. \tag{17.11}$$

Here R denotes the domain of the bulk phases of the system, and Σ denotes the domain of all dividing surfaces of the system. The symbols $\mathrm{d}V$ and $\mathrm{d}\Omega$ denote respectively, volume and area integrations. Evaluating the time derivative on the left-hand side of eqn (17.11), we find that at every position in the bulk phases the equation of continuity must hold, that is,[19]

$$\frac{\mathrm{d_b}\rho}{\mathrm{d}t} + \rho\nabla\cdot\boldsymbol{v} = 0. \tag{17.12}$$

In this expression we have introduced the bulk material derivative[19]

$$\frac{\mathrm{d_b}\psi}{\mathrm{d}t} = \frac{\partial\psi}{\partial t} + (\nabla\psi)\cdot\boldsymbol{v}. \tag{17.13}$$

Eqn (17.11) requires at every point on the dividing surface[19]

$$\frac{\mathrm{d_s}\rho^{\mathrm{s}}}{\mathrm{d}t} + \rho^{\mathrm{s}}\nabla_{\mathrm{s}}\cdot\boldsymbol{v}^{\mathrm{s}} + [\![\rho(\boldsymbol{v}-\boldsymbol{v}^{\mathrm{s}})\cdot\boldsymbol{n}]\!] = 0. \tag{17.14}$$

Here we have introduced the surface material derivative, defined as[19]

$$\frac{\mathrm{d_s}\psi^{\mathrm{s}}}{\mathrm{d}t} = \frac{\partial\psi^{\mathrm{s}}}{\partial t} + (\nabla_{\mathrm{s}}\psi^{\mathrm{s}})\cdot\dot{\boldsymbol{y}}. \tag{17.15}$$

The velocity $\dot{\boldsymbol{y}} = \boldsymbol{v}^{\mathrm{s}} - \boldsymbol{u}$ is the intrinsic surface velocity, and $\boldsymbol{u}$ is the speed of displacement of the interface.[19] The operator ∇_{s} denotes the surface gradient operator, and the double bracket notation in eqn (17.14) is defined as[19]

$$[\![\psi\boldsymbol{n}]\!] = \psi^{\mathrm{I}}\boldsymbol{n}^{\mathrm{I}} + \psi^{\mathrm{II}}\boldsymbol{n}^{\mathrm{II}}. \tag{17.16}$$

Here ψ^M ($M=$I,II) is the value of ψ in phase M, evaluated at the dividing surface Σ, and $\boldsymbol{n}^M$ is the unit vector normal to Σ, pointing in the direction of phase M. Eqn (17.14) is often referred to as the overall jump mass balance.[19] At every point on the dividing surface it describes the time rate of change of the overall surface mass density, as a result of in-plane convection, and exchange of mass with the adjoining bulk phases (represented by the double bracket term). Similarly, we find that for all components in the mixture, conservation of mass implies that in the bulk phase[19]

$$\rho\frac{\mathrm{d}_\mathrm{b}\omega_{(J)}}{\mathrm{d}t}+\nabla\cdot\boldsymbol{j}_{(J)}-r_{(J)}=0, \tag{17.17}$$

where $\omega_{(J)}$ is the mass fraction of component J in the bulk phase, $\boldsymbol{j}_{(J)}$ is the mass flux vector for component J, and $r_{(J)}$ is the chemical reaction rate (in kg per unit volume per second) for component J. At each point on the dividing surface we must satisfy the jump component mass balance[19]

$$\rho^\mathrm{s}\frac{\mathrm{d}_\mathrm{s}\omega^\mathrm{s}_{(J)}}{\mathrm{d}t}+\nabla_\mathrm{s}\cdot\boldsymbol{j}^\mathrm{s}_{(J)}-r^\mathrm{s}_{(J)}+\left[\!\left[\rho(\omega_{(J)}-\omega^\mathrm{s}_{(J)})(\boldsymbol{v}-\boldsymbol{v}^\mathrm{s})\cdot\boldsymbol{n}+\boldsymbol{j}_{(J)}\cdot\boldsymbol{n}\right]\!\right]=0, \tag{17.18}$$

where $\boldsymbol{j}^\mathrm{s}_{(J)}$ is the surface mass flux vector of component J, and $r^\mathrm{s}_{(J)}$ is the rate per unit area at which component J is converted by surface reactions. The bulk and surface mass flux vectors are defined as[19]

$$\boldsymbol{j}_{(J)}=\rho_{(J)}(\boldsymbol{v}_{(J)}-\boldsymbol{v}),\quad \boldsymbol{j}^\mathrm{s}_{(J)}=\rho^\mathrm{s}_{(J)}(\boldsymbol{v}^\mathrm{s}_{(J)}-\boldsymbol{v}^\mathrm{s}). \tag{17.19}$$

The jump component mass balance describes the time rate of change of the surface density of component J, as a result of (in order of appearance in the equation) in-plane surface diffusion, surface reactions, convective transfer between the interface and adjoining bulk phases, and diffusive exchange between bulk and interface.

The principle of conservation of momentum requires that the time rate of change of momentum of a multiphase system be equal to the body forces acting on the material in the bulk phases and interfaces, and the stresses applied on it through its outer boundaries. For the bulk phases this principle implies that at each point the differential momentum balance must be satisfied:[19]

$$\rho\frac{\mathrm{d}_\mathrm{b}\boldsymbol{v}}{\mathrm{d}t}-\nabla\cdot\boldsymbol{T}-\sum_{J=1}^{N}\rho_{(J)}\boldsymbol{b}_{(J)}=0, \tag{17.20}$$

where $\boldsymbol{T}$ is the stress tensor, and $\boldsymbol{b}_{(J)}$ are the body forces per unit mass acting on component J. The principle also implies that on the dividing surfaces we must satisfy the jump momentum balance[19]

$$\rho^\mathrm{s}\frac{\mathrm{d}_\mathrm{s}\boldsymbol{v}^\mathrm{s}}{\mathrm{d}t}-\nabla_\mathrm{s}\cdot\boldsymbol{T}^\mathrm{s}-\sum_{J=1}^{N}\rho^\mathrm{s}_{(J)}\boldsymbol{b}^\mathrm{s}_{(J)}+[\![\rho(\boldsymbol{v}-\boldsymbol{v}^\mathrm{s})(\boldsymbol{v}-\boldsymbol{v}^\mathrm{s})\cdot\boldsymbol{n}-\boldsymbol{T}\cdot\boldsymbol{n}]\!]=0. \tag{17.21}$$

Here $\boldsymbol{T}^{\mathrm{s}}$ is the surface stress tensor, and $\boldsymbol{b}^{\mathrm{s}}_{(J)}$ are the body forces per unit mass acting on component J in the interface. It is common to write the stress tensors in these equations as

$$\boldsymbol{T} = -p\boldsymbol{I} + \boldsymbol{\sigma}, \quad \boldsymbol{T}^{\mathrm{s}} = \gamma\boldsymbol{P} + \boldsymbol{\sigma}^{\mathrm{s}}. \tag{17.22}$$

where p is the thermodynamic pressure, $\boldsymbol{I}$ is the 3D unit tensor, $\boldsymbol{\sigma}$ is the bulk extra stress tensor, γ is the surface tension, $\boldsymbol{P}$ is the surface projection tensor (the unit tensor for the tangential surface fields), and $\boldsymbol{\sigma}^{\mathrm{s}}$ is the surface extra stress tensor. Note that in eqn (17.20) to (17.22) we use a sign convention in which the sign of the bulk stress tensor is chosen to be opposite to that of the pressure tensor, $p\boldsymbol{I}$. Substituting eqn (17.22) in eqn (17.21) we obtain

$$\rho^{\mathrm{s}}\frac{\mathrm{d}_{\mathrm{s}}\boldsymbol{v}^{\mathrm{s}}}{\mathrm{d}t} - \nabla_{\mathrm{s}}\gamma - 2\gamma H\boldsymbol{n} - \nabla_{\mathrm{s}}\cdot\boldsymbol{\sigma}^{\mathrm{s}} - \sum_{J=1}^{N}\rho^{\mathrm{s}}_{(J)}\boldsymbol{b}^{\mathrm{s}}_{(J)} + [\![\rho(\boldsymbol{v}-\boldsymbol{v}^{\mathrm{s}})(\boldsymbol{v}-\boldsymbol{v}^{\mathrm{s}})\cdot\boldsymbol{n} + P\boldsymbol{n} - \boldsymbol{\sigma}\cdot\boldsymbol{n}]\!] = 0, \tag{17.23}$$

where H is the curvature of the interface. The jump momentum balance is a generalized form of the Young–Laplace equation, and describes the time rate of change of surface momentum as a result of surface tension gradients (Marangoni stresses), curvature induced stresses, deviatoric stresses, body forces, and (given by the double bracket term) inertial, hydrostatic, and deviatoric stresses exerted on the interface by the adjoining bulk phases. The Young–Laplace equation is frequently used in the analysis of surface dilatational rheology experiments, performed with droplet tensiometry methods. For complex fluid–fluid interfaces it is often not appropriate to use this equation to analyse experiments, and one should use the generalized form given in eqn (17.23).[1]

The principle of conservation of energy states that the time rate of change of the sum of the internal and kinetic energy of a multiphase system is equal to the work performed on the system by the body forces and stresses, plus the energy transmitted to the system through its outer boundaries, plus the energy transmitted to the interior of the system through radiation. It implies that in the bulk phase we must satisfy the differential energy balance[19]

$$\rho\frac{\mathrm{d}_{\mathrm{b}}u}{\mathrm{d}t} = \boldsymbol{\sigma} : \nabla\boldsymbol{v} - p\nabla\cdot\boldsymbol{v} + \sum_{J=1}^{N}\boldsymbol{j}_{(J)}\cdot\boldsymbol{b}_{(J)} - \nabla\cdot\boldsymbol{q} + \rho\hat{Q}, \tag{17.24}$$

where $\boldsymbol{q}$ is the energy flux vector, and $\hat{Q}$ is the rate of radiant energy transmission per unit mass to the material in the bulk phases. To satisfy conservation of energy we must require also the jump energy balance to hold at each point on the dividing surface, given by[19]

$$\rho^{\mathrm{s}}\frac{\mathrm{d}_{\mathrm{s}}u^{\mathrm{s}}}{\mathrm{d}t} = \boldsymbol{\sigma}^{\mathrm{s}} : \nabla_{\mathrm{s}}\boldsymbol{v}^{\mathrm{s}} + \gamma\nabla_{\mathrm{s}}\cdot\boldsymbol{v}^{\mathrm{s}} + \sum_{J=1}^{N}\boldsymbol{j}^{\mathrm{s}}_{(J)}\cdot\boldsymbol{b}^{\mathrm{s}}_{(J)} - \nabla_{\mathrm{s}}\cdot\boldsymbol{q}^{\mathrm{s}} + \rho^{\mathrm{s}}\hat{Q}^{\mathrm{s}} - \left[\!\!\left[\rho\left(u + pv - u^{\mathrm{s}} + \frac{1}{2}|\boldsymbol{v}-\boldsymbol{v}^{\mathrm{s}}|^2\right)(\boldsymbol{v}-\boldsymbol{v}^{\mathrm{s}})\cdot\boldsymbol{n} - (\boldsymbol{v}-\boldsymbol{v}^{\mathrm{s}})\cdot\boldsymbol{\sigma}\cdot\boldsymbol{n} + \boldsymbol{q}\cdot\boldsymbol{n}\right]\!\!\right]. \tag{17.25}$$

Here $\boldsymbol{q}^{\mathrm{s}}$ is the surface energy flux vector, $v = 1/\rho$ is the volume per unit mass of the bulk phase, and $\hat{Q}^{\mathrm{s}}$ is the rate of radiant energy transmission per unit mass to the material in the interfaces. The double colon in the first term on the right-hand side of eqn (17.25) denotes a double contraction between tensors.

Finally, to satisfy the second law of thermodynamics, we must impose at each point in the bulk phase the differential entropy balance,[19]

$$\rho \frac{\mathrm{d_b} s}{\mathrm{d}t} = -\nabla \cdot \boldsymbol{j}_S + \rho e, \tag{17.26}$$

where s is the entropy per unit mass, $\boldsymbol{j}_S$ is the entropy flux vector, and e is the rate of entropy production per unit mass. At each point on the dividing surfaces we must satisfy the jump entropy balance,[19]

$$\rho^{\mathrm{s}} \frac{\mathrm{d_s} s^{\mathrm{s}}}{\mathrm{d}t} = -\nabla_{\mathrm{s}} \cdot \boldsymbol{j}_S^{\mathrm{s}} + \rho^{\mathrm{s}} e^{\mathrm{s}} - [\![\rho(s - s^{\mathrm{s}})(\boldsymbol{v} - \boldsymbol{v}^{\mathrm{s}}) \cdot \boldsymbol{n} + \boldsymbol{j}_S \cdot \boldsymbol{n}]\!]. \tag{17.27}$$

Here s^{s} is the surface entropy per unit mass, $\boldsymbol{j}_S^{\mathrm{s}}$ is the surface entropy flux vector, and e^{s} is the rate of surface entropy production per unit mass. To satisfy the second law of thermodynamics we must require

$$e \geq 0, \quad e^{\mathrm{s}} \geq 0. \tag{17.28}$$

Now that we have derived the jump balances for the surface mass density, momentum, energy, and entropy, we will focus our attention on the fluxes appearing in these balances.

17.4 Constitutive Equations for Surface Fluxes

In this section we will discuss how constitutive equations for the surface fluxes can be derived using NET frameworks. There are various NET frameworks available capable of constructing constitutive equations for surface fluxes, and the procedures to construct these equations are different for each of them. Here we will limit ourselves to three frameworks: classical irreversible thermodynamics, extended irreversible thermodynamics, and GENERIC.

17.4.1 Classical Irreversible Thermodynamics

17.4.1.1 General Form of the Flux–Force Relations

In the classical irreversible-thermodynamic framework we start with an assumption for the functional dependence of the surface entropy on the system variables.[64] For a multiphase system with complex interfaces we will assume here that

$$s^{\mathrm{s}} = s^{\mathrm{s}}\left(u^{\mathrm{s}}, \hat{\Omega}, \omega^{\mathrm{s}}_{(1)}, \ldots, \omega^{\mathrm{s}}_{(N-1)}, \Gamma^{\mathrm{s}}_1, \ldots, \Gamma^{\mathrm{s}}_n, \boldsymbol{c}^{\mathrm{s}}_1, \ldots, \boldsymbol{c}^{\mathrm{s}}_m, \boldsymbol{C}^{\mathrm{s}}_1, \ldots, \boldsymbol{C}^{\mathrm{s}}_k\right), \tag{17.29}$$

where $\hat{\Omega}=1/\rho^{\rm s}$ is the area per unit mass. From this assumption we find that the material time derivative of the surface entropy per unit mass is given by

$$\begin{aligned}\rho^{\rm s}\frac{{\rm d_s}s^{\rm s}}{{\rm d}t} &= \frac{\rho^{\rm s}}{T^{\rm s}}\frac{{\rm d_s}u^{\rm s}}{{\rm d}t}-\frac{\gamma\rho^{\rm s}}{T^{\rm s}}\frac{{\rm d_s}\hat{\Omega}}{{\rm d}t}-\frac{\rho^{\rm s}}{T^{\rm s}}\sum_{J=1}^{N}\mu^{\rm s}_{(J)}\frac{{\rm d_s}\omega^{\rm s}_{(J)}}{{\rm d}t}\\ &-\sum_n\frac{\rho^{\rm s}\Xi^{\rm s}_n}{T^{\rm s}}\frac{{\rm d_s}\Gamma^{\rm s}_n}{{\rm d}t}-\sum_m\frac{\rho^{\rm s}}{T^{\rm s}}\boldsymbol{w}^{\rm s}_m\cdot\frac{{\rm d_s}\boldsymbol{c}^{\rm s}_m}{{\rm d}t}-\sum_k\frac{\rho^{\rm s}}{T^{\rm s}}\boldsymbol{W}^{\rm s}_k:\frac{{\rm d_s}\boldsymbol{C}^{\rm s}_k}{{\rm d}t},\end{aligned}\tag{17.30}$$

where $T^{\rm s}$ is the surface temperature, and $\mu^{\rm s}_{(J)}$ is the surface chemical potential of component J. The coefficients appearing in the last line of this expression are defined as

$$\begin{aligned}\Xi^{\rm s}_n &\equiv T^{\rm s}\left(\frac{\partial s^{\rm s}}{\partial\Gamma^{\rm s}_n}\right)_{\bar{u}^{\rm s},\hat{\Omega},\omega^{\rm s}_{(J)},\Gamma^{\rm s}_p(p\neq n),\boldsymbol{c}^{\rm s}_m,\boldsymbol{C}^{\rm s}_k},\\ \boldsymbol{w}^{\rm s}_m &\equiv T^{\rm s}\left(\frac{\partial s^{\rm s}}{\partial\boldsymbol{c}^{\rm s}_m}\right)_{\bar{u}^{\rm s},\hat{\Omega},\omega^{\rm s}_{(J)},\Gamma^{\rm s}_n,\boldsymbol{c}^{\rm s}_p(p\neq m),\boldsymbol{C}^{\rm s}_k},\\ \boldsymbol{W}^{\rm s}_k &\equiv T^{\rm s}\left(\frac{\partial s^{\rm s}}{\partial\boldsymbol{C}^{\rm s}_k}\right)_{\bar{u}^{\rm s},\hat{\Omega},\omega^{\rm s}_{(J)},\Gamma^{\rm s}_n,\boldsymbol{c}^{\rm s}_m,\boldsymbol{C}^{\rm s}_p(p\neq k)}.\end{aligned}\tag{17.31}$$

Substituting eqn (17.30) in the jump entropy balance eqn (17.27), and using the overall jump mass balance eqn (17.14), the component jump mass balance eqn (17.18), and the jump energy balance eqn (17.25), we find (assuming $\hat{Q}^{\rm s}=0$, and $r^{\rm s}_{(J)}=0$, for all J)

$$\begin{aligned}\rho^{\rm s}e^{\rm s} &= \frac{1}{T^{\rm s}}\overline{\boldsymbol{\sigma}}^{\rm s}:\overline{\boldsymbol{D}}^{\rm s}+\frac{{\rm tr}\boldsymbol{\sigma}^{\rm s}}{T^{\rm s}}{\rm tr}\boldsymbol{D}^{\rm s}-\sum_n\frac{\rho^{\rm s}\Xi^{\rm s}_n}{T^{\rm s}}\frac{{\rm d_s}\Gamma^{\rm s}_n}{{\rm d}t}-\sum_m\frac{\rho^{\rm s}}{T^{\rm s}}\boldsymbol{w}^{\rm s}_m\cdot\frac{{\rm d_s}\boldsymbol{c}^{\rm s}_m}{{\rm d}t}\\ &-\sum_k\frac{\rho^{\rm s}}{T^{\rm s}}\boldsymbol{W}^{\rm s}_k:\frac{{\rm d_s}\boldsymbol{C}^{\rm s}_k}{{\rm d}t}-\frac{1}{T^{\rm s}}\sum_{J=1}^{N}\boldsymbol{j}^{\rm s}_{(J)}\cdot\boldsymbol{d}^{\rm s}_{(J)}-\frac{1}{(T^{\rm s})^2}\left(\boldsymbol{q}^{\rm s}-\sum_{J=1}^{N}\mu^{\rm s}_{(J)}\boldsymbol{j}^{\rm s}_{(J)}\right)\cdot\nabla_{\rm s}T^{\rm s}\\ &-\frac{1}{T^{\rm s}}\left[\!\left[\rho\left((u+pv)\left[\frac{T-T^{\rm s}}{T}\right]-\rho\left(\sum_n\Xi^{\rm s}_n\Gamma^{\rm s}_n-\sum_m\boldsymbol{w}^{\rm s}_m\cdot\boldsymbol{c}^{\rm s}_m-\sum_k\boldsymbol{W}^{\rm s}_k:\boldsymbol{C}^{\rm s}_k\right)\right.\right.\right.\\ &\left.+\sum_{J=1}^{N}T^{\rm s}\left(\frac{\tilde{\mu}_{(J)}}{T}-\frac{\tilde{\mu}^{\rm s}_{(J)}}{T^{\rm s}}\right)\omega_{(J)}+\frac{1}{2}(\boldsymbol{v}-\boldsymbol{v}^{\rm s})^2-\frac{1}{2}(\boldsymbol{v}^{\rm s})^2+\frac{1}{2}\boldsymbol{v}^2\frac{T^{\rm s}}{T}\right)(\boldsymbol{v}-\boldsymbol{v}^{\rm s})\cdot\boldsymbol{n}\\ &\left.\left.+\boldsymbol{q}\cdot\boldsymbol{n}\left[\frac{T-T^{\rm s}}{T}\right]-(\boldsymbol{v}-\boldsymbol{v}^{\rm s})\cdot\boldsymbol{\sigma}\cdot\boldsymbol{n}+T^{\rm s}\sum_{J=1}^{N}\boldsymbol{j}_{(J)}\cdot\boldsymbol{n}\left(\frac{\tilde{\mu}_{(J)}}{T}-\frac{\tilde{\mu}^{\rm s}_{(J)}}{T^{\rm s}}\right)\right]\!\right]\geq 0.\end{aligned}\tag{17.32}$$

Here the bar over the tensors $\boldsymbol{\sigma}^{\rm s}$ and $\boldsymbol{D}^{\rm s}$ denotes the symmetric traceless part of these tensors, $\boldsymbol{D}^{\rm s}$ is the surface rate of deformation tensor, equal to $\frac{1}{2}(\boldsymbol{P}\cdot\nabla_{\rm s}\boldsymbol{v}^{\rm s}+[\nabla_{\rm s}\boldsymbol{v}^{\rm s}]^{\rm T}\cdot\boldsymbol{P})$, $\tilde{\mu}^{\rm s}_{(J)}=\mu^{\rm s}_{(J)}-\frac{1}{2}(\boldsymbol{v}^{\rm s})^2$ is the velocity modified surface chemical potential, $\tilde{\mu}_{(J)} = \mu_{(J)} - \frac{1}{2}v^2$ is the velocity modified chemical potential of component J in the bulk phase, T is the temperature in the bulk phase, and the vector $\boldsymbol{d}^{\mathrm{s}}_{(J)} \equiv \nabla_{\mathrm{s}}\mu^{\mathrm{s}}_{(J)} - \boldsymbol{b}^{\mathrm{s}}_{(J)}$.

Eqn (17.32) is a bilinear form consisting of products of fluxes and driving forces. For the traceless part of the surface extra stress tensor, and its trace, eqn (17.32), suggests the following functional dependence (using the fact that fluxes depend only on driving forces of equal tensorial order):

$$\overline{\boldsymbol{\sigma}}^{\mathrm{s}} = \overline{\boldsymbol{\sigma}}^{\mathrm{s}}(\overline{\boldsymbol{D}}^{\mathrm{s}}, \overline{\boldsymbol{W}}^{\mathrm{s}}_1, \ldots, \overline{\boldsymbol{W}}^{\mathrm{s}}_k), \tag{17.33}$$

$$\mathrm{tr}\boldsymbol{\sigma}^{\mathrm{s}} = \mathrm{tr}\boldsymbol{\sigma}^{\mathrm{s}}(\mathrm{tr}\boldsymbol{D}^{\mathrm{s}}, \mathrm{tr}\boldsymbol{W}^{\mathrm{s}}_1, \ldots, \mathrm{tr}\boldsymbol{W}^{\mathrm{s}}_k, \Gamma^{\mathrm{s}}_1, \ldots, \Gamma^{\mathrm{s}}_n). \tag{17.34}$$

For the tensors $\boldsymbol{C}^{\mathrm{s}}_q$ $(q = 1, \ldots, k)$ and scalars Γ^{s}_p $(p = 1, \ldots, m)$ eqn (17.32) suggests that we choose

$$\rho^{\mathrm{s}}\frac{\mathrm{d}_{\mathrm{s}}\boldsymbol{C}^{\mathrm{s}}_q}{\mathrm{d}t} = \rho^{\mathrm{s}}\frac{\mathrm{d}_{\mathrm{s}}\boldsymbol{C}^{\mathrm{s}}_q}{\mathrm{d}t}(\boldsymbol{D}^{\mathrm{s}}, \boldsymbol{W}^{\mathrm{s}}_1, \ldots, \boldsymbol{W}^{\mathrm{s}}_k), \tag{17.35}$$

$$\rho^{\mathrm{s}}\frac{\mathrm{d}_{\mathrm{s}}\Gamma^{\mathrm{s}}_p}{\mathrm{d}t} = \rho^{\mathrm{s}}\frac{\mathrm{d}_{\mathrm{s}}\Gamma^{\mathrm{s}}_p}{\mathrm{d}t}(\mathrm{tr}\boldsymbol{D}^{\mathrm{s}}, \mathrm{tr}\boldsymbol{W}^{\mathrm{s}}_1, \ldots, \mathrm{tr}\boldsymbol{W}^{\mathrm{s}}_k, \Gamma^{\mathrm{s}}_1, \ldots, \Gamma^{\mathrm{s}}_n). \tag{17.36}$$

To satisfy eqn (17.32) the mass and heat flux vectors, and structural vectors should have the following functional dependence:

$$\boldsymbol{j}^{\mathrm{s}}_{(J)} = \boldsymbol{j}^{\mathrm{s}}_{(J)}(\boldsymbol{d}^{\mathrm{s}}_{(J)}, \nabla_{\mathrm{s}}T^{\mathrm{s}}, \boldsymbol{w}^{\mathrm{s}}_m), \tag{17.37}$$

$$\boldsymbol{q}^{\mathrm{s}} - \sum_{J=1}^{N}\mu^{\mathrm{s}}_{(J)}\boldsymbol{j}^{\mathrm{s}}_{(J)} = T^{\mathrm{s}}\boldsymbol{j}^{\mathrm{s}}_S = T^{\mathrm{s}}\boldsymbol{j}^{\mathrm{s}}_S(\boldsymbol{d}^{\mathrm{s}}_{(J)}, \nabla_{\mathrm{s}}T^{\mathrm{s}}, \boldsymbol{w}^{\mathrm{s}}_m), \tag{17.38}$$

$$\rho^{\mathrm{s}}\frac{\mathrm{d}_{\mathrm{s}}\boldsymbol{c}^{\mathrm{s}}_q}{\mathrm{d}t} = \rho^{\mathrm{s}}\frac{\mathrm{d}_{\mathrm{s}}\boldsymbol{c}^{\mathrm{s}}_q}{\mathrm{d}t}\left(\boldsymbol{w}^{\mathrm{s}}_1, \ldots, \boldsymbol{w}^{\mathrm{s}}_m, \boldsymbol{d}^{\mathrm{s}}_{(J)}, \nabla_{\mathrm{s}}T^{\mathrm{s}}\right). \tag{17.39}$$

Using the fact that $\mu^{\mathrm{s}}_{(J)} = H^{\mathrm{s}}_{(J)} - TS^{\mathrm{s}}_{(J)}$, where $H^{\mathrm{s}}_{(J)}$ and $S^{\mathrm{s}}_{(J)}$ are the partial surface enthalpy and entropy of species J, the term on the left-hand side of eqn (17.38) can be written as $\epsilon^{\mathrm{s}} + \sum_{J=1}^{N} T^{\mathrm{s}}S^{\mathrm{s}}_{(J)}\boldsymbol{j}^{\mathrm{s}}_{(J)}$, where $\epsilon^{\mathrm{s}} = \boldsymbol{q}^{\mathrm{s}} - \sum_{J=1}^{N} H^{\mathrm{s}}_{(J)}\boldsymbol{j}^{\mathrm{s}}_{(J)}$ is often referred to as the *measurable* heat flux, which is independent of the chosen frame of reference.[33] The energy flux $\boldsymbol{q}^{\mathrm{s}}$ in contrast does depend on the choice of reference frame.

Expanding these relations up to first order in their arguments, we obtain for the surface extra stress tensor:

$$\overline{\boldsymbol{\sigma}}^{\mathrm{s}} = 2\varepsilon_{\mathrm{s}}\overline{\boldsymbol{D}}^{\mathrm{s}} + 2\sum_k L^{\mathrm{s}}_k\overline{\boldsymbol{W}}^{\mathrm{s}}_k, \tag{17.40}$$

$$\mathrm{tr}\boldsymbol{\sigma}^{\mathrm{s}} = \varepsilon_{\mathrm{d}}\mathrm{tr}\boldsymbol{D}^{\mathrm{s}} + 2\sum_k M^{\mathrm{s}}_k\mathrm{tr}\boldsymbol{W}^{\mathrm{s}}_k + \sum_n \tilde{M}^{\mathrm{s}}_n\Xi^{\mathrm{s}}_n, \tag{17.41}$$

where ε_s is the surface shear viscosity, ε_d is the surface dilatational viscosity, and L_k^s, M_k^s, and $\tilde{M}_n^s$ are scalar coefficients, which may depend on temperature and composition of the interface. When the dependence on the structural variables in eqn (17.40) and (17.41) is negligible, and we combine both equations, we obtain

$$\boldsymbol{\sigma}^s = (\varepsilon_d - \varepsilon_s)(\mathrm{tr}\boldsymbol{D}^s)\boldsymbol{P} + 2\varepsilon_s\boldsymbol{D}^s, \tag{17.42}$$

which is the linear Bousinesq model,[19] the surface equivalent of the Newtonian fluid model. For the structural variables we find up to linear order:

$$\rho^s \frac{\mathrm{d}_s \boldsymbol{C}_k^s}{\mathrm{d}t} = 2X_0^s \boldsymbol{D}^s + \sum_k X_k^s \boldsymbol{W}_k^s, \tag{17.43}$$

$$\rho^s \frac{\mathrm{d}_s \Gamma_p^s}{\mathrm{d}t} = Y_0^s \mathrm{tr}\boldsymbol{D}^s + \sum_k Y_k^s \mathrm{tr}\boldsymbol{W}_k^s + \sum_n \tilde{Y}_n^s \Gamma_n^s, \tag{17.44}$$

$$\rho^s \frac{\mathrm{d}_s \boldsymbol{c}_q^s}{\mathrm{d}t} = \sum_m z_m^s \boldsymbol{w}_m^s + \sum_J z_{(J)}^s \boldsymbol{d}_{(J)}^s + z_T^s \nabla_s T^s. \tag{17.45}$$

Again, the scalar coefficients X_0^s, X_k^s, Y_0^s, Y_k^s, $\tilde{Y}_n^s$, z_m^s, $z_{(J)}^s$, and z_T^s, appearing in these expression may all be a function of temperature and composition of the interface. For the mass and energy flux vectors we obtain:

$$\boldsymbol{j}_{(J)}^s = -\sum_K D_{(JK)}^s \boldsymbol{d}_{(K)}^s - \alpha_{(J)}^s \nabla_s \ln T^s + \sum_m \beta_{(J)m}^s \boldsymbol{w}_m^s, \tag{17.46}$$

$$\boldsymbol{q}^s - \sum_{J=1}^N \mu_{(J)}^s \boldsymbol{j}_{(J)}^s = -\sum_J \alpha_{(J)}^s \boldsymbol{d}_{(J)}^s - \lambda^s \nabla_s \ln T^s + \sum_m \beta_m^s \boldsymbol{w}_m^s, \tag{17.47}$$

where $D_{(JK)}^s$ denote the components of the $N \cdot N$ diffusion matrix, $\alpha_{(J)}^s$ is the surface thermal diffusion coefficient for component J, and λ^s is the surface thermal conductivity. The coefficients $\beta_{(J)m}^s$ and β_m^s quantify the coupling of, respectively, the surface mass and energy flux vectors with the vectorial structural fields. Note that eqn (17.46) and (17.47) contain couplings between mass and energy transfer, which are the surface equivalents of the Soret and Dufour effect. When the Soret effect is negligible, and contributions to the mass flux stemming from the structural vectors are as well, eqn (17.46) reduces to (assuming in addition that forced diffusion is negligible)

$$\boldsymbol{j}_{(J)}^s = -\sum_K D_{(JK)}^s \nabla_s \mu_{(K)}^s, \tag{17.48}$$

which is the surface equivalent of Fick's law. For a system with uniform surface composition, with negligible contributions to the surface energy flux induced by the structural vectors, eqn (17.47) reduces to

$$\boldsymbol{q}^s = -\frac{\lambda^s}{T^s} \nabla_s T^s. \tag{17.49}$$

This expression is the surface equivalent of Fourier's law. Substitution of eqn (17.48) and (17.49) in eqn (17.32) shows us that $D^{\mathrm{s}}_{(JK)} \geq 0$, and $\lambda^{\mathrm{s}} \geq 0$.

Eqn (17.32) allows us also to construct constitutive expressions for the fluxes describing exchange between the interface and the adjoining bulk phases. Following along the lines we used for deriving the constitutive equations for the surface fluxes, we obtain (M, $N=$ I,II):

$$\boldsymbol{\sigma}^M \cdot \boldsymbol{n}^M - \rho^M \boldsymbol{v}^M (\boldsymbol{v}^M - \boldsymbol{v}^{\mathrm{s}}) \cdot \boldsymbol{n}^M = \sum_{N=\mathrm{I}}^{\mathrm{II}} \boldsymbol{\zeta}^{M,N} T^{\mathrm{s}} \cdot \left(\frac{\boldsymbol{v}^N}{T^N} - \frac{\boldsymbol{v}^{\mathrm{s}}}{T^{\mathrm{s}}} \right), \tag{17.50}$$

$$\begin{aligned} &\boldsymbol{q}^M \cdot \boldsymbol{n}^M + \rho^M \left[u^M + p^M v^M + \frac{1}{2}(v^M)^2 \right] (\boldsymbol{v}^M - \boldsymbol{v}^{\mathrm{s}}) \cdot \boldsymbol{n}^M - \boldsymbol{v}^M \cdot \boldsymbol{\sigma}^M \cdot \boldsymbol{n}^M \\ &= -\frac{T^M - T^{\mathrm{s}}}{R^M_K} - \sum_J \Lambda^{TM}_{(J)} T^M T^{\mathrm{s}} \left(\frac{\tilde{\mu}^M_{(J)}}{T^M} - \frac{\tilde{\mu}^{\mathrm{s}}_{(J)}}{T^{\mathrm{s}}} \right), \end{aligned} \tag{17.51}$$

$$\boldsymbol{j}^M_{(J)} \cdot \boldsymbol{n}^M + \rho^M_{(J)} (\boldsymbol{v}^M - \boldsymbol{v}^{\mathrm{s}}) \cdot \boldsymbol{n}^M = -\Lambda^M_{(J)} \left(\frac{\tilde{\mu}^M_{(J)}}{T^M} - \frac{\tilde{\mu}^{\mathrm{s}}_{(J)}}{T^{\mathrm{s}}} \right) - \Lambda^{TM}_{(J)} (T^M - T^{\mathrm{s}}). \tag{17.52}$$

Here $\boldsymbol{\zeta}^{M,N}$ are friction tensors, quantifying the exchange of momentum between bulk phase and dividing surface, R^M_K is the Kapitza resistivity, quantifying the resistance against energy transfer between bulk and interface, $\Lambda^M_{(J)}$ are mass-transfer coefficients for exchange between the bulk and interface driven by differences in (velocity modified) chemical potential, and $\Lambda^{TM}_{(J)}$ are mass-transfer coefficients for exchange between bulk and interface, driven by temperature differences. Note that eqn (17.50) to (17.52) are not only constitutive equations for the exchange of mass, momentum, and energy between the bulk phases and the dividing surface, they also act as boundary conditions, coupling the differential equations for the bulk density, velocity, and energy fields, with their respective jump balances.

17.4.1.2 Specific Examples of Constitutive Models

As an example of the type of constitutive equations which can be constructed within the CIT framework, let us consider an isothermal system, with uniform composition, and a microstructure described by a single symmetric tensorial structural variable, $\boldsymbol{C}^{\mathrm{s}}$. Eqn (17.40) to (17.43) then reduce to

$$\overline{\boldsymbol{\sigma}}^{\mathrm{s}} = 2\varepsilon_{\mathrm{s}} \overline{\boldsymbol{D}}^{\mathrm{s}} + 2L^{\mathrm{s}} \overline{\boldsymbol{W}}^{\mathrm{s}}, \tag{17.53}$$

$$\mathrm{tr}\boldsymbol{\sigma}^{\mathrm{s}} = \varepsilon_{\mathrm{d}} \mathrm{tr}\boldsymbol{D}^{\mathrm{s}} + 2M^{\mathrm{s}} \mathrm{tr}\boldsymbol{W}^{\mathrm{s}}, \tag{17.54}$$

$$\rho^{\mathrm{s}} \frac{\mathrm{d}_{\mathrm{s}} \boldsymbol{C}^{\mathrm{s}}}{\mathrm{d}t} = 2X^{\mathrm{s}}_0 \boldsymbol{D}^{\mathrm{s}} + X^{\mathrm{s}} \boldsymbol{W}^{\mathrm{s}}. \tag{17.55}$$

To proceed we must derive an expression for the tensor $\boldsymbol{W}^{\mathrm{s}}$. To arrive at such an expression we first expand eqn (17.29) in terms of the structural variable:

$$s^{\mathrm{s}} = s_0^{\mathrm{s}} + \frac{\nu_1}{2T^{\mathrm{s}}}\boldsymbol{C}^{\mathrm{s}} : \boldsymbol{C}^{\mathrm{s}} + \frac{\nu_2}{3T^{\mathrm{s}}}\mathrm{tr}(\boldsymbol{C}^{\mathrm{s}} \cdot \boldsymbol{C}^{\mathrm{s}} \cdot \boldsymbol{C}^{\mathrm{s}}) + \frac{\nu_3}{4T^{\mathrm{s}}}\mathrm{tr}(\boldsymbol{C}^{\mathrm{s}} \cdot \boldsymbol{C}^{\mathrm{s}} \cdot \boldsymbol{C}^{\mathrm{s}} \cdot \boldsymbol{C}^{\mathrm{s}}) + \ldots, \tag{17.56}$$

where s_0^{s} is the non-structural contribution to the surface entropy per unit mass, and ν_i are scalar coefficients. From the definition of the tensor $\boldsymbol{W}^{\mathrm{s}}$ in eqn (17.31) we find

$$\boldsymbol{W}^{\mathrm{s}} \equiv T^{\mathrm{s}}\left(\frac{\partial s^{\mathrm{s}}}{\partial \boldsymbol{C}^{\mathrm{s}}}\right) = \nu_1 \boldsymbol{C}^{\mathrm{s}} + \nu_2 \boldsymbol{C}^{\mathrm{s}} \cdot \boldsymbol{C}^{\mathrm{s}} + \nu_3 \boldsymbol{C}^{\mathrm{s}} \cdot \boldsymbol{C}^{\mathrm{s}} \cdot \boldsymbol{C}^{\mathrm{s}} + \ldots. \tag{17.57}$$

By substituting this in eqn (17.53) to (17.55), we obtain the following set of equations (retaining only terms up to second order in the structural tensor):

$$\overline{\boldsymbol{\sigma}}^{\mathrm{s}} = 2\varepsilon_{\mathrm{s}}\overline{\boldsymbol{D}}^{\mathrm{s}} + 2L^{\mathrm{s}}\nu_1\overline{\boldsymbol{C}}^{\mathrm{s}} + 2L^{\mathrm{s}}\nu_2\overline{\boldsymbol{C}^{\mathrm{s}} \cdot \boldsymbol{C}^{\mathrm{s}}}, \tag{17.58}$$

$$\mathrm{tr}\boldsymbol{\sigma}^{\mathrm{s}} = \varepsilon_{\mathrm{d}}\mathrm{tr}\boldsymbol{D}^{\mathrm{s}} + 2M^{\mathrm{s}}\nu_1\mathrm{tr}\boldsymbol{C}^{\mathrm{s}} + 2M^{\mathrm{s}}\nu_2(\boldsymbol{C}^{\mathrm{s}} : \boldsymbol{C}^{\mathrm{s}}), \tag{17.59}$$

$$\frac{\mathrm{d}_{\mathrm{s}}\boldsymbol{C}^{\mathrm{s}}}{\mathrm{d}t} = 2\hat{X}_0^{\mathrm{s}}\boldsymbol{D}^{\mathrm{s}} + \frac{1}{\tau_1}\boldsymbol{C}^{\mathrm{s}} + \frac{1}{\tau_2}\boldsymbol{C}^{\mathrm{s}} \cdot \boldsymbol{C}^{\mathrm{s}}, \tag{17.60}$$

where $\hat{X}_0^{\mathrm{s}} = X_0^{\mathrm{s}}/\rho^{\mathrm{s}}$ and $1/\tau_i = X^{\mathrm{s}}\nu_i/\rho^{\mathrm{s}}$. We see that the coefficient τ_1 has units of time. The first term on the right-hand side of eqn (17.60) describes how an applied deformation drives the structure of the interface out of equilibrium, the second term describes the relaxation back to the equilibrium structure. The last term is a nonlinear correction on the linear relaxation behaviour described by the second term. The model in eqn (17.58) to (17.60) is expected to hold only for relatively small departures from equilibrium (deformations up to about 1 %).[39,49] This because it is constructed from linear flux-driving force relations, and a simple Taylor expansion of the surface entropy in terms of the structural variables, eqn (17.56). There are various ways in which we can improve this model. Firstly, we could retain higher-order terms in the expansion of the surface entropy. Alternatively, we could use functional forms for the surface entropy which have been shown to be more accurate.[49,58] We could also replace the material time derivative in eqn (17.60) by a frame invariant derivative such as the upper-convected surface derivative. This equation would then be given by

$$\frac{\mathrm{d}_{\mathrm{s}}\boldsymbol{C}^{\mathrm{s}}}{\mathrm{d}t} - \boldsymbol{C}^{\mathrm{s}} \cdot (\nabla_{\mathrm{s}}\boldsymbol{v}^{\mathrm{s}})^{\mathrm{T}} - (\nabla_{\mathrm{s}}\boldsymbol{v}^{\mathrm{s}}) \cdot \boldsymbol{C}^{\mathrm{s}} = 2\hat{X}_0^{\mathrm{s}}\boldsymbol{D}^{\mathrm{s}} + \frac{1}{\tau_1}\boldsymbol{C}^{\mathrm{s}} + \frac{1}{\tau_2}\boldsymbol{C}^{\mathrm{s}} \cdot \boldsymbol{C}^{\mathrm{s}}. \tag{17.61}$$

Finally, we can recognize that the scalar coefficients in the model are functions of the structural variables. For example, the first term on the right-hand side of eqn (17.61) could be replaced by

$$2\hat{X}_0^{\mathrm{s}}\boldsymbol{D}^{\mathrm{s}} \rightarrow \left(2\hat{X}_0^{\mathrm{s}}\boldsymbol{\Delta} + \hat{X}_1^{\mathrm{s}}\boldsymbol{C}^{\mathrm{s}}\boldsymbol{C}^{\mathrm{s}}\right) : \boldsymbol{D}^{\mathrm{s}}, \tag{17.62}$$

where $\boldsymbol{\Lambda}$ is the fourth-order surface unit tensor. Eqn (17.61) would then change to:

$$\frac{\mathrm{d_s}\boldsymbol{C}^{\mathrm{s}}}{\mathrm{d}t} - \boldsymbol{C}^{\mathrm{s}} \cdot (\nabla_{\mathrm{s}}\boldsymbol{v}^{\mathrm{s}})^{\mathrm{T}} - (\nabla_{\mathrm{s}}\boldsymbol{v}^{\mathrm{s}}) \cdot \boldsymbol{C}^{\mathrm{s}} = 2\hat{X}_0^{\mathrm{s}}\boldsymbol{D}^{\mathrm{s}} + \hat{X}_1^{\mathrm{s}}(\boldsymbol{C}^{\mathrm{s}} : \boldsymbol{D}^{\mathrm{s}})\boldsymbol{C}^{\mathrm{s}} + \frac{1}{\tau_1}\boldsymbol{C}^{\mathrm{s}} + \frac{1}{\tau_2}\boldsymbol{C}^{\mathrm{s}} \cdot \boldsymbol{C}^{\mathrm{s}}. \tag{17.63}$$

Such modifications produce models which can describe experimental data also at much higher deformations than 1 %.[49] In Subsection 17.4.3 we will show how similar nonlinear models can be constructed in a more straightforward manner with the GENERIC framework.

17.4.2 Extended Irreversible Thermodynamics

In the previous section we discussed how constitutive equations can be derived for complex fluid–fluid interfaces within the CIT framework, and included internal variables in the set of independent system variables, to account for the effects of changes in the microstructure of the interface on dynamics of a system. This leads to models which link the (nonlinear) response of an interface to a perturbation, directly to changes in its microstructure, induced by the perturbation. To test such models, we need experimental data on the time evolution of the microstructure during a deformation. Since most interfaces are actually very thin [typically in the range of (1 to 10) nm], obtaining this information is a nontrivial issue. When probing mechanical properties of interfaces in a surface rheology experiment, we would need to combine the rheological experiment with microscopic or (light, neutron, or X-ray) scattering techniques, to measure rheological response and structural evolution simultaneously. Such a combination is referred to as surface rheo-optics,[59] and is still rarely used. So mostly, information on the actual state of the interface during deformation is not available. In such cases the EIT framework provides an alternative approach for constructing constitutive models. In this framework again additional scalar, vectorial, and tensorial variables are added to the set of independent system variables. But rather than including some locally averaged measures of the microstructure, the EIT framework includes the fluxes themselves as system variables. So in EIT, eqn (17.10) is replaced by

$$\{\rho^{\mathrm{s}}, \boldsymbol{m}^{\mathrm{s}}, \bar{u}^{\mathrm{s}}, \rho^{\mathrm{s}}_{(1)}, \ldots, \rho^{\mathrm{s}}_{(N-1)}, \mathrm{tr}\boldsymbol{\sigma}^{\mathrm{s}}, \boldsymbol{q}^{\mathrm{s}}, \boldsymbol{j}^{\mathrm{s}}_{(1)}, \ldots, \boldsymbol{j}^{\mathrm{s}}_{(N-1)}, \overline{\boldsymbol{\sigma}}^{\mathrm{s}}\}. \tag{17.64}$$

In this example we have chosen the trace of the surface extra stress tensor as a single scalar variable, the energy and mass flux vectors as vectorial variables, and the traceless part of the surface extra tensor as a tensorial variable. To create so-called multimode rheological models we may alternatively choose

$$\{\rho^{\mathrm{s}}, \boldsymbol{m}^{\mathrm{s}}, \bar{u}^{\mathrm{s}}, \rho^{\mathrm{s}}_{(1)}, \ldots, \rho^{\mathrm{s}}_{(N-1)}, \mathrm{tr}\boldsymbol{\sigma}^{\mathrm{s}}_1, \ldots, \mathrm{tr}\boldsymbol{\sigma}^{\mathrm{s}}_k, \boldsymbol{q}^{\mathrm{s}}, \boldsymbol{j}^{\mathrm{s}}_{(1)}, \ldots, \boldsymbol{j}^{\mathrm{s}}_{(N-1)}, \overline{\boldsymbol{\sigma}}^{\mathrm{s}}_1, \ldots, \overline{\boldsymbol{\sigma}}^{\mathrm{s}}_k\}, \tag{17.65}$$

where $\overline{\boldsymbol{\sigma}}^{\mathrm{s}}_k$ represents the kth mode of the surface extra stress tensor. With this choice of surface variables we can now proceed along the same lines as in the previous section, where we discussed constitutive modelling in the CIT framework. So firstly, we assume that the surface entropy satisfies

$$s^{s} = s^{s}\left(u^{s}, \widehat{\Omega}, \omega^{s}_{(1)}, \ldots, \omega^{s}_{(N-1)}, \mathrm{tr}\boldsymbol{\sigma}^{s}, \boldsymbol{q}^{s}, \boldsymbol{j}^{s}_{(1)}, \ldots, \boldsymbol{j}^{s}_{(N-1)}, \overline{\boldsymbol{\sigma}}^{s}\right). \tag{17.66}$$

Next, we take the surface material time derivative of eqn (17.66), and substitute it in the jump entropy balance, eqn (17.27). From the resulting expression for the rate of surface entropy production we can again extract flux-driving force relations. Up to linear order we then arrive at the following expressions for the surface extra stress tensor (single mode):[44]

$$\frac{\mathrm{d}_{s}\overline{\boldsymbol{\sigma}}^{s}}{\mathrm{d}t} + \frac{1}{\tau_{s}}\overline{\boldsymbol{\sigma}}^{s} = \frac{2\varepsilon_{s}}{\tau_{s}}\overline{\boldsymbol{D}}^{s}, \tag{17.67}$$

which, of course, can be obtained also directly from eqn (17.60) by setting

$$\boldsymbol{C}^{s} = \overline{\boldsymbol{\sigma}}^{s}, \quad \hat{X}^{s}_{0} = \varepsilon_{s}/\tau_{s}, \quad \tau_{1} = -\tau_{s}, \quad \tau_{2} \rightarrow \infty. \tag{17.68}$$

The parameter τ_s in eqn (17.67) is the surface shear relaxation time. For the traceless part of the surface extra stress tensor we obtain

$$\frac{\mathrm{d}_{s}\mathrm{tr}\boldsymbol{\sigma}^{s}}{\mathrm{d}t} + \frac{1}{\tau_{d}}\mathrm{tr}\boldsymbol{\sigma}^{s} = \frac{2\varepsilon_{d}}{\tau_{d}}\mathrm{tr}\boldsymbol{D}^{s}, \tag{17.69}$$

where τ_d is the surface dilatational relaxation time. Eqn (17.67) and (17.69) constitute the linear surface Maxwell model. In view of its linear nature, this model is valid only for small deformation rates (typically much smaller than $1\ s^{-1}$). We can easily extend this model to higher deformation rates by incorporating higher order terms in the stress. If we choose

$$\boldsymbol{C}^{s} = \overline{\boldsymbol{\sigma}}^{s}, \quad \hat{X}^{s}_{0} = \varepsilon_{s}/\tau_{s}, \quad \tau_{1} = -\tau_{s}, \quad \tau_{2} = -\alpha_{s}/\varepsilon_{s}, \tag{17.70}$$

in eqn (17.61), we obtain

$$\frac{\mathrm{d}_{s}\overline{\boldsymbol{\sigma}}^{s}}{\mathrm{d}t} - \overline{\boldsymbol{\sigma}}^{s}\cdot(\nabla_{s}\boldsymbol{v}^{s})^{\mathrm{T}} - (\nabla_{s}\boldsymbol{v}^{s})\cdot\overline{\boldsymbol{\sigma}}^{s} + \frac{1}{\tau_{s}}\overline{\boldsymbol{\sigma}}^{s} + \frac{\alpha_{s}}{\varepsilon_{s}}\overline{\boldsymbol{\sigma}}^{s}\cdot\overline{\boldsymbol{\sigma}}^{s} = 2\frac{\varepsilon_{s}}{\tau_{s}}\overline{\boldsymbol{D}}^{s}. \tag{17.71}$$

This is the surface equivalent of the single-mode Giesekus model[60] for the bulk extra stress tensor. The corresponding expression for the trace of this tensor is given by

$$\frac{\mathrm{d}_{s}\mathrm{tr}\boldsymbol{\sigma}^{s}}{\mathrm{d}t} + \frac{1}{\tau_{d}}\mathrm{tr}\boldsymbol{\sigma}^{s} + \frac{\alpha_{d}}{\varepsilon_{d}}(\mathrm{tr}\boldsymbol{\sigma}^{s})^{2} = \frac{2\varepsilon_{d}}{\tau_{d}}\mathrm{tr}\boldsymbol{D}^{s}. \tag{17.72}$$

The coefficients α_s and α_d are the surface shear and dilatational mobility parameters, which satisfy

$$0 \leq \alpha_{s} \leq 1, \quad 0 \leq \alpha_{d} \leq 1. \tag{17.73}$$

In the EIT framework we obtain for the surface energy flux vector (when mass transfer is negligible):[44]

$$\frac{\mathrm{d}_{s}\boldsymbol{q}^{s}}{\mathrm{d}t} + \frac{1}{\tau_{\lambda}}\boldsymbol{q}^{s} = -\frac{\lambda^{s}}{\tau_{\lambda}T^{s}}\nabla_{s}T^{s}. \tag{17.74}$$

Here τ_λ is the relaxation time associated with surface energy transfer. Eqn (17.74) is the surface equivalent of the Maxwell–Cattaneo equation[51] for the bulk energy flux vector. When this expression is substituted in the jump energy balance (17.25), we obtain a hyperbolic partial differential equation, which predicts a finite speed of propagation of thermal signals, rather than the infinite speed of propagation which results from using the surface equivalent of Fourier's law (17.49).

Eqn (17.74) could alternatively also be obtained with the CIT framework, starting with eqn (17.45). For an interface with homogeneous surface composition, this expression reduces to (choosing $\boldsymbol{c}^{\mathrm{s}} = \boldsymbol{q}^{\mathrm{s}}$, $\hat{z}^{\mathrm{s}} = z^{\mathrm{s}}/\rho^{\mathrm{s}}$, and $z^{\mathrm{s}}_T/\rho^{\mathrm{s}} = -\lambda^{\mathrm{s}}/(\tau_\lambda T^{\mathrm{s}})$)

$$\frac{\mathrm{d}_{\mathrm{s}}\boldsymbol{q}^{\mathrm{s}}}{\mathrm{d}t} = \hat{z}^{\mathrm{s}}\boldsymbol{w}^{\mathrm{s}} - \frac{\lambda^{\mathrm{s}}}{\tau_\lambda T^{\mathrm{s}}}\nabla_{\mathrm{s}}T^{\mathrm{s}}. \tag{17.75}$$

By expanding the surface entropy in terms of $\boldsymbol{q}^{\mathrm{s}}$, we obtain

$$s^{\mathrm{s}} = s^{\mathrm{s}}_0 + \frac{\delta_1}{2T^{\mathrm{s}}}\boldsymbol{q}^{\mathrm{s}}\cdot\boldsymbol{q}^{\mathrm{s}} + \frac{\delta_2}{4T^{\mathrm{s}}}(\boldsymbol{q}^{\mathrm{s}}\cdot\boldsymbol{q}^{\mathrm{s}})(\boldsymbol{q}^{\mathrm{s}}\cdot\boldsymbol{q}^{\mathrm{s}}) + \ldots. \tag{17.76}$$

Here δ_1 and δ_2 are scalar coefficients. Using the definition of $\mathbf{w}^{\mathrm{s}}$ in eqn (17.31), retaining only terms up to first order in $\boldsymbol{q}^{\mathrm{s}}$, and setting $\hat{z}^{\mathrm{s}}\delta_1 = -1/\tau_\lambda$, we obtain eqn (17.74). We see that when we use fluxes as structural variables in the CIT framework, we obtain constitutive models comparable to those derived using EIT. However, in CIT the set of structural variables is not limited to the fluxes, which may provide more flexibility in constructing constitutive models.

17.4.3 GENERIC

The General Equation for Non-equilibrium Reversible–Irreversible Coupling (GENERIC) is a formulation of nonequilibrium thermodynamics in which the dynamics of a system is described by a single equation, which for multiphase systems takes the form[42,43,45–49,61]

$$\frac{\mathrm{d}A}{\mathrm{d}t} = \{A, E\} + \{A, E\}^{\mathrm{mint}} + [A, S]. \tag{17.77}$$

Here E is the Hamiltonian of the system, S is its total entropy, and A is an arbitrary observable of the system (for example, its total mass or total energy). For multiphase systems A is given by

$$A = \int_R a\mathrm{d}V + \int_\Sigma a^{\mathrm{s}}\mathrm{d}\Omega, \tag{17.78}$$

where a is the density of A in the bulk phases, and a^{s} is the surface density of A on the dividing surfaces. The first term on the right-hand side of eqn (17.77) is the Poisson bracket, which contains reversible contributions to the time rate of change of A.[42,43,61] The second term is the moving interface normal transfer or "mint" term, and it ensures structural compatibility of the GENERIC with the chain rule of functional calculus, when moving interfaces are present in the system.[45–48] The third term on the right-hand side is the dissipative bracket, which contains the irreversible contributions to the time rate of change of A.[42,43,61]

These brackets contain contributions from all surface and bulk variables, and the specific form of these contributions is restricted by a number of conditions. The Poisson bracket must satisfy

$$\{A,B\} = -\{B,A\}, \tag{17.79}$$

$$\{A,\{B,C\}\} + \{B,\{C,A\}\} + \{C,\{A,B\}\} = 0. \tag{17.80}$$

The latter identity is referred to as the Jacobi identity. Furthermore, the Poisson and mint bracket must be constructed such that the degeneracy requirements for the total mass $\mathcal{M}$, the total momentum M_{tot}, and entropy are satisfied (for arbitrary B):

$$\{S,\ B\} + \{S,\ B\}^{\mathrm{mint}} = 0, \tag{17.81}$$

$$\{M_{\mathrm{tot}},\ B\} + \{M_{\mathrm{tot}},\ B\}^{\mathrm{mint}} = 0, \tag{17.82}$$

$$\{\mathcal{M}, B\} + \{\mathcal{M}, B\}^{\mathrm{mint}} = 0. \tag{17.83}$$

The total mass and momentum of the system are defined as

$$\mathcal{M} = \int_R \rho \mathrm{d}V + \int_\Sigma \rho^{\mathrm{s}} \mathrm{d}\Omega, \quad M_{\mathrm{tot}} = \int_R \boldsymbol{m} \mathrm{d}V + \int_\Sigma \boldsymbol{m}^{\mathrm{s}} \mathrm{d}\Omega. \tag{17.84}$$

The dissipative bracket satisfies

$$[A,\ B] = [B,\ A], \tag{17.85}$$

$$[A,\ A] \geq 0, \tag{17.86}$$

$$[A,\ E] = 0. \tag{17.87}$$

For systems with a complex microstructure the Hamiltonian and total entropy are often expressed as

$$E = \int_R \left[\frac{\boldsymbol{m}^2}{2\rho} + \bar{u} + \bar{u}_{\mathrm{c}}(\rho, \bar{\Gamma}, \boldsymbol{C})\right] \mathrm{d}V + \int_\Sigma \left[\frac{(\boldsymbol{m}^{\mathrm{s}})^2}{2\rho^{\mathrm{s}}} + \bar{u}^{\mathrm{s}} + \bar{u}^{\mathrm{s}}_{\mathrm{c}}(\rho^{\mathrm{s}}, \bar{\Gamma}^{\mathrm{s}}, \boldsymbol{C}^{\mathrm{s}})\right] \mathrm{d}\Omega, \tag{17.88}$$

$$\begin{aligned} S = & \int_R \left[\bar{s}(\rho, \bar{u}, \rho_{(1)}, \ldots, \rho_{(N-1)}) + \bar{s}_{\mathrm{c}}(\rho, \bar{\Gamma}, \boldsymbol{C})\right] \mathrm{d}V \\ & + \int_\Sigma \left[\bar{s}^{\mathrm{s}}(\rho^{\mathrm{s}}, \bar{u}^{\mathrm{s}}, \rho^{\mathrm{s}}_{(1)}, \ldots, \rho^{\mathrm{s}}_{(N-1)}) + \bar{s}^{\mathrm{s}}_{\mathrm{c}}(\rho^{\mathrm{s}}, \bar{\Gamma}^{\mathrm{s}}, \boldsymbol{C}^{\mathrm{s}})\right] \mathrm{d}\Omega, \end{aligned} \tag{17.89}$$

where $\bar{u}_c$ and $\bar{u}_c^s$ are the configurational contributions to, respectively, the bulk internal energy per unit volume, and surface internal energy per unit surface. The symbols $\bar{u}$ and $\bar{u}^s$ now denote the non-configurational contributions to these internal energies. Similarly, $\bar{s}_c$ and $\bar{s}_c^s$ are the configurational contributions to, respectively, the entropy per unit volume and surface entropy per unit area. Note that we have assumed here that the microstructures of bulk and interface can both be described adequately by a single scalar variable, and a single tensorial variable. The extension to systems where multiple scalar (or tensorial) variables are needed is straightforward.

With these expressions for the Hamiltonian and total entropy, and the appropriate form of the Poisson, mint, and dissipative brackets (for which in view of their length we refer the reader to ref. 48 and 49), eqn (17.77) yields a complete set of partial differential equations for all bulk and surface system variables, and a complete set of boundary conditions, that couple the equations for the bulk variables with the jump balance for the associated surface variable.[48,49] With regard to this, GENERIC differs from the CIT and EIT frameworks. In CIT and EIT we construct constitutive models using the entropy balance as a guide, which can subsequently be substituted in the jump balances. Eqn (17.77) to (17.89) generate balance equations in which the constitutive model is already incorporated, from which, if so desired, expressions for fluxes can be extracted.

Here we will focus only on the balances we obtain for the structural variables. For the bulk and surface scalar variables we obtain from eqn (17.77)

$$\frac{\mathrm{d_b}\bar{\Gamma}}{\mathrm{d}t} + \bar{\Gamma}\nabla\cdot\boldsymbol{v} - \boldsymbol{G}:\nabla\boldsymbol{v} + \frac{R_1}{T}\frac{\partial\bar{f}_\mathrm{c}}{\partial\bar{\Gamma}} - \nabla\cdot\left[\boldsymbol{D}_\Gamma\cdot\nabla\left(\frac{1}{T}\frac{\partial\bar{f}_\mathrm{c}}{\partial\bar{\Gamma}}\right)\right] = 0, \tag{17.90}$$

$$\begin{aligned}\frac{\mathrm{d_s}\bar{\Gamma}^\mathrm{s}}{\mathrm{d}t} &+ \bar{\Gamma}^\mathrm{s}\nabla_\mathrm{s}\cdot\boldsymbol{v}^\mathrm{s} - \boldsymbol{G}^\mathrm{s}:\nabla_\mathrm{s}\boldsymbol{v}^\mathrm{s} - 2H\boldsymbol{v}^\mathrm{s}\cdot\boldsymbol{G}^\mathrm{s}\cdot\boldsymbol{n} + \frac{R_1^\mathrm{s}}{T^\mathrm{s}}\frac{\partial\bar{f}_\mathrm{c}^\mathrm{s}}{\partial\bar{\Gamma}^\mathrm{s}} - \nabla_\mathrm{s}\cdot\left[\boldsymbol{D}_\Gamma^\mathrm{s}\cdot\nabla_\mathrm{s}\left(\frac{1}{T^\mathrm{s}}\frac{\partial\bar{f}_\mathrm{c}^\mathrm{s}}{\partial\bar{\Gamma}^\mathrm{s}}\right)\right] \\ &- \left[\!\left[\boldsymbol{\Phi}^\mathrm{s}:\boldsymbol{J}_{\Gamma C}:\left(\frac{\boldsymbol{\Phi}}{T}\frac{\partial\bar{f}_\mathrm{c}}{\partial\bar{\Gamma}} - \frac{\boldsymbol{\Phi}^\mathrm{s}}{T^\mathrm{s}}\frac{\partial\bar{f}_\mathrm{c}^\mathrm{s}}{\partial\bar{\Gamma}^\mathrm{s}} + \frac{1}{T}\frac{\partial\bar{f}_\mathrm{c}}{\partial\boldsymbol{C}} - \frac{1}{T^\mathrm{s}}\frac{\partial\bar{f}_\mathrm{c}^\mathrm{s}}{\partial\boldsymbol{C}^\mathrm{s}}\right)\right]\!\right] = 0,\end{aligned} \tag{17.91}$$

For the boundary condition coupling these two equations we find ($M = \mathrm{I,II}$)

$$\begin{aligned}\boldsymbol{\Phi}^M:\boldsymbol{J}_{\Gamma C}^M:&\left(\frac{\boldsymbol{\Phi}^M}{T^M}\frac{\partial\bar{f}_\mathrm{c}^M}{\partial\bar{\Gamma}^M} - \frac{\boldsymbol{\Phi}^\mathrm{s}}{T^\mathrm{s}}\frac{\partial\bar{f}_\mathrm{c}^\mathrm{s}}{\partial\bar{\Gamma}^\mathrm{s}} + \frac{1}{T^M}\frac{\partial\bar{f}_\mathrm{c}^M}{\partial\boldsymbol{C}^M} - \frac{1}{T^\mathrm{s}}\frac{\partial\bar{f}_\mathrm{c}^\mathrm{s}}{\partial\boldsymbol{C}^\mathrm{s}}\right) \\ &= -\bar{\Gamma}^M(\boldsymbol{v}^M - \boldsymbol{v}^\mathrm{s})\cdot\boldsymbol{n}^M + \boldsymbol{D}_\Gamma^M\cdot\nabla\left(\frac{1}{T^M}\frac{\partial\bar{f}_\mathrm{c}^M}{\partial\bar{\Gamma}^M}\right)\cdot\boldsymbol{n}^M,\end{aligned} \tag{17.92}$$

For the structural tensor fields we obtain

$$\frac{\mathrm{d_b}\boldsymbol{C}}{\mathrm{d}t}-\boldsymbol{C}\cdot(\nabla_{\mathrm{s}}\boldsymbol{v})^{\mathrm{T}}-(\nabla_{\mathrm{s}}\boldsymbol{v})\cdot\boldsymbol{C}+\boldsymbol{R}_2:\left(\frac{1}{T}\frac{\partial\bar{f}_{\mathrm{c}}}{\partial\boldsymbol{C}}\right)-\nabla\cdot\left[\boldsymbol{D}_{\mathrm{C}}\vdots\nabla\left(\frac{1}{T}\frac{\partial\bar{f}_{\mathrm{c}}}{\partial\boldsymbol{C}}\right)\right]=0, \tag{17.93}$$

$$\frac{\mathrm{d_s}\boldsymbol{C}^{\mathrm{s}}}{\mathrm{d}t}-\boldsymbol{C}^{\mathrm{s}}\cdot(\nabla_{\mathrm{s}}\boldsymbol{v}^{\mathrm{s}})^{T}-(\nabla_{\mathrm{s}}\boldsymbol{v}^{\mathrm{s}})\cdot\boldsymbol{C}^{\mathrm{s}}-4H\boldsymbol{v}^{\mathrm{s}}\cdot\boldsymbol{C}^{\mathrm{s}}\boldsymbol{n}+\boldsymbol{R}_2^{\mathrm{s}}:\left(\frac{1}{T^{\mathrm{s}}}\frac{\partial\bar{f}_{\mathrm{c}}^{\mathrm{s}}}{\partial\boldsymbol{C}^{\mathrm{s}}}\right)$$
$$-\nabla_{\mathrm{s}}\cdot\left[\boldsymbol{D}_{\mathrm{C}}^{\mathrm{s}}\vdots\nabla_{\mathrm{s}}\left(\frac{1}{T^{\mathrm{s}}}\frac{\partial\bar{f}_{\mathrm{c}}^{\mathrm{s}}}{\partial\boldsymbol{C}^{\mathrm{s}}}\right)\right]-\left[\!\left[\boldsymbol{J}_{\Gamma C}:\left(\frac{\boldsymbol{\Phi}}{T}\frac{\partial\bar{f}_{\mathrm{c}}}{\partial\bar{\Gamma}}-\frac{\boldsymbol{\Phi}^{\mathrm{s}}}{T^{\mathrm{s}}}\frac{\partial\bar{f}_{\mathrm{c}}^{\mathrm{s}}}{\partial\bar{\Gamma}^{\mathrm{s}}}+\frac{1}{T}\frac{\partial\bar{f}_{\mathrm{c}}}{\partial\boldsymbol{C}}-\frac{1}{T^{\mathrm{s}}}\frac{\partial\bar{f}_{\mathrm{c}}^{\mathrm{s}}}{\partial\boldsymbol{C}^{\mathrm{s}}}\right)\right]\!\right]=0, \tag{17.94}$$

with boundary conditions $(M=\mathrm{I},\mathrm{II})$

$$\boldsymbol{J}_{\Gamma C}^{M}:\left(\frac{\boldsymbol{\Phi}^{M}}{T^{M}}\frac{\partial\bar{f}_{\mathrm{c}}^{M}}{\partial\bar{\Gamma}^{M}}-\frac{\boldsymbol{\Phi}^{\mathrm{s}}}{T^{\mathrm{s}}}\frac{\partial\bar{f}_{\mathrm{c}}^{\mathrm{s}}}{\partial\bar{\Gamma}^{\mathrm{s}}}+\frac{1}{T^{M}}\frac{\partial\bar{f}_{\mathrm{c}}^{M}}{\partial\boldsymbol{C}^{M}}-\frac{1}{T^{\mathrm{s}}}\frac{\partial\bar{f}_{\mathrm{c}}^{\mathrm{s}}}{\partial\boldsymbol{C}^{\mathrm{s}}}\right)=\boldsymbol{D}_{\mathrm{C}}^{M}\vdots\nabla\left(\frac{1}{T^{M}}\frac{\partial\bar{f}_{\mathrm{c}}^{M}}{\partial\boldsymbol{C}^{M}}\right)\cdot\boldsymbol{n}^{M}. \tag{17.95}$$

In these expressions $\bar{f}_{\mathrm{c}}=\bar{u}_{\mathrm{c}}-T\bar{s}_{\mathrm{c}}$ is the configurational Helmholtz energy per unit volume, and $\bar{f}_{\mathrm{c}}^{\mathrm{s}}=\bar{u}_{\mathrm{c}}^{\mathrm{s}}-T\bar{s}_{\mathrm{c}}^{\mathrm{s}}$ is the surface configurational Helmholtz energy per unit area. The second-order tensor fields $\boldsymbol{G}$ and $\boldsymbol{G}^{\mathrm{s}}$ describe the coupling of the scalar variables with the velocity gradient, and are defined as

$$\boldsymbol{G}=g_1\boldsymbol{C}+g_2\boldsymbol{I}+g_3\boldsymbol{C}^{-1}\qquad \boldsymbol{G}^{\mathrm{s}}=g_1^{\mathrm{s}}\boldsymbol{C}^{\mathrm{s}}+g_2^{\mathrm{s}}\boldsymbol{I}+g_3^{\mathrm{s}}(\boldsymbol{C}^{\mathrm{s}})^{-1}. \tag{17.96}$$

Here the scalar coefficients g_i and g_i^{s} are functions of, respectively, $\bar{\Gamma}$ and $\bar{\Gamma}^{\mathrm{s}}$, and the scalar invariants of, respectively, $\boldsymbol{C}$ and $\boldsymbol{C}^{\mathrm{s}}$. The specific functional form of these coefficients is restricted by the Jacobi identity. The scalar coefficients R_1 and R_1^{s} quantify relaxation processes for, respectively, $\bar{\Gamma}$, and $\bar{\Gamma}^{\mathrm{s}}$. The fourth-order tensor fields $\boldsymbol{R}_2$ and $\boldsymbol{R}_2^{\mathrm{s}}$ describe relaxation processes for, respectively, $\boldsymbol{C}$, and $\boldsymbol{C}^{\mathrm{s}}$. The second-order tensor fields $\boldsymbol{D}_\Gamma$ and $\boldsymbol{D}_\Gamma^{\mathrm{s}}$ quantify diffusion processes for $\bar{\Gamma}$ and $\bar{\Gamma}^{\mathrm{s}}$. The sixth-order tensor fields $\boldsymbol{D}_C$ and $\boldsymbol{D}_C^{\mathrm{s}}$ describe diffusion processes for $\boldsymbol{C}$ and $\boldsymbol{C}^{\mathrm{s}}$. The second-order tensors $\boldsymbol{\Phi}$ and $\boldsymbol{\Phi}^{\mathrm{s}}$ are coupling tensors, describing a coupling of the exchange between bulk and interface of scalar and tensorial variables. Such couplings are, for example, important in systems with interfaces stabilized by anisotropic particles: when a particle diffuses towards the interface, it may have to adapt its orientation with respect to the interface, before it can adsorb, and hence the exchange of the scalar and tensorial variable are clearly coupled. Finally, the fourth order tensor field $\boldsymbol{J}_{\Gamma C}$ quantifies the transfer of the scalar and tensorial variable between bulk phase and interface.

We see that the balances for the bulk and surface variables take on a similar structure, in which the time rate of change of a variable is given by (in order of appearance in the equations) convective processes, relaxation processes, and diffusive processes. For the jump balances we have additional contributions for diffusive (and convective, in the case of the scalar surface variable) exchange between the interface and the adjoining bulk phases. These expressions are coupled with an expression for the surface extra stress tensor, given by

$$\boldsymbol{\sigma}^{\mathrm{s}}=\boldsymbol{\sigma}_{\nu}^{\mathrm{s}}+2\boldsymbol{C}^{\mathrm{s}}\cdot\frac{\partial\bar{f}_{\mathrm{c}}^{\mathrm{s}}}{\partial\boldsymbol{C}^{\mathrm{s}}}+\boldsymbol{G}^{\mathrm{s}}\frac{\partial\bar{f}_{\mathrm{c}}^{\mathrm{s}}}{\partial\bar{\Gamma}^{\mathrm{s}}}. \tag{17.97}$$

By selecting appropriate functional forms for the Helmholtz energies, the relaxation tensors, diffusion tensors, and coupling tensors we can create a wide range of structural models, which can be compared to experimental data for the mechanical behaviour of complex interfaces, even at high-deformation rates (in excess of 100 s^{-1}).[49] As an example, let us consider an interface in which the microstructure can be described by a single tensor, and for which the surface Helmholtz energy takes the form $\bar{f}_{\mathrm{c}}^{\mathrm{s}}=T^{\mathrm{s}}(\alpha_1\mathrm{tr}\boldsymbol{C}^{\mathrm{s}}+\alpha_2\mathrm{tr}(\boldsymbol{C}^{\mathrm{s}}\cdot\boldsymbol{C}^{\mathrm{s}})+\ldots.)$. Eqn (17.97) then reduces to (setting $\boldsymbol{G}^{\mathrm{s}}=\boldsymbol{0}$)

$$\boldsymbol{\sigma}^{\mathrm{s}}=\boldsymbol{\sigma}_{\nu}^{\mathrm{s}}+2\alpha_1T^{\mathrm{s}}\boldsymbol{C}^{\mathrm{s}}+4\alpha_2T^{\mathrm{s}}\boldsymbol{C}^{\mathrm{s}}\cdot\boldsymbol{C}^{\mathrm{s}}. \tag{17.98}$$

Let us also assume that the relaxation tensor is linear in the structural tensor,

$$R^{\mathrm{s}}_{2\alpha\beta\mu\nu}=\frac{1}{\tau_C}\delta_{\alpha\mu}C^{\mathrm{s}}_{\beta\nu}, \tag{17.99}$$

where τ_C is a relaxation time (note that in general we would choose a fully symmetrised version of eqn (17.99), and if needed, could include a second order term in the tensor $\boldsymbol{C}^{\mathrm{s}}$). If, in addition, we may assume that the deformation does not introduce any gradients in the tensor field, and that there is no exchange of structural information with the bulk phase, then eqn (17.94) reduces to

$$\frac{\mathrm{d_s}\boldsymbol{C}^{\mathrm{s}}}{\mathrm{d}t}-\boldsymbol{C}^{\mathrm{s}}\cdot(\nabla_{\mathrm{s}}\boldsymbol{v}^{\mathrm{s}})^{\mathrm{T}}-(\nabla_{\mathrm{s}}\boldsymbol{v}^{\mathrm{s}})\cdot\boldsymbol{C}^{\mathrm{s}}-4H\boldsymbol{v}^{\mathrm{s}}\cdot\boldsymbol{C}^{\mathrm{s}}\boldsymbol{n}+\frac{1}{\tau_{\mathrm{C}}}(\alpha_1\boldsymbol{C}^{\mathrm{s}}+\alpha_2\boldsymbol{C}^{\mathrm{s}}\cdot\boldsymbol{C}^{\mathrm{s}})=0, \tag{17.100}$$

which is very similar (but not completely equal) to the result we obtained in eqn (17.61), using the CIT framework.

This flexibility and generality is one of the strengths of the GENERIC framework. Through its modular form it can easily be extended to include additional effects, which are not included in eqn (17.90) to (17.95). For example, in some systems we may want to include a coupling between relaxation processes of the scalar variable, and relaxation processes for the tensorial variable, or a coupling between the diffusion processes of these two variables. Such contributions are straightforward to incorporate in the GENERIC.[61] The framework provides an excellent basis for tests of consistency of thermodynamic models, and widespread use can be recommended.

17.5 Conclusions and Outlook

In the field of experimental surface rheology we can currently observe a significant shift from characterization of simple surfactant stabilized interfaces, to more complex fluid–fluid interfaces, stabilized by polymers, proteins, protein aggregates, colloidal particles, or mixtures of these components.[1] There seems to be a general unawareness of the nonlinearities which may occur in the response of these interfaces as a result of applied deformations, even at very small deformation (rates). A detailed study of these nonlinearities is impeded by the lack of constitutive models for surface behaviour. Nonequilibrium thermodynamic frameworks such as the ones we have discussed here are excellently suited to construct such models, and with more focus on this particular subfield, NET could play an important role in the characterization of complex multiphase systems by providing experimentalists with thermodynamically consistent models for the surface extra stress tensor. The experimentally observed coupling between surface stress-deformation behaviour, and mass transfer between the interface and bulk phase, can also be conveniently and consistently modelled within NET frameworks.

Here we have compared the CIT framework (with structural variables included in the set of system variables), the EIT framework, and GENERIC framework. For all these frameworks we have illustrated how constitutive equations can be derived for the surface fluxes (describing transport processes along the interface), and the fluxes describing transfer of mass, momentum, and energy between the interface and its adjoining bulk phases. The latter equations basically serve as boundary conditions, coupling the differential equations for the bulk density, momentum, and energy, with the jump balances for, respectively, surface density, surface momentum, and surface energy.

The EIT framework is particularly useful when experimental data on the structural state of the interface is not available (and the dynamics of the system are such that they can be adequately described using fluxes as system variables, on all relevant timescales). EIT allows us to derive the surface equivalent of constitutive models for bulk fluxes, such as the surface Maxwell or surface Giesekus model for the surface extra stress tensor, or the Maxwell–Cattaneo model for the surface energy flux vector. For the CIT and GENERIC frameworks we have primarily focussed on the derivation of structural models, which couple the response of a system to a perturbation from its equilibrium state, directly to the changes in the microstructure, induced by that perturbation. With these approaches we can construct models for the surface stress tensor which can be applied also at high deformation rates. Structural models have so far not found widespread application in the field of surface rheology. In contrast, in bulk phase rheology these models are much more widespread, and have proven to be quite successful: they have been applied to polymer melts, solutions of branched polymers, immiscible polymer blends, dispersions of (anisotropic) particles, or nematic phases.[61–63] When structural data of complex interfaces during deformation becomes more readily available, the surface structural models we discussed here may prove to be equally valuable for gaining a better understanding of the complex nonlinear dynamics of these interfaces.

Acknowledgements

The author thanks Hans Christian Öttinger, Dick Bedeaux, and Signe Kjelstrup, for many interesting discussions, which inspired much of the work reported in this chapter.

References

1. L. M. C. Sagis, *Rev. Mod. Phys.*, 2011, **83**, 1367.
2. S. Jordens, L. Isa, I. Usov and R. Mezzenga, *Nat. Commun.*, 2013, **4**, 1917.
3. J. B. A. F. Smeulders, C. Blom and J. Mellema, *Phys. Rev. A: At., Mol., Opt. Phys.*, 1990, **42**, 3483.
4. J. B. A. F. Smeulders, C. Blom and J. Mellema, *Phys. Rev. A: At., Mol., Opt. Phys.*, 1992, **46**, 7708.
5. K. H. de Haas, G. J. Ruiter and J. Mellema, *Phys. Rev. E: Stat. Phys., Plasmas, Fluids, Relat. Interdiscip. Top.*, 1995, **52**, 1891.
6. E. Scholten, L. M. C. Sagis and E. van der Linden, *Macromolecules*, 2005, **38**, 3515.
7. E. Scholten, J. Sprakel, L. M. C. Sagis and E. van der Linden, *Biomacromolecules*, 2006, **7**, 339.
8. E. Scholten, L. M. C. Sagis and E. van der Linden, *J. Phys. Chem. B*, 2006, **110**, 3250.
9. E. Scholten, L. M. C. Sagis and E. van der Linden, *J. Phys. Chem. B*, 2004, **108**, 12164.
10. L. M. C. Sagis, *J. Controlled Release*, 2008, **131**, 5.
11. R. Dimova, S. Aranda, N. Bezlyepkina, V. Nikolov, K. A. Riske and R. Lipowsky, *J. Phys.: Condens. Matter*, 2006, **18**, S1151.
12. K. D. Danov, P. A. Kralchevsky and S. D. Stoyanov, *Langmuir*, 2010, **26**, 143.
13. *Foams: Theory, Measurements, Applications*. ed. R. K. Prud'homme and S. A. Khan, Marcel Dekker, New York, 1996.
14. M. A. Bos and T. van Vliet, *Adv. Colloid Interface Sci.*, 2001, **91**, 37.
15. B. S. Murray, *Curr. Opin. Colloid Interface Sci.*, 2002, **7**, 426.
16. P. Fischer and P. Erni, *Curr. Opin. Colloid Interface Sci.*, 2007, **12**, 196.
17. A. A. Doinikov, J. F. Haac and P. A. Dayton, *Ultrasonics*, 2009, **49**, 263.
18. A. A. Doinikov, J. F. Haac and P. A. Dayton, *Ultrasonics*, 2009, **49**, 269.
19. J. C. Slattery, L. M. C. Sagis and E.-S. Oh, *Interfacial Transport Phenomena*, Springer, New York, 2007.
20. H. M. Hilles, A. Maestro, F. Monroy, F. Ortega, R. G. Rubio and M. G. Velarde, *J. Chem. Phys.*, 2006, **126**, 124904.

21. H. M. Hilles, F. Monroy, L. J. Bonales, F. Ortega and R. G. Rubio, *Adv. Colloid Interface Sci.*, 2006, **122**, 67.
22. J.-M. Jung, D. Z. Gunes and R. Mezzenga, *Langmuir*, 2010, **26**, 15366.
23. N. P. K. Humblet-Hua, E. van der Linden and L. M. C. Sagis, *Soft Matter*, 2013, **9**, 2154.
24. P. A. Rühs, N. Scheuble, E. J. Windhab and P. Fischer, *Eur. Phys. J.: Spec. Top.*, 2013, **222**, 47.
25. S. E. H. J. van Kempen, H. A. Schols, E. van der Linden and L. M. C. Sagis, *Soft Matter*, 2013, **9**, 9579.
26. S. E. H. J. van Kempen, H. A. Schols, E. van der Linden and L. M. C. Sagis, *Food Funct.*, 2014, **5**, 111.
27. F. Ravera, E. Santini, G. Loglio, M. Ferrari and L. Liggieri, *J. Phys. Chem. B*, 2006, **110**, 19543.
28. P. F. Noble, O. J. Cayre, R. G. Alargova, O. D. Velev and V. Paunov, *J. Am. Chem. Soc.*, 2004, **126**, 8092.
29. M. G. Basavaraj, G. G. Fuller, J. Fransaer and J. Vermant, *Langmuir*, 2006, **22**, 6605.
30. B. Madivala, J. Fransaer and J. Vermant, *Langmuir*, 2009, **25**, 2718.
31. L. Onsager, *Phys. Rev.*, 1931, **37**, 405.
32. L. Onsager, *Phys. Rev.*, 1931, **38**, 2265.
33. S. R. de Groot and P. Mazur, *Nonequilibrium Thermodynamics*, North Holland, Amsterdam, 1962.
34. D. Bedeaux, D. A. M. Albano and P. Mazur, *Physica A*, 1975, **82**, 438.
35. B. J. A. Zielinska and D. Bedeaux, *Physica A*, 1982, **112**, 265.
36. D. Bedeaux, *Adv. Chem. Phys.*, 1986, **64**, 47.
37. A. M. Albano and D. Bedeaux, *Physica A*, 1987, **147**, 407.
38. I. Prigogine and P. Mazur, *Physica*, 1953, **19**, 241.
39. L. M. C. Sagis, *Soft Matter*, 2011, **7**, 7727.
40. D. Jou, J. Casas-Vásquez and G. Lebon, *Rep. Prog. Phys.*, 1988, **51**, 1105.
41. D. Jou, J. Casas-Vásquez and G. Lebon, *Extended Irreversible Thermodynamics*, Springer, Berlin, 3rd edn, 2001.
42. M. Grmela and H. C. Öttinger, *Phys. Rev. E: Stat. Phys., Plasmas, Fluids, Relat. Interdiscip. Top.*, 1997, **56**, 6620.
43. H. C. Öttinger and M. Grmela, *Phys. Rev. E: Stat. Phys., Plasmas, Fluids, Relat. Interdiscip. Top.*, 1997, **56**, 6633.
44. L. M. C. Sagis, *Physica A*, 2010, **389**, 673.
45. H. C. Öttinger, D. Bedeaux and D. C. Venerus, *Phys. Rev. E: Stat., Nonlinear, Soft Matter Phys.*, 2009, **80**, 021606.
46. L. M. C. Sagis, *Adv. Colloid Interface Sci.*, 2010, **153**, 58.
47. L. M. C. Sagis, *Physica A*, 2010, **389**, 1993.
48. L. M. C. Sagis and H. C. Öttinger, *Phys. Rev. E: Stat., Nonlinear, Soft Matter Phys.*, 2013, **88**, 022149.
49. L. M. C. Sagis, *Phys. Rev. E: Stat., Nonlinear, Soft Matter Phys.*, 2013, **88**, 022150.
50. L. M. C. Sagis, *Eur. Phys. J.: Spec. Top.*, 2013, **222**, 105.
51. C. Cattaneo, *Atti Semin. Mat. Fis. Univ. Modena*, 1948, **3**, 3.

52. J. W. Gibbs, *The Collected Works of J. Willard Gibbs*, Yale University Press, New Haven, 1928, vol. 1.
53. J. van der Waals, 1893; reprinted in *J. Stat. Phys.*, 1979, **20**, 200.
54. D. Korteweg, *Arch. Néerl. Sci. Exactes Nat., Ser. II*, 1901, **6**, 1.
55. A. D. Rey, *Eur. Phys. J. E: Soft Matter Biol. Phys.*, 2000, **2**, 169.
56. J. C. Slattery and D. C. Lagoudas, *Mech. Mater.*, 2005, **37**, 121.
57. E. S. Oh and J. C. Slattery, *Philos. Mag.*, 2008, **88**, 427.
58. G. C. Sarti and G. Marrucci, *Chem. Eng. Sci.*, 1973, **28**, 1053.
59. E. van der Linden, L. M. C. Sagis and P. Venema, *Curr. Opin. Colloid Interface Sci.*, 2003, **8**, 349.
60. H. Giesekus, *J. Non-Newtonian Fluid Mech.*, 1982, **11**, 69.
61. H. C. Öttinger, *Beyond Equilibrium Thermodynamics*, Wiley-Intersience, Hoboken, 2005.
62. R. G. Larson, *The Structure and Rheology of Complex Fluids*, Oxford University Press, Oxford, 1999.
63. C. W. Macosko, *Rheology: Principles, Measurements, and Applications*, Wiley-VCH, New York, 1994.
64. S. Kjelstrup and D. Bedeaux, *Non-Equilibrium Thermodynamics of Heterogeneous Systems*, World Scientific Publishing Co., Singapore, 2008.

Subject Index

References to tables and charts are in **bold** type